Prehistoric Monster Mash

Science Fictional Dinosaurs, Fossil Phenoms, Paleo-pioneers, Godzilla & Other Kaiju-saurs

Allen A. Debus

Text Copyright—2019 by Allen A. Debus. All Rights Reserved.

ISBN: 9781090963697

Dedicated to Karl E. Debus-Lopez and Tanner Wray

Tyrannosaur from *Boy's Magazine*, 1933. A British publication no longer circulated as of 1934; no copyright stated.

Table of Contents

OUT OF THE FANZINES! 1

ONE 6

    REVENGE OF THE 'LOST WORLD' MOVIE MONSTERS 8
    CHAPTER ONE — *EERIE EYRE: "WAR EAGLES" - BEYOND THE TEST REEL* 10
    CHAPTER TWO — *GIANT DINO-MONSTER 'THEORY': VISUALIZING THE REAL 'DINO' IN DINO-MONSTER REEL* 32
    CHAPTER THREE — *REMINISCING OVER DINOSAURUS!* 57
    CHAPTER FOUR — *STOP-MOTION GODZILLA* 61
    CHAPTER FIVE — *EGG ON MY FACE?* 66
    CHAPTER SIX — *NORTH VERSUS SOUTH: FAMOUS & OBSCURE PALEO-MONSTER ENCOUNTERS VIA THE POLES* 72
    CHAPTER SEVEN — *MYSTERY OF THE SPIDER-MEN* 86

TWO 98

    HOPEFUL MONSTERS 100
    CHAPTER EIGHT — *REPLICATING REPTILICUS ... IN PRINT* 102
    CHAPTER NINE — *GREATEST TERROR-SAURS OF SCI-FI & HORROR* 120
    CHAPTER TEN — *KAIJU CONJURE: A RUSSIAN 'GHIDORAH'* 127
    CHAPTER ELEVEN — *MANY FACES OF GFANTIS* 133
    CHAPTER TWELVE — *THE FLYING DEATH: AN AMERICAN 'RODAN' PROGENITOR FROM 1903?* 137
    CHAPTER THIRTEEN — *ZILLA'S NEGLECTED NOVELIZED NEMESIS OF '98—GARGANTUA* 143
    CHAPTER FOURTEEN — *SKULLDUGGERY: A PILTDOWN 'ELEMENTARY'* 150
    CHAPTER FIFTEEN — *RISE OF THE GIGANTIS IMPOSTERS* 164

THREE 170

    APES & GODZILLA (REDUX) 172
    CHAPTER SIXTEEN — *YET ANOTHER TALE OF 'LOST WORLD' PALEO-MONSTERS* 174
    CHAPTER SEVENTEEN — *KOMODO: ANCESTRAL DAIKAIJU* 179
    CHAPTER EIGHTEEN — *KONG VERSUS GODZILLA – A BATTLE OF NOVELIZATIONS* 187
    CHAPTER NINETEEN — *MODES OF SURVIVAL: GODZILLA 'VERSUS' KONG (PART 2)* 207
    CHAPTER TWENTY — *KING KONG DAYS* 214

FOUR 222

    PALEO-BOOKS 224
    CHAPTER TWENTY-ONE — *TRAGG-NIFICENT! A COMIC BOOK OF PREHISTORIC BEARING* 227
    CHAPTER TWENTY-TWO — *AMERICA'S FIRST POPULAR 'DINOSAUR BOOK'?* 236
    CHAPTER TWENTY-THREE — *TWO MOST 'SIGNIFICANT' DINOSAUR BOOKS OF THE DINOSAUR RENAISSANCE* 244
    CHAPTER TWENTY-FOUR — *COMING DARK AGES: FOUR METAPHORICAL 'HORSEMEN' OF THE PALEO-POCALYPSE* 256

CHAPTER TWENTY-FIVE — DINOSAURS IN FANTASTIC FICTION - EXTRAS - III 278

FIVE 288

PALEO-PIONEERS 290
CHAPTER TWENTY-SIX — WHEN BRONTOSAURUS WAS MONSTROUS: 291
THREE PIONEERS' ENCHANTMENT WITH OUR FIRST FILMIC PALEO-MONSTER 291
CHAPTER TWENTY-SEVEN — THAT 'OTHER' HOLMES...310
CHAPTER TWENTY-EIGHT — IRONICAL DINOSAURS—DRAGON BONES OF THE GOBI: ROY CHAPMAN ANDREWS, THE CENTRAL ASIATIC EXPEDITIONS, AND THE MOST CELEBRATED FOSSIL 'CONSOLATION PRIZE' ... EVER!...318
CHAPTER TWENTY-NINE — EDWIN H. COLBERT: WANDERING LANDS, THE TRIASSIC............328
& COLD-BLOODED ARCHOSAURS...328
CHAPTER THIRTY — 'TIME' FOR ALBRITTON'S CATASTROPHISM 339
CHAPTER THIRTY-ONE — A LOOK 'BAKK-' ROBERT BAKKER: REVOLUTIONARY PALEONTOLOGIST..........344

SIX 362

ODD EVOLUTION 364
CHAPTER THIRTY-TWO — FAR OUT EVOLUTION—OMG! 365
CHAPTER THIRTY-THREE — MOREAU: H. G. WELLS'S "EXERCISE IN YOUTHFUL BLASPHEMY" 376
CHAPTER THIRTY-FOUR — WELLS VERSUS VERNE: PARTICULARS IN PALEO-FICTION.............388

SEVEN 394

DINOSAURIANA 396
CHAPTER THIRTY-FIVE — "IT'S ALIVE"!!...399
CHAPTER THIRTY-SIX — A RECORD OF SPACE-TIME 416
CHAPTER THIRTY-SEVEN — A PALEOART 'REVOLUTION'? 423
CHAPTER THIRTY-EIGHT — COLLECTING PALEO-PAPER 429
CHAPTER THIRTY-NINE — INTO THE VALLEY OF DINOSAURS 441

EIGHT 452

G-FEST PANELS 454
CHAPTER FORTY — ALL-TIME GREATEST PREHISTORIC MONSTERS OF SCIENCE FICTION: A PANEL EXPERIENCE 455
CHAPTER FORTY-ONE — TWO MALIGNED DINO-MONSTER FILMS 463

NINE 470

CHICAGOLAND'S PALEO-WORLD 472
CHAPTER FORTY-TWO — PAUL SERENO STRIKES ... AGAIN & AGAIN..............................474
CHAPTER FORTY-THREE — DINOSAURS COME TO LIFE 483
CHAPTER FORTY-FOUR — SAY "YES" TO MICHIGAN'S PALEO-TREASURES! 490

TEN 498

EXO-PALEONTOLOGY 500
CHAPTER FORTY-FIVE — AGAINST THE GRAIN: OFF A COMET INTO A VALLEY OF EXO-DINOSAURS 502
CHAPTER FORTY-SIX — A COSMIC ICE AGE EXTINCTION: (LEARNING BY 'O-S'-MOSIS).....................510
CHAPTER FORTY-SEVEN — STEINER'S (AND GRILLMAIR'S) UNORTHODOX ICE AGES THEORY................520

ABOUT THE AUTHOR **535**

ABOUT THE COVER **535**

Out of the Fanzines!

In June 1981 a curious 'dinosaur' short story appeared coauthored by Jack Dann and Gardner Dozois titled, "A Change in the Weather."[1] This story was not exactly presaging our extinction if human-caused climate change continues such that we'll end up dead as dinosaurs if we don't strive to do something about it.Instead, their story focused on the maddening *ubiquity* of dinosaurs that was becoming overwhelmingly apparent in modernity. My 'take away' was that its authors perhaps noticed how prevalent or even 'relevant' dinosaurs were fast becoming in popular culture—symbolically and in 'imagetext.' In their story, dinosaurs—pterodactyls and fin-backed *Dimetrodons*—get in everybody's way, and in the end literally rain out of the skies like 'cats and dogs.'

While "Change in Weather's" protagonist is absolutely fed up with this Mesozoic mess, not everybody would have shared his annoyed reaction. For the dinosaur renaissance with its newsworthy, burgeoning emphasis on dinosaur physiologies and phylogenies, fossil embryos, eggs and babies, as well as Late Cretaceous mass extinctions, was heating up rapidly then—linking lives and deaths of dinosaurs in a kind of revolutionary social 'movement'—and like many others I was totally fascinated. Unlike that crazy 1981 short story, living dinosaurs weren't in fact causing traffic congestions or making persistent nuisances of themselves. However they *were* increasingly invading books, magazines, toy store shelves, movies and televised programs, meanwhile becoming assorted 'collectible' paraphernalia at unprecedented rates.

And I absolutely loved it!

There's an accepted term reserved for today's adults (usually males) who grew up during the late 1950s, '60s and early '70s addicted to watching monster and sci-fi horror movies & television shows, while building monster models and collecting monster magazines, records and toys. This term is Monster Kid, that facet of the baby boomer 'generation' to which I proudly belong. I still watch Svengoolie's (i.e. Rich Koz's) program aired in Chicago on Saturday nights. But others—albeit lesser numbers of compatriots—also grew up 'dinosaur' at the same time. We played with various kinds of dinosaur toys and read dinosaur books. We were thrilled by museum displays featuring prehistoric animal skeletons, restorations and painted scenes from deep time. We watched sci-fi flicks where 'live' dinosaurs appeared onscreen to ravage mankind. Some of us even collected fossils, or no less imaginatively wondered what if dinosaurs someday were to literally rain out of the skies like cats and dogs.

So join the club! I propose a new term for 'us' … at the ripe old age of 65 years—I am *also* a self-proclaimed, inaugural 'Dino-Monster Kid.' This new book, *Prehistoric Monster Mash,* is compiled for us Dinosaur Kids, as well as certain Monster Kids who reveled in monsters of the Mesozoic persuasion!

Some readers may get a misconceived impression that the contents of this book—comprising quite a range of paleontological and dino-related topics from Jules Verne through the dinosaur renaissance heyday—seem, well, 'disconnected.' Although I've intended this as a *prehistoric* monster mash-up (with a nod to Bobby (Boris) Pickett), some may view this compilation more so as a monstrous 'mish-mash,' given the broad topical range. (Not that there's anything wrong with that?) However, as a ranking dino-phile and Monster Kid, I've always felt that the general topic of 'dinosaurs' encompasses everything from popular consideration of dinosaur science through dinosaur movies and, yes—even Godzilla (well, at least from 'his' earlier filmic history). It's just how I'm 'wired.' I've outlined this perspective in several of my McFarland books—perhaps most definitively in my *Dinosaurs Ever Evolving* (2017).

Nearly every chapter herein was previously published as a fanzine article (illustrated with numerous images and photos that unfortunately cannot be reproduced here). Within each of *PMM's* ten major sections I've bundled chapters addressing interrelated themes or topics, beginning, for example, with "Revenge of the Lost World Movies." And so, while it may seem that there's little rhyme or reason among major heading contents in this collection, here's some insight into my method of madness.

First, each chapter has *something* to do (however tenuously) with prehistory and ancient life represented either in real or fictional form. Secondly, each chapter reflects a topic that at one time or another became of passionate interest to me, which I suppose makes this another nostalgic excursion. Despite this sense of tangentiality, savvy readers may detect considerable 'interconnectivity' between chapter contents. From beginning to end, there's (more or less) a 'movement' from science fictional aspects, topically, toward matters increasingly founded in science fact. Also, this book contains some of my most recent previously unpublished musings. (As an aside, my very first bit of paleo-themed writing that perhaps one might say 'went somewhere' or 'meant something' at the time happened to be my Junior year Deerfield High School English research term paper—graded in 1971—titled "The Oxford Meeting of 1860," concerning a famous evolutionary debate.)

Topics covered range from the utterly fantastic to reporting of the genuinely scientific. This book encompasses Mary Shelley to Paul Sereno, mad scientists hellbent on Frankensteinian science and awful evolution, 'invasions' of continental ice sheets to those from outer space, a sense of the 'apocalyptical' in paleo-sci-fi, as well as the promised interspersed mashup of real, 'reel,' and other types of fictional prehistoric monsters! You won't find scientific discussion of *Tyrannosaurus* here, for example, or a full chapter on whether *Brontosaurus* was warm-blooded. You can find those topics covered expertly in other books. Nope—this time I'm working on the fringe of prehistoric life in popular culture, where I've arrived late in life, now as a senior. Why such an odd concatenation? Delve within and see. This isn't a science book: some chapters are more introspective than others, and most are retrospective. In compiling this particular collection, readers may glean further original perspectives (e.g. extending above and beyond ideas expressed in my previous books) on how, over time, prehistoric life—with dinosaurs at the masthead—clawed out of stiffly written, obscure science journals into mainstream popular culture. And yet a volume like this can only represent a cross-section of what's out there in the 'dino-verse.' There's only so much exploration and consideration to such matters one may manage in a lifetime.

A late 2010s trend has been the posting of a bevy of all manner of dinosaur and dino-monster related "clips," videos and "VLOGs" on YouTube. Rather than joining the fray, I'm from that prior book-reading generation. I just don't expend that much time 'surfing' through those videos and articles online—however elaborate and interesting some of them might be. Fanzines and books may be going out of style (and sales of this book will never make me rich), but therein shall remain my preferred medium of choice. Hence this 'new' in-print offering.

So much of life is about the destination, short-circuiting the journey. Given college coursework which included geology and paleontology classes, fossil collecting forays of the past and loads of armchair/avocational pursuits of the past half century, I only consider myself an 'amateur' paleontologist at best. This entry therefore represents one more semi-nostalgic romp—pondering books, movies, models, toys, records, miscellaneous dino-monsters and dino-paraphernalia, as well as certain distinguished people of paleontology that one Dinosaur Kid encountered along the way, during the 'movement'—through time and space.

Lastly, before undertaking our first 'episode,' I hope readers would forgive my reliance upon less consistent, flexible footnote presentation throughout, including some within-text referencing as originally presented in printed fanzine versions, and also my decision to dispense this go-around with an Index or Bibliography. In several cases I've revised chapters if certain details had been inadvertently omitted from the original fanzine pieces. Editing is an extended process. Often subtle details are omitted from first appearance of an article, while typos 'barnacle' in. So, in content, these chapters supersede the original versions of those articles. I'm not a tech/computer-savvy guy, and it is challenging for me to create a book on Createspace.. I receive lots of help along the way, but any and all errors in the final product are mine.

It is my hope that this 'montage' will entertain, particularly, lovers and aficionados of our classic dino-monsters and their exalted paleo-necromancers. So please enjoy *this* ... (most likely) my 'last' dinosaur book!

Allen A. Debus
Hanover Park, IL
June 2019

Note to Preface: (1) In *Playboy Magazine*, although I encountered it in a 1990 anthology titled *Dinosaurs!* edited by Jack Dann and Gardner Dozois.

ONE

Drawing suggestive of a "War Eagles" scene setting by Mike Fredericks (2019) made especially for this book, used with permission. See Chapter One for more.

Revenge of the 'Lost World' Movie Monsters

Scream-inducing dino-monsters confront mankind most frequently in movie theaters and frighten from our television screens. While the public generally is aware what dinosaurs are, (unlike us Dinosaur Kids who actually *do* know!) they don't particularly care for them. Unless there's a new blockbuster movie release laden with cool special effects to see, then they'll pay their admissions fee regardless. *King Kong, Godzilla, Jurassic Park* and *World* ... we simply cannot escape their lure or their clutches! But if you think we've got it bad, just consider those devoted souls who obsessively aspire to share their visions of the dinosaurian world, even in a timeline strictly forbidden by contingency and inevitable law of extinction where menacing dino-monsters live alongside ourselves, onscreen! Usually, in order to suitably suspend disbelief, humans must somehow be 'transported' to lost world places in order to confront creatures displaced out of geological time. There's much originality and ingenuity applied on behalf of writers and directors in devising and creating these lost world settings in order to stoke our imaginations.

Entire books have been written about dinosaurs in the movies—some nearly as entertaining as the movies themselves. In this section though, let's consider several classic dinosaur film titles as well as a few lesser known flicks—never completed or 'lost'—that have haunted us and their producers for, in some cases, over a century. For every great movie there's usually a script that dazzles, sometimes founded on a foundational novel or short story. In the case of dino-monster movies, aspects of paleontological science become an essential ingredient at the core.

So how has paleontology infested movieland? Here's an offering of several examples, insights that will reappear recurrently throughout other chapters in other sections of this book.

These chapters originally appeared in the following publications:
Chapter One—*Mad Scientist*, no.20, Fall 2009, pp.24-35.
Chapter Two—*Scary Monsters,* no.88, June 2013, pp.107-119.
Chapter Three—(Previously unpublished).
Chapter Four—*G-Fan*, no.79, Spring 2007, pp.28-31.

Chapter Five—*G-Fan*, no. 89, Fall 2009, pp.24-26.
Chapter Six—*Mad Scientist*, no.33, Summer 2018, pp.32-42.
Chapter Seven—(Previously unpublished).

Chapter One — *Eerie Eyre: "War Eagles" - Beyond the Test Reel*

Willis O'Brien (1886-1962) is renowned as a great "ideas man" for fantastic films; many uncompleted. Although he was a marvelous artist and long regarded as a consummate stop-motion animator, he wasn't the consummate pitchman or self-promoter for those ideas. Producers fretted over his perfectionist work ethic, perceived by some as plodding progress. Stop-motion animated effects—so arduously achieved—might not be completed punctually, according to contract. Financial budgeting therefore could be problematic, even though for O'Brien the *art* of filmmaking was always first and foremost. Furthermore, there is a mindset that O'Brien was taken advantage of. In twilight years, he may have been regarded as a 'fuddy duddy,' no longer able to move fast enough for unscrupulous people in the business completing projects 'on the cheap.' Irving Block, for example—effects man for *The Giant Behemoth*—said, "He got screwed, he didn't get his due."

As Don Shay outlines in his 1982 *Cinefex* 7 biographical account, "Willis O'Brien: Creator of the Impossible," O'Brien had a lifelong interest in boxing. Perhaps his additional interests in prehistoric animals stemmed from conversations he'd had with paleontologists he had guided in the Crater Lake region, "coupled with ... discovery of the fossilized remains of a sabre-toothed tiger." (Shay, p.6) So, perhaps unsurprisingly, throughout his movies, animated effects and unpublished stories, battling (e.g. - boxing) prehistoric monsters, (or a huge gorilla boxing with a human, as in *Mighty Joe Young*) is a recurrent motif.

Fans of movies carrying O'Brien's indelible mark of technological and artistic authority—*King Kong* (1933)*, The Son of Kong* (1933), *Mighty Joe Young* (1949), and *The Lost World* (1925)—can still rave about what he was able to do when the creative 'force' was with him. However, several 'dinosaur' films he conceived ("Creation" (1929-32), "War Eagles" (1938-39), "Gwangi" (1941-42), "Valley of the Mist" (1944), and "King Kong vs. Prometheus" (1960)) never reached fruition, that is, *according to his design and under his watch*. And of these titles, circumstances of only one, "War Eagles," could change dramatically in the future. We'll consider what this movie idea—War Eagles—was all about, but only after pondering how O'Brien's other filmic orphans eventually arrived on silver screen, albeit in roundabout ways.

Perhaps the most storied of O'Brien's uncompleted dinosaur movies was RKO's Creation, *yes*, the film that, as noted by film historians, through a fortuitous chain of events, paved the way to *King Kong*. Creation, dubbed by Mark F. Berry (*The Dinosaur Filmography*, McFarland, 2002) as the "most famous dinosaur movie never made," merged ideas from Arthur Conan Doyle's *The Lost World* (1912)—produced by First National Pictures in 1925—with to lesser extent, Edgar Rice Burroughs'1918 novel, *The Land That Time Forgot*. (It is the submarine device, to be described shortly, which seems borrowed from Burroughs.) Creation isn't another dinosaur on the loose type plot. Rather, the film plays on that ever popular 'humans–on-the-lost-prehistoric-monster-island' theme. O'Brien with Arthur O. Hoyt (*The Lost World's* director) conceived the story, while a detailed scenario with dialog was scripted by Hoyt and Beulah Marie Dix.

Essentially, the story is about survival in a savage, primeval land still populated by giant animals long extinct elsewhere. In the face of ever present, raw-edged danger and isolated from civilization, true personalities surface. Hoyt viewed this production as a surefire masterpiece, claiming, "This picture cannot be compared with any other ... being produced today. It is neither drama nor comedy, and yet it contains both elements. It is high adventure-melodrama of a highly imaginative kind. Only through the medium of the motion picture is it possible to attempt verisimilitude. The stage cannot attempt it—the novel must leave much to the reader's imagination. On the screen, we have the perfect illusion—these terrible monsters breathing, fighting and bleeding in mortal combat among themselves and with the people." (*Spawn of Skull Island*, by George E. Turner, *et. al.*, 2002, p. 200)

The ensuing summary of Creation's plot is primarily derived from James Van Hise's *Hot-Blooded Dinosaur Movies* (1993, pp. 84-5), and Jeff Rovin's *From the Land Beyond Beyond: The Making of Movie Monsters You've Known and Loved* (1977, p. 23). A hero, 'Steve,' who loves pretty, but spoiled and wealthy 'Elaine Armitage,' embarks on a yachting adventure led by Elaine's father. Steve argues with his onboard rival, a disreputable suitor named 'Hallett,' just as a Chilean submarine captain hails the crew, warning them of an approaching typhoon. As the waters grow turbulent, Steve, Elaine, Hallett and the rest climb into the submarine, evidently to safety—but of course this is a fantastic movie—so there's no guarantee everybody will survive what comes next. Following a seaquake of tremendous magnitude, a volcanic mountain rises from the seabed. The submarine enters a cavern within the new island where, through portholes, characters spy prehistoric monsters. Shipwrecked, survivors surface on an island lake situated within a volcanic caldera surrounded by a primeval jungle; they access the shore in a life-raft.

The castaways are astonished to see a live *Brontosaurus* herd and battling three-horned *Triceratopses*. In a later scene, a twin-horned, mammalian *Arsinoitherium* menaces men crossing a log overhanging a gorge, shaking off several to their doom, below. Magnificent pterodactyls wing down, snatching two crewmen hanging on for dear life. (This 'draft' log scene was re-staged using different monsters for *King Kong*.) Back at their camp, Steve, who survives the prehistoria attack, begins arguing with Hallett over Elaine (who now understands that Steve loves her more). Frayed tempers erupt from weeks of subsistence living, resulting in poor judgement. Hallett fires his gun at a Brontosaur, which vengefully destroys their hovel. Then trigger-happy Hallett callously shoots a cute and inquisitive juvenile *Triceratops* ... in the eye. The justifiably enraged mother in turn chases Hallett, toppling a tree over him, then gores him with her horns. So who, or which, is the monster? Audiences probably would have applauded Hallett's avenged demise.

Steve and Elaine recover the wireless radio from the submarine wreck, only to then stumble upon ruins of an ancient temple, where prehistoria abound, evidently protecting wealth of a former, lost kingdom: precious gems—an idea which wended its way into *Son of Kong*, 1933. A winged *Pteranodon* threatens to carry Elaine aloft (another scene retooled for *Kong*). Rescued by Steve, the pair are pursued by a plated *Stegosaurus*, which chases them into the path of a *Tyrannosaurus*. The two monstrosities struggle until (as later filmed by Disney in 1940, i.e. *Fantasia*), *Stegosaurus* succumbs. Steve and Elaine, who have been observing from a crevice, now see several juvenile Tyrannosaurs hungrily approaching the Stegosaur corpse. They flee through a hole in the temple wall conveniently opened by one of the crashing dinosaurs. Returning to the lake, they manage to fix the radio and signal an SOS. The lake is now boiling due to an impending volcanic eruption, causing panicked dinosaurs to stampede all around them. But a Chilean airplane pilot rescues Steve and Elaine just prior to the island's destruction.

Considerable effort was expended to create Creation, all at considerable cost, with a whopping budget of $1.2 million. Test reels featuring O'Brien's and Marcel Delgado's stop motion animation magic were done (at least one, the *Triceratops* chase scene, survives today). Many fine "continuity sketches" and artwork (by Mario Larrinaga, Byron Crabbe and Ernest Smythe) were completed detailing the look and atmosphere of certain settings and scenes.

According to Don Shay commenting in *Cinefex* no.7 (1982, p.29):

> "A full year was spent on the extensive preparations required to mount the unprofitable 'Creation' project ... Mario Larrinaga, Byron Crabbe and Ernest Smythe prepared scores of large, detailed renderings ... Marcel

> Delgado was hard at work on the model dinosaurs, a number of which had never before, nor since, appeared on the screen. Two brief test reels were to be filmed ... One, involving destruction of the storm-swept yacht ... the 'Creation' test (was) taken from the sequence in which 'Hallett', portrayed by actor Ralf Harolde, shoots the baby *Triceratops* and is then pursued by its enraged mother. The test footage ... represented a quantum leap in technique over the comparatively primitive composite work in *The Lost World*."

Film historians like to mention that Creation's pre-production efforts marked significant advances in the making of fantastic films. For instance, thanks to technicians, not only did jungle sets appear more lush and improved relative to those in *The Lost World*, but producers also planned to accommodate recording of sound. Now, fifty percent more individual shots had to be taken per second (24 frames per sec., vs. 16 per sec., as in *Lost World*). This is because, as Rovin explained, "The mechanics of an optical soundtrack—the transcription of audio to lines which are printed on the film itself and then translated back to sound by the projector—dictated that the film be run at twenty-four frames per second." (pp. 24-25) Added camera work only tacked on more time and expense. Comparing animals appearing in *Lost World* vs. those outfitted for "Creation"—many of these later having roles in *King Kong* and *Son of Kong*—Delgado's sculpting techniques had considerably improved. Indeed, Creation became a proving ground for a much more lavish and, yes, unforgettable production.

Although "fascinated by the technical aspects," by late 1931, Merian C. Cooper decided, however, that "after nearly a year, ('Creation') was going nowhere." (*Spawn of Skull Island*, 2002, p. 97) As Van Hise comments, "Like *The Lost World*, the story is just a thin showcase for the dinosaurs and isn't very interesting without the added touch of fine special effects." Cooper claimed that Creation was "Just a lot of animals walking around." (But BBC scored success over half a century later with their seminal 1999 production, *Walking With Dinosaurs*.)

Because production costs were quickly mounting, as painful as it was to Hoyt, O'Brien and sculptor Delgado, Cooper pulled the plug, redirecting efforts toward his new conception, perfecting a now acclaimed masterpiece, a fixture in popular culture, movie lore and a "household name": *King Kong*. Creation's death knell had sounded, inevitably and figuratively transforming into the famous gong, summoning Skull Island's giant prehistoric gorilla.

But, despite his 1933 successes with *King Kong* and *Son of Kong*, and, (apart from considerable personal anguish suffered during this period) more agonizing experiences would haunt O'Brien. In 1938, O'Brien began work on War Eagles, while his Creation cohort—Hoyt—vainly if not stubbornly, strove to keep the magic alive, the Creation concept on the masthead, although switched to

Columbia Pictures. So (aside from *King Kong*, which differs substantially plotwise) did progenitor Creation eventually break through to the silver screen, albeit in some vastly derivative form? Well, from what little we know, Hoyt's later project, The Lost Atlantis was essentially a reworking of Creation's scenes and plot elements. Was O'Brien's and Hoyt's 1929 brainchild forever 'sunk'? Quite possibly, when we consider the incentive, and workmanship leading to and beyond Hoyt's The Lost Atlantis, also begun in 1938, in a tepid sense fruition was achieved.

Lost Atlantis was backed by the special effects maven Fred Jackman (*Lost World's* technical effects director) and was intended to be a masterpiece of movie magic! Twenty minutes of test animation was produced on a $20,000 advance budget for Columbia executive Harry Cohn. Marcel Delgado visited the production facility during preparation of two test reels. He observed about 20 women creating dinosaur puppets and that their handiwork—the dinosaur models—were "cuter (looking) than they should have been." Delgado also felt their models "lacked the reptilian ferocity and leathery skins that he considered essential." Colorful advance advertising displayed a mother and pair of juvenile horned *Monocloniuses* E.g. "Agathaumas"); a theropod (*Ceratosaurus* ?) thunders from temple ruins, ala Creation.

In the September 1939 issue of *Popular Mechanics* (printed about the time that support was withdrawn from the film) one learns that, perhaps tinged with hyperbole, "two girls who worked over a month building the creatures threatened to quit because they said the half-completed monsters crawled through their dreams." (p.130A) After Cohn reneged on the deal, two years later Walter Lantz (Universal's cartoon icon and creator of "Woody Woodpecker"), and Edward Nassour took a stab at the vanishing Lost Atlantis project. Lantz aspired to insert test dinosaur footage into one of his cartoon shorts, but that effort wasn't successful. *Spawn of Skull Island's* (2002) authors (George E. Turner, Orville Goldner (with Michael H. Price and Douglas Turner) discussed circumstances with Lantz in 1973. Lantz said, "Utilizing a method of model animation invented by Nassour, we made one reel 'including a fight between dinosaurs, ... the models were small, the human figures standing only two inches high, and the project was abandoned because it proved too expensive and time-consuming." None of this Lantz-Nassour footage has been found.'" (p.221)

A bit of recent skullduggery on the part of several film historians now (doubtfully) suggests that test reel dinosaur footage prepared for Lost Atlantis *may* have made its way into another dinosaur movie, *The Lost Continent* (Lippert Pictures, 1951). In *FilmFax Plus* (no. 105, Jan/March 2005, p.122) William Fogg mentioned rather suspicious circumstances surrounding stop-motion animation witnessed in *Lost Continent*. Fogg states "...750 feet of 35 mm black and white film featuring prehistoric animals would be completed in six weeks and delivered to Lippert Productions for the sum of $9,000." And according to the contract,

"The animation in this film is to be compared to some of the footage of prehistoric monsters that was seen a few weeks ago in the screening room at the studio..." Hmmm—six weeks is a suspiciously short time in which to produce high quality stop-motion animation scenes.

In 2002, authors of *Spawn of Skull Island* suggested:

"In fact, we have grown to suspect that some of the Lantz-Nassour footage *may have* reached the screen. In 1951, Robert L. Lippert produced *The Lost Continent*, an almost good yarn about scientists and soldiers at large on an uncharted island. They are pursued by an angry *Brontosaurus* and observe a battle between *Triceratopses*, and one of the party is gored by a *Triceratops*—all incidents that the Hoyt scenario had depicted. The only special-effects credits are assigned to Augie Lohman, a specialist in mechanicals, and Ray Mercer, who handled the titles and optical printing. Lohman told George Turner that the creatures were not of his making, and that he didn't know who built or animated them. The animation is smooth, but the creatures *are* rather 'cute' and lacking in intimidating presence (consistent, perhaps, with Delgado's description) ... Are these the dinosaurs from The Lost Atlantis? Your guess is as good as ours, but we suspect they are." (p.222)

Furthermore, as Mark F. Berry suggests in his *Prehistoric Times* article (no. 79, Fall 2006, pp.22-23), "So ... could that mysterious footage 'seen in the screening room' be the old 'Lost Atlantis' footage? Could *The Lost Continent* animation be a mixture of old and new animation?" Well, if it is, then O'Brien's and Hoyt's long-lost Creation project may have indeed bore (meager) fruit after all.

So, let's review. It's as if RKO's aborted Creation project re-materialized in two other films: (1.) although conceptually quite different in story outline, *King Kong* borrowed several of Delgado's spectacular Creation models and certainly profited from enhanced technical effects used to create the atmosphere of a lush, miniature jungle setting, and (2.) through an 'intermediate' stage—so conceptually similar to Creation—several of Lost Atlantis's stop-motion dinosaur scenes may have been assimilated into *The Lost Continent*.[1]

Analogously, O'Brien's next major planned dinosaur production, which also withered on the vine, conceptually "re-materialized" *twice*—Gwangi.

The name, "Gwangi," is most familiar to us today because it was incorporated into a completed film in 1969, titled *The Valley of Gwangi* (Warner Bros., Seven Arts). But, few of you may realize that O'Brien's Gwangi was originally conceived nearly three decades earlier, in 1941.

According to *Spawn of Skull Island's* authors (p.222), contemporary news reports carried the waggish tale of tiny horses roaming in the Grand

Canyon area; a local Indian legend that huge monsters dwelled there also fueled filmic fire. Accordingly, Gwangi's 1941 plot will also sound more or less familiar. Essentially, cowboys discover a herd of miniature, dog-sized horses, which turn out to be of prehistoric ilk ("Eohippus"). The cowboys are with a Wild West show, owned by a woman named 'J.T.' (The cast of characters also included 'Tuck' - J.T.'s lover, a mysterious native, and an eccentric scientist.)

They soon realize that these strange horses represented a terrific gate attraction. Pursuing the horses into a remote valley near the Grand Canyon, they discover living dinosaurs and winged reptiles. One cowboy is swept off his horse by a pterodactyl, then rescued. A large *Allosaurus* (known as the legendary, titular "Gwangi") storms the scene. Thwarting the lassos, it engages battle with a *Triceratops*. Three-horn dies, but when Gwangi is knocked out in a landslide the dinosaur is secured by rope and transported back to the carnival in a makeshift wagon. Later, Gwangi is attacked by lions, which it kills, and then escapes to a village. The huge Allosaur is pushed to its demise over a cliff by a truck.

Marcel Delgado sculpted the Allosaur-Gwangi model prototype. O'Brien (with artist Jack Shaw) finished several oil paintings beautifully illustrating action scenes between cowboys and fighting dinosaurs. By 1942, miniature sets and six glass paintings designed for accentuating live action had been completed. Three actors were cast for lead roles. But when a privately commissioned Gallup Poll revealed that the public wouldn't be so interested in yet another dinosaur movie, and that people couldn't pronounce the name, "Gwangi," despite the "excellent script," interest in the movie's vast potential began to fizzle. RKO had already invested $50,000 in this production, which was abandoned following a studio reorganization. Even efforts to enlist Ruth Rose Schoedsack (*King Kong's* most influential script writer), in 1943, couldn't garner adequate support.

And so, during the height of World War II, Colonial Pictures was unable to secure the necessary financing to complete the film. In retrospect, Ray Harryhausen stated "The disappointment to O'Brien must have been tremendous as this followed closely the disintegration of his War Eagles at MGM." (Berry, p.439) As noted by authors of *Spawn of Skull Island* (p.224), "some of O'Brien's miniature sets and other artifacts were rediscovered during the 1980s in a loft at what is now Culver City Studios—tiny monuments to a monumentally lost cause."

Don Shay remarks, following Gwangi's demise, by the mid 1940s, "... hard times set in. No one seemed interested in the kind of work Obie did best." (*Cinefex 7*, p.53) Yet, Gwangi *was* made, in a large sense, conceptually not once, but twice. The first time was in the form of a film titled *The Beast of Hollow Mountain* (Nassour Studios, 1956), which O'Brien participated in. *Hollow* was less hallowed than the second production, which was an homage to O'Brien.

According to Neil Pettigrew, the second generation Gwangi theropodous dinosaur was born from the 'skeleton' of the first, constructed over the metal armature used for the 1941 version. For the 1956 'take' on the theme, O'Brien was enlisted, although not in full capacity. This movie, titled *The Beast of Hollow Mountain*, once more featured cowboys in a desert (Mexican) setting, but the plot is far less ambitious. This time there is no Wild West show—other than the movie itself—and the dinosaur isn't provokingly captured with lassos, so challenging to accomplish using stop-motion animation. The titular "Beast," presumably an *Allosaurus,* denizen of the Hollow Mountain is also a solitary dinosaur; there's no astounding dinosaur battle. In contrast to O'Brien's vision, *Beast's* Director Edward Nassour strove for expediency. While *Beast* is indeed a separate story from Gwangi, combination of ranchers and dinosaur in a remote desert setting at least approximates western 'flavor' that O'Brien intended to capture fifteen years earlier.

    A canonical synopsis of *Hollow Mountain* appeared in Neil Pettigrew's *The Stop-Motion Filmography* (1999). "As low-budget monster-on-the-loose pictures go, this is well-staged and a lot of fun, but it is difficult to watch it without thinking how much better it might have been had Willis O'Brien been allowed to handle the special effects. He wrote the original story and sold it to producer Edward Nassour, getting a promise that he would be hired to do the effects. But Nassour reneged on his promise, barred O'Brien from the studio set, made drastic revisions to the story and gave the effects contract to Jack Rabin and Louis DeWitt." (p.68) Inevitably, O'Brien's input was, as many have opined, unfairly credited, relegated to a mere, "from an idea by Willis O'Brien."

    Recently, Edward Nassour's son, "Ed," provided an enlightening interview for William Fogg's *FilmFax Plus* article (no. 105, pp.66-73). Acknowledging that O'Brien was a "great idea man," Ed stated that while he may have preferred using O'Brien to do the effects,"it was obvious that O'Brien would not be able to get the work finished in time for the release date... O'Brien might have been angry, but this 'barred from the set' stuff never happened." (To the contrary, Don Shay had commented in his 1982 *Cinefex* 7 article (p.62), that "... Obie was not even permitted past the studio front office.)

    Ron Lizorty, who contributed a letter to *FilmFax Plus* (no. 107, pp.38,40) challenged several of Ed Nassour's conciliatory claims. Lizorty declared, "... I've been told by ... FX craftsmen who confided in me that Obie was taken advantage of by Nassour ... Just think if Obie had been allowed to follow through with his artistic expertise and animation—using the exquisite Marcel Delgado "Gwangi" model. We'd have another masterwork—and in color yet!" Seconding this opinion, Pettigrew summed circumstances in *The Stop-Motion Filmography*, claiming, "... *Beast* ... is hampered by its inability to exploit the dramatic potential of its effects and is often clumsily directed by Nassour. More critically, others have dismissed *Beast* as a "childishly interpreted ... cowboys vs. dinosaurs movie" lacking "...finesse O'Brien would have brought to the table." Ultimately, however, *The Beast of Hollow Mountain* bombed at the box office. It took (Ray) Harryhausen to realize full possibilities of O'Brien's vision of combining western and prehistoric genres when he made *The Valley of Gwangi* 13 years later." (p.74)

So *The Valley of Gwangi* (1969) honored O'Brien's vision and talents, spearheaded by a master animator himself, O'Brien's protege'—Ray Harryhausen. On both story and technical bases, *The Valley of Gwangi* is a finer production than *Beast of Hollow Mountain*, (despite several technical innovations and 'firsts' evident in the latter). Mark F. Berry noted in 2002: "Flowing throughout the film is a current of good versus evil imagery ... *Gwangi* is punctuated by visual reminders of humanity's disrespect for the natural world ... *Gwangi* contains some of Ray's most dynamic, energy-packed animation, resulting in some of the best dinosaur action ever put on film." (pp.400-402) Because most readers are probably familiar with 1969's superior *Valley of Gwangi*, and also because here I am emphasizing O'Brien's dinosaur film ideas that were either 'lost' or *not* completed, I won't expound.

Then, in 1944, O'Brien, with wife Darlyne, conceived another idea for a movie involving dinosaurs, once more invoking a western theme. Titled, "Emilio and Guloso," later—"El Toro Estrella," but eventually, by April 1950, retitled "Valley of the Mist"—the (latter) venture was to have featured the combined talents of both O'Brien and Ray Harryhausen (who collaborated on the successful *Mighty Joe Young* (RKO, 1949). Here, twelve-year-old Mexican lad, 'Emilio,' raises a bull calf to maturity. Emilio is heartbroken when his father sells the bull to 'Senor Garzon'—owner of a bullring. So Emilio decides to capture another prize animal to offer in trade for 'his' bull; this "Gwangi-like" animal turns out to be a giant reptile, which according to Indian legend lives in the nearby mountains. Using cattle, Indians aid Emilio in baiting the huge lizard. In the original "Emilio and Guloso" story version, the captured dinosaur then escapes, winding up in a bullring where it kills several bulls; Emilio's bull defeats the dinosaur. Emilio sadly soothes his mortally wounded bull, as it dies. There were two story versions, however. In the first, the dinosaur breaks free on its own; in the second, when Senor Garzon refuses to return the bull to Emilio, boy frees dinosaur.

In the later, 1950 version of the story, the dinosaur portion of the plot was expanded. Instead of a single dinosaur, Emilio and his Indian helpers are confronted by a pterodactyl flock, then witness a battle between an *Allosaurus* and *Triceratops*. The *Allosaurus*, of course, is the Gwangi-like animal that is captured, transported into town, and ultimately engages in bullfighting. Indians unleash the dinosaur; Emilio's bull breaks free to save the day, emerges victorious in battle, and this time isn't killed by the dinosaur. Harryhausen recalls how promising was this movie premise, claiming in 1998, "Jesse Lasky, Sr., was going to produce it (for Paramount), and the screenplay was written by Jesse's son, Jesse Jr. We had a good screenplay, but at that time we couldn't raise the money. I don't know why."

After languishing for over two decades, in 1975, a version of the movie including prehistoric animals was released by Nassour Productions. Well, sort of. In his *Dinosaur Filmography*, (pp.96-98) Berry scrutinizes the peculiar result, a 45-minute long featurette, titled, *Emilio and His Magical Bull*. It will suffice to state here, merely, that the film—entirely animated using "replacement animation"—is disappointing in its projection of prehistoric "monsters." No, Harryhausen wasn't involved. Instead, Edward Nassour relied on "amateurs" assisted to lesser degree by Henry Sharp and Henry Lyon, who had accomplished fluid scenes of the Allosaur for *Beast of Hollow Mountain*. Although the Nassours were probably striving for a "fairy-tale quality," *Magical Bull* often misses the mark. Think of this production as the sort of children's fare achieved for those beloved 1960s Christmas specials 'toons,' like *Rudolph the Red-Nosed Reindeer*, or that other one about "Suzy Snowflake," except suffused with an odd "Valley of Gwangi" ambience. Strange indeed, and surely not what the O'Briens once intended back in 1944! Nonetheless, Berry praises the modest effort: "... it's an interesting and occasionally inspired attempt at an art form that has too seldom been tackled." (p.96) Steve Archer reproduced O'Brien's delightful (undated) treatment for "The Story of Emilio and Guloso" in his *Willis O'Brien: Special Effects Genius* (McFarland, 1991), pp. 90-95.

As James Van Hise notes, throughout the 1950s, O'Brien kept conjuring story ideas for movies involving dinosaurs that could be stop-motion animated. For example, following Valley of the Mist, which he never saw in its final disappointing form, he dreamed up tales incorporating prehistoric monsters that never went far beyond the drawing board stage, "The Eagle" and "Last of the Labyrinthodons." In the former, an injured eagle cared for by a rancher is blamed for devoured livestock—when the real culprit is a dinosaur. As in Valley of the Mist, the pet eagle kills the marauding dinosaur. O'Brien's lengthy treatment for "The Eagle" was reproduced by Steve Archer in *Willis O'Brien*, (pp. 103-112). Meanwhile, "Labyrinthodon" was about a vessel-destroying sea monster. Another titled "Matilda, or the Isle of Women"—written for the Marx Brothers, which never went beyond the typed page—incorporated a friendly *Brontosaurus*. "Matilda" is fleshed-out in Archer's 1993 book (pp.122-129).

Yet, during this late phase of his career, O'Brien did lend his technical expertise to a few completed projects—Irwin Allen's *The Animal World* (1956) and, teaming with stop-motion animator Peter Peterson, for *The Black Scorpion* (1957) and *The Giant Behemoth* (1958)—the latter involving a gigantic dinosaurian sea monster that attacks London. (Harryhausen was the principal animator for *Animal World*'s 20-minute prehistoric sequence; O'Brien functioned as "Supervising Animator.") *Black Scorpion* and *Giant Behemoth* were made on shoestring budgets. As Rovin notes, "...many of the sequences ... (were) shot in

(Peter) Peterson's garage" (p.48). Such productions were a far cry from what he aspired to do with his remaining years, yet they would be O'Brien's last (major) hurrah!

O'Brien still had major irons in the fire though, as he aspired to recapture glory of his youth through Irwin Allen's remake of *The Lost World*, and, in what could have been his crowning glory, "King Kong vs. Frankenstein," later retitled "King Kong vs. Prometheus." Although O'Brien sought to turn his career full circle—back to former days of filmic prominence—instead his career was slipping into a downward spiral! Film historians delightfully declare how *The Lost World* remake (1960) turned out to be such a fiasco. And in the same breath they take potshots at how O'Brien was cheated out of his 'Kong vs. Frankenstein' idea. I don't fully share these opinions, however, and at risk of going a bit 'tangential' on you, here's why.

In *Lost World's* case, O'Brien had indeed been invited to serve as special effects supervisor. But producer Irwin Allen had seen how long it had taken for O'Brien and Harryhausen to complete the (costly) stop-motion camera work for *Animal World*. And so, instead, for *The Lost World's* remake Irwin Allen decided to use another kind of effect that had been used quite successfully in (one of my favorite films), *Journey to the Center of the Earth* (20th Century Fox, 1959). Here, Allen relied on live reptiles to which fins, horns and other appurtenances had been glued. Horrors—"gagged up lizards" as O'Brien ridiculed the process! Consequently, O'Brien's involvement and interest in the new *Lost World* diminished considerably. Mark F. Berry suggests that O'Brien "did almost no work on this film." As Harryhausen stated to Berry, "He didn't, no; they just used his name."(p.252) Okay—now here's the thing—when my grandparents took me to see *The Lost World* in the summer of 1960, I was completely mesmerized and quaking in my boots at sight of the huge reptiles roaring and tussling on the wide screen! So, to me, even though O'Brien may have been involved only at most in name, rather than spirit, only, *Lost World* (1960) was highly effective and therefore holds a special place in my heart and mien.

Yes, those who exalt the stop-motion variety of dinosaur special effects quite often are most likely to snobbishly 'pooh pooh' any other kind of effect (other than cgi, not available during O'Brien's lifetime), as if stop-motion is somehow the only noble, or painstaking artistic way of vivifying prehistoric (and other gigantic) monsters. (I've sculpted dozens of miniature prehistoric animals from wire armature and heat-hardening compound, some even adorned with troublesome feathers, and so I *know* how tedious and time-intensive the process is of just simply making a *static* monster 'prototype.') So, after O'Brien's bout with "gagged-up lizards," came what monster film historians like to point out was the most humiliating form of appropriation, 'thanks' to a 'villain' producer named John Beck working in cahoots with the Toho Motion Picture Company. The two most definitive accounts of the tale are provided in Steve Ryfle's

Japan's Favorite Mon-Star: The Unauthorized Biography of 'The Big G' (1999, pp.80-81), and Donald F. Glut's *Classic Movie Monsters* (1978, pp.331-333).

Don Glut (p. 332) reproduces six of O'Brien's drawings showing how the hairy, dome-headed and sometimes cyclopean-looking "Ginko" menace (of which more will be said shortly) was to have appeared. In his *Willis O'Brien: Special Effects Genius* (pp. 80-81,1991), Steve Archer reproduces thirteen faces drawn by O'Brien showing possible Frankensteins. Darlyne O'Brien stated that the '13 faces' art is an example of how O'Brien would test different characters "to see what he would like best." But these "...were taken to Japan and ... never returned to us."

O'Brien's King Kong vs. Frankenstein story outline—to have relied upon stop-motion effects, filmed in color—is very intriguing. Because this Frankenstein monster differed in appearance from the 1931 'Boris Karloff' classic Universal version, having "elephantine skin, a stubby, bald head and long, ape-like arms, and standing over 20 feet tall and weighing 34 tons ..." it would have posed a deadly foe and nemesis for the similarly proportioned Kong. In story plotting, this new monster would have been assembled and stitched together from large African animals, such as elephants and rhinoceroses. According to Glut, O'Brien intended to name it the "Ginko." Of course, these two mythic monsters would have inevitably encountered one another in a gargantuan battle to the death, perhaps rivaling Kong's prior 1933 struggle with the Tyrannosaur. One Daniel O'Shea, an RKO attorney, gave O'Brien the thumbs-up for this terrific idea, and then things completely devolved from there.

As summarized by Ryfle, the plot was to have featured 'Kurt Frankenstein,' a grandson "...descendant of the Frankenstein clan (who) creates a new monster in a laboratory hidden in an African jungle. The monster escapes, kills its creator, and destroys the laboratory, and when news of the creature makes its way to America, several savvy promoters set out to capture it and bring it back to San Francisco to put it on display. Simultaneously, another group sets out to Skull Island to capture King Kong." Well, yes, that's the general idea, however, Glut notes there were at least two versions to the story, the first being O'Brien's and, secondly, the version by script writer George Worthing Yates.

In the O'Brien version, as pitched and sold to O'Shea, Carl Denham retrieves Kong from Skull Island in order to "stage a boxing match between him and some other monster in San Francisco." The second monster turns out to be the new titanic 'Frankenstein,' or Ginko. In one scene, a female gymnast walks a tightrope held between the Ginko's massive paws. However, the rope breaks; Kong, being protective of the injured girl, believes the Ginko did this maliciously, so he attacks the abomination. The pair battle their way furiously though downtown San Francisco until both creatures topple off the Golden Gate Bridge. To O'Brien's core idea, Yates added the element of Kurt Frankenstein intending to create an invincible army of promethean monsters! But, in Yates' version, the "crafty" creature proves to be Kurt Frankenstein's mastermind—not the other way around.

Perhaps only innocently trying to steer things along in the right direction, O'Shea brought producer John Beck on board, who in turn had hired Yates. It was Yates who re-titled the "Ginko" beast in his script as "Prometheus." As late as November 2, 1960, word on the street—as projected in a *Variety* "blurb"—was that "... Beck and O'Brien ...signed a contract for production of a sequel to *King Kong*," to be made during the summer of 1961. But Beck couldn't find financial backing in Hollywood, so, naturally, he harnessed other options. Ultimately, tempted by cash as the story goes, Beck sold the "Prometheus" story to Toho, without notifying O'Brien. Toho was interested in the rights to Kong (not the Prometheus end of things). And of course, Toho matched a truly titanic, suitmation Kong with their flagship monster in *King Kong vs. Godzilla* (released in Japan as *King Kong tai Gojira* on August 11, 1962), which (except for one quick battle scene on Mount Fuji) did not rely on stop-motion effects accentuating these suitmation monsters. Toho's *King Kong Escapes* (1967) is closer to the "Frankenstein vs. Prometheus" idea; so perhaps O'Brien got ripped-off twice. Indeed, Toho seemed smitten by O'Brien's ideas and during the mid 1960s even reeled off films like *Frankenstein Conquers the World*, and *War of the Gargantuas*.

Beck is to blame, of course, for the outcome. Consider aged O'Brien anxiously awaiting news from Beck ... that sadly never came. According to Don Shay (p. 69), "Obie phoned him regularly and was repeatedly assured that he would be notified when anything developed." All the while Beck was insidiously transforming his brainchild into something entirely different. Toho wanted rights to RKO's Kong, but was less enamored with O'Brien's story ideas and his technical expertise and artistry. O'Brien was snuffed out of the picture when 'his' Kong became drafted into Toho's script. Yes, unscrupulous Beck is cast as the villain, but seriously folks ... is the result so infamously bad? (And here I realize that I'm about to feel as welcome as a Cubs fan straying into U.S. Cellular Field on Chicago's southside.)

As noted by film historians, O'Brien must have felt inconsolable. As the story goes, if gagged-up lizards weren't bad enough, surely a suitmation Kong would have been far, far more disheartening to witness. O'Brien died on Nov. 8, 1962, that is, *before* being subjected to Toho's publicity-fueled extravaganza, because the U.S. version wasn't released until June 1963. And yet, Godzilla *is* the "Modern Prometheus." So, on a metaphorical level, and moreover, considering his other 'lost' filmic opportunities, even if improperly credited, O'Brien's original "Kong vs. Ginko" movie idea reached fruition in a most enduring fashion after all, playing in scores of matinee theaters and drive-ins to throngs of youngsters of the time (such as this writer and scores of kids then unfamiliar with the name, Willis O'Brien), watching with jaw-dropped-to-the-floor alacrity. I won't rhapsodize here, but at nine years of age, yes, strangely enough *King Kong vs. Godzilla* truly was one of the most influential films of my youth (and therefore in a sense still is).

And now on to War Eagles, drafted in 1937's wake of the *Hindenburg* dirigible disaster in New Jersey, as well as the publicized loss of heroic ace pilot Amelia Earhart, her plane downed somewhere in the Pacific.

For decades, few outside the motion picture industry's inner sanctum—or others 'in the know'—had heard of War Eagles. Thanks to Don Glut, who wrote the first popular article about this never completed film in 1966 for *Modern Monsters* (Oct.-Nov.), War Eagles soared into public consciousness once more. A slightly revised version of Glut's article was recently reprinted in his *Jurassic Classics* (McFarland, 2001). A decade following the ill-fated "Creation," O'Brien and Cooper teamed expertise for another movie concept involving dinosaurs, titled War Eagles, planned as a Technicolor MGM production. This lavish creation, which studio brass hoped would surpass even *King Kong*'s artistry differed considerably plot-wise and in scope from the western themes O'Brien would turn to.

Thematically, War Eagles was another in the spate of war-disaster/angst genre' allegories popular around the turn of the 20th century (leading through the World War I era), in which readers thrilled to the prospect of New York City's cataclysmic destruction. In such fare, Germany was typically the aggressor nation. Perhaps, H. G. Wells (apocryphally) thought, upon sight of the Manhattan skyline, "What a ruin it will make!" Accordingly, Wells scripted novels such as *The War in the Air* (1908); in *War of the Worlds* (1898), Martians annihilate New York City.

I am uncertain which of the two, Merian C. Cooper vs. O'Brien, dreamed up the War Eagles story originally. Mark Berry suggests it was Cooper's story; Van Hise suggests O'Brien conceived it; Rovin credits Cooper, as do authors of *Spawn of Skull Island*. Steve Archer attributes the story to Cooper. But, muddying the waters, O'Brien's second wife, Darlyne may also have been a 'conspirator.' "She would help him create ideas to write the stories, and she worked on them with him. She typed most of the work, and he would do sketches and that's how those ideas became an entity. And then they would rush them over to Cooper, and he would say ... no! - ... or ... ah ha!" (Archer, *Willis O'Brien*, p.89)

However, Cooper did present a proposal dated September 20, 1938 to MGM brass—a seven-page outline titled "White Eagle or War Eagles"—a story he believed (as then related to actress Fay Wray) would be even "bigger than Kong." Much of this outline and much other illuminating information is incorporated within David Conover's 2011 book (edited by Philip J. Riley), *War Eagles: The Unmaking of an Epic—An Alternate History for Classic Film*

Monsters. Moreover, Glut notes that "War Eagles" became yet another (doomed) "Obie" project.

Part of the problem though is that Cooper's picture as conceived was overly ambitious, indicatively, as several exciting elements were necessarily dropped from his original outline in later scripts. Conover states, Cooper (with writer James Creelman) were "… having a hell of a time trying to figure out how to make it all work …Cooper … may have overplayed his showman's hand by trying to do so much in one picture. When one reads his original synopsis, the rescue of (i.e. the beautiful Viking princess) Naru from the clutches of the apemen and the dinosaur menace feels like the story's climax and the wedding that follows the very happy ending. But the protagonists *still* have to save New York!" (Conover and Riley, p.35)

Essentially War Eagles, as scripted and revised by authors such as Harold Lamb, James Creelman and eventually by Cyril Hume, concerned the last (modern) Vikings, a lost race who have tamed giant eagles, riding them into combat versus living dinosaurs. In earlier screenplay versions, the Vikings are encountered in Arctic insular confines, while in Hume's later October 10, 1939 screenplay (with changes Feb. 16, 1940) a pilot flying solo, by then renamed "Slim" Johansen, crashes his plane instead in Antarctica. In Cooper's original 1938 treatment, the reckless pilot intends to fly "from north to south, crossing both the North and South poles" as part of a stunt advertising a product named "Bi-Car-Brome." Slim eventually crashes in an Antarctic prehistoric volcanic valley.

War Eagles is another "lost race" story, (like H. Rider Haggard's novel, *She*), a popular theme especially during the late 19th century. In a superseded version of an earlier script, history professor 'Hiram P. Cobb' relies on 'Jimmy Mathews' (a biplane stunt pilot) to verify his theory, that the Vikings retreated to a "blind spot" near the North Pole centuries ago. Flying intrepidly northward through freezing skies, they crash after colliding with a large white object (which later turns out to be one of the titular 'eagles'). Instead of a frozen, icy tundra, however, the two survivors find themselves upon a warm, temperate landscape clothed in vegetation. Next, surmounting a plateau, Cobb and Mathews discover a Viking village—thus validating Cobb's theory after all! The two outlanders are accepted into the primitive community and Mathews is attracted to 'Naru.' Although to impress (if not pacify) studio brass who were concerned about Cooper's sluggish pace of production, an "Allosaur Capture Sequence" in which ape-people subdue an allosaur was scheduled to be filmed by late 1938 as (second) test footage, this dramatic scene was also not incorporated into Hume's final script.

Meanwhile, Cobb and Mathews are startled to find among the local fauna, a species of gigantic, 15-foot tall snow eagle, individuals of which have been tamed by Viking warrior 'cavalry.' Their aquiline 'steeds' are even equipped with "...saddles made from the hides and horns of the huge ceratopsian dinosaur *Triceratops*." (Glut, p.195) When Jimmy Mathews ropes the eagle that had collided with his biplane earlier, he is carried precariously into the sky; yet Mathews tames it in flight. In order to subsist, the Vikings must occasionally ride their eagles into the Valley of the Ancients, where they scrounge for supplies. They also airlift sacks of the valley's fertile soil to the plateau to grow their crops. While it would seem more sensible for the Vikings to simply live in this more luxuriant Valley, rather than commuting back to their village atop the harsher plateau, they're obstructed from doing so because of an infestation—savage allosaurs patrolling the Valley. (O'Brien was clearly attracted to the theme of dinosaurs living anachronistically in lost world, hidden 'valleys.')

Soon Mathews is enlisted in a flight mission to the Valley of Ancients. Unfortunately, slaughter ensues below, as the Vikings encounter a pack of hungry Allosaurs, which had been attracted in turn by a Brontosaur that attacked Mathews. Defending themselves from the vicious Allosaurs, Vikings are devoured. Despite battling the dinosaurs courageously with razor-sharp beaks and mighty talons, several of the magnificent eagles, tethered helplessly, are also killed. Mathews and a few warriors manage to get airborne on eagles, thereby casting a rain of spears and arrows down upon the Allosaurs, which in turn flee. But danger lurks both on the ground as well as in the skies, for (in one of the later, increasingly complicated drafts scripted by Hume) Mathews' eagle tangles with a winged *Pteranodon* that is eventually dispatched. (In another draft a pterodactyl swarm presumably menaces the aquiline 'airforce.')

Jimmy then repairs the radio in the damaged biplane, only to become startled by news of an impending attack on New York City by a nation (Germany) using an invincible weapon capable of neutralizing electricity. This new weapon is carried into combat aboard massive dirigibles. In one version of the story, Cobb, Mathews and their Norsemen allies "...fly south and encounter the enemy's aerial armada over New York, and defeat it after a huge battle." (Van Hise, p. 88) In another version as outlined by Glut (p.195), "With great ambition in mind...," it is the Vikings who "...decide to invade new frontiers. They settle upon attacking no less than New York City! The titanic flying monsters from a lost world invade the metropolis, only to be fought off by the current mechanized wonder of civilized man—the dirigible." (Steve Archer describes it as a "flying fortress," p.37.) Thus, as Glut adds, "...a battle between two ages is waged in the skies above Manhattan." Steve Archer records there were "...various versions of the script, dated Sept. 20, 1938, with additional script changes dated April 20, 1939..." (p.41).

However, perhaps Cyril Hume's October 10, 1939 script ("with changes Feb. 16, 1940") is now best known to aficionados as 'canonical' thanks to tireless researches conducted by Conover. Hume's delightfully engaging script happens to be the only complete version I've read. As opposed to previous War Eagles script versions, geographically, Hume's story's setting is in Antarctica rather than the Arctic. There are other key differences as well. For those who haven't had the pleasure yet, overall, the plot idea is rather like 1957's *The Land Unknown*, except considerably embellished in every degree and nuance. For instance, Hume trimmed the plot and chopped characters Cobb and Matthews. He diminished "menace" in the Valley of the Ancients by eliminating those apemen antagonists and, to ease studio concerns, striving for a modicum of political correctness, excised blatant references to Germany and Nazis (although they're still implied). But Hume instilled character charisma and crafted a smoother plot, consistent with Cooper's ideas.

Allosaurus becomes chief monster menace confronted in the Valley. Exciting scenes were envisioned involving these Jurassic jihads on the 'lost' (but found) society of Vikings. The enlarged eagles are flown into battle against allosaurs who threaten the protagonist and his girl—Naru. A *Triceratops* battle erupts versus an allosaur in a "terrific battle," as described in the script, with Slim and Naru perched precariously above on a tree limb watching in terror. But the tree crashes to the ground pinning Naru when Tops shoves Allosaur into it. Then Hume directs, "*Allosaurus* kills *Triceratops*. *Triceratops* falls across log, lifting it enough to free Naru's leg."

Afterward in an ensuing vengeful act, Slim eradicates the entire Allosaur herd by engineering an avalanche of boiling lava upon them from a nearby active volcano. While dinosaurs had already succumbed to a naturally occurring volcanic eruption in 1925's *The Lost* World film, this War Eagles sequence scripted by Hume also presages later scenes in movies such as *One Million B.C.* (1940), 1960's *The Lost World, Rodan the Flying Monster* (1956), and *Godzilla—1985*—all in which dino-monsters are fatefully burned to a crisp in burning lava, for sake of saving human lives. After the Allosaur annihilation, Slim and Naru are wedded according to Norse tradition, but their marital bliss is rudely interrupted by harrowing geopolitical events, broadcast via radio—salvaged from Slim's plane!

It seems the—ahem—thinly veiled, "Generalissimo of That Nation" has ordered an aerial attack upon New York City. But Slim's intrepid Norseman allies vow to assist him by riding their Eagles into aerial battle against the invader. Slim triumphantly rides "White Eagle"—his once indomitable 'steed' which he tamed—into battle. Of course, the Norse 'air force' thwarts the attacking horde—led by a gigantic dirigible in the clouds carrying advanced weaponry—using a neutralizer ray that nullifies electrical power on the ground. A squadron of attacking war bombers is also on the way. The United States seems powerless to defend itself! Then a dramatic dog-fight battle ensues between bow and arrow-armed Norsemen riding War Eagles, with talons slashing invading interceptor planes. Eventually Slim, aided by his comrades, destroy the dirigible and its super-weapon. Hume concludes with a spectacular vision as White Eagle, Slim and Naru are perched atop the shoulder of the Statue of Liberty. "Slim stands beside White Eagle, one arm around Naru … BOOM into a big magnificent CLOSEUP WHITE EAGLE, as he screams in triumph, and wildly beats his great shining wings."

Fortunately, Conover also uncovered Harold Lamb's extensive pseudo-scientific backstory, titled "Notes on the Great Eagles and Their People," in part suggesting plausible origins for those very large (war) eagles, including the Pleistocene bird of prey *Harpagornis* whose wings may have spanned 10-feet. This creature has been referred to as "Haast's Eagle."

Considerable resources were expended on the "War Eagles" front, prior to its studio surrender. In fact, of all O'Brien's lost projects discussed here, War Eagles was perhaps furthest along, or nearly as much so as was Creation. As he had done before for *King Kong*, Marcel Delgado sculpted several puppets (scaled to '1 inch equals 1 foot') with armatures for this footage (i.e. Allosaur and Eagle models) and other scenes that weren't filmed. Don Shay records how Delgado "experienced tremendous difficulty in trying to glue real feathers onto his miniature eagles. George Lofgren, a taxidermist on staff at MGM, was brought in to solve this problem, which he did by providing treated bird skins with the original feathers still attached." (p.54)

Van Hise (p. 88) claims that by early 1939, 400 feet of test stop-motion footage featuring Valley of Ancients scenes ("...Vikings on their War eagles dive-bombing the Allosaurs before dismounting and finishing the attack on the ground...") was completed. Conover avers that a first test reel was filmed sometime between October 7, 1938 and December 15, 1939 (pp.65, 75), and shot in Technicolor. Conover also comments that while another elaborate, color test sequence was scheduled in late 1939 involving apemen capturing an allosaur, this second effort was never filmed—at least in entirety (p.99).

Given that aerial combat showcasing soaring, dive-bombing eagles would have been reliant upon thin wires, difficult to stabilize between shots, such work proved tedious indeed. The allegedly ten-minute animated test sequence has been lost to history. Numerous paintings featuring striking scenes were prepared (many by Joe Duncan Gleason, Leland Curtis, Harry Stevenson and Jack Shaw), which were used to construct 'vast' miniature movie sets. Several drawings and paintings showing intended scenes survive today. Also, "...five bird armatures—two lying and three sitting ... survive in a private collection." With so few relics, it's almost as if the project never happened.

But the best documentation of how far the work progressed comes directly from Ray Harryhausen, who visited O'Brien on the set in 1939 (and who owned a 'souvenir'—one of the Viking man puppets—the armature for which was later used for a scene in *Mighty Joe Young*). As recorded by Rovin (p.57) Harryhausen:

"... first called him when he was making 'War Eagles' at MGM. A school friend of mine knew him, and she said he was a very understanding man, and just to call him up. So I called ...told him that I'd been making dinosaurs all my life. I also said that I would be enormously honored if he would view them. So he kindly invited me to the studio ... Of course when I saw his office with all these hundreds of beautiful drawings for 'War Eagles,' I almost went out of my mind. The walls were *covered* with them! Some were painted in glorious water color and others in oil, but the majority of them were drawn in a bold charcoal technique. It wasn't easy to recover from this sudden and overwhelming exposure to all I had dreamed about for years...." And in the Foreword to Carl Macek's 2008 novel, *War Eagles* (to be discussed below), Harryhausen added "'...'Obie' was using three office rooms ...I vividly remember one big oil painting of the War Eagles perched on the crown of the Statue of Liberty and another painting of a huge dirigible surrounded in the air by the War Eagles. I was absolutely thrilled to be able to see the preliminary work on what Merian C. Cooper had promised in his original treatment of 'War Eagles' to be "Bigger than *King Kong*!'"

Although aspiring to become even "bigger" than Kong, Cooper's and O'Brien's War Eagles was never finished. On Sept. 1, 1939, when World War II was declared (in Europe), Cooper—himself a World War I flying ace pilot—left MGM, patriotically joining the Chinese Flying Tigers in their intensifying fight against Japanese forces. So War Eagles was shelved, and upon his return following the war, Cooper's interest in the project flagged. Perhaps he sensed the public would be weary of a film with a contrived climax founded upon geopolitical tensions and its peculiar super weapon. As opposed to battles orchestrated on miniature sets, Cooper had engaged in real fighting and by 1945, mankind had invented far more terrible super-weapons than the story's "electrical neutralizer." Also, with jet aircraft, V-2 missile launches and other terrible war engines introduced to the Pacific and European theaters, well, by then dirigible technology may have gone over with audiences like a lead balloon; dirigibles no longer have seemed quite as threatening.

Steve Archer concurs that "With its elaborate action sequences—a step beyond Kong ... cancellation of War Eagles must have been an enormous disappointment to O'Brien. The film could have paved the way for future fantasy films from O'Brien and Cooper, made on the same scale. One can only speculate the effect it might have had on O'Brien's career and the technique of model animation, given the progress he had made in his work on *Lost World* and *King Kong*." (p. 41).

According to Mark Berry, during the 1980s, following Harryhausen's retirement, perhaps due to that vivid impression from his first meeting with

O'Brien during filming of "War Eagles," stop-motion animator Jim Danforth discussed "... with Charles Schneer the possibility of rekindling War Eagles." As we know that prospect never materialized either, but is there a movement afoot to 'rekindle' the War Eagles effort on some level after so many decades?

Yes, perhaps! And the first sign of that is Carl Macek's novel published in 2008, *War Eagles* (Angel Gate), "Inspired by an original concept by Merian C. Cooper," with a Foreword by Ray Harryhausen. Notice that O'Brien wasn't so credited in the novel, so I initially wondered whether he'd been slighted once more, or whether James Van Hise may have been wrong about the original story *idea* being O'Brien's after all. However, the book is *dedicated* to both Cooper *and* O'Brien. Author Ray Bradbury praised Macek's novel: "Now at long last, Cooper's dream has been brought back to life, thanks to visual effects legend Ray Harryhausen and his friends. I love the idea and celebrate it! Like everything else with Cooper's name attached, this book is wonderful." And (perhaps prophetically?) Harryhausen added, "I have always hoped to honor my two great heroes by producing a film version of Cooper's remarkable project. This book is an important first step in that direction."

Macek's novel follows the story Van Hise outlined. Names of story characters have been innocently changed. Essentially, 'J. P. Brandt,' an Air Force test pilot has been drummed out of the military, in 1939, for disobeying orders during a test flight, brashly putting President Roosevelt at risk during the maneuver. Consequently, Brandt is enlisted in another flying stunt, to take an experimental airplane, the "UB14," from pole to pole, gaining publicity all the way for its inventor. Over Arctic regions, their plane is damaged by a massive white eagle. "Outlander" Brandt survives the crash only to be rescued by Norsemen; he is cared for by a beautiful blond lady named 'Naru,' daughter of 'Chieftain Raud.'

While being assimilated into Viking culture, Brandt and Naru develop feelings for one another. Brandt is amazed by the white eagles ("*ofrior orn*"), which the Vikings have tamed. Brandt urges a mission to visit the plane wreck to find the whereabouts of his two flight comrades, 'Kolinsky' and 'Will.' Suddenly, the patrol is attacked by a pack of hungry *Utahraptors*, (similar in size and ferocity to the "raptors" which were featured in 1993's *Jurassic Park*, disturbed from their dispatching of a plated *Stegosaurus*. A terrible, blood-soaked battle ensues! Brandt manages to recover an army pistol from the wreckage, which he uses to disperse the carnivores, and Viking survivors fly back to their village. Afterward, Brandt tames a massive white eagle for himself, evidently the one that caused the UB14 crash. Riding his charge, Brandt locates Will and Kolinsky. As it turns out, Kolinsky works for the Polish Secret Service; he was assigned to discover Nazi bases situated in the Arctic from aboard the UB14. There is one such base on the Norsemen's island, a yard where dirigibles bearing a new Nazi weapon—the electrical neutralizer—are awaiting orders to attack

New York! Brandt and his new allies fly southward, joining in battle over New York, paralyzed in the grip of Nazi technology. The U.S. Air Force is unable to launch their planes in retaliation against the attacking dirigibles. But the Vikings *cum* Brandt, upon aquiline steeds, prevail over Germany's offensive.

The *War Eagles* novel may be followed by additional productions, preserved more or less in the traditional story outline. As we learn from Australian writer Robert Hood's website (roberthood.net), a graphic novel devoted to the War Eagles ever-shifting story may soon be published. And, believe it or not, in 2010, yes, the movie *War Eagles* will be released from a screenplay by Andy Briggs! This is in 'pre-production,' the production company itself being "Ray Harryhausen Presents." Oh, surely this will not be a stop-motion variety of film. My guess is that, instead, producers will avail themselves of cgi technology. *War Eagles'* climactic dirigible scenes will probably have a "Sky Captain" 'feel' to them (i.e. as in *Sky Captain and the World of Tomorrow*, 2004). As in *Sky Captain's* case, and viewers of the Walt Disney, Pixar film *Up* (2008) might agree, dirigibles are definitely 'in.' By the time you read these words in *Mad Scientist*, some seven months hence, you may have even seen trailers advertising *War Eagles!*[2]

Hot special-effects are dime a dozen now, perhaps diluting the modern viewer's appreciative sense—that is, compared to the day when just about *any* silly special effect or fantastic costume gave reason to pause. A commonality, or ('fatal flaw'?) of O'Brien's aborted dinosaur films is that they were usually devised without prior (existing) famous or 'classic' story to guide plot and action, upon which the Studio's conviction could be staked. Several of these rough draft film ideas eventually bore fruit in some form, but tortuously only after several iterations, the most famous of all being Creation's transformation into *King Kong*. O'Brien's *The Lost World* (1925), having the least such difficulty, was of course based on the most famous (classic) dinosaur novel of all, Arthur Conan Doyle's *The Lost World* (1912). Now, with Macek's *War Eagles* novel 'out there,' perhaps it is finally time to film the story as it should be told.

Although O'Brien continued to complete movie projects through the 1950s, perhaps the acclaimed *Mighty Joe Young* (1949) was his swan song. Afterward, as noted by Don Shay, "If the father of stop-motion animation was to continue on in the field he had brought into being, he would have to adapt—and adapting—for O'Brien, was to be a painful process." (p.62) By 1960, the industry had essentially rejected O'Brien, and yet his creations live on in spirit.

So far, War Eagles is the *only* one of O'Brien's more considerably fleshed-out, yet 'lost' 'dinosaur films' to miss the silver screen in one way or another. However, on a wing and a prayer, War Eagles may soon fly ... beyond the test reel!

Notes to Chapter One: (1) While such conjectural allegations may seem intriguing, they've not gone unchallenged. Don Glut, for instance (in personal communications) vehemently opposes the gist of these speculations; (2) In retrospect, while updating this manuscript in October 2018 … obviously this remark as originally penned during early 2009 didn't ring true.

Chapter Two — *Giant Dino-Monster 'Theory': Visualizing the Real 'dino' in Dino-Monster Reel*

Science fiction and horror writers and movie directors must surmount a difficulty in conceiving original fare for their audiences. Our exalted 'imaginers' must be sufficiently capable of suspending disbelief with characters and universes they've created, and the overwhelming, cosmic problems which must be resolved such that the story will entertain. And while they usually aren't trying to produce a comedy (unless it's, say, *Abbott and Costello Meet Frankenstein*), how many times have we seen these old sci-fi genre films and snickered at certain scenes or ridiculously scripted lines based on what we know today. Or if we just happen to know more science than the 'average Joe'—possibly including the benighted film director—we can spot scientific inconsistencies that were so *conveniently* glossed over. "That's impossible!" we might find ourselves yelling at the screen. Or, "That violates the laws of physics!" Yes—true—but the history of science has proven that we'd best be careful to judge. For quite often what is considered impossible eventually enters the realm of the possible through the dark magic of scientific method.

So, certainly, to some degree, maybe our rude snickering is unfair. And yet, we really can't help ourselves, can we?

When it comes to movies involving invading prehistoric monsters, which are usually cast as 'science fictional' in nature (often imbued with elements of horror), in order to satisfyingly suspend disbelief, paleontologists are summoned to address two pressing matters. First, they must sufficiently document the existence of a gigantic mysterious beast that defies comprehension and natural laws, a creature that would otherwise seem impossible. And they must explain why this supposedly "prehistoric" monster is alive in modernity. Hey – if it's truly prehistoric, then by definition it shouldn't be alive today! (A third movieland question, left largely unexplored in this article, regards how the monster may be destroyed, but often that's where other scientific experts enter the picture, besides the paleontologists.) Scientific men seek rational explanations. Hence the dino-monster must be "explained" rationally, if even on the basis of flimsy evidence and doubtful logic – just the kind of stuff we love! Science fiction without the science often would translate into fantasy horror. The films under consideration here were all intended as science fiction, albeit tinged with terror.

The Deadly Mantis:

Using this platform to delve into a few of the tricks used in movies to suspend disbelief over paleontological matters, let's examine Universal's *The Deadly Mantis* (1957) as an introductory exercise. This unusual film crosses the 'big bug" genre' of the 1950s with distinct paleo-themes. (Opening narrative also projects Cold War angst, although without those usual scenes of a detonating nuclear warhead that appear in the American films of this period.) The trouble begins when a natural geo-disturbance causes glacial Arctic ice containing an enormous Praying Mantis to surface and thaw. The Mantis emerges from icy entombment wholly alive, and begins menacing mankind, at first mysteriously. Discovery of a broken, 5-foot long "hook" – mottled green in color – at a USA military plane crash site leads to scientific investigation over the nature of the beast. What is this unusual specimen?

When pathologist Professor Anton Gunther (Florenz Ames) can't identify the "appendage," intervention of paleontologist Professor Nedrick Jackson (William Hopper) is sought. Jackson happens to be an authority on the "Oligocene carnivory," which essentially means he has expertise on carnivores (all kinds, we might suppose) that lived globally some 30 million years ago, or so. It sounds like a fairly broad field. Jackson also has a reputation (reminiscent of Baron Georges Cuvier, two centuries ago) of being able to reconstruct prehistoric creatures with no more to start from than only a few bones. Good thing he's available to consult on what will later turn out to be a huge bug then!

Okay – before calling Terminix, paleontology comes to the rescue! Through careful script writing, Jackson's initial observations lead curiously and quite nearly to a 'correct' conclusion within a few minutes. After remarking that the specimen doesn't resemble bone, yet is more like gristle or cartilage (usually associated with bone), Jackson ridiculously says, that it can't be from an animal because it's not bone, and that is because every known species of animal has a bony skeleton! This would have most savvy viewers wondering where exactly Dr. Jackson obtained his Ph.D. Then he adds, "Even birds have bony skeletons." Really – how fascinating! Jackson must have noticed that phenomenon during a Thanksgiving dinner. He goes on to claim that through this thinking-out-loud exercise they're making "considerable progress." Oh stop, please! If the script writers had only been clear that not every kind of *animal* has a bony skeleton, the science would have seemed more intact. Or Jackson might have simply missed saying the word "vertebrate" in the first line cited above. Or maybe he missed some key classes back in his college days.

Yet things are moving along here as discussion turns to invertebrates, like insects. Conclusions are swiftly flying by the seat of our pants. Jackson, while noticing a detail on the specimen through a magnifying glass states emphatically, that insects "… have a fold where muscles are attached," referring to general insect anatomy. Dr. Gunther muses, "It's a possibility." "An insect?" asks General Ford (Donald Randolph). Yes possibly, according to Dr. Jackson, that is, through "process of elimination." But a further, simple analytical test of fluid from the end of the hook would reveal whether the "blood" contains red corpuscles. If red corpuscles are not present, then this would support a theory that the specimen is from an insect, although they would still not know which species. General Ford praises Dr. Jackson, even suggesting that he's beginning to sound a little like Sherlock Holmes. Okay! 'If only these people could listen to themselves,' we find ourselves screaming! No claw or appendage from an insect could e-v-e-r be THAT huge. Physical scaling laws bound by gravity and physiology defy that possibility in the present age. True – in the geological past there were prolonged periods, such as during the Paleozoic Era, when atmospheric oxygen exceeded that of the present day and insects then did grow to larger sizes than in the present – such as the dragon fly with the 2-foot wing span that will be referred to shortly in a later movie scene. But that's a long shot from where Dr. Jackson is leading the inquiry.

Basically, everything is concluded in the next important scene, which takes place in Dr. Jackson's office at the Natural History Museum. This is where Jackson, whose thinking is facilitated by Marge Blaine (Alix Talton) takes us along a frenetic ride of pseudo-logic leading to his conclusion that the hook/appendage came from a Mantis. Marge conveniently recognizes that the shape of the appendage resembles that of the spur on the leg of a grasshopper or cricket (which are herbivorous, not carnivores). Nearly every assertive statement uttered in this scene becomes a definitive, if not foregone conclusion, no doubt to cut to the chase and reduce the movie length. Also, if the science-speak is said rapidly enough, then audiences won't have time to contemplate what's going on or be able to properly critique it. Now Jackson almost sounds momentarily doubtful. "If that's from an insect, it's from the biggest creature that ever lived." In a brief sigh of comic relief, Marge replies, "Are you sure you're feeling alright?" (lololol) Well, he probably isn't, but for sure he's really in the moment.

Now things come together almost too rapidly for viewers to keep up. In keeping with Cold War monster movie themes, it must first be proven to the audience that Science can deduce what this particular creature is, thus facilitating our ability to identify it when it appears, leading to its ultimate destruction. And that's just what's about to happen. Next, Jackson receives a phone call – the fluid test results have come in; the blood test reveals there were NO red corpuscles.

Ah, so it's an insect after all. Then how could such a cold-blooded creature be alive, resurrected in the frozen Arctic? Jackson first shows us a 90-million-year-old fossil insect (certainly not a real specimen) preserved in amber, thus implying the monster must be ancient, which is a setup for what will be stated, shortly. Why is the living monster on the rampage so huge? Jackson refers to a fossil slab over his mantle showing the 2-foot wide dragon fly. See how big they could become! "Best you can do?" challenges Marge. Jackson replies, well he didn't say they were looking for a mere dragon fly. After all, the monster is carnivorous (thank goodness Jackson specializes in "carnivory"), and therefore, yes, he's already decided – therefore, it must be a Praying Mantis! Wow! Well, fossils "prove" that prehistoric dragon flies could only grow to two feet, but absence of negative evidence allows Jackson to suggest that prehistoric praying mantises could have grown to considerably larger sizes.

Jackson now is on a roll, stating that the field of paleontology isn't narrowed to what we understand … fortunately paleontologists can "stretch understanding to take in the universe." Huh? Okay, more "science-speak." Shaking off that gem, next there is a discussion of frozen Woolly Mammoths discovered in Siberia. Jackson speculates contrary to conventional wisdom that the Mammoths actually were *alive* when discovered, recovering from suspended animation as the surrounding ice melted. Stating that the Mammoths were not cadavers "proves" if not highly suggests that, likewise, the Mantis could have also been preserved in a frozen state from some earlier geological age. Furthermore, Jackson extrapolates that if the Mantis was frozen it likely was held in a state of suspended animation. "You believe this?" Dr. Gunther rightfully queries. Jackson replies, "Well, I don't *disbelieve* it," and furthermore, until someone comes along with a better theory he's sticking to it. We're even shown an anatomical picture of a Praying Mantis in case audiences aren't familiar with this insect. The odd, almost maddening part of the matter is that despite his tortured use of scientific reasoning, he turns out to be … correct. And, predictably, the giant Mantis appears dramatically only to be destroyed.

So in conclusion, what a torturous, convoluted, if not downright flimsy chain of logic was used to 'prove' the existence of the monster that will inevitably menace mankind.

If no attempt is made to suspend disbelief in some inventive way, like through the nerdy role of a "Dr. Nedrick Jackson" as in *The Deadly Mantis*, then the story is more akin to outright fantasy rather than being science fictional in tone. There's nothing inherently wrong with either approach, except that in movies from America's Cold War period, use of explained science and technology became part of the formula for destroying giant mysterious invading monsters. But before certain weapons can eradicate, first scientists must use scientific reasoning and methods to *prove existence* of the monster. Only after the monster is explained on the basis of sound rationale (accounting for its particular powers, attributes, properties and abilities) may it be destroyed. This successful 'formula' mirrored America's approach to prevailing globally, both at the end of and following World War II through scientific invention and use of technology and industry. So why not use analogous strategies for defeating (non-human) monsters as well. Eventually, scientists learned that exposure to radiation (particularly fallout from nuclear weapons) was harmful, and so this led to another intellectual crisis (as well as yet another paradigm shift in monster film-dom).

Proving existence of the monster, however strange, imagination-defying or impossible they may seem lies at the root of 1950s and early 1960s paleontologically themed giant monster movies. Scientists seek natural explanations for creatures that seemingly defy known physical laws or even comprehension; therefore in these movies, sometimes, the "rules of what is 'naturally' possible must be bent." The giant prehistoric Mantis is certainly a stretch (bad pun!) and problematic when it comes to convincing viewers how a 5-foot long hook belongs to what Marge describes as a "cute little bug." But when it comes to movie dinosaur monsters (or dino-monsters), there are usually fewer mental gymnastics to go through before accepting what is happening on the screen as 'real-reel.' And so here let's focus on the essential ploy used in such period films, that is, how directors went about their task of proving to audiences *existence* of the giant dino-monster. There's a time-honored trick they usually resorted to, which was inaugurated by Jules Verne but reached maturation with Sir Arthur Conan Doyle. Remember—part of the traditional sci-fi formula is that before dino-monsters can be defeated—they first must be proven to exist, and that their existence must not be beyond the realm of the possible. Let's see where this idea stemmed from and how it played out in our favorite Cold War era giant dino-monster movies.

The Lost World:

In a recent e-mail communication, I extemporaneously referred to myself as a "dino-monster scholar," and so accordingly I'll try to live up to that lofty, self-serving title here by exploring another paleontological aspect related to suspending disbelief in sci-fi movies. I particularly enjoy dino-monster works of fiction and movies – those in which the science of paleontology is bent ever so gently (or sometimes rudely wrenched aside!), as to unleash throes of imagination. For example, three of our most beloved sci-fi novels of the past rely upon or incorporate to significant degree, geological and paleontological elements. Briefly consider, for example, Jules Verne's *Journey to the Center of the Earth* (1864, 1867), Edgar Rice Burroughs's *At the Earth's Core* (1914), and

Arthur Conan Doyle's *The Lost World* (1912); yes, all of which were later filmed with varying degrees of success.

Even when Verne's tale was written, geologists certainly realized that it would be virtually impossible (there's that word again!) to actually journey inside the Earth all the way to its center, or the core. And yet, through elegant story writing and plotting, Verne and, later, Burroughs masterfully suspended disbelief, (taking sleight of hand advantage of what was then at most, discarded pseudo-science), making it verily seem that adventures they wrote about were at least, well, entirely plausible when indeed they were not. The key is that they made us *want* to believe!

By contrast, because much of the surface of our planet was still relatively unexplored by 1912, the premise of Conan Doyle's tale (i.e. dinosaurs still living in a remote corner of the world) might have seemed rather more in the realm of, at best ... *possibly* "possible" at the time it was written. Yet he introduces a trick to suspend disbelief, one replicated by subsequent authors and directors through the next half century, one which in slightly different vein was introduced by Verne 50 years earlier. A "trick"?

Yes, and a rather simple yet suggestive one. In Conan Doyle's serialized 1912 novel, we learn that the first modern explorer to ascend the "lost" plateau in the Amazon is fictional character Maple White, whose notebook was retrieved during a subsequent mission conducted by Professor Challenger. Within this notebook are sketches of supposedly living animals Maple White witnessed on the plateau summit, including plated dinosaur *Stegosaurus*. A facsimile of the notebook drawing, showing the dinosaur staring at Maple White, was 'reproduced' in *The Strand Magazine*, drawn for the issue by artist Harry Rountree, although 'signed' for authenticity in the lower right corner by 'Maple White.' Seeing is believing, and contemporary readers of the novel would have had no recourse but to be tempted to 'believe' there was a person named Maple White who was actually on the plateau where he saw this living animal! Well, at least seeing the drawing published adjacent to Conan Doyle's remarkable words facilitated imaginative fancy. As in politics, or rather like Agent Mulder of *The X-Files*, we're overly prone or conditioned to accept information confirming what we wish to believe, "even in the face of facts," even if it's circumstantial. (quote from Richard Stengel, *Time*, Oct. 2012.)

A dozen years later, First National Pictures filmed an adaptation from the novel as a silent movie, *The Lost World* (First National Pictures, 1925), starring Wallace Beery as Professor Challenger. Notably, director Harry O. Hoyt included scenes where viewers not only glimpsed Maple White's drawings in the notebook, but also witnessed mounted prehistoric animal skeletons in London's "Zoological Hall." These scenes are visual extrapolations of passages selected from the novel, where Conan Doyle's Challenger presents physical evidence 'proving' to *Daily Gazette* reporter Edward Malone, as well as intrigued page-turning novel readers, the veracity of Maple White's observations. Both Conan Doyle and Hoyt suspended disbelief sufficiently, so as to lure us in, hook, line and sinker. At this point, we really want to thumb our noses at those fools who scoff at Challenger, an esteemed scientist who wields powerful conviction on this point. Buttressed by "evidence" presented by Conan Doyle and Hoyt through fictional Challenger, we simply *prefer* to believe that dinosaurs still live on the plateau, and so we shall.

Incidentally, Jules Verne did something similar in his *Journey to the Center of the Earth* novel, showing us 'real' fossils embedded in the walls of labyrinthine underground caverns, before these frightening 'prehistoria' come alive! First National's suggestive use of *Lost World's* genuine fossils wasn't the first-time dinosaur skeletons were used as movie props in motion pictures (i.e. *Adam's Rib*, 1923)—nor would it be the last. There's another innovation heralded by First National; namely, eye-catching promotional poster artwork featuring a giant theropodous dinosaur smashing elements of a modern city underfoot.

The Beast From 20,000 Fathoms:

Next, we see the trick re-utilized in 1953's *The Beast From 20,000 Fathoms*, involving release of a giant dino-monster from glacial ice following a nuclear warhead test detonation in the Arctic. Although following the blast, physicist Dr. Nesbitt (Paul Christian) sees a living giant monster moments before becoming injured in a snowy avalanche, medical personnel believe he's suffering from "traumatic hallucination." Nesbitt also claims because he's a scientist that his observations should override any psychological evaluation—but to no avail. There were no monster footprints left near where Nesbitt was rescued (although these could have quickly become covered in a storm). Soon the medical staff at least partly convinces Nesbitt that he was delirious at the time.

That is, until a ship, the *Fortune*, sinks at sea, with crew claiming they spied a sea serpent just before the wreck, an incident which Nesbitt reads in a newspaper account. Then he's quickly off to see Dr. Thurgood Elson (Cecil Kellaway), the "foremost paleontologist in the world," at the Department of Paleontology. Elson's researches are facilitated by his pretty assistant, Miss Lee Hunter (Paula Raymond). Immediately, the scene entices audiences. We see the mounted skeleton of some kind of sauropod (e.g. "brontosaur-like") animal and a large image of the American Museum *Tyrannosaurus rex* skeleton (suggested in a panel mount), both props from 1938's *Bringing up Baby*, in shots behind Nesbitt. Shades of Verne and Conan Doyle!! Why—if you melded the skull of

"Rex" with the body of the sauropod, you'd essentially have the 'resurrected' body of the Rhedosaurus Nesbitt saw in the Arctic.

In this scene infused with paleo-ambiance, Dr. Elson pooh-poohs Nesbitt's sighting of the monster and doubts his idea that such a creature might have hibernated like a bear for over 100-million years, until heat from detonation of the Bomb melted entombing ice. It's too unlikely that an animal could survive for such an incredibly long time. Lee Hunter chimes in, offering the story about "dead Mastodons" frozen in Siberia (i.e. although so far, no frozen Mastodons have been discovered anywhere—only Mammoths). She suspects Nesbitt's theory could have merit, especially because she's a "sympathetic bystander who has a deep abiding faith in the work of scientists; otherwise she wouldn't be one herself." Nesbitt's intriguing story seems worthy of investigating further—ahem, in her apartment.

Hunter has gathered together "all the sketches of the known prehistoric animals" for Nesbitt to examine. So the ability to define what the creature is (which by now has destroyed two ships and shortly a lighthouse tower), depends on deciphering paleoart. Ironically, as we'll see shortly, the unrecognized answer to the riddle existed plain as day in Dr. Elson's museum hall all the while. Sifting through the pictures, Nesbitt remarks that he didn't realize there were so many kinds of prehistoric animals, to which Hunter replies that they haven't even reached the Cretaceous Period yet. This is incorrect, however, because Nesbitt has already sorted through two Charles R. Knight painted restorations of "Trachodon" and *Tyrannosaurus* engaging a *Triceratops* family in battle, all of which are Cretaceous animals. Maybe Hunter isn't so geologically minded after all. Few audience viewers probably saw that only a subset of the pictures Nesbitt pages through were established samples of paleoart, as opposed to having been prepared for the movie by production artists and staff.

Then, suddenly, he notices a picture that resembles the animal he saw. Hunter quickly shows him another which is strikingly even more so like the monster he witnessed! By golly they've found it! Now the monster can be named – it's the Rhedosaurus. Of course, this is a drawing made for purposes of the film by production staff. Now Hunter and Nesbitt reason that if one of the sailors who survived the sea serpent sinking can also pick out the same picture (even though they've only seen the monster in the water with its legs submerged), then they could "prove" to Dr. Elson such a monster lives and that Nesbitt's outlandish theory is correct after all. When shipwrecked sailor Jacob Bowman (Jack Pennick) dramatically identifies the same picture, this concurrence is too uncanny. "What further proof do you need?" Nesbitt implores. Elson, chuckling at the thought that a Mesozoic animal could be alive today, is suddenly convinced, doing a 180-degree turnabout. Furthermore, Elson declares the monster is 100-million years old and a "direct ancestor" of the mounted pseudo-'sauropod' skeleton displayed before them.

Now that the monster has been identified and named, they begin planning how to stop it, leading to its inevitable destruction. But first, Elson must witness the creature firsthand in its natural habitat—the submerged canyons where its only fossils have been found. (One wonders how the fossils were found and excavated from such a deep underwater deposit, an unlikely and expensive proposition.) So in a later scene, Elson enters a diving bell. Upon spying the creature below the waves, we further learn that it is a "Paleolithic survival," that its "...dorsal is singular, not bilateral," and that the "capital suspension is cantilevric." "But the most astonishing thing about it is..." And at that moment, poor Elson expires. From his few gibberish observations, however, one may rightly question whether he might have already been suffering from the bends.

So clearly, use of dinosaur imagery—some real and some cooked-up for occasion of the film—facilitated Nesbitt's recognition of the Rhedosaurus, the critical first step required in a sci-fi flick before the offending beast can be exterminated. The mounted skeleton of the Rhedosaurus' direct ancestor is not a real dinosaur known to science. But it looks sufficiently like dino-skeletons, which many people in the audience would have seen examples of in natural history museums, to pass for a genuine dino-monster. (And incidentally, there was an Australian dinosaur sauropod genus named the *Rhoetosaurus*, a name which was pronounced much like Harryhausen's invention.)

Gojira:

Conan Doyle's strategy—the 'trick' was further employed in the decade's most pivotal, heartfelt, tightly wound and powerful giant dino-monster movie, Toho's *Gojira* (1954). *Gojira* was inspired both by RKO's *King Kong* and *The Beast From 20,000 Fathoms*. But it was instilled with much originality as well, derived from Japanese wartime and postwar experiences with nuclear bombs. Let us examine how the paleontologist in this story, Dr. Yamane (Takashi Shimura), identifies the sea monster that becomes known to the world as Godzilla. First, there is a backdrop to this film, stated and accepted by the audience. Godzilla claims status as a legendary sea creature of the "old days," when women were sacrificed to appease it, inhabiting Odo Island's coastal waters. The creature is also briefly witnessed by sailors on doomed freighters which have suddenly burst into flame, the first of which is named the *Eiko-maru*.

But can the sinkings instead more logically be resultant of an undersea volcanic eruption? That idea is quickly dismissed. The islanders put things together – as the fishermen aren't catching anything lately; it must be some large animal causing this. Ceremonial exorcism of the beast only seems to stir things up, as one eyewitness, "Shinkichi" sees a huge animal that night storming through their village. Following destruction by hurricane gusts, and given the local shipping catastrophes, Dr. Yamane is summoned to conduct a thorough investigation of the island. But while addressing officials before embarking for Odo, Yamane prophetically cites the case of "alleged" Snowman footprints in the Himalayas as examples of what Nature has hidden from us. "After all, Earth has many deep abyssal pockets that may contain secrets." A highly liberal thinking paleontologist!

When the investigation commences, Yamane quickly takes note of groundwater contaminated with radiation on the same side of the island where a huge footprint is found, as well as a trilobite embedded within this footprint. Trilobites became extinct about. 250-million years ago, but strangely this specimen is not fossilized. Startlingly, it's fresh! Then the monster appears but comes ashore on the other side of the island, where it is witnessed by dozens of people, including Yamane who takes a photograph. Afterward, an enormous trackway is seen along the shore leading back to the sea. This is a bipedal monster, unlike the Rhedosaurus. So it would seem that the monster's existence has been very well documented during this visit to Odo Island.

But there is still the pressing matter of explaining how such a creature could be alive today. Yamane is tasked with the job of suspending disbelief concerning this remarkable creature for the viewing audience. True—we've all seen the monster—but how could this be accepted as a plausible prehistoric beast? Much of what we need to know comes in the next pivotal sequence, where Yamane lectures to worried officials back in Japan. The news is grim, and what he has to say is bolstered by paleoart—once again—images of prehistoric life.

In a darkened room, Yamane shows slides of prehistoric animals; first a crudely drawn *Brontosaurus*, and then – significantly a painting showing two animals clearly 'borrowed' or swiped from Rudolph Zallinger's *Age of Reptiles* (1947) mural displayed at Yale University's Peabody Museum. Here amidst a volcanic field, we see a striped *Plateosaurus* of the Triassic Period, and more importantly a bipedal, pot-bellied *Tyrannosaurus rex*, dating from the Cretaceous Period—one of the principal creatures whose anatomical features were borrowed for creating the Godzilla suit. Dr. Yamane suggests, however, that like Godzilla, these creatures stem from the Jurassic Period, only *two million* years ago. Yamane needs to consult his geological timetable, however, as the Jurassic Period is defined by rocks dating from between 205 to 144 million years old. While refinements in this chart have occurred since 1954, even then the "Late Jurassic" was set around 150 million years ago.

Zallinger's *Tyrannosaurus* also has a single subdued row of integumentary, triangular scales situated along its spinal column, not unlike those seen on the Rhedosaurus. (As an aside, note that in Dr. Elson's aforementioned reference to the "singular, not bilateral dorsal" feature may be in reference to the Rhedosaur's low ridge of triangular spinal scales that 'became' far more prominent, jagged and ostentatious a year later in *Gojira's* Godzilla. Also, it is interesting that the 'ancestral' 'sauropod-ish' skeleton on display in Elson's museum has no bony, stegosaurian osteoderms along its backbone. So Rhedosaurus's (possibly derived) dorsal feature may be integumentary.

Next, Yamane theorizes how this creature came to be. He doesn't have any fossils belonging to this species, but suspects that during and after the Jurassic a creature "somewhere between" marine and terrestrial reptiles evolved a transitional "intermediate" form. Okay, but why is this animal alive today?

Yamane continues, after accepting its folkloric name of "Godzilla," and estimating from the photo that it is about 150-feet tall, which would defy bounds set by gravity. Yamane suspects that it was disturbed from its particular marine niche following recent hydrogen bomb testing in the Pacific. This conclusion is supported by "strong evidence," such as the trilobite (which he claims has been extinct for only two million years), and the radioactive sandy sediment found in its shell. Also, Geiger counter readings show presence of strontium 90, a product of nuclear bomb detonation. The sand that came from Godzilla, which is characteristic of Jurassic "red clay" marine deposits, absorbed an immense of radiation. Therefore, Godzilla must have absorbed this radiation as well. To metaphorically cap his summary, in a later scene we see Yamane contemplating the folly of mankind in his study adjacent to a miniature model of a *Stegosaurus* skeleton—the dinosaur that had plates along its spine (in this case shown aligned as a single row). Showing an image of a real dinosaur with plates arranged along its spinal column would fortify audience belief in a fictional creature, Godzilla, which has several rows of plates on its spine. This skeleton also foreshadows that of Godzilla's (plated) skeleton briefly witnessed on the seabed after Dr. Serizawa's (Akihiko Hirata) oxygen-destroyer has been successfully deployed.

Now Japan proceeds to destroy the monster, although Yamane would prefer to study its resistance to radioactivity. Yamane concludes the film prophetically, warning of other formidable godzillas that could be stirred up if mankind continues conducting nuclear tests. How good a job has Yamane done in explaining the monster? Has he suspended disbelief sufficiently?

Well, Godzilla is even larger than Rhedosaurus, and its enormous, bipedal proportions simply defy laws of physics. Yamane didn't explain any of that. After the monster begins breathing fire over the city, this curious adaptation is never explained by our resident paleontologist either. If the transitionally evolved Godzilla species is a direct mutation from the Bomb, he didn't say so explicitly. How does a marine creature like this 'breathe'so efficiently on land as well as underwater? No traces of gills are visible. But, while he also cannot explain how such a creature could survive in spite of its high degree of radioactivity, to his credit, Yamane did offer to *study* this death-defying characteristic. Even though Yamane was able to establish the monster's origins, he remains powerless to suggest how it may be destroyed, (a task left to Dr. Serizawa). Scenes where Tokyo is destroyed by the horrible phantasm visually absorb and mentally engage audiences to such high degree that there is no need for further suspension of disbelief that might be ordinarily achieved through continued pseudo-scientific lecturing, prattling, etc. of how such a creature may live yet defy physical laws. Recalling Hiroshima's arresting scenes of destruction, we simply must accept and grieve.

The 1956 Americanized version of the film, *Godzilla, King of the Monsters*, casting Raymond Burr in a spliced-in role as reporter Steve Martin, is not nearly as good. And science supporting 'belief' in the monster is much watered down, if not worse. For instance, Yamane declares that Godzilla is 400-feet tall, making it that much more of an impossible beast! Much of the rest repeats what is learned in the Japanese version (including its presentation of 'real' dinosaur paleoimagery intended to bolster belief in the "reel" monster), although maybe some things were lost in translation. But this time Yamane states that Godzilla was "resurrected" due to repeated hydrogen bomb experiments, which can be interpreted to mean anything from mutation to simple (re-)awakening from beneath the waves following a disturbing blast (after "rare phenomena" of nature allowed the species to "'reproduce itself," presumably while submerged, through geological ages until the present day). Or did he imply some sort of supernatural process? Then at the film's conclusion, Steve Martin remarks that the menace was "over" without suggesting Yamane's angst over what might happen if nuclear bomb testing were to continue (which of course happened).

Gigantis the Fire Monster:

The very nature of sequels is that the need for proving existence of the monster is minimized (because we've already seen it 'alive'). Also, explaining background science of the monsters can be presented much earlier in the film, instead of having the usual big, convoluted buildup deciphering the scientific mystery. Usually in such cases the focus hinges on addressing why it is alive (again!), and above all, how may it be destroyed (again). In Toho's *Godzilla Raids Again* (1955), there isn't too much emphasis placed on explaining the nature of a second Godzilla. This is indeed a second specimen as we can see from the more buck-toothed fangs and the slightly different sounding growl. But there is added intrigue over a new dino-monster which Godzilla battles to the death. But if audiences already accept that a Godzilla once appeared (i.e. in *Gojira*), then appearance of a second Godzilla *and* another species of giant dino-monster adversary is much easier to swallow.

This time, instead, two paleontologists (Drs. Yamane and Tadokoro) consider a new adversarial creature, the "Anguirus" which has been reported. It has been witnessed on a small Pacific island (sparring with the 2nd Godzilla) by two pilots, who share their observations with a scientific team. The pilots are shown dinosaurs drawings, a pile of paleo-pictures (yes—including another rendition of Zallinger's *Age of Reptiles Tyrannosaurus*), and they flip through pages in fictional "Dr. Preterry Hawdon's" illustrated dinosaur book. And then one of the pilots excitedly finds a picture of the quadruped Anguirus, (pasted into) the book, which as Dr. Tadokoro (Masao Shimizu) explains is an "ankylosaur" that was/is 150 to 200 feet tall, a geological contemporary of Godzilla. It would be unbecoming of me to challenge a true expert such as Dr. Hawdon in such affairs, but note this height would have made Anguirus as tall or taller than the first Godzilla. So either the new live specimen is smaller than its discovered fossilized specimens, or Hawdon meant 150 to 200-feet "long" instead of "tall." Also, borrowing from a prevalent former idea concerning the American plated *Stegosaurus*, Anguirus allegedly contains at least three different brains, adding to its aggressive nature. This time, most of the explanation on logically explaining the beast(s) addresses the new monster, Anguirus. Where has it been hiding? In undersea caves in the ocean abyss, created by geological upheaval.

Both species supposedly lived 70 to 150-million years ago, when most representatives of their species became extinct. However, Dr. Tadokoro claims that hydrogen bomb testing "woke Godzilla," and then also "awakened the ankylosaur." Dr. Yamane, sounding despondent and depressed, adds that Godzilla is "sensitive" to light, presumably due to "memories" of H-bomb testing exposure. This light sensitivity trait becomes a key plot element later on. For the second Godzilla is attracted to the coast by ignited gasoline tanks. Meanwhile, Anguirus seems relatively invulnerable to Godzilla's fiery breath. Yes, once more Japan is under a greater threat than the prospect of nuclear weapons.

As was the case with *Gojira, Godzilla Raids Again* was modified for American audiences, being re-released as 1959's *Gigantis the Fire Monster*. "Are there darker and more sinister secrets on Earth?" a narrator introduces rhetorically. Apparently! Much of the film proceeds as before, although here, Dr. Tadokoro's dubbed-in voice expounds in a slightly different and rather more convoluted vein. The new dino-monster is also referred to a "gigantis" variety of dinosaur, also known as an "anguilosaurus, or the Anguiris, "killer of the living." And strangely enough, it may "come alive after years of hibernation due to radioactive fallout," Tadokoro reads from the children's dinosaur book. Also, it says Anguirus could wipe out the human race. Perhaps most importantly for what is to follow, Anguirus happens to stem from a breed of "fire monsters," with emphasis on the aspect of fire. Did I say *fire*?

Now, in the American version, Dr. Yamane shows a 1:40 minute-long film showing the provenance of the two fire monsters, emphasizing their fiery nature. (Incidentally, to date, the source of stop-motion animated dinosaur footage in this clip remains a mystery.) This is essentially a "life through geological time" clip which has been developed "as science has been able to reconstruct it." This strange video purportedly explains why these monsters have a common fire derivation. Unfortunately, this addition to the American version of the film makes Dr. Yamane look rather feckless.

Since Paleontology's infancy two centuries ago, the "time-honored" means of conveying knowledge of the geological past in books, dioramas and natural history museums was through conveying imagery showing Life through Time. This was how people in the western world generally came to understand what past geological ages were like and what life they contained. The aforementioned "Age of Reptiles" mural by Zallinger for example, was the 20th century's most magnificent example of a life through time portrayal. So it is perhaps not too surprising that in an "Americanized" version of a Japanese dino-monster movie where scientists are striving to put the rational "spin" on two otherwise distorted and impossible creatures, that a traditional way of explaining where such beasts came from would be in deference to the standard for portraying the geological past, as in a pseudo life-through-geological-time filmic presentation.

And so the film clip (and we do not even know if it was made to match already scripted dialogue or whether if the script was tweaked to align with stock footage sequences), focuses on significance of fire in the creation of the titular "fire monsters." There is a heightened emphasis on explaining provenance of the so-called "fire monsters." The narrator of the film clip discusses Earth's fiery origins, boiling gases resulting in hotter and more primitive forms of life that were born, adapted to fiery matter. Indeed, "their very existence was based upon the element of fire." Oops—did he say "element" of fire? Now the science is sounding downright alchemical, or 'medieval.' Anyway, dinosaurian creatures were evolving that breathed and apparently survived in fire. Indeed, "fire was a part of their organic makeup." Cosmic rays bombarded the Earth causing extinctions and forcing the fire-adapted monsters to go underground to survive.

Then Earth went through major upheavals; volcanic disturbances occurred and lava erupted from Earth's core. When the monsters "came out of hiding" after nuclear bomb tests filled the atmosphere with radiation, because they were conditioned to being exposed to fire, they were practically indestructible. Okay, we get it …. Fire!!

Well, once reanimated in Japan, one fire monster kills the other, and the survivor (i.e. the second Godzilla) later becomes encased in glacial ice. So ice prevails over fire…. (until *King Kong vs. Godzilla* when the 2nd Godzilla melts his way out of the glacier).

Rodan, the Flying Monster:

In 1956, Toho unleashed one of its most endearing giant monsters in a film later billed as *Rodan, the Flying Monster* for its 1957 American release. Although I haven't seen the Japanese original, it generally follows the American version, with exception of the added introductory footage of two nuclear bomb tests added to the beginning, mirroring how *The Beast From 20,000 Fathoms* began. This footage offers a pseudo-explanation for what will be witnessed shortly. A narrator ominously questions whether our planet can sustain such staggering blows as the Bikini Atoll hydrogen bomb test ….without causing retaliation by a horror undreamed of?

When occupants of a small mining town are terrorized by giant insect larvae living in the mine shaft, it isn't obvious (yet) that we're dealing with another set of prehistoric horrors, or that their origins in present day stem from dispersal of radiation. That former realization comes from observations made by "Shigeru" (Kenji Sahara), who witnessed an awful scene deep within a coal mine cavern, divulged in a flashback sequence as he recovers from a head injury. Shigeru's terrifying recollections of a gigantic winged egg hatchling devouring the enormous larvae ("Meganulon") leads to biologist Professor Kashiwagi's (Akihiko Hirata) scientific determinations. Kashiwagi appears to be an 'expert' on Mesozoic reptiles.

Meanwhile, planes are being destroyed over the Pacific by high altitude, supersonic "UFOs." No doubt Toho must have latched on to America's flying saucer craze during this period. And when a blurry photograph of a large winged creature is obtained that looks like a gigantic "bird" with a clawed foot, Dr. Kashiwagi has a strong hunch. After examining the photo, Kashiwagi immediately wants his assistants to produce his file on "transitional saurian forms of the Mesozoic and Carboniferous periods," particularly information on the pterosauria.

With Shigeru's memory suddenly restored, the means by which he is able to "convince" a scientific team as to what he actually saw—an existing giant dino-monster—is again accomplished through reviewing an assortment of paleoart pictures. This time, it is a drawing of a winged pterodactyl known as *Pteranodon* that Shigeru seizes. Henceforth, conclusions begin rapidly piling up! Dr. Kashigawi goes on a science-speak roll. Possibly facilitated by his examination of the blurry photograph, he becomes completely convinced the animal is real, despite its unlikelihood. But many questions must still be answered. After all, this is a science fiction movie and before mankind may plan the dino-monsters' destruction eventually, somebody must explain why these creatures are out and about. That's simply the order of such things. After all, prehistoric beasts should *not* be living in our present.

Now much speculation is offered, masquerading as fact, which everybody is buying hook, line & sinker once again. The scientists seem to be rapidly leap-frogging from one half-baked conclusion to another. Yet the most maddening thing is that, ultimately, they're right! Perhaps these animals never really died, Kashiwagi proposes. Maybe they only "slept" through countless millennia, deep within the bowels of the Earth. It is deduced that because the

Rodans are reptiles, then like snakes they must hibernate underground after feeding. Maybe they got caught down there. (As in Verne's "Journey," geological circumstances – including a nearby volcano, the ideas of deep time, a subterranean setting and prehistoria living within have been marvelously melded.)

A scientific team ventures back into the cavern, obtaining giant egg shell fragments. After examining the specimen through a microscope, Kashiwagi declares excitedly, "It's egg shell alright … (and) it's reptilian!" Furthermore, its cell structure is "primitive," therefore Late Cretaceous in age. Next, using their modernistic room-sized "electronic computer," measurement of the shell curvature permits calculation of the flying creature's physical dimensions. The results are staggering. The creature, a "carnivorous pterodactyl closely related to the extinct *Pteranodon*—(i.e. species "rodan")—which may have been fully grown upon hatching and capable of flying right away had a 500-foot wingspan. These wings could generate typhoon magnitude force winds, and the creature (it later turns out there are *two* adult pterodactyls) must be the cause of the recent UFO sightings. Not only that, but the creature must have weighed 100 tons. Yet it somehow flew! Keep in mind that until now Shigeru is the only survivor who has seen one of the flying monsters up close. So we've concluded that such a giant mysterious dino-monster exists.

But we aren't ready to summon attacking military forces yet, because the scientists still haven't addressed a key question that must be considered. For if these *Pteranodons* are considered extinct, why is this "Rodan" species flying around the Pacific? Kashiwagi offers more outright speculation disguised as a "theory of his own" which is good enough to work with. Paraphrasing: millions of years ago an egg was hermetically sealed and buried in a volcanic eruption, perpetuating the germ of life until atomic explosions fractured the Earth's crust, thus admitting air and water. Infiltration of warm water caused the egg to hatch, and also "explains the giant insects." This is perhaps why the file on the Carboniferous Period was sought in an earlier scene, because during the Carboniferous Period, insects and bugs did actually grow to larger sizes, sustained by increased atmospheric oxygen levels. But the Carboniferous ended some fifty million years prior to the beginning of the Mesozoic Era (when oxygen levels were even less than today) and therefore long before pterodactyls evolved.

In *Rodan* we see the 1950s Cold War formula exemplified once more. The giant dino-monster's existence and otherwise impossible physical dimensions and properties are 'proven,' facilitated using paleoart imagery of real prehistoric animals buttressed with paleontological authority. Only then can explanations for why the creatures happen to be alive today be concocted, fortified by an ominous introductory specter of radiation dispersal caused by nuclear bomb testing.

Paleosaurus—The Giant Behemoth:

In appearance, the quadrupedal dino-monster known as "Paleosaurus," featured in *The Giant Behemoth* (1958) would appear to be a near ripoff of Harryhausen's Rhedosaurus. Even the respective story elements curiously converge. But Paleosaurus also borrows from the Japanese line of dino-monster films, as it is a radioactive beast. So how is disbelief suspended sufficiently so that viewers can enjoy this film that much more while enjoying the monster's rampage through modernity?

The plot seizes upon recognized threats to the oceanic world: fear and worries over bio-magnification of pollutants up the food chain from contaminated plankton, (a phenomenon known to science). The essential part of the theory of this dino-monster's powers is that radiological elements may be biologically concentrated in creatures at the top of the food chain. In this way, a large carnivorous dino-monster would receive the lion's share of radiological elements concentrated within its bodily tissues. During a lecture, marine biologist Steve Karnes (Gene Evans) prophetically warns of something awful that someday could rise up from the sea to "strike back at us," given man's continual dumping of nuclear wastes into the sea and atomic bomb tests generating radioactive fallout. And something does, not unlike the Behemoth of biblical prophecy.

At first, existence of such a monster is deduced from several lines of indirect evidence. After several mysterious deaths, with victims showing acute symptoms rather like radiation burns seen in those exposed to the nuclear bomb at Hiroshima, followed by an enormous fish kill, and presence of a pulsating goo washed ashore that burns human flesh and sea monster sightings, Dr. Karnes' plagued mind is racing toward a most peculiar theory. As discussed with atomic energy commission scientist Dr. Bickford (Andre Morell) he doesn't want to overlook any evidence. Clearly, however, Karnes is losing his objectivity, as he's becoming wedded to a radiation theory, or what Bickford later calls the 'Monster or Behemoth theory.' "Something came out of the ocean," Karnes declares.

So, as is customary in such proceedings, samples are obtained and delivered to a well-equipped laboratory for testing. In this case the samples are dead fish, which are then subjected to a device that analyzes the radiation signature as distributed within the tissues. A high concentration of radioactive particles is found in one specimen. Next, Karnes boards a ship where he directly spies a "Nessie-looking" type of sea monster through his binoculars swimming through the waves. As they near the sea-monster, high levels of radioactivity are generated, so they've established a solid correlation between radiation and a sea-monster 'source.' Ships are damaged subsequently, testifying to the great strength of this gigantic creature. Meanwhile, Bickford has performed some nebulous "independent confirmation from his own laboratory." Strangely, the glowing substance in the radiograph was found to contain "cells from the stomach wall of an unidentified species." Well, that unidentified species is not just another run of the mill marine sea monster. It can walk on land too, as "proven" by a photograph of a truly enormous footprint. Now, more than ever, Karnes and Bickford need to discuss matters with – you guessed it! Britain's "best man in paleontology." Yes—they're rapidly jumping to conclusions, but of a kind that can only be verified by a scientist who specializes in the study of prehistoric life!

Thus Karnes has directly observed the creature, but before it may be destroyed he also needs to know whether the monster may be pseudo-prehistoric as well. Once again, as before, the paleontologist, Dr. Sampson (Jack MacGowran) relies on paleoart and other forms of paleoimagery fortifying a message to viewers that his pronouncement of the monster as a variant of prehistoric beast is absolutely true. When Karnes and Bickford enter Sampson's office area, they stand before the leg bones of a sauropod. (As opposed to the mounted skeleton in *Beast From 20,000 Fathoms*, this time we only see a partial skeleton.) But shortly we also see an enlarged photograph of another mounted specimen suggesting there is also a skeleton of the *Iguanodon* displayed in the room, as well as an array of other smaller fossils.

From the photo of the huge footprint, Sampson deduces that the creature belongs to "the old Paleosaurus family – like this specimen we've got here" (i.e. pointing to the partial skeleton)… only much larger. How convenient! He estimates the creature may be 150 to 200-feet long. And, following in tradition of his profession, Sampson also produces a picture from his file showing the "whole creature." Having a picture of the living animal as reconstructed from fossil bones, importantly provided under a paleontologist's authority, makes it seem more genuine and prehistoric. To further reinforce connections between the living animal with the nature of its "scientific" basis, in a later scene when Karnes and Bickford are apprising government officials about the monstrous threat, another drawing of the Paleosaurus can be seen in the conference room on a chalkboard.

When informed that the creature is alive, Sampson grows excited. Now he wants to lead an expedition to find it. Sampson explains that it may be headed to the Thames River estuary, where these creatures would go to die in the shallows, which is where their fossil bones have been found. He also opines that he always believed the Paleosaurus lineage was still alive somewhere, as all those reports of alleged sea-monsters could only have been founded upon sightings of the "tall, graceful neck of Paleosaurus" riding through the waves. But why is it still alive today? Sampson can only meekly offer that "No form of life ceases abruptly," which isn't a definitive explanation. It turns out that Paleosaurus was electric like an eel, which allows it to project its radiation through deadly emanations. Strangely, it also can't be detected on radar. Sampson never explains how the monster can breathe underwater without apparent gill structures. Presumably, not unlike those other amphibian dino-monsters we've encountered before, Paleosaurus has an ability to retain huge amounts of oxygen to remain submerged for prolonged periods. Although the living anachronism Paleosaurus has prehistoric, paleontological ties (e.g. most likely to Rhedosaurus), Samson never reveals the geological age its distant ancestors belong to.

And so London is attacked for the second time in a dino-monster film, (i.e. the first time being in First National's *The Lost World*). Ultimately, knowledge of the Paleosaurus' radioactive condition facilitates a means for killing it. However, by dispatching it underwater with a nuclear-tipped device to accelerate its inherent radiation poisoning, soon the decomposing cadaver will release its radiological components which will then disperse along the Thames and back into the sea. Terrifyingly, another huge fish kill is then reported along the eastern Atlantic seaboard, from Maine to Florida. An implication is that more Paleosauruses will soon invade America.

Gorgo:

Gorgo is rather different. By the time of "Gorgo's" entry, American audiences were becoming quite used to the idea of amphibious dino-monsters that were of astoundingly gigantic size wreaking havoc on metropolitan centers. And also there were those common ties to nuclear bomb testing, although in the film, *Gorgo* (1961), the origins of Gorgo and its mother (whom I have referred to as "Ogra") are not linked to nuclear weapons testing. Instead of radioactivity, mysterious seafloor paroxysms cause the appearance of two huge dino-monsters off Nara Island in the Irish Sea. Following an underwater eruption, two divers, Joe Ryan (Bill Travers) and Sam Slade (William Sylvester) pull their boat into harbor at Nara Island, where they soon shall make an amazing discovery.

A Viking longboat figurehead of dragon-like configuration held in private collection of an archaeologist is referred to as "Ogra," suggesting that the titular dino-monster is a beast of legend, which has been historically sighted in coastal waters. Soon on a treasure dive offshore, Ryan and Slade spot the real dino-monster that caused another man to die from fright. The seamen lack paleontological training, so that when Joe asks Sam what he thinks they saw, Sam can only reply that he doesn't know. But whatever it was he never wants to see it again. Later in the evening, (the younger) "Gorgo" mounts the shore but is repelled with burning torches. Following a harrowing experience in a diving bell, Gorgo is captured in a fishing net. After reporting the creature's capture, two Dublin paleontologists, including Prof. Hendricks (Joseph O'Connor) proclaim that the specimen should be delivered to their institution where it can be studied because it is of "enormous scientific value." But Ryan and Slade prefer to maximize their profits. And so instead the creature is transported to London where it will be exploited, exhibited in Dorkin's Circus. At Dorkin's, the dino-monster is renamed "Gorgo," abbreviated from name of the mythical Gorgons, (and is not based on the real tyrannosaurid name, *Gorgosaurus*).

As of yet, despite appearance of two inquisitive paleontologists, Gorgo's quasi-prehistoric nature remains uncertain. The purpose of these scientists is diminished in contrast, say, to the more prominent roles of Dr. Yamane in *Gojira* or Dr. Elson in *The Beast From 20,000 Fathoms*. At first, in *Gorgo* we learn more about the dino-monster's origins from news reporters and a publicist rather than directly from the paleontologists. We hear on a news reel, for instance, that the monster was probably released from an undersea cavern from an unprecedented eruption. Later, paleontologists lament that the creature was essentially stolen upon its delivery to London, stating they have lost a unique opportunity for studying evolutionary biology. They urge caution because (like Rhedosaurus) the creature might carry disease and parasites. Then a publicist declares that Gorgo – the most Earth-shattering discovery of the 20th century, "should have been extinct ten million years ago." However, one may wonder why it absolutely *should have been* extinct at that time.

But the paleontologists' role becomes more defined in a later scene, where the two scientists meet with Ryan and Slade to show them a preliminary finding they've made. And this is where the movie theme shifts from exploitation to vengeful parental destruction, ultimately leading to regained freedom. The two scientists divulge that Gorgo is a mere juvenile of its species, and that its parent is truly colossal – nearly 200 feet tall! And for our purposes, they resort to use of paleoimagery to bolster their argument, suspending disbelief with a drawing showing reconstructed skeletons of a related species in both juvenile and adult forms. The problem here is one of gross misidentification.

These skeletal drawings do not in the least conform to Gorgo's suitmation design. Bipedal Gorgo stands vertically upright (like a human) and has a proportionally large head and clawed hands. But the creatures shown in the drawing, which are both clearly dinosaurian in nature – although this is not stated, are stooped over with more horizontally inclined vertebral columns and heads and arms that are proportionally smaller. In fact, the skeletal drawings are more like those of a real, herbivorous dinosaur named *Camptosaurus*, which only grew to 25 feet in length. Interestingly, while there are no mounted skeletons of any dinosaurs shown in Prof. Hendricks' institution, a pair of skeletons are displayed at the Smithsonian Institution – both adult and juvenile specimens of the *Camptosaurus*. But Gorgo appears to be derived from a flesh-eating, theropodous type of dinosaur, not from a herbivore (even though all non-avian dinosaurs were extinct long before ten million years ago).

Anyway, the stern, prophetic warning from the paleontologists is that the parental form may soon emerge from the sea. But, ironically, the paleontologists are a primary cause of the calamity that is about to ensue! The fact that Hendricks had earlier suggested that during the voyage to London Gorgo's skin should be kept wet, allowing skin particles to wash into the sea, permits the mother to trace her offspring's scent all the way to London, causing the usual havoc and inevitable destruction.

And so London dies once more.

Reptilicus:

Speaking of inevitable mayhem and destruction … Americanized version of *Reptilicus* is even far stranger. Here we have a case of a mysterious soft tissue "fossil" or relic that creates its own self-reconstruction through a biological regeneration process. Imbued with prehistoric origins by scientists in this film, *Reptilicus* (1961), this particular creature appears more draconian than any kind of more familiar species of dino-monster (until introduction of Toho's Ghidrah).

Miners prospecting for copper in Lapland deep in the Arctic bring up some soft tissue, issuing blood, thawed by the drill bit. Specimens include a fleshy substance with a leathery skin and some fossilized bones. The team leader, "Svend," (Bent Mejding) suspends drilling, intuitively realizing they'll need to consult a paleontologist. And so Professor Martens (Asbjorn Anderson) and Dr. Peter Dalby (Poul Wildaker) enter the picture. An ominous closeup shot shows the tissue pulsating eerily.

Arriving on the scene, Martens and Dalby determine the bones belong to a large creature which had been frozen underground for a long time. And here, to suspend disbelief, Dalby refers to the traditional tale of those Woolly Mammoths frozen in Siberia (where this sort of thing happens all the time). However, this new find is unique because the bones appear to be reptilian. The specimens are delivered to Denmark's Aquarium (instead of a paleontological science hall or wing). And while we see no skeletal reconstructions of dinosaurs within this building, Dalby and Martens are adeptly striving to reconstruct the mysterious dino-monster from its fossil bones and soft tissue remains. One paleontologist claims, "…reconstruction is impossible" because there may be insufficient bones – although this creature would be expected to be unlike any other. A clue is that the bones appear to be cartilaginous, like those of a shark. Good thing their lab is situated in an aquarium then! They're really sounding confused, but fortunately Svend delivers additional bones with which to facilitate their reconstruction.

While the paleontologists scramble to restore appearance of this monstrously sized creature previously unknown to science, their answers are about to unfold before their very eyes. Astonishingly, the key to it all turns out to be a marvelously preserved segment of tissue from the creature's tail tip. From this specimen, the paleontologists magically deduce that the creature was 90-feet "or more" in length. Also, it must be a "giant dinosaur" that lived 70 to 100-million years ago, even though not a shred of evidence is offered to support these conclusions. Thus the emphasis is clearly being placed on the perfectly preserved tail tissue section as opposed to the other bones. Soon the old fossil will create its own self-restoration.

Now things quickly unravel. When Dalby carelessly leaves the cold storage room door ajar, the specimen thaws overnight. But rather than decomposing, the tissue is alive and growing. While frozen underground in Lapland, its cellular tissues have been maintained in a suspended state of animation for millions of years. Not only that, but the tail segment is regenerating itself into ….. *what*? At this stage, they can only suggest that whatever the tail will grow into will surely be "reptilian."

Now the two paleontologists have documented that the creature is a prehistorical kind of creature, and that (at least part of it so far) is alive. So naturally it is time to clumsily add the military component of the equation. Enter General Grayson (Carl Ottosen), a benighted soul who at first can't figure out why he's in the movie. But we know, right? (Wink, wink.) Yep, it's nearly time to kill the prehistoric monster, which by the way has some catching up to do. And besides, where is that paleontological presentation of the paleoimagery (i.e. drawings of dinosaurs, etc.) which we've seen in all the giant dino-monster films discussed here?

As in *Gorgo,* newspapers seize the reins. A reporter instead of a scientist names the creature – "Reptilicus." And headlines begin circulating: "Prehistoric Monster growing in tank" and "Incubator tank feeds monster from past." The prehistoric nature of this beast clearly seems to have been established. As it grows, Martens observes its huge bony scales (which later turn out to be invulnerable to cannon shells), and an acid slimy gland secretion, spewed from its mouth. Suddenly, the creature undergoes an exponential growth spurt, possibly stimulated by electrical charges in the atmosphere during storms, and escapes. Nobody left alive has seen the whole animal yet, however. But an unseen trackway leads to the sea—so it must be amphibious. (Reptilicus even uses its wings to fly in scenes expunged from the English version of the film. See back cover of this book.)

Now here's the scene you've been patiently waiting for. Martens suggests to Grayson, who is holding a stack of prehistoric animal drawings, that Reptilicus is a "cross between one of *these* and an amphibious reptile." And the picture visible in Grayson's hands happens to be that of a sauropod – like a *Brontosaurus*, shown standing adjacent to a diminutive man, for scale. The brontosaur doesn't in any resemble winged, draconian Reptilicus! This presentation of real dinosaurian imagery is clearly the least significant or convincing of any such similar scenes among classic Cold War era dino-monster movies. It has no real purpose here. It's a throw-in, added probably because other predecessor movies—those aforementioned here—feature these "illuminating" moments. Martens elaborates with his latest theoretical mumbo jumbo, declaring "Nature went through a long period of experimentation ... I believe Reptilicus is one of Nature's attempts to bridge the step from reptile to mammal." Not only that, but Martens somehow pins down its geological origin to 70-million years ago. Actually, however, as demonstrated by the fossil record, that "bridge" had already taken place about 180-million years before Reptilicus' suspected 70-million-year-old age of origin. In reply, one of the military brass stereotypically remarks that whatever it is we must destroy him. And even though Reptilicus is maybe mostly reptilian and not essentially a sea creature, they'll have to continue their search underwater. Hmmm—impeccable logic, eh?

By the time of Paleosaurus' prior arrival nuclear bomb testing had become a knee jerk reaction indirectly resulting in, or otherwise 'proving' the inevitable appearance and existence of a giant, unnatural dino-monster. However, radiation hasn't caused Reptilicus' appearance. The most important aspect of Reptilicus' unnatural history is that it cannot be bombed, (including presumably with a nuclear warhead), because if Reptilicus' body is blown to bits, each piece will regenerate into separate Reptilicuses. Prof. Martens is asked how to destroy the creature, but alas, channeling Dr. Yamane, he does not know. They eventually figure out how to poison it in one piece, although another chunk of thawed Reptilicus flesh is already ominously pulsating, regenerating on the ocean floor. Fortunately for us all, no filmic sequels were ever made. And so, unlike others under consideration here, this film lacks a clear, foreboding message for mankind, other than, perhaps, be careful when drilling for copper in Lapland. Or, more likely, fossil tissues can be dangerous if they're regenerated into full-bodied living creatures (something we learned in *Jurassic Park*).

Conclusion:

Films addressed in this article cover the Cold War period when such giant dino-monsters and assorted species of 'prehistoria' were still taken most seriously. These movies came with stark messages for mankind. Early on, producers thought such movies were proper vehicles for delivering stern messages, and the formula stuck. At best such creatures are *faux*-prehistoric, or prehistoric anachronisms since they're living in modernity—so technically they can't be "prehistoric" in the now. All of them are biologically impossible according to natural laws. No known vertebrate animal could hibernate for millions of years. The Japanese variety are too huge and would collapse under own weight, or their organs wouldn't properly function. Radioactive monsters couldn't possibly live (e.g. Paleosaurus and Godzilla), even if mutated. Rodan and Reptilicus have wings but are too large to fly. Furthermore, it seems that in the Japanese films, radiation isn't usually the accepted cause or culprit of the dino-monster's great size. (Although along with Paleosaurus and Rhedosaurus, nuclear bombs instigate their appearances and their wrath). Their giant sizes being somehow *resultant* of radiation exposures became grafted into popular culture and general ideology of 1950s giant monster movies through several American directors (like Bert I. Gordon, or David Weisbart who produced 1954's *Them!*). Also, several of the dino-monsters are amphibious, yet lack signs of gills or other aquatic adaptations (like early whales). Their invulnerability to conventional weapons defies comprehension. Besides Reptilicus' penchant for fresh cows on the hoof, dietary habits of these giant creatures are left to imagination.

Because aberrations such as Rhedosaurus, Godzilla, Anguirus, Paleosaurus, Rodan, Gorgo and Reptilicus are such flagrantly different sorts of "dinosaurs," extra emphasis must be placed on explaining their occurrences and unnatural history. In cases of movieland dinosaurs that are "real," that is, more closely aligned with those known to science, as in films like *The Lost World* (1925), *Jurassic Park* (1993), *King Kong* (1933), fewer pains and concocted stories are needed to explain why these supposedly extinct creatures are still alive. In such cases, scientific intrigue simplifies to something like this, 'Yes – it's a genuine, living tyrannosaur. But then it must be explained how has it (a.) survived, or (b.) been resurrected into modernity?' The answers usually are, 'Gosh, we aren't scientists, but we just ever so conveniently found it on a lost world, which is naturally where one would expect to find such animals.' Or, 'yes, we are scientifically trained people, and so we recreated it in a laboratory using biochemistry.'

Also, by the mid 1960s, real dinosaurs had attained a new unprecedented level of popularity among America's masses, which continued to grow during throughout the next half-century. It just wasn't necessary to introduce or explain what a dinosaur was anymore and suggest or imply using assorted paleoimagery spliced into the film that if real dinosaurs existed (e.g. some with curious plates, horns or ones that walked bipedally and had large carnivorous heads), then suspending disbelief, something else could *also* exist that was *allied to* real dinosaurs yet was not so much like any real kind of prehistoric animal known to science. Of course, following Reptilicus, audiences probably had fully wised-up to the fact that these giant dino-monsters and those that would soon follow, (e.g. Gamera, Gappa and Yongary), clearly were not at all genetically allied to *real* prehistoric animals, such as those appearing in Zallinger's famous *Age of Reptiles* mural.

Usually in sci-fi and horror flicks, disbelief in what is about to transpire at the gruesome hands of a mad scientist, cursed with deranged capabilities, is satisfactorily suspended by use of apparatus in a La-BOR-a-tory, bedecked with splendorous, electrically sparkling Strickfaden devices, or bubbling flasks of chemicals in flasks and glassware, or even physics inventions (e.g. Time Machines, Trekkie - transportational devices, radiation death ray emitters, etc.). But with paleo-themed sci-fi/horror monster movies circumstances are necessarily different. Here, different pseudo-scientific lines of evidence must be called upon to suggest the veracity of what is unfolding before our very eyes. And during the Cold War era period, movie producers and directors recognized the 'formula' and value of presenting examples of real paleoimagery (dinosaur skeletons and paleoart drawings) on screen to help prove reality of giant fictional dino-monsters that would soon plunder civilization.

Chapter Three — *Reminiscing Over Dinosaurus!*

"Alive! After 70 million years! Roaring! Walking! Destroying!"
(Ad line for *Dinosaurus!*)

Sometime in the early 1960s (post-September 1961), our dad began a semi-regular ritual that continued through the next decade—taking my brother Rick and I to see the early Saturday afternoon prehistoric monster movie at the show. One of the earliest of these was a delightful and masterful—that is, to my then 2nd or 3rd grade psyche—movie titled, *Dinosaurus!* (Fairview Production/Universal-International, 1960). I departed exuberantly from the theater that day, in a daze thinking I'd just seen the most utterly fantastic special effects projection of 'real' dinosaurs since those witnessed in RKO's *King Kong*, (which by then I'd seen televised two or three times). As was common back in the day, one had no way of knowing when you might have opportunity to see a movie released at the cinema ever again—televised (unless you went back to the theater for the next matinee). In my case I didn't see the film again until over half a century later when it was broadcast on horror movie host Svengoolie's Chicagoland program one Saturday evening (e.g. during the mid- to late 2000s), with occasional repeated airings over the next decade.

By then I *had* read about the movie in two books published between 1999 and 2002 (Neil Pettigrew's *The Stop-Motion Filmography*, and Mark F. Berry's *The Dinosaur Filmography*, respectively). In effect, I had re-acquainted myself with the movie ... twice ... without having actually seen it again ... yet. But reading their written summaries only whetted my appetite for that exalted day when *Dinosaurus!* might air again someday, which then did happen only a few years later when Rich Koz (e.g. Svengoolie) must have 'telepathically' received my urgent alpha-cerebral transmissions pleading for its broadcast.

My *adult* reaction to *Dinosaurus!*? Well, on the positive side it still held charm. Remorsefully, anticipated feelings of nostalgia were swept asunder by harrowing regret for all the years gone by in-between—quite a contrasting kind of perspective relative to my first viewing! For if I felt that differently about this once beloved film, then how drastically had my mindset changed over the years? The answer should have been glaringly obvious but had eluded detection. I was beguiled not realizing as I should have known that few fondly remembered 'children's movies' remain as enchanting for us in maturity. That evening watching Svengoolie's *Dinosaurus!* broadcast I could finally declare, 'lesson learned.'

Dinosaurus! is largely ignored today too. Few would consider it to be a 'classic' movie. Why? The story (i.e. screenplay by Dan E. Weisburd and Jean Yeaworth—from an original idea by co-Producer Jack H. Harris) isn't too far-fetched for this genre, and actors perform reasonably. Authorities concur that actor Gregg Martell's characterization of the resurrected Neanderthal Man is rather exceptional. But the movie's special effects pale beside magnificent cgi-animation inaugurated in 1993's *Jurassic Park* (and many fantastic films thereafter). Therein lies root of today's problem. *Dinosaurus!'s* two dinosaurs—*Tyrannosaurus* and *Brontosaurus* just don't cut it anymore. So is such disregard tantamount to a 'referendum' against old-school stop-motion animation?

Here's an outline of the overall plot, for now skipping lighthearted scenes.

Not unlike Hollywood's Frankenstein Monster, during a storm, lightning revives well-preserved corpses of two dinosaurs—*Tyrannosaurus* and *Brontosaurus*—and a caveman recovered from a near-shore deposit in the Caribbean. There are subtle indications that this 'Triassic trio' may have been alive together—Flintstones-like—prior to burial. Further, against all odds and logical sense, they were discovered "frozen" on the seabed. Well, a dismal effort *is* made to explain this mysterious *in situ* condition. For soon handsome good guy "Bart" posits, perhaps counterintuitively, that dynamiting during dredging operations "… must have blasted through the rock that entombed them. Some compressed gas must have caused the freezing, I guess."

Meanwhile, one greedy, heinous islander "Mike" wants to procure the caveman specimen and sell him on the U.S. mainland for lots of cash. The boy character "Julio" doesn't like Mike—especially when he dastardly destroys his plastic Marx toy dinosaurs. Julio befriends Caveman and their brontosaur 'steed,' while being chased by the hungry tyrannosaur. Inevitably the two dinosaurs fight, with carnivore triumphing. The wounded brontosaur later drowns in quicksand. Sacrificing his life, Neanderthal heroically rescues Julio and a kind woman named Betty. Frightened islanders seek refuge in an abandoned shoreline fortress, with a surrounding gasoline 'moat' ablaze. "Bart" bravely battles tyrannosaur from within a "steam' shovel. Its swinging metal arm forces dinosaur over a cliff—thus plunging the final prehistoric relic into the sea. (We've since seen analogous 'manned-machine versus tenacious monster' climactic scenes subsequently in 1986's *Aliens* and also in 1993's *Carnosaur*.)

Dinosaurus! offers continual 'hang-on-to-your-seats, boys & girls' action, tension and drama—plus bits of genuinely funny comedy performed by the Caveman to boot. So how *couldn't* this be a great memorable film, not merely a personal favorite! Why is it so overlooked, (yes, even by Svengoolie in recent years)?

The matter becomes more perplexing when considering some of the familiar names 'behind the movie' (e.g. besides actors who have relatively few credits and are largely unknown today). First, two famed science fiction writers, Alfred Bester and Algis Budrys, hammered out an initial story outline and idea; Budrys's 'synopsis' ran up to 600 pages long! According to Bill Warren, "Dan Weisburd was brought in to write the screenplay, (be-) cause Budrys was all wrote out." And those featured dinosaurs weren't created by hacks. After consulting with Willis O'Brien, the models were sculpted by Marcel Delgado—of 1933/RKO *King Kong* fame! According to Mark Berry (author of *The Dinosaur* Filmography, 2002), O'Brien "helped convince Jack Harris to use the stop-motion technique." (p.91) To my brightly lit youthful eyes then, Delgado's *Tyrannosaurus* model seemed the best such made *ever* appearing in a movie. Other stalwart pros, Wah Chang, Phil Kellison and Dave Pal, did animation photography, (Victor Delgado, Marcel's brother, worked on armature-building and "mechanicals"). Not a shabby behind the scenes lineup!

And the script, especially inspired scenes written by Jean Yeaworth for Martell's Neanderthal persona, is generally decent. Further, while most of the acting in this movie is at least passingly good, Martell's performance is consistently praised. Michael Klossner, author of *Prehistoric Humans in Film and Television* (Mcfarland, 2006, p.225), even declares that "Gregg Martell's work as the Neanderthal Man in *Dinosaurus!* is the finest performance ever by an actor playing a prehistoric person."

However, let's probe aspects of the movie which critics have denounced. First—those two dinosaurs. Their menacing visages were projected through both application of smaller stop-motion models as well as several larger props (e.g. the *Brontosaurus* neck and head), built for close-up, interactive scenes with human actors. Scenes when these larger, wire-controlled head and neck props intercut with stop-motion animation appear joltingly 'noticeable,' rather than seamlessly spliced-in, as the larger models don't closely match respective smaller ones. Neil Pettigrew (author of *The Stop-Motion Filmography*, 1999) complains that movie jungle sets through which dinosaurs move aren't as lush or of high quality as those appearing in *King Kong*—sci-fi movie maven Bill Warren refers to the sets as "highly unimaginative" and "boring." Pettigrew also isn't pleased with the "exaggerated doll-size of the (dinosaurs') eyes." (p.176) Now in adulthood, I cannot disagree.

Several stop-motion animated scenes involving the tyrannosaur are generally well performed and state-of-the-art, for that time sufficiently 'convincing' (i.e. when, after awakening, 'Rex' kills a man on the shore; its attack on a bus; the tumultuous battle with brontosaur; and in the climactic steam-shovel fight). But as Bill Warren notes (in his *Keep Watching the Skies—the 21st Century ed.*, (2010) the tyrannosaur constantly "shrieks … eventually sounding comic. Its tongue wabbles up and down much of the time, and the sculpted-in expression of furious rage makes the tongue-waggling seem even sillier." (vol. 1, pp.233-34) Warren offers scant praise for that cool steam shovel versus 'Rex' fight.

Brontosaur's shuffling movement scenes appear even more problematic. Pettigrew stated Delgado's brontosaur model was "massive and caused difficulties for the animators." (p.178) Furthermore, Warren has another insight—that the models "were so unstable (they are very light for animation models) and the work had to be done so fast there's little even a master animator could do." Furthermore, Warren notes that the models don't hold their positions well in scenes "resulting in a true jerkiness … the cumbersome models never seem real, and the complete lack of flexing in the body of the brontosaur makes it look like a big toy being pulled along by a string." (vol. 1, p.233) Furthermore, that small brontosaur puppet was also reused for a pivotal scene in *The Twilight Zone* ("The Odyssey of Flight 33"), in which it looked more convincing, but only because "… the action of the model was so limited … perhaps because more time was available to the animator." (p.234)

And for the clincher—despite what it had going for it in a positive sense, overall, as noted by Warren (p.235), *Dinosaurus!* rated "pretty lousy" in *Psychotronic Video* magazine. Ouch!

Although not of Academy Award-winning caliber—beyond Gregg Martell's acclaimed acting performance: lighthearted and comical in one extended scene and heroic when he stood up for others in peril—the rest of the cast performed reasonably well—at least acceptably for a run of the mill genre movie involving anachronistically rejuvenated dinosaurs. But still they're "colorless," opines Warren, mindful that most of them were relatively inexperienced.

Popcorn, fighting dinosaurs, a club-wielding/brontosaur-riding caveman, another boy who played with the same dinosaur toys that I did … and even a pretty girl to be rescued from the bad guys! As a Saturday afternoon matinee in the early 1960s—this was perfect kiddie fare—and that's how I choose to remember *Dinosaurus!* Not like the experts. Here's hoping Svengoolie shows it again sometime soon.

Chapter Four — *Stop-Motion Godzilla*

With a plaque now commemorating an unlikely California address where American scenes for *Godzilla: King of the Monsters* were once filmed, we should now consider memorializing another far less publicized film 'studio,' of sorts—in Chicago where the first stop-motion Godzilla films were filmed in 1959 and 1964. Where exactly? At Don Glut's house on Magnolia Lane near Wrigley Field. Yes—and now you can watch them yourself on DVD. Never heard of these 'kaiju' movies? Please read on.

Okay, don't get me wrong. This isn't exactly Harryhausen-caliber work, but nor is it 'Gumby versus Dinosaurus' either. (Actually, there was a 'claymation' dino-character in the old Gumby series, but I digress.) But, even though many dino-manic fans, many of whom read circa 1960 issues of *Famous Monsters of Filmland* have known of Don's youthful films for a long time, until recently we'd never had opportunity to judge their quality. Let me offer a little rundown on these films and what they signify. Don's sometimes crude efforts were always serious-minded; today, despite industrial advances in computer graphics and 'kaiju-mation,' magically performed from the friendly confines of his basement, backyard and bedroom, Don's early efforts still broadcast considerable charm.

As a youth, Don made a total of 41 short amateur films, of which 8 were of the dino-monster variety. With model railroad sets, Marx playset pieces and various dinosaur models available then or of his own making, and armed with trusty 16 mm camera, he accomplished what many 'monster kids' of the time – including myself although a decade later than Don, aspired to. He made monsters like King Kong and Godzilla destroy cities and it all must have been very cool and fun to do then, as it is to watch now.

Fourteen years before earning his B.A. degree in cinema from the University of Southern California, with his encouraging mother Julia holding the camera, young Don avidly tried his hand at making his first dino-monster film, Diplodocus At Large (1953). This is about as basic as it gets, with Don's hand and arm inserted into an Ollie the Dragon hand puppet 'attacking' a toy city constructed out of Lincoln Logs and Plasticville model buildings. Fortunately, the dinosaur is destroyed by a flying saucer before things get ... too out of 'hand.' Good thing he '*digit*-ized' this one for us though.

Then in his next amateur effort, Earth Before Man (1956), instead of puppets, it was waxy plastic Miller dinosaurs pulled by almost 'invisible' strings through aquarium rocks settings, or Mesozoic scenery borrowed from Marx playland sets. As Don notes in commentary on his *I Was a Teenage Movie Maker* DVD, the Miller dinosaurs would tend to melt in the hot sun in scenes filmed outdoors (for better lighting). But an 'earthquake' replete with smoldering volcano soon ended their sweltering misery. Notably, Don added a waxy gorilla figure to Earth Before Man's menagerie. No doubt, here, Don was simulating the huge presence of another, certain huge Hollywood ape whose exploits had then recently been reissued to theaters around the country (*King Kong*, 1933), although scenes from Irwin Allen's and Ray Harryhausen's *The Animal World* (1956) were also being (crudely) emulated.

Juvenile efforts soon gave way to the first dino-monster experimental film Don made enhanced with stop-motion animated effects, Dinosaur Destroyer (1959), a title borrowed from the title of a 1949 *Amazing Stories* story and also a comic book title. Dinosaur Destroyer was sort of a melding of two Harryhausen films *20 Million Miles to Earth* (1957), and *The Beast From 20,000 Fathoms* (1953), but featuring a monster resembling and inspired by Godzilla's precursor, the Rhedosaurus, which Don clearly revered. Admittedly, Don lacked knowledge as to how such movie effects were made, so he set out to (re-) invent them himself, for the entertainment of his family and friends. And he more or less succeeded. For his clay pseudo-Rhedosaur hatches from a (chicken) egg (taken onboard by rocket ship from Venus back to Earth). Then the rapidly growing 'rhedo-chicken' attacks his model railroad set. After climbing a water tower, the monster is boiled alive (in burning model glue).

So after dabbling in sci-fi themes once more in films such as The Time Monsters (1959), featuring a *Tyrannosaurus/Triceratops* tussle and a battle between a Kong-like ape and a horned *Ceratosaurus*, Don moved on to The Fire Monsters (1959), obviously indicative of his recent screening of *Gigantis the Fire Monster* (1959), in which Gigantis and Anguirus waged battle through the streets of Chicago. Adeptly using forced perspective for certain scenes, Don boldly took his two monster models to a downtown bridge where with tripod-mounted camera, he filmed their movements, which surely must have been an odd sight for pedestrians and spectators. Eventually, fire-breathing Gigantis climbs the Prudential Building, then Chicago's tallest building, before falling to its death.

Shortly thereafter, for a school project Don filmed Age of Reptiles (1959-1960), although he added a Godzilla-like conclusion titled 'what if the dinosaurs lived today'? This particular segment resulted in a lesser grade on his school assignment, but no matter now, for today's viewers can still enjoy scenes of Don's finback fire-breathing *Spinosaurus* destroying Chicago, as well as Don being eaten by his clay monster. Evidently, Chicago was to Don, as Tokyo is to Toho.

Following another short, 1961's Time is Just a Place, next, he tackled the *King Kong* theme, essentially performing his own version of the classic RKO film, although disguised with the title, Tor, King of Beasts (1962), (with rocket ship sequences inspired from *When Worlds Collide* and *Commander Cody*). This film is really rather amazing, (but still not his best)! Using his camera, Don recorded scenes from a televised broadcasting of *King Kong* directly from his TV screen. Then he used this footage to exactly reproduce these scenes for Tor, King of Beasts. The orchestrated results, such as Kong-like Tor's fight with a *Tyrannosaurus*, are most impressive. Don takes Tor through all of Kong's paces; while he scores little for originality with this entry, he gets an 'A' for effort. Poor Tor falls (as did Gigantis earlier) from the Prudential Building to his death.

However, in my humble opinion Don's best kaiju film (also his last stop-motion movie) was his dino-monster rumble, Son of Tor (1964), the last of his amateur dino-monster films. This time, Carl Denham with fellow intrepid explorers return by plane to Tor's island, only to confront a regular monster fest of famous dino-creatures. Gorgo, Ymir, Godzilla, Tor's son 'Torro,' and Paleosaurus (i.e. from *The Giant Behemoth* (1959)) and the Rhedosaurus all make cameo appearances (as well as other dinosaur 'props'). After Gorgo devours a pterodactyl, Godzilla battles Ymir. Godzilla and Ymir topple over a cliff. Torro must defeat a horned *Styracosaurus* while a huge aquatic *Kronosaurus* nibbles a rear plane wing. Carl/Don rescues Torro from a tar pit. But after the volcano erupts, Torro, in turn, saves Don, allowing him to return to civilization. The volcano stirs up the pot, forcing a handsome Rhedosaurus and Paleosaurus to flee for their lives. Torro isn't quite as lucky, succumbing to a smoldering flow of chili con carne. These models all look good and move well, pointing to a possibility that Don was well prepared for honing his stop-motion wizardry into a career, maybe even forming a production company focusing on original American kaiju, filmed using stop-motion rather than suitmation. (Partially, this *has* come to pass with his 1996 movie, *Dinosaur Valley Girls*, which didn't incorporate any kaiju.)

Don had convinced himself that Hollywood had boundless sums of money with which producers made movies, and that, therefore, the dino-monsters he had seen in movies like *King Kong*, and *The Beast From 20,000 Fathoms* were achieved using gigantic electronic robots! Knowing he couldn't accomplish such feats in his kitchen, he set out striving to achieve similar effects using clay model miniatures. Eventually his settings spilled out of the kitchen, to the Lincoln Park Conservatory, the Addison rocks along Lake Michigan, several downtown locations and the Palwaukee Airport. By the early 1960s, he had very much impressed several Hollywood special effects people, including 'Forry' Ackerman, and 'Mr. Big' himself—Bert I. Gordon, already famous for classic 1950s movie productions such as *The Amazing Colossal Man*, and *The Spider*! Gordon was especially interested in a forced perspective scene filmed for Tor, King Beasts, involving a 'brontosaur' chasing a man up a tree, an effect which Don had achieved by filming his stop-motion dinosaur moving along a Chicago park picnic bench.

Whereas it's somewhat debatable as to who invented 'the calculus' (Newton or Leibniz), I'm afraid Don's secondary invention of stop-motion dino-monster animation came far too late for him to claim priority over individuals like Marcel Delgado and Willis O'Brien. But in a very short period, Don's dino-movie making skills increased dramatically from knocking toy buildings over with a hand puppet. Only a decade later, Don had resolved many of the problems of shooting stop-motion scenes. While at first he *deliberately* made the scenes appear jerky by jiggling the camera, because that's how the old RKO and the 1950s Harryhausen effects appeared 'live' on television, Don eventually learned how to make his monsters move seamlessly and with less strobe.

Through experience, he learned other basics such as shooting from the tabletop in an *upward* direction, rather than shooting downward, so as to achieve the illusion of scale enormity. His volcanic effects improved as well, with lava 'chemistry' evolving from ketchup to flowing chili con carne mixed with burning model glue, and, inevitably, incorporating homemade gunpowder. Chocolate syrup was used for blood. He not only 'reinvented' but made full use of rear projection screens, and experimented successfully with two-screen shots, just like the pros at RKO. He even inserted footage of live reptiles posing as huge dino-monsters into several scenes, several of which he adorned with glued-on fins, spikes or other 'accouterments.' (To me however, dolphins filmed at Brookfield Zoo don't make good stand-in 'ichthyosaurs,' as was intended in Tor, King of Beasts.) Don painted Godzilla's fins white to accentuate intervals when the monster exhaled flame from his toothy maw. It's all there to see on his DVD, which you may order from Amazon.com. And there are 33 other amateur fantasy and sci-fi themed movies on the DVD as well, many of which were made during his 'rebellious' college years.

Have you ever imagined how a stop-motion Godzilla would appear in extended movie scenes? Well, why not see for yourself because this was already done over five decades ago by dino-monster enthusiast, Donald F. Glut! Plus, there are many other dino-scenes besides, not mentioned here.

References: (1) Donald F. Glut. *Jurassic Classics: A Collection of Saurian Essays and Mesozoic Musings.* McFarland & Company, Inc., 2001, chapter titled "Amateur Animation: Dinosaurs and Ancient Apes," pp. 97-108; (2) Donald F. Glut, *Classic Movie Monsters*. Metuchen, NJ: Scarecrow Press, 1978. pp. 357, 403; (3) Donald F. Glut, *I Was A Teenage Movie-Maker: The Book,* McFarland & Company, Inc., 2007.

Chapter Five — *Egg on My Face?*

At risk of being unceremoniously drummed out of *G-Fan* (or having 'egg on my face'), I confess there are scenes in the TriStar *Godzilla* movie that I enjoyed. But we'll come to that shortly.

First, a 2009 Perspective:

The idea for this chapter came to me during the 2009 G-Fest, where I participated in the event's final Panel discussion, concerning the future of the Godzilla film series. Eventually, Godzilla In Name Only ("GINO's") name (aka "Zilla") surfaced, when it struck me that the adult monster's design and body outline rather mimics the creatures of the successful *Alien* series. I agree with the consensus that "GINO" isn't the Godzilla we've become accustomed to, for a number of significant reasons. But the *major* reason (its greatest failing) is that such a gigantic monstrosity could not sprint so quickly ("greater than 80 knots"), outmaneuvering military helicopters. Shouldn't it move ponderously just as if there was a man walking inside a monster suit? We're talking about *daikaiju*, for crying out loud.

Each kind of movie monster has its noticeable, condemning design flaws, while on the receiving end each viewer has his or her levels of tolerance as to what they'll accept. In the art of suspending disbelief, some sci-fi creators 'get it' (like Toho's masterminds did with 1954's *Gojira*) while others simply do not. Additionally, there's a distinct tendency for directors and producers of 'remakes' to not show respect for, or even understand essence of, the "original material," as one Panel member put it. Consequently, during the late 1990s, TriStar's *Godzilla* became the most reviled of all the giant monster movies; even *Reptilicus* was worthy of lesser scorn. And yet, if you read the favorable reviews posted at IMDB (the International Movie Database), my prediction is that the next generation of movie monster fans will come to accept GINO as a not so bad a film, after all. (But will TriStar's entry ever eclipse *Gojira's* score at IMDB? Perish the thought.)

Besides 1979's *Alien*, the *other* monster series that Dean Devlin and Roland Emmerich strove to capitalize on with their monster designs was the popular *Jurassic Park* industry. This is evident in sequences involving those smaller 'Zilla-raptors' shot in a miniature mockup of Madison Square Garden. These are the scenes I liked most. Movie heroes and characters are menaced by GINO's offspring, which hatch from hundreds of eggs laid in the demolished theater. Here lies the real threat to civilization. Horrors—what if they escaped!? Then there would be more than 40,000 GINOs menacing mankind in a single year.

No—those hatchlings aren't dromaeosaurid raptors, like *Jurassic Park's Velociraptors*. They aren't even supposed to be 'dinosaurs' under any strict cladistical definition, as they evidently materialized asexually from their gargantuan parent—a mutated Iguana. They're "born pregnant." Asexual reproduction is a (non-mammalian) trait manifested under certain extreme conditions. But, in a 'way cool' sort of fashion, they do appear dinosaurian!

As discussed in Rachel Aberly's *The Making of Godzilla* (1998), considerable effort went into designing GINO, and many lavish props were made to facilitate the resulting special effects. Their monster suit design isn't too bad overall, that is, if they didn't try to pass it off as Godzilla, and if they hadn't unfortunately decided to force the creature to defy known or mysterious 'laws' of "daikaiju physics." But those little guys, the baby GINOs, score much more highly in the 'scale' (a pun!) of monster-dom.

Eggs factor hugely (another pun!) in the Japanese monster movie industry. Consider *Mothra*, for example, or the original *Rodan*. For most of you, perhaps, GINO's eggs left you with just a hollow feeling, but to me, given my appreciation for *Jurassic Park*, well, they worked me 'over easy.'

The real horror in TriStar's *Godzilla* isn't so much the huge parent smashing buildings in New York City, but rather the prospect for eventual doom—mankind's inevitable extinction—if even a dozen or so of those baby Zillas escaped their Madison Square Garden hatchery. This is more of a *Jurassic Park*-ish, 'DNA genetic power' gone out-of-control kind of theme (akin to *Alien* & its sequels), as opposed to a (solitary) Japanese monster-on-the-loose motif. Arguably, the producers didn't excel in driving home this idea. "If those escape and multiply, in a very short time a new species will emerge. One that could replace us as the dominant species of this planet," as the "worm guy" Nick Tatopoulos (Matthew Broderick) prophetically warns in the movie (and in H. B. Gilmour's children's novelization from the Devlin/Emmerich sceenplay (1998, p. 120). Yet again, through bland acting, there's no real sense of urgency or conviction in this declaration. A failing. In *Godzilla* (1998), Devlin and Emmerich ineffectually interlocked fears over threats posed by genetic and atomic power.

In the end, one egg does survive the Madison Square Garden arena bombing. While this could have ripened into a sequel, there was no impetus to do so. In Gilmour's novelization we learn, "A crack formed in its shell. It was a very small crack. It might have been made by the vibrations of the trains, or the jackhammers of the cleanup crew high overhead, or other forces outside the egg. But it might also have come from within, from something inside the shell struggling to break free and live" (p. 137). Certainly, life always finds a way! And Baby Zilla escapes, grows monstrously sized, and soon has its own Saturday morning cartoon series. Yet, here, the most monstrous element of this fortuitous survivor is dismissed; the second-generation GINO has lost the means to reproduce asexually. So, in the late 1990s cartoon series, the producers dispensed with the *Jurassic Park* Frankensteinian/DNA theme, instead swerving back to more familiar ground; a giant monster allied with mankind fighting other giant antagonistic monsters on the loose.

But those Zilla-raptors are a total scream. We have over 20 minutes of them on view in TriStar's *Godzilla*. They're clearly derived from *Jurassic Park*, such as that scene in the laboratory when that cutesy looking baby *Velociraptor* hatches before Alan Grant's and Ellie Sattler's startled eyes. Yet, of course, *Velociraptors'* reproduction threatens to continue, unchecked, despite genetic safeguards bioengineered into dinosaurian DNA. Mankind would have as nearly an unholy time ridding itself of intelligent, mad-slashing "Raptors" run amuck over the planet, as it would from fending off 40,000 grown Zillas.

As Rachel Aberly divulged in *The Making of Godzilla*, "Most of the babies were actually suits, individually tailored for specially trained puppeteers.... One of the design factors that enabled the Baby Godzillas to stand and saunter was their tail, which featured a springy bungee cord running down the length of it. This allowed the tail to move naturally with the body and counterbalanced the weight of the suit..." (pp.105-106) Eventually, suitmation scenes were augmented, married to CGI for more dramatic effect. The baby Zillas, consequently, are far more realistic and convincing than, say, is the raptor-like genus *Deinonychus*, bloodily molesting actors in Roger Corman's *Carnosaur* (1993).

I'm not aspiring to become the founder of a renegade "We love GINO" society whatsoever. The Zilla-Raptor scenes are the *only* aspect of TriStar's *Godzilla* that I genuinely appreciate. What bothers me most about TriStar's *Godzilla* is not GINO's appearance, but the cloying, annoying, lighthearted sense of comic relief, of which there was no shred in *Gojira* (1954). Curiously, my wife, Diane, loves TriStar's *Godzilla*. Why? She rates it 3 stars out of 4 for very reasons that I loathe; she enjoys the humor and comic relief. Plus—prior to America's 9/11—it was entertaining to see all the faux-destruction.

Apparently, it takes nothing less than a disastrous event of historical proportions—a catastrophe such as a nuclear bomb explosion, or, say, a major terrorist attack—to dampen audience enthusiasm for humor in city-wrecking, giant monster films. What a difference a (real) catastrophe makes. Every moment in *Gojira* is imbued with a dramatic, heart pounding sense of urgency.

Throughout TriStar's *Godzilla,* released three years prior to 9/11, contrast the main characters' general sense of 'going through the motions' and sprinkled silliness with post-9/11 *Cloverfield's* (2008) far more effective, pervading sense of dread! (Diane hated *Cloverfield*.)

Whereas *Gojira* bears dark, somber messages, GINO sought to capitalize on dino-monster movie blockbusters such as *Jurassic Park*. It's hard to be completely original these days when there's a monster 'suited' for every individual taste and personality type.[1]

Now for a Fresher 2015 Perspective—Beyond the Movie:

Stephen Molstad's 'non-godzilla' 1998 tie-in GINO novelization, *Godzilla,* is pretty darn good overall. Despite insipid humor sprinkled throughout, passages in the book flow much more coherently and evenly than as seen in those sappy movie scenes. Inherently, the novelization is about environmentalism, particularly (underdeveloped) fears of overpopulation as represented by the hatching of hundreds of "baby gojiras" – born pregnant, in Madison Square Garden, threatening mankind's extinction should they escape. Concerns over radiation are also emphasized, as Nick Tatopoulos (protagonist and narrator) is a dedicated "worm guy" biologist – studying worms exposed to radiation from the Chernobyl site for signs of anomalous growth. He feels this is a dignified profession, citing Charles Darwin's 19th century detailed studies of worms, documented in the latter's 1881 book, *The Formation of Vegetable Mold*. Fortunately, all those gojira babies are destroyed when a USA military plane launches Tomahawk missiles, reducing the Garden to a smoldering volcanic "caldera," from which the vengeful adult GINO-Gojira erupts in hot pursuit.

TriStar contemplated a movie sequel, as Molstad inserted passages at the end of his novel in which Nick not only has a nightmarish vision about an egg that wasn't destroyed in the Garden, as seen in the movie, while adding a "rumor" that the adult Gojira had *also* laid a second, earlier clutch of eggs on a Caribbean island during its journey to New York, which hatched and would result in a "Cretaceous-period Park."

What doesn't work so well in the book are references and descriptions to the titular monster as a 'Godzilla' beast—which it is clearly not—as well as the silly lapses of humor pervading throughout its pages. (e.g. "I think we're going to need bigger guns." "It's 'Gojira,' you moron!" Or even that tag line innuendo, "Size does matter," and quips about American vs. French coffee.) Instead of "Godzilla," the lead monster and movie title should have been named something like "King Velocimodo," "Megastegoraptor," or "Overpopulus Rex." Why? Because on so many levels this monster is simply *not* Godzilla, and that remains its central, antagonizing problem! Rather than assimilating with the Godzilla story and theme, it misguidedly and brashly attempts to outdo *Jurassic Park*. (Those baby gojiras are reminiscent of *JP*'s velociraptors.)

There's nothing wrong or bad about the creature's monstrous (non-godzillean) appearance in itself, but this monster does things that any self-respecting Godzilla – or "Gojira" as it is referred to in the book – would never do (or be capable of). It is referred to as a "megalithic lizard," and paleontologist Dr. Elsie Chapman rather unconvincingly attempts to align it with a (theropodous) dinosaurian ancestry.

In the novelization, keeled plates on Gojira's bony shoulder blades "flap" like vestigial wings. Much worse, is its "lightning-fast" high velocity running speed, as well as its incongruous kangaroo-like hopping motion, allowing it to leap "a couple city blocks at a time." It burrows! And yet despite its 500-ton weight it can also tiptoe, "stealing cat-quietly" up a street. And there are too many boring, insipid jokes about proper pronunciation of Nick's last name that might elicit laughs from a crowd relatively unfamiliar with the original solemn story. The naming of Mayor Ebert, whose publicity manager is named "Gene" (as in "Siskel") panders to former noted film critic Roger Ebert; the 'comedy rule of three' isn't employed effectively, yet annoyingly with repetition as in certain passages like when characters fleeing in a seriously banged-up taxi think they must have finally 'lost' the angry pursuing adult monster.

If the relatively unknown novelization is decent, then how might I react to another screening of the 1998 movie? Might I now see it through 'rose-colored glasses'? I've long felt ambivalent about the 1998 movie. Over a decade ago, in 2003, I even commented that I liked the movie and that the monster's appearance was "invigorating," stating in *Scary Monsters* magazine that "even monsters must evolve! It's a law of nature, and of monsterdom as well!" Sure, but now considerably older and (hopefully) wiser, I say, clearly this monster film isn't about Godzilla, and it should have been edited and retitled accordingly for optimal viewer satisfaction. Stuffing this peculiar creature into an established Godzilla mythos was a misguided procrustean maneuver, when an original storyline for TriStar's new monster (but not named Godzilla!) should have been written instead. And by the way, it isn't that the monster designs (both adult Gojira and babies) aren't good; they're all really 'okay.' The problem lies more with how the giant creature moves onscreen, which is often counter-intuitive for its size. Also, maybe just a quibble, but no living animal could have laid *that* many eggs at one time, all of such great size.

Of course, the quite readable novelization is a direct function of movie script quality, even if Molstad smoothed out some of its rougher spots. Nick Tatopoulos' line in the film speaks for the entire production …. as he muses, it's the dawn of a 'completely incipient (piscivorous) species': which simply is *not* Godzilla or of its esteemed lineage. But GINO, or 'Zilla' as some refer to it, might have made a great opponent, say, for the Cloverfield monster. With that said, the CGI effects are splendid throughout, but those quasi-romantic/relationship scenes between Nick and reporter Audrey seem too drawn out. Watching it again one early August afternoon in 2015, my wife gave it a solid 7 stars out of 10, adding "This is a great-looking monster." I, on the other hand assigned *two* ratings. Qualified as an alleged 'godzilla flick,' it only deserves 3.5 out of 10; but changing the title and trying to forget that Zilla is supposed to have anything to do with the 'real' God-Zilla, I can honestly give it 6.2.

Note to Chapter Five: As of July 14, 2015, TriStar's *Godzilla* scored 5.3 stars out of 10 at IMDB, but with a whopping 142, 178 reviewers, followers and/or supporters & fans.

Chapter Six — *North versus South: Famous & Obscure Paleo-Monster Encounters via the Poles*

We must first realize that because, for many decades, it seemed a quite reasonable, foregone conclusion that dinosaurs and their (e.g. presumably "cold-blooded") reptilian brethren would not survive or adapt to life in a cooler, polar region of the globe—even in the warmer Mesozoic Era—discoveries of the 1960s and beyond, pointing counter-intuitively, could not be anticipated. Woolly Mammoths—yes, but dinosaurs—a resounding 'no.' But then, in his popular books, vertebrate paleontologist Edwin H. Colbert startled dino-aficionados of the time with news of large 'iguanodont' tracks found by geologists in August 1960 on West Spitzbergen, east of north Greenland, well within the Arctic Circle. Colbert also fortified plate tectonics theory when he identified bones and hallmark skull of a strange, Early Triassic reptile, *Lystrosaurus,* found in the coldest place on Earth, Antarctica, in late 1969. Had continents moved, or had these animals lived in such places? Both it would seem.

As an emerging "dinosaur renaissance" unfolded during the mid-1980s, discovery of a rich bed of dinosaur bones found in northern Alaska became immediately enlisted in the warm-blooded dinosaur controversy, which by then was woven into the fiery dinosaur mass extinction debate. In 1985, some paleontologists claimed that the Alaskan remains falsified the asteroid theory of Late Cretaceous mass extinctions because if dinosaurs were already naturally pre-adapted to harsh conditions of the far North then they should've been able to cope with a pervading impact ash cloud churned up into the atmosphere by the impact that blocked out the Sun's light for several months.

Several years later paleontologists discovered dinosaur bones in southern Australia deposits in regions of the globe that during the Mesozoic were cloaked in darkness for months, experiencing sub-freezing temperatures. Then in 1994, a large, impressive Early Jurassic theropod genus—*Cryolophosaurus*—was discovered in Antarctica's Transantarctic Mountains. Here was indication that at least some dinosaurs—adapted to these conditions—may have been 'warm-blooded' after all! In 2017, came announcement of the discovery of a 'long-neck,' Late Cretaceous sauropod genus, *Morrosaurus*, found in Antarctica. Paleontologist Steve Brusatte stresses that "… the fact that it is much more primitive suggests that Antarctica may have been a refuge for archaic dinosaurs during the run-up to the asteroid impact." (*Prehistoric Times* no.120, p.57)

Before treading further, one must briefly contrast geology of the northernmost and southernmost polar regions. It will be apparent that, not only as geographical knowledge of these two disparate regions emerged, but also more recently, in consequence of evidence supporting climate change, stories soon to be summarized here made a 'transition' of sorts—favoring North over South. At least by 1723, cartographers had established that Arctic polar ice overlay an ocean, not land. Whereas the Arctic pole is a sea covered by ice (melting at redoubled pace every year), the South Pole—first reached by man in 1911—and most southern continent is a forbidding, desolate place strewn by inactive volcanoes covered by many thousands of feet of ice. More recent paleo-tales have centered on Antarctica, rather than the Arctic. It would seem that both severely cold, relatively inaccessible places made them likely, lonely, lost world-opportunistic locations for creative writers to situate paleo-monsters of every possible persuasion!

One last important connection to be made before delving into specific titles is that, pop-culturally, paleo-monsters often are bound to Earth's interior, partly because their remains (or 'living' selves) are found within caves, icy tombs or rock deposits, deep *within* the Earth. Think of movies *Reptilicus* and *Rodan, the Flying Monster* for example, or of Jules Verne's *Journey to the Center of the Earth*. Discoveries of our dino-monsters were only found after Man had blasphemously 'penetrated' the planet—our Mother Earth. Beyond metaphor and symbolism here, one such fascinating real discovery was made in the late 19th century, when in 1878 Belgian coal miners disinterred skeletons of (at least) five *Iguanodons* found 1,100 feet in the deposit! To astronomers, farther away in Space implies deeper in Time; analogously, to imaginers of resurrected paleo-monster stories, probing within Earth's subterranean realm may unveil planetary prehistory. So, perhaps unsurprisingly, some of the earliest sci-fi/horror tales incorporated caves and caverns infested with living 'prehistoric' monsters. (Two meriting parenthetical mention here are Arthur Conan Doyle's "The Terror of Blue John Gap," (1910,) and C. J. Cutcliffe Hyne's "The Lizard" (1898). But see my *Dinosaurs in Fantastic Fiction* (2006) for more on these and others.)

It's certainly intriguing to ponder … 'what lies lurking upon polar icy realms, in fossil form?' Sci-fi master, Stephen Baxter, offered a provocative 'what if' scenario in his 2003 novel, *Evolution*. In one segment of this life through time novel, he described dinosaur fauna of Antarctica that fortuitously survived the Late Cretaceous asteroid mass extinction, persisting until *10 million years ago*. Not so implausible, methinks. Will we someday find their remains too, perhaps frozen, although eerily held in suspended animation?

(My 'rule of thumb' here is that because several of the sci-fi films we've come to know are derived from famous fantastic novels, in such cases I will *briefly* discuss those films *following* the classic literature, under respective "Literature" sections. Films that seem to have no prior direct basis in published literature will only be discussed under respective "Film" sections.)

North – Films:

In 1957's *The Deadly Mantis*, a paleontologist trods through a tortured chain of logic enabling the 'correct' conclusion that an enormous Praying Mantis insect has been released from Arctic ice, only to wreak havoc upon modernity. An implication is that monster Mantis hails from a much earlier geological age. It struck me that it may have been intended as a 'relic' paleontological survivor from Earth's Carboniferous Age (e.g. the "Coal Age"), when atmospheric circumstances supported evolution of enormous terrestrial arthropods. Or, paleo-connections aside, the huge insect may be simply symbolic of the Bomb; yet, plot-wise, radiation isn't of any filmic consequence here! As Bill Warren comments in his *Keep Watching the Skies* (2010, Vol. 1), *The Deadly Mantis* seems "partially designed as a showcase for the Arctic security system of the United States." (p.220)

However, while paleo-Mantis isn't a dino-monster, perhaps the very first *dino*-monster preserved for millennia in ice to plunder a modern city was Superman's super-sized foe—the Arctic Giant. Released as an animated short feature in 1943, the Man of Steel's gargantuan nemesis was discovered frozen, as have several Woolly Mammoths, in western Siberia. The block of ice containing the complete specimen with soft part anatomy intact is transported to Metropolis via "refrigerated freighter," where it is displayed in the Museum of Natural History. But the refrigeration unit in the museum is accidentally shut down, causing enclosing ice to rapidly melt—thus freeing the gigantic low-finned 'tyrannosaur' from suspended animation and unleashing its wrath upon mankind. Only Superman can prevent an utter calamity, while rescuing Lois Lane from its clutches!

Gamera, the giant flying, spinning prehistoric turtle, was also freed from Arctic ice, as originally debuted in 1965's *Gamera the Giant Monster*. Here, an aircraft sortie near the North Pole leads to detonation of a nuclear bomb. American combat jets strike an unidentified aircraft which crashes, extinguishing in a mushroom cloud. Gamera, invulnerable to nukes, is soon dutifully on the rampage! The device of nuclear explosion leading to revival of a prehistoric dino-monster had already been utilized though, 14 years earlier in *The Beast From 20,000 Fathoms* (1953).

In "Beast," a quadrupedal dinosaurian is released from Arctic ice following an experimental powerful nuclear weapon test—devised around the time when the USA was developing its hydrogen bomb. So the giant reptilian, virulent prehistoric germ-carrying "Rhedosaurus," hibernating for one hundred millennia, represents Nature's vengeful dark side of man's tendency to probe into the 'forbidden' atomic realm of matter and energy. A prophetic harbinger of what is to come. The titular, highly infectious (to humans), stop-motion animated Beast attacks New York City before succumbing to a radioactive isotopic injection. Today, one wonders why the movie's Operation Experiment explosion had to be detonated within the Arctic Circle, as opposed to elsewhere—other than the fact that the icy region is remote. Certainly, the USA has a history of experimenting with nuclear explosions in southwestern states and in the Pacific—but never, to my knowledge, the Arctic. (*Beast* does connect to Ray Bradbury's 1951 short story "The Foghorn," plot-wise, through the famous lighthouse destruction sequence.)

The unforgiving Arctic and its unlikely prehistoria were targeted in an Asian suitmation epic, 1977's *The Last Dinosaur*. Masten Thrust is a petroleum magnate—the world's richest man who loves hunting the biggest game, present or past. He's stoked by the proposition of 'bagging' a dinosaur trophy; said dinosaur is alive in the Arctic, according to evidence obtained by an oil prospecting team survivor from a prior mission. So Thrust's band of explorers penetrate an Arctic volcanic basin in a "Polar-Borer," (a sci-fi gimmick derived from Edgar Rice Burroughs' 1914 novel, *At the Earth's Core*), arriving at a warm subterranean valley populated by prehistoric denizens (a 'primitive' tribe of people, and two dinosaurs that fight in slow motion—*Tyrannosaurus* and *Triceratops*). The story is a 'lost world-ish' tale, although casting Thrust as a 'Captain Ahab' character, obsessed with hunting the 'last dinosaur,' or as Mark Berry provocatively refers to it, Thrust's "… dinosaurian Moby Dick."

North – Literature:

Although Jules Verne had prior literary inspirations, explorers in his 1864 novel *Journey to the Center of the Earth* ("*JCE*" the undoubted 'grandparent' of all such later derivative subterranean tales), take their first intrepid steps into the bowels of the Earth through a volcanic vent ("Skartaris" on Mt. Snaeffels) situated on a western peninsula in Iceland. Although its loftiest ramparts are certainly frozen in winter, geographically, the main body of Iceland is just shy of the Arctic Circle. Verne's novel is the founding masterpiece of the genre', however, readers might be far more familiar with the classic 1959 film starring James Mason and Pat Boone based on the novel. In both cases 'prehistoric' species are encountered deep within Earth's crust.

Eventually, after triggering a volcanic eruption near the 'Central Sea' in both novel and film, the explorers are thrust upward on a fiery magma plume through a volcanic vent, to Earth's surface. Incidentally, in his *opus*, *Keep Watching the Skies*: *21st Century Edition,* Bill Warren queries "Why do these lost worlds always wait to self-destruct when outsiders from Our World/Time first arrive? Is there something in our auras that sets off latent pyrotechnics? Perhaps it's only a series of extended coincidences? A cosmic conspiracy? And perhaps it's unimaginative, hack writers who can think of no other way to wrap things up." (p. 837) Well, all I can say is that Verne—who first rigged this 'eruptive' sci-fi gimmick—was by no means a "hack."

In Verne's original novel, most prominently the explorers witness a great sea battle waged between two Mesozoic aquatic sauria—*Ichthyosaurus* and *Plesiosaurus*. In other passages (added to the 7th 1867 edition), the band arrives upon a plain of ancient fossil bones. The novel is so enriched with many older paleontological/geological ideas, too many to recount here—although I encourage readers to consult Chapter One of my 2006 book, *Dinosaurs in Fantastic Fiction* for more on Verne's novel! His wonderful novel merits far more consideration than I can provide here, and only tangentially. The 1959 film's overt suggestions of the 'prehistoric' comes with those gigantic *Dimetrodons* scrapping along the shore of the Central Sea; these do not appear in the novel. And that giant mushroom 'forest' appearing in the film is a vestige from artist Edouard Riou's illustrated edition of Verne's novel—although represented therein as a relic Transition Forest 'stemming' geologically from the Coal Age.

But the next best-known foray into the Earth via the North Pole, precipitating encounters with prehistoric inhabitants of Earth's inner world—Pellucidar, famously recruited Tarzan of the Apes, in Edgar Rice Burroughs' 1930 novel, *Tarzan at the Earth's Core*. The idea underlying Pellucidar borrows from 'holes at the poles' hollow Earth pseudo-science, a then-decades old 'theory' warranting only a brief digression here.

Although the erroneous idea for polar openings into Earth's interior extends back to the 17th century, one individual in particular, John Cleves Symmes, became famously associated with the concept during the early 19th century. According to David Standish, author of *Hollow Earth* (2006), "After Symmes, ideas about the hollow earth, in real life and in fiction, became inextricably linked with the poles and polar expeditions, right down to the present." (p.58) A corollary of the idea, based on tepid evidence from southern sea explorations, was that once one penetrated through an outer frigid pole, a warm environment would be found at depth … inside inner Earth. Edgar Allan Poe was obsessed with the idea, writing tales inspired by it—such as *The*

Narrative of Arthur Gordon Pym: none of Poe's holes-at-the-poles stories concerned paleontological confrontations. In 1828, one of Symmes' staunch disciples—Jeremiah N. Reynolds drummed up interest in a voyage, one aspect of which could have been possible discovery of an Antarctic polar opening; for political reasons that mission never got underway. When an expedition was finally launched (*sans* Reynolds), departing in October 1829, of course no 'Symmes Holes' were found, although interestingly, naturalist James Eights found the first fossils from Antarctica "embedded in icebergs sheared from the Antarctic mainland" (Standish, p.91), indicating a former temperate climate. As we'll see later, while most of the early intrigue eventuated on southern 'holes,' in theory, the concept also applied to the northern pole area as well.

In Burroughs' prior 1914 novel, *At the Earth's Core*, access to Earth's inner world of "Pellucidar"—warmed by a small interior 'central' star, was initially gained through a non-polar latitude. Because Pellucidar cooled much more slowly than Earth's outer crust, plants and animals residing therein are older geologically, 'primitives' relative to fauna and flora of the outer domain. A dreaded and dominant species ruling in Pellucidar are the intelligent Mahars from the Saurozoic, (mysteriously allied genetically to pterodactyl genus *Rhamphorhynchus* of outer Earth's Jurassic age). But there are progenitors of modern men coexisting with 'prehistoric' mammals and dinosaurs living there too. The rationale for this anachronistic congregation of prehistoria is that lack of geological catastrophes inside Earth (which otherwise marked extinction and evolutionary radiation boundaries on Earth's outer crust) somehow permitted organisms from 'different' ages to live concurrently in Pellucidar.

But why recruit Tarzan of the Apes for the 1930 adventure, and why did his expedition elect to try a northern 'hole,' an impossible feat that Burroughs expertly suspends disbelief in? Tarzan's group learned of the polar openings in an earlier Burroughs novel—*Tanar of Pellucidar* (1929). Essentially, without further ado, a hero of the 1914 novel, David Innes, is in need of rescue from a havoc-raising pirate band, named the Korsars that inadvertently sailed over the polar rim, as explained in *Tanar*. Only a mighty hero such as Tarzan might be able to take on the weighty task in such a primitive jungle world, one that would appear "like heaven" to him. This time though they descend into Pellucidar in a 1,000-foot long Zeppelin constructed of ultralight, strong metal. After landing, threats come fast afoot, an ecology teeming with great Pleistocene mammals—sabertooths and humanoid Sagoths; giant flightless birds; great pterodactyls; strange dinosaurs and a pack of armed 'dinosauroids'! Yet somehow the rescue party survives and accomplishes their mission. True—prehistoria encountered by Tarzan and company aren't indigenous to polar conditions, although in this novel and the next one to be outlined, below, they're confronted only after penetrating through the icy northern realm.

Relying on Symmes' 'theory,' in esteemed Russian geologist Vladimir Obruchev's 1924 little known novel, *Plutonia*, a geological expedition also makes it to the inner part of our planet, via a north polar route, where prehistoric creatures still thrive. This band of explorers tread using snow sledges across the tundra. Then, almost imperceptibly, on a gradual, "endless descent," venture into Earth's interior. Once they arrive, however, as in Verne's novel, the explorers discuss geology throughout their journey, while dispatching prehistoric animals ("living fossils") with their rifles. The inner world is lit by a small 'central' star which is designated "Pluto." At first their quarry and observed creatures are of mammalian ilk, yet as they progress farther, more 'ancient' fanua are encountered. Essentially, for most of their journey they're retreating backward in time, from the 'Pleistocene' toward Mesozoic life—and relic-surviving dinosaurs! Spying a *Triceratops*, a savvy explorer states, "That means we're in the Cretaceous Period now! ... And the further downstream we go, the more of these monsters we'll probably meet." Menaced by a sea of marine 'lizards,' huge ants, volcanic eruptions and a tribe of 'Stone-age' people, the explorers eventually wend their back up top, only to find Russia at war with Germany.

Speaking of war—a very Cold War—paleontologist Jose Luis Sanz mentions two other stories (unread by this author) in his 2002 book, *Starring T. Rex!* (pp.59-61). One of these sounds particularly interesting—a novel by Alan Comet titled *The Waking of the Past* (1955), in which communist scientists revive a horde of prehistoric monsters from surrounding icy tombs, which then attack Europe. I am inserting Sanz's description of this novel here:

> "It is likely that the various discoveries of mammoths in the Siberian permafrost were one of the factors responsible for the process of synchronization between humans and dinosaurs within the fantasy discourse. All that is required is to swap the proboscideans (mammoths, elephants, mastodons, etc.) for dinosaurs. This is the approach taken in the story by Spanish science fiction writer Alan Comet (the nom de plume of Enruqie Pascual) in his 1955 novel *El Despertar del Pasado* (*The Waking of the Past*). It is the 21st century and the Soviet Union has been defeated, leaving a few communists hiding out in an underground base in Siberia. Professor Komarow, who revives human beings by inserting a small electrode into their brains, manages to bring back to life thousands of 'monsters of the past' that were sleeping in the ice. The communists set this terrifying horde loose on Europe in order to defeat their enemies. In the end, the United States air force puts a stop to it, and also manages to return the 'monsters' to Siberia and put an end to the Soviet peril once and for all." (p.59)

Then in 2008, Carl Macek published a novel titled *War Eagles* entirely inspired by Merian C. Cooper's late 1930s, ill-fated effort to produce a film by that name. Whereas an abandoned movie script specified a pilot landing near the South Pole where dinosaurs are encountered, Macek instead reoriented his protagonist landing in the Arctic (during the World War II era), where dinosaurs such as the *Utahraptor* and stegosaurs have survived to modernity. Those titular 'war eagles' are enlarged eagles which fly their Nordic companions into battle against Nazis and menacing theropods. (See Chapter One for more on War Eagles.)

South—Films:

In *The Land Unknown* (1957), explorers travel to Antarctica where they descend via helicopter—damaged by a collision with a pterosaur—3,000 feet below sea level to an ancient jungle (cyclorama) forest populated by prehistoric creatures, including a pot-belly suitmation tyrannosaur and a (mechanical) elasmosaur. Giant (live) lizards also roam about. While the steamy jungle visuals and the tyrannosaur seem inspired by Rudolph Zallinger's then world famous paleoart mural *The Age of Reptiles*, overall, the film itself seems rather Lovecraftian, perhaps in a disconcerting 'mountains of madness' mien. Perhaps, however, given the potentially heat-generating, subglacial, volcanic nature of the vast southern continent, may we as yet anticipate an undiscovered "warm-water oasis" there, buried in the ice, teeming with various prehistoria?

"Magma," the giant prehistoric walrus, as described by Jeff Rovin in his *Encyclopedia of Monsters* (1989) appeared near the South Pole in Toho's 1962 film, *Gorath*. Two versions of this sadly comical-looking monster appeared, one being a rubber suit worn by Haruo Nakajima; another was a 3-foot long model. But I must digress here while also acknowledging another—this time stop-motion animated—'prehistoric walrus' ('Walrus Giganticus') which menaced intrepid heroes near the North Pole in 1977's *Sinbad and the Eye of the Tiger*. So, at least in the battle of gigantic paleo-walruses, North ekes out a tie with South, (thanks to Ray Harryhausen). When it comes to walruses ... "big" may be equated with 'prehistoric': isn't that the usual monster formula?

South—Literature:

The earliest south polar *published* story entry I've read was published in 1901—Frank M. Saville's novel, *Beyond the Great South Wall*. In this neglected tale, explorers trek to Antarctica—in search of the lost Mayan race, which presumably fled southward centuries ago—where they fortuitously encounter the "last dinosaur," a rampaging, vengeful brontosaur. This creature, named "Cay," eventually expires in a volcanic eruption. But a much more enduring 'lost world-ish' story, Edgar Rice Burroughs' *The Land That Time Forgot* ("LTF"), was published in 1918. Unlike Verne, who presented a paleontogically-ordered array of prehistoric representations (in fossilized, living or imagined form) in *JCE*, Burroughs analogously, yet mysteriously condensed his paleo-creatures into *single organisms* evolving throughout their individual lifetimes, 'marching' upward via a 'ladder' of evolution from cellular matter toward becoming Stone-Age peoples. This all happens on an expansive insular terrain, a lonely forbidden place known to inhabitants as "Caspak." Where exactly is Caspak? Although uncharted, according to prolific sci-fi writer Lin Carter, "Far to the south, near Antarctica, lies mysterious Caspak." (In the tale, Caspak was named by a prior 18th century explorer as "Caprona.")

LTF's story protagonist, Bowen Tyler, notes "…I have experienced a cosmic cycle, with all its changes and evolutions for that which I have seen with my own eyes in this brief interval of time—things that no other mortal eye had seen before, glimpses of a world past, a world dead, a world so long dead that even in the lowest Cambrian stratum no trace of it remains. Fused with the melting inner crust, it has passed forever beyond the ken of man other than in that lost pocket of the earth whither fate has borne me and where my doom is sealed."

Captured on a German World War I U-boat sub, Tyler and his navigators are lost at sea, sailing southward through unfamiliar waters beyond sighting of icebergs, until a verdant land—ringed by misty cliffs rising perpendicularly from the sea—is spotted where they put in. Soon the party is menaced and preyed upon by a host of prehistoric creatures, including a vicious tyrannosaur and a marine plesiosaur—Tyler and a female, Lys, survive. The novel was scripted into a 1975 movie, of same title, starring Doug McClure. Novel and film were continued in sequels, (released 1918 and 1977, respectively) each titled *The People That Time Forgot*, in which stolid men return to Caspak. In both the 1975 and 1977 films, audiences were treated to a not too bad Mesozoic menagerie, with props and puppets all based on a variety of real dinosaurians animated in a rather convincing primeval setting. However, the former movie has far more entertainment value because more such effects effort went into *LTF*. At the end of the movie *LTF* (perhaps all too predictably—yawn!) a volcanic eruption dooms the German submarine.

Following Burroughs' example, another writer—John Taine (who gained fame as a noted mathematician under his real name—Eric Temple Bell), sent an exploring party to Antarctica, where they encounter living, colossal godzillean-sized dino-monsters in his 1929 novel, *The Greatest Adventure*. Under his pseudonym, Taine wrote several science fiction novels; he evidently was stoked by prospect of organic evolution—a theme of *The Greatest Adventure*, perhaps his masterpiece, as outlined below.

After bones and other remains of (otherwise) extinct creatures found freshly preserved in the Antarctic are brought to his attention, Dr. Eric Lane heads an investigatory expedition there. But, once they're spied in-the-flesh, these colossal brutes are "off," biologically "abnormal," even "evil" and "unearthly": they're nothing known to science, not conforming to the known fossil record. "They are like bad copies, botched imitations ... of those huge brutes whose bones we chisel out of the rocks from Wyoming to Patganoia ... Every last one of them is a freak." Yet, eerily, they're correlated with mysterious archaeological inscriptions discovered on a huge cement wall built "as hard as diamond" sealing an immense crypt, upon which are pictograms, showing monsters representing."... different ages of the Earth's history." It seems that in a Frankensteinian fashion, eons ago intelligent beings experimentally created 'unnatural' life that then evolved on its own from spores, into the obscene dino-creatures witnessed and trapped here. Later these creators "perished in a war to destroy their own creations," lest they pollute the entire planet. As if evolution, once started, is somehow preordained (which is *not* the case).

On the heels of Taine, a decade later, came H. P. Lovecraft's *At the Mountains of Madness*. Intriguingly (like Verne before them), Burroughs, Taine and Lovecraft each dabbled in science fictional aspects of how life progressed through geological time, and recording of past life—the geological paradigm of their time and a 'majestic' theme that's usually 'lost' or faded when such tales are converted to film for popcorn munching audiences. Reading such tales, it's certainly a good thing Antarctica is so far away from North America, visited by only the most intrepid of explorers—and so icy cold—or civilization would be screwed!

Although great dino-monsters don't appear in the flesh in Lovecraft's 1930 tale, peculiar Antarctic fossils—including dinosaur bones are discovered early in the story, and eventually other information—bas-relief carvings and inscriptions, allowing human investigators from Miskatonic University to reconstruct the horrifying, hidden prehistory of our planet. Like Taine's prior novel, *Mountains of Madness* isn't a 'hollow Earth' story—although explorers do comb subterranean passages of a colossal, cyclopean city, a "Palaeogaen megalopolis" structure within a mountain. Eventually, they arrive at a conclusion that "The things once rearing and dwelling in the frightful masonry in the age of dinosaurs were not ... dinosaurs, but far worse ... They were the Great Old Ones that had filtered down from the stars when the Earth was young."

Following World War II, with its frozen landscape increasingly mapped, surveyed by air and under scientific scrutiny, few (to none) additional published stories involving identification of fictional Antarctica paleo-creatures were written—until the 1990s when the trend fired up once again. In fact, the next three came out near the end of the millennium, beginning with Leigh Clark's *Carnivore* (1997). Certainly, when compared to prior literature discussed here,

Carnivore is a minor work. I read most of it during a Chicago to LA flight two decades ago. First, a fossilized egg is discovered preserved in an Antarctic glacier. This is collected and stored adjacent to some *radioactive* waste drums at an isolated outpost. (Hmmm. Anyone see a problem there?) Then an EPA person arrives to inspect, and miraculously survive the awful proceedings. For once the hatchling springs forth (surprise!), 'ta da' ... it's a tyrannosaur. Growing rapidly, the rad-tyrannosaur is hungry and it begins gobbling people with reckless, bloody abandon! It's like a rampaging mini-godzilla, on the loose—and in the end, unkillable.

Marc Cerasini's 1998 novel, *Godzilla At World's End* deftly pays homage to Lovecraft (as well as Burroughs and Verne), as climactic scenes are staged in the Antarctic realm, deep within Earth—during a "new age of monsters." This is a fun novel, in which Godzilla teams with humanity to destroy the crystalline Ancient Ones hold on terrible monsters tasked to destroy Earth. In the end, Godzilla battles the Ancient Ones' ultimate monster—Biollante, leaving both creatures trapped at the "center of the Earth forever!" But as one of the humans remarks to his team, "Godzilla might look like a hero now, boys. But I wasn't much older than both of you when I saw the terrible things he is capable of doing." (p.315)

Another late entry from the last century was James Rollins' 1999 novel, *Subterranean*, in which Antarctic investigators face evolutionary oddities—relic egg-laying monotreme forms untouched by Late Cretaceous asteroid collision ash cloud, surviving—isolated for 66 million years inside Antarctic tunnels and caverns within Earth. (Some species of which are, kind of 'channeling' Wellsian 'morlocks.') These creatures, collectively, represent a "... whole ecosystem's environmental niches (which) have been filled with various monotreme species." (p.246) So they're transitional between reptiles and placental mammals. In one instance, fleeing humans encounter "Some flying predator! A descendant of the pterodactyl, maybe." (p.269) True—these aren't dino-monsters, per se, so I won't hoist this one further; but it's yet another example of man confronting strange living evolutionary descendants from prehistoric times in a polar region. A good 'evolution gone awry' yarn too, recalling Burroughs' *LTF*!

Finally, in David Sakmyster's and Rick Chesler's *Jurassic Dead* (2014), riffing a line from *Jurassic Park*, a tyrannosaur long-buried in Antarctic ice proves "... that death, like life, will find a way." When the 'Rex' thaws, it reacts like a rabid zombie-saur, attacking men with wild abandon, thus converting each human into a zombie as well. Parasitic enzymes in saliva transmitted via biting cause revivification of corpses. In fact, it may have been these microbes that caused extinction of the dinosaurs, via rampant cannibalism, 66 million years ago—*not* an asteroid collision. And perhaps in a nod to Taine's much earlier novel, an environmental activist does his best to prevent the plague—a 'walking dead' zombie apocalypse like no other imagined!—from spreading, and the enraged, undead near-indestructible tyrannosaur from escaping the southern continent. (Thus, in this latter instance, elements of *JD* do mirror those in *Carnivore*.) After "instigating" a "failsafe," climactic volcanic eruption, our heroes flee in a plane, escaping the pursuing tyrannosaur ... *but wait,* there are sequels in the series. (For more on the sequels see Chapter Twenty-four.)

There may be other films and examples of fantastic literature missed that can be encompassed in this North vs. South standoff, but those outlined here serve my purpose.[1]

Conclusion:

What trends are noticed in this sketch? Well, first we see that 'North' is represented by slightly more films than 'South,' and the former were mostly released longer ago. But for 'Literature,' especially for those originally published in English and during more recent decades the apparent gap has closed. In 2004, audiences were treated to a pretty good (non-dinosaur) film, *AVP: Aliens vs. Predators* (novelized by Marc Cerasini), in which representatives of the Predator race hunt those 'insectoid' parasitic Aliens for sport in a great 'Rubik's Cube' of a pyramidal maze, long concealed under Antarctic ice. Why any trends at all? After all, if you believe in any of those pseudo-geological theories of "pole shift," in which Earth's crust can easily and sometimes *suddenly* slide around its molten core, then relative position of the poles relative to the Sun's warming rays would be quite variable.[2]

Dino-monsters have been a mainstay of fictional ideas since inception of the paleontological sciences into popular culture two centuries ago. Jules Verne vaulted a variety of prehistoric animals (*sans* dinosaurs) into the lore of science fiction in 1864, but it was Burroughs who introduced dinosaurs and pterodactyls in full force to adoring fans of fantastic fiction, beginning with his *At the Earth's Core,* and "Caspak" trilogy. The hardly accessible polar realm (both North and South) satisfied tenets of veritable "lost worlds," because as long as inherent disbelief could be suspended, then there were conceivable ways to revivify paleo-creatures there ... in some way or another. As long as modern Man hadn't already plundered creatures living there and the fragile indigenous ecologies, then surely monsters long extinct everywhere else might still live at the poles—right?

Another pattern is that for both North and South *Literature,* usually it is the older examples (through the 1930s), tend to emphasize life through geological time/evolutionary ideals. (An exception is the aforementioned segment from Stephen Baxter's 2003 life through time-directed novel—*Evolution*.) However, movies either ignore or downplay evolutionary themes (e.g. relative to the novels they may be derived from). Movies highlight visual aspects of monster-on-the-loose drama.

It would seem that having an entire largely unexplored *continent* (i.e. Antarctica) on which to hide prehistoria upon (or within) has become more attractive to sci-fi writers of late, than explaining how Mesozoic saurians, etc. might be dislodged from rapidly melting Arctic Ocean ice, or thawing frozen tundra regions subjected to (hotly debated) human-caused climate change. Consider—if even the polar bears are swimming increasingly further distances from floe to floe, then the North polar region might seem a little less mysterious, romantic, and perhaps far more uninviting a setting for staging prehistoric pandemonium.

Furthermore, we note that sci-fi's first 'lost world' kind of places weren't situated in primitive insular jungles (e.g. Arthur Conan Doyle's Amazonian plateau, or Kong's Skull Island), but instead accessed via chilly polar places as in Verne's (North) and Saville's (South) early entries. Polar regions were increasingly under scientific scrutiny then, (apart from Symmes' holes), and thus to some degree, the mystery and intrigue of polar adventure coupled with what *might* be found there, or within, must have captured writers' imaginations.[3] (Consider as well—in Mary Shelley's 1818 novel *Frankenstein: or the Modern Prometheus*, largely accepted as the first modern science fiction/(horror) novel, the Monster eventually wends his way to the cold, haunting northerly latitudes, pursued by that other 'monster,' his creator.)

Finally, there was a 1988 televised episode written by Michael Reaves for the series, *Monsters*, in which intelligent dinosauroids—anticipating their own extinction 66 million years ago—place themselves within a radioactively decay-controlled suspended animation machine, near Reno, Nevada[4] but during a harsh winter storm. Unfortunately for these aggressive dinosaurians (but *most* fortunately for the human race!), they happen to emerge in a frigid arctic blast of cold—a climatic condition to which they're not adapted, thus quickly dooming their bid for planetary resurgence. Yeah—but this has nothing to do with prehistoric monsters of either polar north or south even though the extreme cold gimmick simulates prevalent temperatures presiding there.

So, there you have it—have I chilled you to the marrow?

Notes to Chapter Six: (1) The undistinguished titular monster in *Reptilicus* (1962) was found as part of a frozen fossilized limb, underground, but at a location rather far from the Arctic. Although 'uncharted,' James Gurney's hidden land of Dinotopia does not appear to be near either Pole; (2) John White, *Pole Shift: Scientific Predictions and Prophecies of the Ultimate Natural Disaster*, (1980); (3) For more on such discoveries, see William H. Goetzman's *New Lands, New Men: America and the Second Great Age of Discovery* (1986); (4) It took several years for me to notice, but with apologies, I now stand corrected on an error in my *Dinosaurs in Fantastic Fiction* (2006), p.200, note 15, where it was incorrectly stated that those titular sleeping dragons emerged in an "Arctic environment," rather than (correctly) in a non-polar, yet wintry Reno. At the time of that writing, it had been over a dozen years since I'd seen the episode and dim memory failed.

Chapter Seven — *Mystery of the Spider-Men*

As a devotee of the 1933 RKO film *King Kong*—which I first (remember having) witnessed (with my father) televised in the very late 1950s—mysteries surrounding its production drew attention. And perhaps *the* biggest controversy evidently still raging concerning this classic monster movie concerns what is usually referred to as the "spider pit sequence." Were such scenes ever filmed involving giant spiders and arthropods devouring crew members who fell into the misty jungle gorge—shaken from that log bridging the ravine by Kong just prior to his epic battle with the Tyrannosaur—or not … and *if* filmed, did this giant bug footage escape the proverbial 'cutting room floor'?

I eventually saw photo 'stills' (one appearing in the March 1964 *Famous Monsters of Filmland*) showing a giant spider as well as Mario Larrinaga's, Marcel Delgado's and Byron Crabbe's pre-production drawings showing scenes supposedly shot within the pit landscape. By 2002, there already appeared to be utter conflict as to what had actually happened with this horrific scene. Spider pit photos and sketches reveal this scene was clearly on the minds of 'Kong's' production team during early filming stages, (i.e. at least into December 1932, roughly three months prior to its official March 2nd release date). But then the scene got chopped, somehow. Decades later, I was unprepared for venom and bile over what happened to the alleged footage, seeping through pages of a popular culture movie fanzine I subscribe to—*Filmfax Plus,* spread over a period of several years, once the 'pit' hit the fan(-zine).

Why would this seem important to me—personally? For our purposes it's not just because we're talking about his majesty, Kong, but also because since I first read of the spider pit sequence I've thought of those nasty bugs as additional representatives out of prehistory that somehow escaped extinction on 'Skull Island'—as did its indigenous dinosaurian fauna. So their appearance spoke to me as remnants surviving from the Paleozoic Carboniferous age, when insects (like the 'giant' dragonfly—*Meganeura*) and other arthropods, known from fossils, actually did grow to gigantic sizes during their lifetimes—resultant possibly of excess oxygen retained then—within Earth's primeval atmosphere.

To my knowledge it was Russian geologist Vladimir Obruchev who may have introduced giant-sized insects menacing intrepid explorers in a lost-world-ish, primeval setting—in his 1924 novel, *Plutonia*. His novel featured giant (e.g. 3-foot long) ants, "rulers of the Jurassic world," further suggesting they represented a presumed enlarged prehistoric insect variety founded loosely on fossils. But then another far more famous novel invoking giant primeval bugs was published eight years later.

During the mid-1980s I read Delos W. Lovelace's 1932 novelization penned from a *movie script* version of *King Kong*, including several scenes with prehistoric creatures that don't appear in the final film. Muddying waters a bit, the book also incorporates gripping passages describing that 'men-shaken-off-the-log' tableau, where sailors were cast down into a buggy oblivion beginning at the end of *King Kong's* Chapter 11, and continuing into Chapter 12. But more on this *written* scene later.

By then, I had also heard verbatim from Donald F. Glut during one of his annual treks back to Chicago around Christmas time that Merrian C. Cooper had told him directly (in 1966) that the spider pit scene was filmed but cut prior to film release. Glut documented this brief conversation further in his 2001 book *Jurassic Classics*. This was a more elaborated reiteration of his firsthand account mentioned in Glut's earlier *Classic Movie Monsters* (1978). Glut had been informed by Cooper that the scene interrupted the movie's fast flow and placing—so he voluntarily removed it. Writer Gary Vehar refers to this as "scene-tightening." (*Filmfax Plus* no.118, July/Sept. 2008, p.111) More on this later as well.

Most of us would have thought that such casual research and explanation into the matter would have concluded things quite satisfactorily. But then, strangely, in 2002, came release of George E. Turner's (with Dr. Orville Goldner, et. al.) book, *Spawn of Skull Island*, with introductions by Ray Bradbury and Ray Harryhausen. Startlingly—there on page nine was Bradbury's claim, penned for the book in 2001, that he had seen the spider pit scene (which lasted only about "four to five seconds") when he'd witnessed it presumably at a theater. Bradbury stated "The spider was down below, at the bottom of the pit where the ... men fell in. I can't prove it—but I know I saw it. The memory is just too vivid. You can't make that up." For me, it's hard *not* to trust Bradbury.

So who is right—what actually happened? Film writer and Kong expert Gary Vehar attempted to settle things in his detailed 2008/09 three-part *Filmfax Plus* article series, "Lost Nightmare: The Mystery of the Missing Spider Pit Sequence." *Filmfax Plus* magazine editor Michael Stein was probably not fully prepared for the wrath, ire and despair this well-intentioned piece generated for readers to mull over. Interestingly, Vehar divulges others, besides Bradbury, who also claimed to have seen this elusive footage. In his survey of the evidence Vehar made other fascinating claims.

Clearly there were at least *plans* to film something that could be described as an encounter within a bug-infested gorge for an initial ~10-minute test reel, comprising in total 138 scenes—one of which was described in production notes as "gigantic insects and a huge spider attack the sailors at the bottom of the gorge ... Monstrous lizards and insects, a gigantic spider and an octopus-like creature swarm out to devour the men." (*Spawn of Skull Island*, pp.117-118) Furthermore, notes indicate that men falling into the chasm landed in "soft mud." (p.118)

Then an enormous spider crawls up a liana vine 'rope' toward Driscoll who has sought refuge higher up in a cave. (In final edit, that menacing upwards-crawling spider was switched instead to an unfinished, giant two-legged 'komodo dragon-esque' reptilian.) Vehar states that authors of *Spawn of Skull Island* (including Goldner who worked on the Kong set) held that the spider pit scene was not a "legend," and was "treated as fact"—meaning that it was actually filmed for that test reel. (However, in *Filmfax Plus* no.120 (p.56) Vehar skeptically states comic book collector Warren Reece claimed in 1985 that Goldner told him that the spider scene *never* was filmed.)

Then, after Ruth Rose replaced script writer James Creelman, scenes already incorporated into test reel footage were also retained and transcribed as part of her 'final' script. Thus, Vehar claims, we may "safely surmise" that the sequence had already been filmed. Rose's version is that which Delos Lovelace would have then relied upon to write the late-1932 Grosset & Dunlap novelization, which may have been on its way to publication before those bugs (and other scenes—e.g. a *Styracosaurus* and a battle with *Triceratopses*) got canned from the movie in post-production—or at some other juncture several weeks later.

But other scenes that were either considered 'salacious' or too gruesome for viewing by American audiences were cut from the film, only to be later retrieved and re-inserted half a century later. So, likewise, why can't the infamous spider pit scene be found somewhere and re-inserted? It hasn't happened so far, although based on what some *Filmfax Plus* writers claimed such an outcome might not be beyond realm of possibility. Were removed spider scenes burned in a 1950s RKO film library fire, or did Cooper himself destroy the footage? Alternatively, as others suspect, a spider pit sequence was *never* filmed. So was it filmed—then cut at some juncture—at a conceptual stage, never filmed at all, versus final editing? And what were the decisive reasons for taking either course of action? Budgetary limitations, pacing & flow issues, censorship issues, tight deadlines ...? Details pro & con—beyond what we'll fully address here, were all laid out in *Filmfax Plus*. And then there's Bradbury's remarkable assertion to mull over. Lately the matter has really been 'bugging' me!

Vehar states: "first public mention of the spider pit sequence occurred before the film's release ...in the January 15th 1933 edition of the *Los Angeles*

Times." (*Filmfax Plus*, no.119, Winter 2008, p.51) In this *Times* article it was divulged that Cooper "… insists that he has not made a horror picture. In fact, he threw away one of his best scenes (in which men are hurled into an abyss alive with prehistoric spiders) in order to keep the adventure and fantasy notes dominant." (p.51) And yet this remark may not be conclusive, for it seems Cooper's claims about when and why the footage was chopped 'evolved' through decades. Was it cut *before* January 15th as the *Times* article suggests?

So next, Vehar cites conflicting recollections and a bound copy of Cooper's final movie script gifted to Samuel A. Peeples, regarded as a "well-respected" former screenplay writer. Peeples claimed that notation within his script copy indicates the scene was never shot, but then goes on to state that he had seen footage of a scene showing the bottom of the gorge involving the bugs. Then he remarks that Cooper told him that "… the sequence was never filmed, as the front office at RKO was certain it would be censored." (*Filmfax Plus* no.119, p.53) Confusing—not all of these assertions can be true! Vehar suggests Peeples may simply have misunderstood and that instead maybe "… the scene was filmed but never included in an edit of the film, as RKO was certain it would be censored."

In answering another query, an undated handwritten letter that may be observed in Peter Jackson's 2005 *King Kong* DVD, Cooper stated that he "…personally cut them out of the rough studio print … they stopped the story." So we may have returned to what Glut had informed me years ago. But what is a "rough studio print"? This filmmaking term refers to a preliminary filmed version *before* adding sound or scoring, (in this case later performed by Max Steiner, who *did* claim to have seen the footage). Vehar adds "It is often at this stage that scenes that played well on paper are discovered to be lacking or contrasting to the overall tone, and are deleted." (p.54)

But Cooper also said that "during the last week of January" (1933) he screened a preview of the film in San Bernardino. It was very well received except for when the spider pit scene projected. "The screaming on screen was matched … by the screaming of the audience in the packed theater, a great many of whom left … It stopped the picture cold … so the next day at the studio I took it out myself. O'Brien was heartbroken: he thought it was the best work he'd done, and it was …," Cooper stated, thus fashioning the 'legend.' (p.54) Yet even here Vehar notes that the *LA Times* article indicated the spider pit sequence had already been cut one or two weeks *before* that late January 'legendary' previewing in San Bernardino.

Is it then possible, neglecting Cooper's hyperbole, the preview screening was intended only for RKO executives, some of whom objected to the spider scene? Ultimately, according to Cooper as stated in a Feb. 10, 1966 letter to Jack Polito, the scene was not part of the "finished picture," implying it was at one time part of a preliminary version of the film, such as in a test reel. (p.57) But in another letter, dated Oct. 23, 1966, Cooper 'clarified' how "The spider sequence was *never seen* in any theater; I cut it out before Steiner scored the picture." (p.58, italics per Cooper)

In 1985, one member of the animation team, Bert Willis, recalled animating the spiders and bugs in the small gorge set. But then he stated the "Pennsylvania Board of Censors demanded that the scenes be taken out ... those things were too horrible to show." (*Filmfax Plus* no. 120, p.54) Unfortunately no records exist either to support or falsify Mr. Willis's specific claim. Meanwhile, another soul—this time one J. R. Ayco—came forward informing Forrest J Ackerman that he had apparently seen the spider pit sequence in a Kong version upon its 1933 release in the Phillipines. Such statements would support Bradbury's astonishing 2001 remark! How would this have been possible though if the spider pit sequence was never musically scored (while the rest of it had been)? It's been theorized that some of these individuals, having glimpsed some of the scene's production visuals—drawings and photo stills—may have been stricken with bouts of long-after-the-fact 'false memories.'

Others believe that the test reel included a spider pit sequence already filmed before Ruth Rose's script had been finalized, which was both incorporated into her script *and* Lovelace's novelization. Now let's consider key passages in the 1932 novel, evidently based on screenplay versions incorporating a spider pit sequence. Although Lovelace's winter 1932 'novelized' description of the gorge and spider pit scene probably differed in certain respects from any test reel footage that was filmed, it is extremely well written and detailed! Unlike Cooper's assessment of the alleged test footage, Lovelace's page-turning passages do not slow the story's pace—they create tension and excitement! That section of the book certainly captivated me the first time I read it decades ago.

In the novel, the 'gorge' scene begins, following a battle between Kong and a trio of three-horn dinosaurs, with the men being chased toward the ravine and the log bridge by an enraged *Triceratops*. (There's back-story leading to that plot stage, but I won't digress.) Carl Denham has glanced down to see the living things below and shouts to Jack Driscoll, "That place down there is the breeding spot for the rottenest thing on this island. Look." They see an enormous spider, with a "keg"-sized abdomen, and also a large tentacled "octopus-insect." With menacing 'three-horn' glaring at the men from behind, they clamber onto the log to escape ... but Kong appears on the other side, thus trapping them on the middle of the log, suspended precariously over the gorge. Driscoll and Denham slide to safety into rock fissures below the log on each of their respective sides of the ravine. Maddened Kong begins shaking the log; two men fall off, into the abyss. A large lizard is there to gobble-up one of them, while the other unfortunate, landing in soft mud, suddenly is covered by a swarm of those hungry "great spiders." As others slip off the log, soon the octopus-insect creature is battling the spiders for its share of the prey.

Driscoll has been observing the awful situation, but when he looks down he notices a large spider climbing a heavy vine toward him. Horrifically, it "…got so close that its soft exhalation was audible to the mate (i.e. Driscoll) as it plunged back into the ravine, reaching futilely at other vines." Driscoll slices the vine and the spider tumbles downward. That move essentially ended, as written, what has become known as the "spider pit sequence," which works very well in the novel.

Readers know that in the movie this spider had been substituted for that large two-legged komodo-esque lizard, missing its hind legs. The lizard sculpt had not been completed, presumably because, with time being of the essence, and with the spider suddenly edited out of the action at a late production stage, animation on the reptile had to quickly proceed even if it meant filming and splicing using an unfinished 'puppet.' (Lovelace's promotional novel omitted the elevated train sequence in New York that does appear in the movie, offering clues as to when novel was completed relative to the film.)

Then almost miraculously, *Filmfax Plus* reported on the cover of its 134th issue (Summer 2013) "Lost King Kong Spider Scene Found!!" Too good to be true? The alleged 'discoverer' was Paul Niemiec and his article "Back to the Spider Pit!" really had me going. Here, Niemeic was writing mainly about his brother's recollection of having witnessed the spider pit footage at a Denver Art Museum exhibition in 1987. This show, "Hollywood: Legends and Reality," comprising movie art memorabilia, was originally featured at the Smithsonian, but traveled to Denver by late August. Mr. Niemiec offered a detailed account of the items on exhibition, but (spoiler alert!), as recounted to him by his brother—based on his memory, because *Paul was not able to attend the show.* Hmmm—a bit of a letdown in itself!

Paul described, rather in-depth, on what his brother saw at the *King Kong* display where he witnessed the spider pit footage in a "curtained-off" theater "containing 12 to 18 seats." The "Holy Grail of Lost Scenes" unveiled at last! But our excitement is deflated in relating that Paul's brother divulged there really was no spider pit 'sequence' at all, that is, "as many have imagined it based on pre-production" design artwork. (p.63) Furthermore, contrary to prior claims, the footage *was* musically scored. Here's Paul's summary of what his brother saw that long-ago day.

A *Triceratops*—model puppet seen in "Creation," blocks rearward passage from the log's relative safety. As Kong shakes the log:

> "One by one, the sailors are dislodged and fall to their death, dashed against the hard ground at the bottom of the ravine. But one sailor's fall is caught by a nest of vines or webbing strung from one ravine wall to another, avoiding death by brutal impact. Stuck fast, he can not free

himself. The vines are coated with sticky webbing. The sailor struggles, kicks. From a dark recess, a giant spider emerges, a spider the size of a jeep. It feels the vibration in its web and investigates. Sensing a meal, the spider surges forward. Pinned in his cave by Kong, Driscoll can neither aid his friend, nor injure the spider. The ensnared sailor, seeing that he is to be eaten alive by this nightmare beast, can only scream as the spider's mandibles close over its torso. The rest of the film unspools as normal."

Paul Niemeic illustrated his article with original drawings showing some of the sequence. He estimates the series of shots cut from the movie comprising the 'sequence' ran from 25 to 50 seconds. That 'octopoid-insect' creature did not appear in the salvaged scene. Neither did a giant crab, nor "giant carnivorous plants with tentacles," per surviving production drawings. And the spider seen in this footage isn't the one noticed in a famous production still, but a "plain black giant spider."

Niemeic then reasonably suggested that "rogue copies" of the film, those including the spider pit sequence ("reel # 7") got distributed errantly, thus explaining how Ray Bradbury (and a few others) were able to see a theatrical version of *King Kong* containing the spider pit scenes. As films got damaged during repeated cinema showings, fresh replacement copies were ordered from RKO and sent—but in very rare cases some of those prints included reel #7. Ultimately, in *Filmfax Plus* no. 137, (pp. 6, 99), after doing some background checking of his own, the highly skeptical Richard Aldrich stated that in regard to Niemeic's entertaining article, the fanzine had become "victim of a hoax." The mystery remains!

Based on my *Filmfax Plus* readings, apparently no one has well documented another sighting of the spider pit footage since Paul Niemiec's brother may have done in 1987. However, in issue no. 135 (Fall 2013), another writer, Paul Schiola—a Denver area resident and professional sculptor—came forward also claiming that like Paul's brother, he also had witnessed the spider pit footage "loop" shown at the Denver Art Museum in 1987.

In an October-November 1966 issue of *Modern Monsters* (which later was reprinted in his aforementioned *Jurassic Classics* volume), Don Glut delivered the canon: what was known about the spider pit scene, while debunking theories about where it went and why, concluding:

> "The so-called censored spider sequence is a typical example of the kind of misinformation that has, over the years, circulated about this classic film. ... We can be confident that the spider scene was, in fact, shot. For example, it appears in the *King Kong* novelization ... Yes, there really was a giant-spider scene cut from *King Kong*, but, contrary to most accounts, this scene never made it into theaters; moreover, it was *not* deleted by the

censor. According to Merian C. Cooper, whom the author briefly questioned about the missing spider scene in 1966, the reason for its deletion is ... action around this part of the film is rather quickly paced. Kong is being chased, the crew members are running, much is going on and quite rapidly so. But the spider scene ... well, it moved so slowly that it interrupted the overall pacing of the film at that point. For that reason only, the scene was cut—voluntarily by Cooper and the film's editor, and not by the censor." (*Jurassic Classics,* pp.190-192)

Writer Don Shay echoed as much in his well-researched article "Willis O'Brien Creator of the Impossible." (*Cinefex 7*, Jan. 1982, p.41)

Outright 180-degree dissension rankled over whether the scene was ever filmed at all—prompting rancor within Letters column pages of *Filmfax Plus*! First, Ron Lizorty objected to Vehar's claim of the possibility that the spider pit sequence was ever shot. Lizorty said that Orville Goldner "confirmed what Marcel (Delgado) had told me earlier; that before any animation was attempted, word came down from the front office (Cooper?) to scrap the scene," based on time and money matters. (*Filmfax Plus*, no. 120, Spring 2009, p.6)

But then tension escalated in Michael Stein's next issue no. 121 (Spring 2009), with a letter whimsically titled "King Kong-troversy?" submitted by Jack Polito, replying to Lizorty. Polito reasoned: "Time and again Cooper was quoted by many people over the years that he felt the spider scene slowed the flow of the action and therefore he cut the scene out of the film before it was scored. He has never wavered on that fact. Steiner confirmed this with me. So, how can a scene stop the flow of action if it was never in the film in the first place?" (Purportedly, Steiner claimed never to have scored the scene, although he did see the test reel spider pit footage.) Furthermore, Polito contended the scene was never released in any version distributed to theaters.

Then the Letters section (titled "Re:Edits—Opinions, Ideas, Corrections & Complaints from our Readers") of Issue no. 122 erupted in a flaming storm of controversy. This time Lizorty, who like Polito also has film and artistic modeling credentials, had sent in two letters, disagreeing vehemently with Polito. In his second letter ("Kong-tinued") Lizorty stressed O'Brien's genius and expertise in the making of Kong—the tragic monster (e.g. over Cooper's) and reiterated that the spider pit scene was never filmed.

More back & forth bitter banter appeared in the subsequent *Filmfax Plus* (no.123, Spring 2010). This time Vehar and Lizorty engaged, although their contributions were bucked to the rear of the Letters column. Lizorty opined, "I'm certain some of your readers look forward to an end(?) to the spider pit controversy." In particular Vehar noted that:

"... Polito says Cooper told him it was filmed and cut. Lizorty says Delgado told him otherwise. He also says Goldner told him it wasn't filmed, but he certainly did not say that in *The Making of King Kong*. Goldner's book states in absolute terms that the spider pit was part of the test reel and also strongly implies that it was cut by Cooper and his editor when assembling the final edit. ... Polito says Steiner was surprised to see it had been cut ...So who do we believe?" (pp.118, 121)

By issue no. 125, Polito implored *Filmfax Plus* to stop giving Lizorty a "platform" for the "war of words." No such luck. For in issue no. 136 (Winter 2014) Lizorty himself questioned veracity of the Niemeic Denver Art Museum claims: "If the 'spider pit' footage was screened (there), it may have been a recreation made by a pro or semi-pro FX person." (p.106) Meanwhile editor Michael Stein struck a conciliatory tone: "... the Denver exhibit took place 28 years ago ...The thought also crossed my mind that the mysterious footage may have been loaned to the Denver Museum solely for that exhibit by an independent, local collector who requested anonymity ... (pp.26, 106)

And yet years after *King Kong's* initial 1933 release it *was* possible to see the authentic Kong spiders and *two* spider pit scenes!

First, in their article "Is that what I think it is?" (*Mad Scientist* no.14, Fall 2006, pp.9-14, Martin Arlt and Hal Lane noted that a 1940 movie, *You'll Find Out,* teaming Bela Lugosi and Boris Karloff, inserted a scene in which real life band leader Kay Kyser—also one of the movie's characters—spies original *King Kong* animation models—including two of Kong's would-have-been 'giant' spiders, displayed on shelves and (the spiders affixed to) a wall to make that scene appear creepier. (As an aside—Arlt and Lane note that the famous Kong brontosaur made its last stand in the 1938 film, *Bringing Up Baby* starring Cary Grant and Katherine Hepburn, when it perished alongside the large bronto-skeleton when it crumbled to the museum floor in a pivotal scene.) Glut also mentions another film, RKO's *Genius at Work* (1946). As in *You'll Find Out,* "... the spider model and other creatures from *King Kong* and *The Son of Kong* can be identified on a background shelf of a movie studio set." (*Classic Movie Monsters,* p.305) But in this strange pairing of movie set scenes, the inanimate spiders were merely used as lurid props.[1]

In other sci-fi films though, what is thought to be (most or at least the armature of) one of the original giant spiders once intended for Kong has been used *either* suggestively or through animation to fuel abject horror! In the first category, Glut has identified 1965's *Women of the Prehistoric Planet* in which "the spider was not animated ... It simply 'jumped' when a technician manipulated an unfortunately obvious 'invisible' wire." (p.305).

For the latter category, there's animated footage of a giant 'Kong' spider in 1957's *The Black Scorpion*. In this film—with fluid stop-motion animation of each 'bug' orchestrated by Willis O'Brien and Pete Peterson—a meek effort is made to link the huge creepy crawlers to the 'prehistoric' realm. A scientist declares that such monsters were only found in a fossil state and thought to be extinct, and previously known only from the "Tri-asian Era." (Did he mean 'Triassic'?) After the heroes (one character played by Richard Denning who has already fallen for co-star Mara Corday's character) descend into a volcanic cavern, thus encountering the 'lair' of the buggy monsters (& here a stop-motion animated 'bird' or is it a pterodactyl that flies past them). We cringe as a giant spider emerges from a 'trap door' to menace a boy stowaway. This 'spider'—as noted by Arlt and Lane—is the smaller one visible on a shelf in *You'll Find Out*. It is equipped with pincers, revoltingly chirps like a bird and is usually regarded as one of the O'Brien/Delgado models eliminated by Cooper from the 1933 Kong spider pit 'sequence' (although its armature may have been re-tooled and 're-skinned'). If not exactly qualifying as a 'spider pit,' certainly the monster scorpion's lair could at least be dubbed a 'spider cavern.'

In 1960's *The Lost World*, an enlarged spider confronts a native woman on the prehistoric plateau. Enormous spiders also plague humans in *Planet of Dinosaurs*, *One Million Years B.C.*, and *Valley of the Dragons*. And spiders appearing relatively 'gigantic,' although minus the requisite 'prim-*eval* (with emphasis on the 'evil'-sounding part) "pit" habitat, were introduced in a host of other 'classic' sci-fi/monster films, perhaps most notably in *The Incredible Shrinking Man, Tarantula, The Spider,* and *Son of Godzilla*.

Then in 2005 Peter Jackson's *King Kong* Universal Studios team created a convincing ~five-minute sequence showing how some originally contemplated version of the spider pit sequence may (or could) have appeared. Without further ado, this impressive bit of animation is available on Youtube. There are other clip variations on Youtube, prepared by others, including one in which footage showing *The Black Scorpion's* trap-door spider has been spliced into Peter Jackson's prior animation effort.

With such wrath on display, it would seem that *Filmfax Plus* contributors had leaped out of the spider pit into the proverbial 'snake pit'! Yet in the end perhaps there was conciliation after all as Michael Stein concluded: "Although the mystery surrounding the Kong controversy remains unresolved, the dialog has been more than stimulating on both sides of the argument." (*Filmfax Plus* no.136, p.106)

I'm no expert on the spider sequence and other than Glut I've consulted with no one else who claims to have 'inside' knowledge on the 'lost' footage. There's much more material on this unusual topic available online. But that goes beyond scope and thrill of enlightenment discovery I sensed spread over many months perusing pages of America's foremost, endearing pop-cultural movie fanzine—*Filmfax Plus*.

Note to Chapter Seven: (1) There is considerable discussion online and footage on Youtube concerning whereabouts of the Kong models, reappearing in other films, subsequently—although only used as 'museum props,' not animated. The 30-foot long worm bug 'larva' seen in *The Black Scorpion* may also be another of the original creations made for Kong, but bumped from the film. Beyond giant spiders and *Bringing Up Baby's Brontosaurus,* an array of dinosaur puppets and the *Arsinoitherium* made for "Creation," *The Lost World's* "Agathaumas," and others featured in *King Kong* and *Son of Kong* may also be seen in *Genius At Work, You'll Find Out,* and *Once Upon A Time*'s 'museum' scene (starring Cary Grant).

TWO

Illustration by Harry Rountree, for Arthur Conan Doyle's The Lost World (1912); The Strand Magazine, Chicago Public Library periodicals collection See Chapter Fourteen for more.

Hopeful Monsters

The title of this section was borrowed from a phrase coined by Jason Croghan that he introduced during a 2018 G-Fest panel (which will be described more fully in Chapter Forty).

Not unlike those lost monsters from Kong's Spider Pit sequence, such a variety of 'prehistoric' animals have appeared as *monsters* in movies, television programs, comic books or science fiction/horror novels. We've seen everything from fossil invertebrates and monster sharks to club-wielding cavemen, although the usual suspects remain theropod dinosaurs—the ferocious 'meat-eaters.' This particular section strays from the norm, however, as we bring an assortment of winged creatures into the spotlight, emphasizing a few other fictional forces of nature too. True—prehistoric life in popular culture runs the gamut, and so what's offered here can only be a mere sampling, a 'prehistoric monster mashup' of what explorers of the genre may quickly encounter. While many creators may have 'hoped' for top billing on movie rights, royalties from tie-in toy sales, or predominance on human psyche, not all succeed. We may remain 'hopeful' that such monsters shall never appear … in our limited range of reality.

Referring to Chapter Eleven, I might add that perhaps the *best* Gfantis painting/image I've seen so far is that by artist Chris Scalf, whose brilliantly orange-red hued rendition was 'unveiled' in poster format for the 2015 G-Fest event, (where I met this artist). Check his website!

Chapter Fourteen was rather inspired by a 3-page outline I discovered my father (Allen G. Debus) had written decades earlier—possibly during the mid-1950s, maybe around the year I was born—when he was employed as a chemist at Abbott Laboratories. His intended subject then was titled "The Great Piltdown Hoax—Case History of the Greatest Scientific Fraud," and I don't know whether he ever finished the article. Dad wrote, prophetically: "There are few more ardent quests than that of man to find out more of his own past. The imperceptibly slow evolution of the human race in the dim prehistoric eons is only slowly being uncovered by scientists who have searched in every quarter of the globe for the remains of ancient men … Why it was done and who did it are possibly questions which will never be answered, but uncovering of the hoax will stand forever as an example of the interrelation of one science with another." In his outline, my dad was going to discuss Henry Fairfield Osborn's perspectives in 1918 as expressed in *Men of the Old Stone Age*, "conflicting" with those of Charles R. Knight's published in *Prehistoric Man* (1949), as well as the chemical fluorine tests exposing Piltdown Man an absolute fraud.

I recognize that my dad would have been around the same age that I was when I began writing short articles for fanzines and newsletters. But did my birth obviate his ability to continue with popular avocational writings? Like those unanswered questions over unmasking of the Piltdown hoax, I'll never know.

These chapters originally appeared in the following publications:
Chapter Eight—Composited from two *separate* articles. *G-Fan* no.70, Winter 2005, pp.14-19; and *Mad Scientist* no.26, Winter 2013, pp.12-19.
Chapter Nine—*Scary Monsters* no.48, Sept. 2003, pp.107-114.
Chapter Ten—*G-Fan* no.80, Summer 2007, pp.26-30.
Chapter Eleven—*G-Fan* no.101, June 2013, pp.67-69.
Chapter Twelve—Previously unpublished.
Chapter Thirteen—*G-Fan* no. 71, Spring 2005, pp.70-75.
Chapter Fourteen—*Fossil News,* vol. 10, nos. 3-5, Parts 1-3, March – May 2004, (respectively, pp. 14-18; 14-17; 15-18).

Chapter Eight — *Replicating Reptilicus ... In Print*

Alternative Kaiju:

The name 'Reptilicus' entered my consciousness sometime in the late 1960s, although I never saw the film until the late 1980s. This movie evidently wasn't considered prime-time fare for Chicagoland networks. Programs sensitive to the monster movie genre' such as our local "Creature Features" and "Svengoolie" never seemed to air it. Until after receiving Don Glut's *The Dinosaur Scrapbook* (1982), I had no idea what *Reptilicus* was about. One thing for sure, this monster had a cool-sounding name!

Glut's "Scrapbook" may have been the first book to contain a reasonable description of the movie and its titular monster. There on pages 134-135, I spied a photo of a rather cheesy-looking creature, rather resembling a dragon with vestigial wings, as well as three contemporary comic book covers inspired by the 1962 movie. One such comic series was titled "Reptisaurus the Terrible." Glut stated "Reptisaurus looked basically like a crimson Reptilicus until the seventh issue (October 1962), wherein he suddenly and quite mysteriously appeared with a new body and a snout horn to go along with it."

Of the movie itself, which I've since come to love for its remarkable foibles, Glut succinctly remarked: "The regenerated monster has impenetrable scales, a serpentine neck, ridiculously small limbs, and dragon-like wings. It commands the land, sea and air in its attack on humankind, spitting out poison that kills its human victims on contact, while crowds of smiling, unpaid extras flee from the monster through the streets." (*Scrapbook,* p.134) Well, that at least partially explained why it was hardly, if ever, aired on television. And yet, strangely, I thirsted for more about it!

Reptilicus was evidently a film which ranked even below the deplorable *Robot Monster* (1953), which I'd it seen televised several times in the early 1960s. Wow, I thought, that's really, well ... different. Yet, it wasn't until after our last child was born that I finally managed to see the movie broadcast on a cable station. The Americanized version was shown again at the mid-July 2011 G-Fest, where I recall my wife and I highly enjoying the show.

Over two decades following my initial exposure to *Reptilicus*, accessibility to viewing and learning about this film has markedly changed. Now a cult classic, *Reptilicus* is available as a DVD product, no less, and is critiqued at length in movie books—such as Mark F. Berry's *The Dinosaur Filmography* (2002). What's more, one can surf the internet for *Reptilicus* collectibles, which is how I obtained a little known, yet much weathered treasure: a copy of Dean Owen's promotional paperback tie-in novel, *Reptilicus* (A Monarch Movie Book) published in June 1961. On the front cover the book is billed as "A high-tension story of a prehistoric monster no modern weapon could kill." Sounds like a 'can't miss page-turner,' eh? Then on the back cover we see a rare photo of the winged, *airborne* (yes, airborne!) creature, flying over an urban landscape with the words "Spawn of Hell" printed at lower left.

From reading forgotten wonders like Dean Owen's novel, one may glean how *Reptilicus*, the movie, was perhaps *supposed* to proceed, instead of how it woefully turned out. *Reptilicus*, the novel, must be categorized as an R-rated monster story (even more so than Carson Bingham's 1960 novel *Gorgo*). It is a charming, if not likeable story, although lacking the strange non-sequiturs, incongruities and over-the-top acting performances, not to mention the bizarre acting and dialogue plaguing the made-for-English film version. (Here I said "plaguing," although now must admit that it is for all these odd reasons that I so love the made-for-English audience film version.) Alright, let's consider this pairing of 'alternate kaiju,' primarily as conveyed via Dean Owen's high-tension novel, and then we'll get to those tie-in comic books in this chapter as well, (i.e. as opposed to hearing more about the movie (for which further information may be sought in Mark Berry's excellent aforementioned book.)

Can't say I'm much of a Dean Owen fan, since *Reptilicus* is the only book of his that I've read. Nevertheless, on the novel's Author Profile page, we learn that Owen (a pseudonym for Dudley Dean McGaughy) is "... one of the most popular of present-day authors." Owen authored more than forty Western and suspense novels, including two Monarch Movie Bestsellers—*Konga,* and *The Brides of Dracula*. By the time it came to pound out pages for *Reptilicus* on his typewriter, Owen certainly had the experience for churning out pulp. The novel is credited as based on an original story by Sidney Pink, with screenplay by Ib Melchoir. However, as Bill Warren, author of *Keep Watching the Skies* (Vol. II, 1986), related, Pink sued both Monarch Books and American International for "unauthorized use of his name in publication of the book ... As a result, Pink was held up to 'public contempt and ridicule.' Book ... contained passages of 'such lewd, lascivious and wanton desire as to inflame unsavory and lascivious desires in the reader."

Now let's delve into *Reptilicus*—the novel.

Owen hooks readers from the get-go with a racy scene involving one of the film's stars, 'Svend Alstrup,' the handsome blond Dane. In the novel he's quite the ladies' man. But remember, this was 1961, and the steamy stuff Owen could describe in his novel would have never made it past movie censors, especially in the case of a film intended for a younger audience. Early goings-on between men and women in the novel in a way foreshadows all the carnage that is to come. Bedroom acrobatics aside, let's explore the origins of this 'alternate kaiju' Reptilicus.

Admittedly, it's a *looong* stretch tagging Reptilicus as dinosaurian, and Mark Berry must have included an entry for this creature in his book under the most tenuous terms. From its outward appearance, Reptilicus resembles a dragon, but for purposes of the scientific blather surrounding its discovery in the movie and novel, we'll count this curious kaiju as a 'dino-monster' (a term I believe I've coined).

Remains attributed to a prehistoric monster (a giant 'dinosaur') are pulled up by a copper mining drill bit in the Arctic Lapland of Denmark. The fossils include bones and flesh partially thawed by friction of the rotating drill, still retaining blood, (literally staining the hands of Svend in the film's opening scene), and figuratively foreshadowing Svend's remorse and guilt for deaths later caused by the monster he discovered and helped to regenerate. In the novel, General Grayson refers to the soon to be rampaging Reptilicus as "… nothing but a chunk of prehistoric meat." Professor Otto Martens of Copenhagen's Aquarium is consulted over the mysterious fossils and flesh.

To Martens, this is a remarkable find, because it is the first non-mammalian frozen cadaver found. Svend notes the bones are hollow, suggesting an avian link (thus suggesting its later flying abilities). Martens concurs, commenting, "Very strong bones, but also very light—to enable the creature to take to the air," which would indeed be peculiar for an animal estimated to be ninety-feet long! Later, after Grayson observes the rampaging, regenerated monster which has grown tufts of coarse black hair in between its overlapping scales, Grayson asks Martens whether Reptilicus is a "… cross between some sort of dinosaur and a reptile." Martens explains to Grayson that "Reptilicus is one of nature's attempts to bridge the step from reptile to mammal." Of course, in the movie, the monster is referred to as a 70 to 100-million-year-old dinosaur. Yet, I repeat, it is a dragon.

Following a sampling of sins of the flesh involving Svend, his foreman's wife and then even Professor Martens' youngest daughter, Karen (played by the pretty brunette in the film), Martens is stricken with hubris at sight of the amazing preserved tissues. Martens declares, "I could become the first man in history to regenerate a prehistoric monster." Dr. Peter Dalby replies, "Think of the chaos such a monster could cause in the world today." But Martens cautions he would only let the regeneration process go so far, before terminating it.

Yeah—right! So, in Frankensteinian fashion, Martens and colleague Dr. Dalby decide to place an eight-foot long frozen segment of fossil flesh inside a 'regeneration incubator,' where it will be bathed in special nutrient fluids. (In the movie, this frozen piece is the tail end of the animal.)

As Martens explains to news reporters, "regeneration" is the ability of living tissue to heal itself or to repair an injury, like when a lizard, which has lost its tail regrows that appendage, or when severed starfish and flatworms regenerate themselves from sliced-off arms and pieces. Now, however, Martens has visions of grandeur, exclaiming to reporters that he'll attempt to regenerate a "*complete* prehistoric creature." One reporter, Klaus Peterson, suggests that, given the creature's reptilian nature, when the creature is regenerated, it should be named "Reptilicus martenius." Martens is amused, replying, "Splendid, splendid, but I think we should shorten it to 'Reptilicus."

Although the scientists speculate it would take months to a year to thaw the flesh properly to promote regeneration, Dalby errs by leaving the refrigeration room door opened, defrosting the hunk of flesh. Meanwhile, Lise, Martens' older daughter, notes that the hole left by the drill bit has already begun to heal … overnight! Now triumphant, Martens dreams of the day he'll share the Nobel Prize with Dalby. But Lise is dreading developing circumstances: "Father, I'm frightened. Don't let it live. Kill it—now! Before it is too late." Gripping her shoulders, Martens tries to settle her fear. "Get hold of yourself, child! This is a time for rejoicing, not hysteria. Think what this means, Lise. *Think*. For eons that piece of flesh has been frozen—in suspended animation. Peter (Dalby) let it thaw—accidentally. But we don't have to worry about that." And then channeling his inner mad scientist persona, Dr. Martens exclaims, "Because it lives. *It is alive!*"

Now, with Reptilicus growing in the incubator, for the vague reasons that "the regeneration of a giant prehistoric monster … might become a menace to one or of the United Nations," military and UNESCO representatives are called in—American scientist Connie Miller and bad-mannered Brigadier General Mark Grayson, respectively. (It's as if they've been watching too many Godzilla movies.) Connie notes Grayson has a roving eye for her as well as the attractive Lise; in the film, the two ladies are played by voluptuous blondes who are far better 'behaved' on screen than in the novel we are summarizing here.

Soon, Reptilicus, appearing "unlike any other known dinosaur," has doubled in size. But Grayson, who is dejectedly "playing nursemaid to a hunk of flesh," seems bored with his duty. Meanwhile, Martens and Dalby seem troubled by their 'experiment.' Martens admits "…there have been moments of doubt … Peter. It is the age-old argument. Can a scientist worry about the consequences?... Did Teller worry because the H-bomb could destroy the world? … A scientist experiments, no matter what. Then he does his worrying … I am going ahead, come hell or high water, as the Americans say." Dalby, a voice of reason, only answers, "I hope to God you can keep control" (i.e. of the experiment). But control is quickly lost. Soon after a vestige of the regenerated head forms, electrical discharges from a lightning storm exponentially accelerate Reptilicus' growth. Horrified, Dalby and Martens consider destroying the no longer dormant creature writhing in the tank, but they are too late. And so, awakened from prehistory in Frankensteinian fashion, Reptilicus claims its first victim—Dr. Dalby—and then escapes the Aquarium's confines into the stormy night.

General Grayson and Connie are called in to view the wreckage and, shortly, a report of a monstrous creature seen on a nearby Danish farm reaches them:

> "For the first time Grayson … saw the hellish, murderous creature that had seemed so harmless in the Aquarium tank. The creature had grown overnight … It now towered some 80-feet into the air. The body was long and slender. It looked … like a gigantic reptile with a long pliable tail. The entire body was covered by huge, overlapping scales. Tufts of coarse black hair grew between the scales. The rear legs on which the creature stood were sturdy and squat, the forelegs grotesquely small, but with long sharp claws. On either side of the body were large wings—like the wings of a bat … The skull of Reptilicus seemed to be a cross between the deformed, hideous triangular-shaped head of a rattlesnake and the malicious features of a vampire bat."

Noting the poisonous venom dripping from its fangs, Grayson dutifully orders the machine guns to open fire, all to no avail. Following Reptilicus to the beach, a flame-thrower douses the monster, giving off a "nauseous odor of singed hair and flesh," driving him into the sea.

There are two crisis moments for Grayson following his initial encounter with Reptilicus. The first comes when patrol boat Commander Nielsen is ordered to drop depth charges onto Reptilicus' body when it's discovered on the sea floor, regenerating in shallow water. Grayson is jubilant, knowing that they're bombing Reptilicus to bits. But Connie rains on his parade, warning "Stop them, Mark. Stop them at once! Don't you see … if Reptilicus should be blown to bits he would regenerate … each separate piece would grow into a new animal." Grayson calls a halt to the underwater depth-charging, but not before an injured foreleg is ripped from Reptilicus' prostrate body. Now vengeful Reptilicus rises from the sea, snatching the boat in its jaws, while shore artillery units evacuate. Professor Martens, who has a conscience after all, realizes his sin in restoring Reptilicus to life and suffers a heart attack from the awful sight, but lives. Author Dean Owen writes, "A visiting Englishman, surveying the damage, said somberly, 'Even the Blitz was nothing like this. Frightful, frightful.'" Reptilicus may have been spawned to a certain degree from Cold War angst, but the carnage described in the novel was certainly fueled by the ravages of World War II.

While Grayson ponders where Reptilicus may next strike and how to effectively wage war against such a creature, Reptilicus' injured flesh regenerates and its wings develop fully. Now Reptilicus is airborne, on a rampage in the Baltic region. Reptilicus attacks Hamburg and devastates Stockholm. However, theorizing that Copenhagen's Aquarium is Reptilicus' instinctual home, where it was effectively (re-)born in the incubator tank, Grayson prepares for its eventual reappearance there. Relating her fear of Reptilicus to the heroic Grayson, Connie Miller contrasts the monster with the hydrogen bomb. "A bomb I could face … But this thing—this Reptilicus—is something from another world." (i.e. a 'prehistoric' world) Then she falls into his arms and the two make whoopee behind a locked office door.

Reptilicus, impervious to anti-aircraft fire, begins its aerial assault on Copenhagen. Connie whimpers, "Just to think of that monster flying around up there—it's almost as if the end of the world was at hand." The populace is out of control, buildings are crumbled. A rumor that a hydrogen bomb will be dropped on Reptilicus heightens panic. And then, Grayson decides to use "multiple flamethrowers" on Reptilicus. "A terrible scream of rage burst from Reptilicus as the wings caught fire." The flamethrowers burn away the webbing from its wings, and now Reptilicus, no longer capable of flight, is terrestrial.

A terrible scene of annihilation is described as screaming men and women flee the approaching Reptilicus over Knipplesboro Bridge, while a terrified bridge engineer throws the switch to raise the bridge. In the movie, this ranks as one of the most ludicrous monster-on-the-loose scenes ever filmed, and the one where Don Glut noted smiling extras jogging past the movie camera. But Dean Owen's authorial talents do not fail his audience, for this is truly one of the novel's high-tension moments. Shortly afterward, Reptilicus destroys a trolley full of passengers. Meanwhile, Reptilicus is becoming ever more irritated and enraged by all the fire-power brought to bear against its scaly hide. The creature continues to weave a path of destruction through Copenhagen, its vision impaired from first encounter with those flamethrowers.

Another high-tension moment, framing the story's second 'crisis moment' as well, opens with debate over using a nuclear weapon to kill Reptilicus. Martens overrules Grayson, claiming, "You want to blow him into a thousand pieces? Each minute particle of the creature would be capable of growing into a new Reptilicus … How can you find them all? There will be bits of flesh blown into the lakes, the streams and into the sea." Grayson humors the old man, ordering his soldiers to strip Reptilicus of its scales, that is, if only the monster could be incapacitated using multiple flamethrowers. Martens, having made his point, leaves the room. But then, after Martens leaves, Grayson divulges to Svend that he will use "the bomb" on Reptilicus anyway, once it has been driven to "open country."

However, it is Svend who comes up with the best idea for destroying Reptilicus—using a hypodermic containing sodium pentathol to knock him out, which is prepared by pouring the chemical into a bazooka shell on the university grounds. Let's skip past the unnecessary scene where Lise is romantically 'assaulted' shall we, and move to the story's climax—beginning with the most ludicrous dialogue in the novel—when Grayson tells Captain Brandt that the shell will have to be fired into the soft skin of the monster's mouth. Brandt replies "That means you'd have to fire point-blank." Grayson adds, "And at very close range." No kidding.

Reptilicus meets its own end in the town square. Poor Brandt is crushed under the monster, allowing Grayson to gain his point-blank shot, and the soldiers busily strip Reptilicus of its scales. Then the flamethrowers douse Reptilicus with flame. "Flesh popped and sizzled, seeming to melt under the searing tongues of flame. The fat erupted in a great geyser of flame. The eyes burned away, the skull became an inferno. Gradually the creature seemed to melt. Sizzling fat ran into the gutters and the pavement cracks ... All that remained were the piles of scales." Grayson makes no motion to bomb the scales. Meanwhile, at sea. "The great forelimb that had been blown from the body of Reptilicus by the charge now lay in a secluded cavern. It had almost doubled in size ... The claw itself seemed to shiver with life as it slowly and ominously started to close upon itself." At the movie's conclusion, just before the scene of the claw is shown, Grayson ironically remarks, "It's a good thing that there's no more like him."

And so, mankind is held, symbolically, contemptibly within clutches of a great 'doomsday dinosaur' inadvertently of our own making—essentially, a "monster no modern weapon could kill"!

Herein, I've emphasized the published story (novel) rather than the movie. Partly, this is because one may sometimes glean more about meaning and intent of a monster story then through (only) watching the film. Also, while many readers may have viewed *Reptilicus*, few likely have read Dean Owen's long forgotten novel, which despite its unsavory passages, remains, at plot level, rather faithful to the movie script. As Bill Warren noted, *Reptilicus* the novel is a conceptual merging of three prior films: *Godzilla, The Thing From Another World* and even *Rodan*.

Due to delays associated with the aforementioned lawsuit, Owen's novel came out months before the film. And from reading amusing accounts of *Reptilicus* written by Warren and Mark F. Berry, readers might have an impression that perhaps it would have been best had the film never been released. (Yes, here's another case where, in my opinion, the novel was far better than the film, which is *not* to say that the movie wasn't very entertaining.)

Now ... three items of note about the *film*; Reptilicus vomits cartoonish-looking green acid slime on its victims, a poor swipe of Godzilla's fiery radi-

breath. And in American prints, scenes of the Rodan-esque flying Reptilicus were deleted, supposedly because they were too crudely done and laughable. Besides the Danish women playing roles of Lise and Connie Miller in the movie (Ann Smyrner and Marla Behrens), the most memorable feature about the English version of the movie is how the actors recited their lines 've-ry slo-w-ly' so as to achieve proper enunciation. As Berry noted of this draggy delivery, "Professor Martens says … syllables oozing like snails from his lips." Evidently, screenplay writer Ib Melchior anticipated the movie would be dubbed, so he instructed actors, "When you speak, be sure to speak distinctly and move your mouths distinctly so that we can dub this." Unfortunately, the result was a side-splitting affair, with lines such as, "I have … ne-ver seen … bone frag-ments … like this … be-fore." Many are the happy times I've doubled over in hilarity listening to General Grayson, played by Carl Ottosen, care-fully, robotically e-nun-ci-at-ing his lines in this particular film version. Scripted humor performed by the comical actor playing the lab maintenance worker 'Peterson' in the movie (not the same character as the reporter in the novel) was far less amusing than the unintentional humor.

Special effects in *Reptilicus* are poorly done, perhaps among the poorest for any film of this genre in that period of movie-making. But I won't harp on this outstanding deficiency here. In fact, overall, the movie was so bad that American International initially refused to distribute it. But now it's available on DVD! My advice would be to read the book first though.

Charlton's Winged Demon of the Early 1960s:

Reptilicus/Reptisaurus was arguably the original "triphibian" monster of the comic book realm, as it walked on land, flew and swam! And Charlton's version of the dino-monster, spanning a brief, nine comic book run from 1961 to 1963, was far more versatile and protean than the creature—often subjected to ridicule—witnessed on the silver screen. Acting seen in the picture released in America, is, to say the least, er uh …*well*, quite "unique." But I've had many a grand time watching this film with its purportedly 50-million-year-old monster, one that I never got around to seeing until the mid 1990s! *Reptilicus* (the Americanized movie) has often been harshly mocked and maligned, (although it is certainly worthy of many giggles – merely the "fun factor" at large). But now let us direct our attentions to the short series of comic books published by Charlton and issued fueled by "hype" surrounding the film's monstrous debut (USA – June 1961).

Charlton's issue no. 1 of *Reptilicus*, attributed to Sidney Pink's original story and Ib Melchior's screenplay, appeared in August 1961. While the story mirrors events portrayed in the Americanized version of the film fairly closely (where the monster was not seen flying), certain aspects are neglected while others are played up to further degree. Most noticeably, the Reptilicus monster's *flying* capabilities are greatly accentuated, as in Dean Owen's script novelization published in June 1961. And Charlton's monster version never vomits acid slime at its adversaries. Although the movie monster itself has been reviled and subjected to mockery, conversely, as it appears in the comics the 100-foot long, indestructible flying Reptilicus somehow seems, well, shall we say, more 'reasonably' presented. From a pure dino-monster fan perspective, I wish the movie Reptilicus had been as good as the comic book version. But then, arguably, the film wouldn't have been quite so memorable.

The plot of Reptilicus issue # 1 generally follows the movie script, beginning with the Lapland copper miners' discovery of an extremely well preserved, yet foul-smelling hunk of scaly animal flesh, severed from the tail of some unknown prehistoric animal. Svend Alstrup, in charge of the mining operation radios Professor Martens at Copenhagen's University, who in turn flies out to examine the curious fossil. Dr. Martens concludes that the animal to which the flesh once belonged was reptilian, or "… the most fearsome reptile which ever existed!" Eventually, after returning to the University, Alstrup is introduced to Martens' two ravishing daughters (who incidentally look much better on camera than on the comic book page).

While Martens hopes to study the fossil flesh under refrigerated conditions, an assistant accidentally leaves the laboratory room door ajar. This would ordinarily cause flesh to rot, but instead the warmth incongruously causes preserved cellular tissues to grow! The flesh, sliced from the tail of the animal by the mining drill, uncannily begins to heal from the wound. Then, during a storm, a lightning strike obliterates the tank in which the flesh is regenerating. Afterward, Dr. Martens discovers that his specimen now known as "Reptilicus" has disappeared. In fact, giant footsteps leading from the demolished building indicate that a monster is already on the loose in the countryside.

Meanwhile, two UNESCO officials arrive – General Grayson and a Miss Miller. Grayson is summoned from his golf round when news of the laboratory disturbance circulates. Although at first he wants nothing to do with the matter, Grayson is ordered to lead a military reconnaissance to search for the creature. Rifle fire, bazookas, and tank shells do not hinder Reptilicus, however. Military planes drop napalm and rockets on the bellowing beast, to no avail. Reptilicus retreats into the forest, but is later seen flying over the Atlantic. Having witnessed the horror, Martens determines that Reptilicus is one of nature's attempts to "bridge the gap between fish, reptile and bird," while Grayson concludes Reptilicus is "half snake, half lizard, and all nightmare!"

As in Owen's novelization, Reptilicus flies over the Swedish border to Stockholm where it is attacked by military jets, although such weapons prove useless, other than for driving it back to Denmark. Here, flame-throwers do not deter the monster, and officials consider using a "Big Bomb" (i.e. nuclear) to destroy it. But if this happens, warn Martens and Grayson, the animal would be blown to bits and each separate shred would regenerate into a new Reptilicus "clone." Meanwhile, a secret weapon is being improvised – a shell containing narcotic that will be fired from a bazooka into the monster's bloodstream. With Reptilicus' corpse lying in the city square, the victors exult. Dr. Martens dares to proclaim, "One chance in billions produced Reptilicus! Such a creature can never come into being again!" However, in the last panel, ominously, another scary scaly chunk of flesh is seen floating in the northern waters…..

Well, that is where the movie ends too. The *Reptilicus* movie, novel and comic book are in a larger sense 'congruent,' that is, at least in terms of overall plot. So now you may be wondering what exactly happened *next* to Reptilicus in the comics, eh? Read on.

In Charlton's Reptilicus no.2, (Oct. 1961), "Reptilicus in the African Jungle," the giant winged monster returns. It turns out that the formless flesh flotsam floated southward toward warmer African waters, where it grew. Also, Dr. Claudius, an American physicist embittered by how the USA military turned one of his atomic discoveries—a nuclear powered turbine—against mankind, lives in a nearby jungle hut with his pretty daughter. Dr. Claudius shuns modern civilization, refusing to facilitate man's capability to self-exterminate using nuclear weapons. Then Peter Blinn arrives from the State Department intent on convincing Claudius that his nuclear turbine could be instead benevolently used in the "backward nations" of the world to help end poverty.

Meanwhile – a second incarnation of Reptilicus washes ashore! As the monster devours food ravenously, it grows to immense size. On outspread wings Reptilicus terrorizes the nearby African village. When the villagers plead to Dr. Claudius for help, the good professor is incredulous, although Blinn recognizes from their description of the witnessed monster that – incredible as it would seem—there could actually be a second Reptilicus now on the rampage! Blinn heads off on a safari with warriors to locate the creature, taking along a .375 magnum rifle – just in case. Of course, the gun shots only irritate the snoozing beast and soon Blinn and the natives flee for their lives. After spying the flying monster, Dr. Martens suggests that it may be an "… unlikely cross between a pterodactyl, a brontosaurus and a colossal snake."

The Chieftain proposes that the unwanted monster should be killed in the mystical Valley of Smoke, a volcanic region at the heart of the Cold Mountain where poisonous vapors are emitted. And so discouraging and wounding the "evil" monster with spears and rifle shots, Reptilicus is steadily driven back toward the Valley of Smoke, where great beasts go to die. When an earthquake causes the enraged Reptilicus to slip into a sulfurous, volcanic bog, Blinn takes full advantage of the situation, pumping a slug into its broken wing. Not unlike what happened to the flying pair of Rodan at the end of the 1956 Toho film, (or the "fire monster" near the end of Irwin Allen's 1960 film, *The Lost World*) Reptilicus apparently succumbs within the fiery volcanic emanation. The episode ends ominously, "Meanwhile, what of Reptilicus? Had he at last found a grave that would hold him ... was this death final, or would he rise again some day to terrify mankind?" Well – what would *you* guess?

Something strange happened to Reptilicus between release of issues nos. 2 and 3. The monster's name changed ... to "Reptisaurus." This may have been done for copyright reasons (or possibly for artistic, creative license). Although initially Reptisaurus and Reptilicus monsters highly resemble each other, eventually Reptisaurus's form evolves into a more imposing creature. In issue no.3, one important difference, however, is that Reptilicus' prominent toothy fangs are missing in Reptisaurus – so perhaps the two are separate species after all. On the cover, one of two reptisaurs shown has the fangs while the other doesn't – sexual dimorphism or an artistic error? It seems that one of the flying creatures acquires the name Reptisaurus, while the other is referred to as its "mate." Regardless, the series continued in what was to be the first issue of "Reptisaurus," labeled as issue "no. 3" in the blended Reptilicus/Reptisaurus series.

Issue no.3 (i.e. although this time of Reptisaurus), titled "Reptisaurus the Terrible," issued in January 1962, finds Reptisaurus on the rampage. But on the splash page, Reptisaurus is likened to a great "mutant" pterodactyl. Clearly the editors were still wrestling with Reptilicus', oops – uh, I mean "Reptisaurus'" natural history and geological origin. When a great age of ice descended upon the dinosaurs, Reptisaurus' kind also became extinct, frozen in entombing sediment. Furthermore, in issue no.3, it is atomic bomb testing that awakens two "reptisaurs" from under the ooze and muck of the Caspian Sea. So here one may conclude that the last true "Reptilicus" monster (cloned from fossil frozen flesh salvaged by a mining drill) expired in the Valley of Smoke, while Charlton's second generation (or perhaps species?) of winged dino-monsters, known as "Reptisaurus," did indeed originate into modernity via nuclear bomb testing.

Rescued from oblivion, two reptisaurs fly at high altitudes where they're noticed by outer space aliens from Jupiter intent on invading Earth. Interplanetary war ensues, with the Reptisaurus crunching flying saucers in its jaws. Next, the "Jupos" further engage Reptisaurus from the ground with "Rollo" tanks firing blazing heat rays, but the monster carries away one doomed tank in its mighty talons, dropping it from lofty skies. Reptisaurus is then stunned by a high energy gamma ray burst, and crumples to the ground. Now the female reptisaur is stirred from her egg nest, yet she too is "neutralized." The "Jupos" take photos of themselves triumphantly gloating over their comatose "trophies."

Military fighter aircraft now patrol the area, which quickly join a losing aerial battle against the superior flying saucers. No sooner have the Jupos claimed Earth for their own when suddenly Reptisaurus painfully recovers, angrily seeking revenge. Ordinarily, as so often happens in the comics, it would seem that only Reptisaurus and his mate could turn the tide of victory over for Earth's inferior forces. However, while the parents lie relatively helpless, it is the young reptisaur hatchlings, which nip and tear at the heels of the jittery Jupos that chase the disconcerted aliens away, back into space forever. In the final frame, the reptisaur family led by "Reptisaurus – scourge of the universe," is seen retreating toward a cave leading into the bowels of the planet where they may sleep for perhaps 100,000 years.

Or maybe not.

Regrettably, the plotline of "Reptisaurus the terrible" no.4, issued in April 1962, story titled "Reptisaurus meets his mate," is rather silly. After sleeping for two years in the subterranean cavern, two reptisaurs emerge, one on Baffin Island in the Arctic and the other in Sudan on the African continent. Soon the reptisaurs attack cities separately, while mankind remains powerless to stop them on every shore. Eventually, the two reptisaurs sense each other via emanating (ultra-) sounds, even from many miles away. While Reptisaurus ("the first such monster to blight the sight of man") hurtles at supersonic speeds over the Atlantic toward the US, the second reptisaur is felled by a lightning strike in an Ozark swamp. The pair of winged monsters continue to sense each other's call, converging toward one another over Florida's Everglades. While military officials watch aghast, a vicious dino-monster aerial 'dogfight' ensues, but – surprise—as things turn out this is only a mating ritual. The tale concludes, "They were courtin'! That's all! Every time that big one slammed the smaller one, that was saying 'I love you' Monster style…. " Not exactly a "bride of Frankenstein" affair, but certainly a case of giant monster love, which must have seemed far less palatable to pre-adolescent daikaiju fans then, than giant monsters speaking to each other (as in Toho's *Ghidrah the Three-headed Monster*).

Reptisaurus the Terrible no.5 (June 1962) is a bit of comic relief, rather extending the romance theme of issue no.4 (as if we hadn't had enough of that). Having already found a mate in issue no.4, it would seem that the male Reptisaurus becomes somewhat of a "philanderer" in issue #5, chasing a floozy Chinese celebratory Dragon. Reptisaurus is enjoying his meal near a peaceful Pacific atoll, which of course is then nuked by USA military. The Bomb annihilates the island, but this only fazes Reptisaurus, who now angrily attacks an observation plane. Reptisaurus then files a course over Red China, where it is attacked once more. Poor Reptisaurus, nobody loves him… that is until the winged monster spies a Chinese holiday parade featuring a fire-breathing celebratory dragon that is "… mechanized, articulated … most life-like in every way!" Reptisaurus is immediately smitten! Both *animatronic* dragon and

Reptisaurus are soon parading *together* through streets of the Chinese town. Flames emitted from the dragon's snout (by a technician intending to dissuade Reptisaurus) do not deter the living flesh and blood beast.

But machine gun fire and military jet strafing chases Reptisaurus away to the skies above; it retreats to the Himalayas. But true love will not be deterred by such weaponry, and so Reptisaurus triumphantly returns to claim his 'consort,' although this time the Chinese military insidiously loaded tons of munitions ready to explode via remote control. Love unrequited in the worst way imaginable! So when Reptisaurus nuzzles closely, the dragon explodes, and the disillusioned monster returns to the "roof of the world" to snooze until his next episode.

The Cold War theme returns more prominently in the 6th issue (August 1962), titled "Reptisaurus vs. the Red Star," Reptisaurus awakens, rejuvenated, from his icy throne atop the Himalayas. Buzzed by a communist jet plane, soon atomic weapons are dropped over the roof of the world – fiery mushroom clouds ascend over the great mountain chain. But Reptisaurus isn't destroyed and soon chases the fleeing jets back over communist soil. Entangled in cables, Reptisaurus' future and fate appears grim. (It's rather incongruous that a monster which cannot be destroyed by nuclear bombs can be simply ensnared and immobilized in a netting of ordinary steel cables! But remember, this is a comic book.)

When the Soviets boast of their capture of Reptisaurus, illustrated reports appear in USA newspapers, prompting Washington military brass to rescue the monster. An American official proclaims, "Reptisaurus hasn't intentionally done harm to mankind! He strikes back when we hurt him ... but he is not by nature vicious! ... I should like to see him freed for his own sake." In what is described as a "marvel of organization and military efficiency," American paratroopers are furtively deployed, who cut the cables. A veterinarian recommends feeding the weakened beast with high protein nutrition and vitamins. Like Popeye gulping down his spinach, soon Reptisaurus is soon strong again and takes vengefully to the skies, in pursuit of military planes carrying the red star insignia, a symbol the dino-monster has learned to hate, yet not fear. Ultimately, Reptisaurus trashes what must be all the key communist government buildings in Moscow.

Afterward, men in New York City streets remark that the dino-monster "... must be sick and tired of the way he gets treated abound here. Poor, pitiful creature!" Meanwhile, USA military brass congratulate themselves for their decision to unleash Reptisaurus upon the Soviets.

Back-slapping aside, Reptisaurus' appearance transformed dramatically for issue no.7, (Oct. 1962), "Reptisaurus Returns," and considerably for the better! The old, former Reptisaurus was thin, scraggly – looking, like some sort of weird snake with scrawny wings and legs. Whereas before, as in issue # 6, Reptisaurus was four-limbed (i.e. wings and legs), now suddenly Reptisaurus had sprouted a sturdy (extra) pair of monster arms. Its leathery wings are much broadened, seemingly more capable of flight. And Charlton's refined dino-monster was redesigned with great, rounded dorsal fins, a full set of sharp teeth including two prominent fangs, (as opposed to edentulous). It also now sported a ceratosaur-like nasal horn and featured more imposing body texture and facial contours. This new scarier looking version resembles a true dragon (although Reptisaurus didn't breathe fire)! And even the new story line was improved.

The splash page of "Reptisaurus Returns" features a dramatic battle with mankind. Introductory text suggests that these invulnerable "reptilian mutants" were kept alive since the Mesozoic age through an ice age in suspended animation, and (as stated before) were awakened by a nuclear explosion. These were a "menace that mankind could neither understand nor destroy."

Reptisaurus' new mate doesn't resemble the one previously shown in issue no.4. In fact, the new mate almost appears to be a different species, with a pteranodon-like crest atop its noggin. Also, they're different colors. Together they swim the oceanic depths, foraging. Then, stretching leathery wings, they lift to the skies above, headed toward Africa's steamy, primitive jungle where they shall soon nest. A biologist named Dr. Harden wants to observe them in their natural habitat, even though political strife is brewing on the nearby coast, in Katanga province. But onward goes intrepid Harden with pretty wife, Lois, in tow. When their strong, handsome guide makes a play for Lois, her husband socks him in the jaw, proving that Dr. Harden isn't just a nerd after all. Tinged with sweet metaphor, Lois' and Dr. Harden's love swells just as they discover the enormous Reptisaurus nesting site.

Dr. Harden plans to destroy the eggs, lest the planet would soon be overrun with a giant invulnerable species (making us wonder whatever happened to those reptisaurus young that had hatched by the end of issue # 3?). Harden stealthily swipes one egg to study later in his laboratory. But reptisaurs are crafty monsters and so when they return, they're able to repair the remaining eggs that Dr. Harden had cracked with an axe, thus sparing the embryos developing within.

Enraged, the (presumably) male Reptisaurus flies toward the haunts of man, where it engages itself in the Katanga war. Swooping downward into the city like Rodan in the Toho film, jets, rifles and bazookas do not deter the monster! Then after laying waste to the town, the monster thoughtfully returns to its mate, while we see Harden in his laboratory, with a human-sized reptisaur hatchling peering from inside a cage. Meanwhile …. back in the dark African jungle, those eggs that Harden compromised end up producing viable young as well. The tale spawned by "Reptilicus" is far from over!

Well – you can probably guess what happens next.

In "Reptisaurus the terrible" no.8, (Dec. 1962), titled "Brood of the Beast," reptisaur hatchlings lie in wait in the steamy jungle setting. Meanwhile, Dr. Harden and his assistant experiment. They seek a means of killing the captive creature—as the adult pair remain a "constant threat to mankind." Subjecting the smaller lab specimen to freezing conditions, simulating the glacial age that supposedly killed all the large reptilian Mesozoic forms, only slows down its metabolism. Then – Eureka! – Dr. Harden theorizes that if the young monster were to be encased in ice that also contained radioactive nuclides, that this might do the trick. (But even if it works, one wonders how he plans to encase two gigantic adult reptisaurs in radioactive ice down in a hot African jungle!!)

However, suddenly, a storm rips through electrical poles. With lab electricity short-circuited the young reptisaur is freed from its cage. As the monster threatens Dr. Harden, he conveniently stumbles upon his "sound machine" that emits hypnotic, ultra-high frequency waves. When Harden blows through the specially engineered whistle, strangely enough, this effect calms and entrances the monster, allowing the doctor to safely return it to its cage. How convenient!

Well, when it comes to destroying giant dino-monsters, an ultra-high frequency sound emitter (essentially a dog whistle) isn't anywhere near as cool as, say, the "oxygen destroyer" that consumed the original Godzilla. But, whereas the latter device turned Toho's Dr. Serizawa 'mad' in a way that turned out to be good for mankind, this dog whistle thingie causes Dr. Harden to become "mad" in the typical, old-fashioned bad way – for now Dr. Harden's fevered mind rebels. Now he wants to *command* the reptisaurs with the whistle so that he can control the world." Crazed, demonic laugh—Ahahahahahah!! But as radioactive ice experimentation proceeds, a major, obvious flaw manifests. So much radioactivity is emitted, generating a heat flux as well, that ice encasing the beast melts. As beast awakens, Harden's assistant kills the monster with an overdose of radiation fired from some kind of gamma gun.

The team flies to Africa, with Harden's ulterior motive being to test the dog whistle on adult reptisaurs, hoping to "hypnotize" them to do his evil bidding. Little did he suspect to encounter an entire nest of ghastly reptisaurs! The whistle indeed works on the large male just as Harden hoped it would. But when Harden's assistant escapes his bonds (e.g. Harden had tied him up), the assistant desperately radios for a nuclear bomb strike over the reptisaurs' swamp. This of course kills Harden and presumably the younger reptisaurian brood, but with the hypnotic spell broken, the two adult reptisaurs are shown flying away "toward the horizon."

Charlton strove once more to give us a final Reptisaurus the Terrible segment, which was issued as a special edition during the summer of 1963, titled Idol of the Aztecs. This time Reptisaurus is equated with the Aztec flying serpent god, "Quetzatacl," (probably a misspelling of Quetzalcoatlus). Plagued by yet another mad scientist (reprising core elements culminating issue no.8) who wields an electronic transmitter paralyzing device instead of the dog whistle. Following a nuclear strike, Reptisaurus flees across the Atlantic into a Central American jungle. This coincidentally happens just as a merciless Aztec tribe "lost" for centuries, emerges from underground hiding, awaiting arrival of the great legendary "thunder bird." Upon its triumphant descent from the skies, so their High Priest Xochipilan claims, the Aztec nation shall be led to "victory."

Eventually an expedition led by a famed hunter and explorer to find the monsters starts to unravel, with the handsome guide making a play for the arrogant leader's wife (a common plot element in Charlton's books). But this happens just as the party nears the primeval swamp where the two adult reptisaurs are resting. The explorers are captured by the Aztec warriors and are about to be sacrificed on the temple altar to their god, when suddenly the male Reptisaurus dramatically appears. "And Reptisaurus came, a monster from the past worshipped by a people from the past! He came on wings of thunder, the omen, the talisman they had waited generations for ... a sign that would make Central America and Mexico run red with blood of battle."

Reptisaurus smashes the temple, thus freeing the explorers although their leader has now gone completely insane from the ordeal. In the final frame, we see Reptisaurus and his mate slumbering in their jungle ooze, dreaming back "... through the ages to the dim past when the world was young and he and his kind ruled the Earth!"

Charlton was concurrently running two additional series of books also featuring giant monsters based on movies that had then recently been released (i.e. *Gorgo* and *Konga*). Of the three comic book tie-in series, Reptilicus/Reptisaurus's was the shortest run. Both Reptilicus/Reptisaurus and Gorgo of course shared a pseudo-prehistoric provenance. I have outlined the scope and meaning of the Gorgo series elsewhere[1], but it would appear as if Charlton's editors were going for highly analogous messages with their Reptilicus/Reptisaurus series. These messages are not nearly as accentuated as in the Gorgo series where that dino-monster becomes almost a Gaian, planetary symbol as well as a (reluctant) capitalist monster, defending the Free World.[2] Gorgo and his mother take the brunt of many nuclear detonations, thus thwarting threat of communism and impending World War III—prevalent fears during the Cold War/nuclear "arms race" era. Outer space alien creatures Gorgo fought were not much better, psychologically, than the power-hungry despots, as characterized in the books, ruling the Soviet Union and Red China.

Like Gorgo, Reptilicus/Reptisaurus takes on an inadvertent USA/world defender role as well. Reptisaurus even makes a cameo in Gorgo no.12 "Monsters' Rendezvous" (April 1963), forming an alliance in defeating would-be Venusian conquerors, rather like when Rodan teamed up with Godzilla, for instance to fight the outer space alien winged Ghidrah. Interestingly, although by October 1962 the appearance and name of the Reptilicus monster had morphed to the more imposing figure of Reptisaurus, the winged monster appearing in Gorgo # 12 is the old 4-limbed, snaky-looking Reptilicus form.

Compared to Gorgo's triumphs, the winged monster is only second banana, rather like Rodan was Godzilla's accomplice in Toho's productions. Somehow Reptilicus never realizes or attains Gorgo's symbolic status, nor does it ever seem quite as indestructible or vengefully powerful.

My impression is that Charlton didn't really know what to do with the Reptilicus monster, that is, beyond issue #1, which followed the movie script. It's as if they were groping for interesting stories, but found themselves borrowing key plot elements and themes from the Gorgo series (which must have been more successful supporting its greater longevity). A stunning physical transformation into the Reptisaurus witnessed in issue no.7 finally got their monster out of its mating/dating friend-zone woes. But by issue no.9 the storyline simply seems to have grown tired and stale. Gorgo clearly had an edge over Reptilicus in relating Cold War era themes, and the mad scientist bit was also growing repetitive and uninspired. Reptilicus/Reptisaurus never championed the Free World to the extent that Gorgo and his mother heroically did.

But then in the movies, audiences surely had far more empathy for the Gorgo/mother pairing than for acid slime-vomiting Reptilicus. Just as the movie *Gorgo* exceeds the filmic *Reptilicus* in many respects, Charlton's Reptilicus and Reptisaurus the Terrible pale by comparison to the Gorgo comic book series.

Notes to Chapter Eight: (1) Founded on a solo panel slide presentation I gave on the Gorgo Charlton comic series at G-Fest in July 2012, I incorporated much of my talk in Chapter Twelve of my 2016 book, *Dinosaurs Ever Evolving: The Changing Face of Prehistoric Animals in Popular Culture;* (2) Godzilla's role in film, as well as Gorgo's and to lesser extent Reptilicus' in the Charlton series may sometimes seem "heroic," in the sense that, certainly from human perspective, heroic deeds are being accomplished. After all, this is partly what qualifies and defines them as "daikaiju." But in such cases these dino-monsters may not be intent on *deliberately* performing actions that would seem heroic. They're just responding to highly adverse situations and circumstances, sometimes inadvertently. If we could translate and speak "monster talk" with them time and again after they've just spared the world from utter annihilation, they'd probably deny (i.e. in typically heroic fashion) any suggestion that they've responded like mankind's "super-heroes," saying, "Aw, shucks,

Little human Person – it's just what I do." Nevertheless, beaming with pride, we'd simply reply, "*Hey man*, that's sounds like 'hero' talk"!

Chapter Nine — *Greatest Terror-saurs of Sci-fi & Horror*

".....Satan laughing spreads his wings........"
('War Pigs,' Black Sabbath, 1970)

Not unlike the shadowy, bat-winged demon haunting A Night on Bald Mountain (*Fantasia*, 1940), the visceral nature of winged pterodactyls (i.e. 'pterosaurs'), or 'terror-(ptero)-saurs,' evokes primeval and preternatural horror! Terror-(ptero)-saurs, heralded, winged masters of prehistoric skies, have been known to science for two centuries. In the genre of science fiction and horror, they also rank highly among our most treasured saurian monsters. Quite apart from those lumbering, land-lubbing fantasy dino-monsters, sight of *flying*, often malevolently-minded 'dinosaurs' in monster/sci-fi films is a special treat. Although paleontologist H. G. Seeley aptly referred to pterosaurs as 'dragons of the air' in a 1901 treatise, here I'm not referring to mythical winged dragons, fictitious giant birds—such as the giant two-headed 'Roc' of Arabian folklore, griffins, Lewis Carroll's fanciful 'Jabberwock,' Harpies of Greek mythology, or even the huge insect Mothra—but those awe-inspiring fantasy *pterosaurs* captivating our imaginations in the popular culture of science fantasy, fiction and horror.

A list of the greatest fantasy 'terror-saurs' of all time would naturally include old standbys such as those supersonic Rodans, the stop-motion animated *Pteranodon* dispatched by King Kong on Skull Island, bat-winged pterodactyls which nearly gobbled Raquel Welch in *One Million Years B.C.*, the pterosaur hive encountered by Professor Challenger's expedition to *The Lost World* in Arthur Conan Doyle's classic novel, *Jurassic Park III's* (2001) caged *Pteranodon* family, James Gurney's illustrated Dinotopian pterosaurs 'airlines,' the 40-foot long, fire-breathing dragon-like pterosaur 'Vermithrax Perjorative' from the 1981 movie *Dragonslayer*, and even two lesser Japanese entries, *Gappa the Triphibian Monster*, and three-headed, electricity-spitting Ghidrah/King Ghidorah. However, any short list of winged 'terror-saurs' would be incomplete without mentioning sci-fi's *greatest* pterosaurs—an intelligent, evil race known as Mahars created by Edgar Rice Burroughs for his "Pellucidar" series of tales, creatures which so far haven't been given justice on screen.

Pterosaurs are now far better understood scientifically than was the case in Seeley's time, a century ago. Understandably, their natural history was confusing, as pterosaurs seemed as "... monster(s) resembling nothing that has ever been seen or heard of upon the earth, excepting the dragons of romance or heraldry." During the 19th, century pterosaurs cruised through a remarkable voyage of intellectual evolution. Like many of today's winged monsters of film, several early scientists speculated that, instead of being designed for flight adaptation, pterosaurs were aquatic swimmers. This was because, instead of seeing wings preserved on their fossil slabs, pterosaurs' long 'wing-fingers' were incongruously thought to be 'oars.'

But there were also early naturalists who correctly recognized from their skeletal remains that pterosaurs flew, although at the time the closest living relatives to extinct pterosaurs were regarded as cold-blooded reptiles. Accordingly, pterosaurs were depicted as leathery-winged, "epicene creatures, half-bird and half-bat." Influential illustrations of reconstructed pterosaurs published in 1817 and later in 1836 reinforced visions of pterosaurs-as-prehistoric-bat-like creatures with broad wings connected to the ankles (a flight organ of bats known as the 'uropatagium'), spanning the inside area between the legs and connecting to tail. In many 19th century restorations pterodactyls' hideous, demonic forms haunted ancient shorelines. In these old depictions, from their lofty vantage point pterosaurs—harbingers of terror, presided over horrific scenes of savagery, bloodshed and titanic struggle.

While the image of pterosaurs as cold-blooded, bat-winged creatures persisted for a century (i.e. such as the inaccurately portrayed *Pteranodon*, which snatches Fay Wray in *King Kong* (1933), or pterodactyls battling for Raquel Welch-as bait in one of *One Million Years B. C.'s* thrilling scenes (1966), by the late 1970s such interpretations dramatically shifted. Today, pterosaurs are known to be winged marvels of ancient skies. No longer depicted as prehistoric 'bats' (or, for that matter, flipper-powered penguins adapted for underwater flight), pterosaurs are recreated accurately, aloft, using state of the art scientific knowledge and computer graphics and robotics expertise.

Astonishingly, in the early 1970s scientists discovered an enormous species of pterosaur, named *Quetzalcoatlus*, which had a 40-foot wingspan, not quite Rodan-sized yet monstrously proportioned. Description of the airplane-sized, evidently warm-blooded "Q" fueled scientifically-founded restorations of the winged genus battling the most terrible carnivores of the late Cretaceous period. So, as modern dinosaur specialists might concur, over 66-million years ago a very real analog of a 'Rodan' really did combat ancient, bipedal Godzilla-type nemeses. Gigantic vertebrates, seemingly approaching dimensions of some of the classic terror-saurs in monster and sci-fi films, really could fly after all!

To this writer's knowledge, pterosaurs made their sci-fi debut in Jules Verne's *A Journey to the Center of the Earth* (1864), although these were not the featured 'prehistorians' in his epic. Although never witnessed by the intrepid explorers, they are referred to as "birds reconstructed by the immortal Cuvier," which "flap... stupendous wings aloft in the dull strata of subterranean air." In illustrated 1867 editions of the novel, Verne's artist Edouard Riou figured several winged pterosaurs flying over the Central Sea's 'antediluvian' shoreline.

Pterosaurs gained ground in Arthur Conan Doyle's *The Lost World* (1912), stealing two major scenes in the novel. First, in a descriptive passage, Professors Challenger, Summerlee and company stir up a rookery of pterodactyls:

> "From this crawling flapping mass of obscene reptilian life came the shocking clamour which filled the air and the mephitic (i.e. offensive), horrible, musty odour which turned us sick..... Their huge, membranous wings were closed by folding their forearms, so that they sat like gigantic old women, wrapped in hideous web-coloured shawls, and with their ferocious heads protruding above them."

Quickly retreating, the explorers are lucky to escape the airborne menace, alive. In the novel's climax, at a public meeting well attended by critics and hecklers, Challenger releases a captured pterodactyl from a crate to prove that prehistoric animals have survived into modernity atop the 'lost' South American plateau. The astonished crowd gazes upon, "The face of the creature .... like the wildest gargoyle that the imagination of a mad medieval builder could have conceived." Then, whereas in the silent 1925 film it is a brontosaur which flees into the Thames River, instead—in the novel, Challenger's pterodactyl escapes through an open window into the misty night. While a stop-motion *Pteranodon* was animated for a key scene in the 1925 silent film the novel's magnificent scenes, described above, were unfortunately omitted.

Most of us probably caught their first glimpse of a fantasy pterosaur in *King Kong*, notably the winged genus *Pteranodon* which Kong kills high atop a cliff on Skull Island. The scene was also described by Delos W. Lovelace in the 1932 novel:

> "(Driscoll) pulled himself over the ledge in time to see Kong seize a great pterodactyl and begin its destruction. This time the affair was not a fight. It was too one-sided for that. The reptile had swooped down to the white form on the ledge. Kong had turned about in time and, seizing it as its long talons reached for Ann, angrily tore the creature to pieces."

Before Kong's encounter with the great pterodactyl, however, we must turn to Burroughs' primeval visions. Burroughs, creator of an 'inner earth' named "Pellucidar," was evidently fascinated by winged saurians. For instance, besides his Mahars to which we shall return to presently, Burroughs wrote the first scene involving aerial combat between an airplane and a giant pterodactyl, (mimicked in later films such as *Rodan*) described in his novel, *The People That Time Forgot*, (1918). Pilot Tom Billings records:

> "The first intimation I had of it was the sudden blotting out of the sunlight from above, and as I glanced quickly up, I saw a most terrific creature swooping down upon me. It must have been fully eighty feet long... with an equal spread of wings... It was coming straight down toward the muzzle of the machine-gun and I let it have it right in the breast; but still it came for me...."

Fortunately, Billings survives to tell the tale of his 'dogfight' with a Mesozoic saurian. Another extraordinary scene pitting biplane pilot Dr. Burt Winslow vs. pterosaur (i.e. '*Pteranodon*'), described in Don Glut's *Frankenstein in the Lost World* (Tome 6 in the New Adventures of Frankenstein series, 2002) was inspired by Burroughs' novel.

Winged 'terror-saurs' of fantasy film and literature are rarely portrayed accurately, but then isn't that the fun of it all? They're often shown way too large, and anatomically incorrect for the real genera they're supposedly based on. Aside from its tremendous size, take a look at Kong's toothy, stop- motion-animated *Pteranodon*, for instance; *Pteranodon* were really toothless. And the *Pteranodon* giving Raquel Welch a 'lift' in *One Million Years B. C.* has 'bat-like,' instead of pterosaur-like fingers. Then, it's really hard to tell which real genus of pterodactyl Rodan is intended to be. (Although in the 1957 movie scientists claim Rodan is a living, gigantic relative of *Pteranodon*—species "Rodan." For a survey of these and other sensational fantasy pterosaurs of the movies, page through chapter 5 of Don Glut's *The Dinosaur Scrapbook*, 1980. And for a listing of real generic names of pterosaurs known to science, see Appendix One to Glut's *Dinosaurs: The Encyclopedia*, Supplement 4, (2006).)

Edgar Rice Burroughs' incredible Mahars have been depicted similarly through the years. The Mahars, rulers of Pellucidar, are supposedly a stock of pterosaur named *Rhamphorhynchus*, which lived during the Late Jurassic and Early Cretaceous periods, about 150-million years ago. *Rhamphorhynchus* was identified in the 1880s by American paleontologist Othniel C. Marsh, who investigated its wing impressions preserved on a fossil slab. Although merely a crow-sized species based on its fossilized remains, long-tailed *Rhamphorhynchus* has been conveniently enlarged beyond human proportions for dramatic, pop-cultural effect, as in Burroughs' Pellucidar tales. Thus, in the earliest portrayals by artist J. Allen St. John (i.e. for a 1922 edition of *At the Earth's Core*), or as portrayed by artists Joe Kubert, Russ Manning and Alan Weiss decades later in 1970s comics based on Burroughs' 1930 novel *Tarzan At the Earth's Core*. Even in their strange parrot-like monster costumes featured in a 1976 Amicus film starring Doug McClure and Peter Cushing, the Mahars always appear much larger than life! And far scarier too! (Readers can see more striking Mahars imagery than can be provided here by other renowned artists by searching for the "Pellucidar Gallery" Internet website.)

In both 1977's *The People That Time Forgot* and *The Land Unknown* (1957), air-pilots have unfortunate 'first encounters' with pterosaurs, damaging the aircraft, thus obviating means of ready escape from foreboding lost world-ish settings.

Where did Mahars, the most vile and bloodthirsty terror-saurs of all time, dwell? They resided not in prehistory nor in outer space or even in skies above, but in skies *below*—that's right, *within* the Earth in a land known as Pellucidar. In Burroughs' imaginative 1914 story, two explorers bore through 500 miles of the earth's crust in a mechanical drilling 'iron mole.' Emerging on the inner surface world of Pellucidar, the explorers discover, astoundingly, that the earth is hollow, as its spherical shell rotates about a tiny radiant star with orbiting 'moon,' forming an '*intra*-stellar' planetary system. Primitive and bizarre life forms inhabit this primitive world of a 'million years ago.' During the earth's formation, because the inner crust cooled more slowly than the outer crust, the evolutionary state of the inner world is 'primitive' relative to our side of the planet. Therefore, Pellucidar became inhabitable at a much later geological period (relative to events of the earth's outer crust). The Mahars are one late-surviving 'Saurozoic' life form which have managed to subjugate all other races of ape, monster and prehistoric creature. Even in 1914, the idea of a 'hollow earth' setting for Burroughs' scientific romance was decidedly 'old hat,' but Tarzan's creator certainly was a master at suspending disbelief!

At the Earth's Core reads much like a *Planet of the Apes* type fantasy. The two outer world protagonists (David Innes and Abner Perry) are whisked by the iron mole into a prehistoric, stone-age world inhabited by Sagoths, a race of primitive, gorilla-like beings controlled by the Mahars. The Mahars, described as hideous, powerful, eight-foot-long winged reptiles, torment and prey upon Pellucidar's weaker human tribes, the Gilaks—a Cro-Magnon-like race. Sagoths are used to restrain Gilaks. Mahars are depicted by Burroughs' artist, J. Allen St. John, as monstrously-sized 'Rhamphorhynchoids.' They cannot hear or 'speak' audibly, but instead communicate with the Sagoths using a mysterious sixth sense transmitted telepathically through the 'fourth dimension.' All remaining Mahars are female, because long before the winged race learned how to procreate artificially so as to prevent their overpopulation using a closely guarded chemical technology termed the Great Secret. The Mahars are usually accompanied by another less intelligent, 'awe-inspiring,' winged menace—forty-foot-long pterodactyls known as Thipdars. Like the pterosaur Kong slew, Thipdars are much larger-than-life *Pteranodons*.

David Innes, horrified by the Mahars' cruelty, vows to end their reign of terror. Gilaks are slain in their amphitheaters, much like slaves and Christians were put to death in gladiatorial events or eaten alive by savage animals within the walls of the Roman Colosseum two millennia ago. The sadistic Mahars experimentally dissect live human victims without anaesthetics. Innes also observes Mahars mesmerizing captive humans—bred and fattened like cattle, who are then dismembered and eaten alive in a temple. From a hidden vantage point, Innes helplessly observes a tortured, hypnotized female. First, "....one of the girl's arms was gone—gnawed completely off at the shoulders—but the poor thing gave no indication of realizing pain, only the horror in her set eyes seemed intensified. The next time they appeared the other arm was gone, and then the breasts, and then a part of the face—it was awful."

After forming a Gilak tribal alliance, led by the protagonists, the Mahars are defeated in battle. Then, in a scene perhaps inspired by H. G. Wells' *The Time Machine* (1895), Innes eventually ascends to the surface in the "iron mole" to gather books and records necessary for quickening the Gilaks' cultural development. But Innes is horror-stricken after realizing that instead of traveling to the surface world with his bride-to-be, Dian the Beautiful, a creepy Mahar stowaway (most cleverly disguised as Dian) is onboard!

Burroughs' enchanting world of Pellucidar inspired seven additional literary 'sequels,' six written by Burroughs, the final entry being penned by John Eric Holmes in 1976. Thipdars and Mahars had recurring, (although, until Holmes latest novel) steadily diminishing roles throughout the series. In *Pellucidar* (magazine ed. - 1915, 1st hard cover ed. 1923), Innes returns with the Mahar stowaway to the inner world. Now Emperor of Pellucidar, he unites the Gilak tribes and wages military assault on the neighboring Mahar cities. Nevertheless, Innes reflects, "The task of ridding Pellucidar of these hideous creatures is one which in all probability will never be entirely completed, for their great cities must abound by the hundreds and thousands in the far-distant lands....." Following *Pellucidar* came Burroughs' *Tanar of Pellucidar* (1st book ed. 1930), perhaps the most popular sequel of all, *Tarzan at the Earth's Core* (1st book ed. 1930), *Back to the Stone Age* (1st book ed. - 1937), *Land of Terror* (1944), and *Savage Pellucidar*, (published posthumously as a 1st book ed. 1963). Besides his heroics in Burroughs' novel, Tarzan often battled Mahars and thipdars in comics staged in Pellucidar settings.

Then, in the grand finale', John Eric Holmes' engaging *Mahars of Pellucidar* (1976), psychologist Chris West 'beams down' by wire to the inner world, facing Mahars and Thipdars in a bloody struggle for survival. Here we learn that some Mahars are male and that their telepathic power is a form of "radio sense." We also tour the Mahars' egg-incubator rooms, and once more witness Mahar sadism in their temple of despair. West is himself subjected to psychological experimentation conducted by Mahar scientists before escaping to the forests beyond the city walls. West and his party confront many prehistorian dangers along the way, at one time finding themselves in the midst of a climactic battle waged between an enraged three-horned *Triceratops* and a monstrous *Tyrannosaurus*. Like Innes before, West triumphs over the Mahars and the Sagoth horde. In my opinion, "Mahars of Pellucidar" stands on its own merit and would make a terrific screenplay. According to author Jeff Rovin, Holmes' sequel, "Red Axe of Pellucidar," never saw publication.

Jeff Rovin lists Edgar Rice Burroughs' Mahars, monsters which have haunted me since childhood, among the horrific bestiary known to human foes, including many deemed prehistoric in nature in his fascinating account, *The Encyclopedia of Monsters* (1989). Theirs' is a well-deserved entry, as the time-honored Mahars rank foremost among horror fiction's classic variety of winged monsters! Now can somebody please write a decent script and design better costumes for their next screen appearance?

In the span of years since this chapter was originally published as a fanzine article, monstrous pterodactyls continued their plague upon humanity in 2015's *Jurassic World* and after a prolonged hibernation Rodan returned to the fray in *Godzilla: King of the Monsters* (2019).

Chapter Ten — *Kaiju Conjure: A Russian 'Ghidorah'*

Ghidorah! A name conjuring images of one of Toho's most gigantic and formidable kaiju! And most impressive-looking too.[1]

It was a startling early 2007 *National Geographic* press release that swept King Ghidorah's (also known as 'Ghidrah's) incredible bat-winged form to mind. "Two-headed reptile fossil from age of dinosaurs found," announced discovery of a 120-million-year-old marine reptile hatchling with twin necks and heads—extraordinary! At only 3 inches long *Choristodere*, as it was christened, is miniscule compared to Ghidorah. And dear Choristo-dere lacked wings, multiple tails, wings and couldn't' fly (except underwater). However, visions of Toho's famous three-headed extraterrestrial dragon stirred by an actual case of a multi-headed Mesozoic dragon soon had me pondering particulars and further facets of what is kaiju?

Cases of multi-headedness (polycephaly) have been scientifically documented in a variety of mammals and reptiles. Such unfortunate creatures are sometimes considered monstrous for their inescapable genetic abnormalities, yet they clearly aren't kaiju, that is, per J.D. Lees's definition published in *G-Fan* no.78. And while two-headed deformities occur infrequently, three-headed cases are extremely rare, (like the three-headed turtle found in Taiwan in 1999).

But what about other multi-headed monsters, created by Man, namely those of myth, legend and sci-fi? What are these beasts, and can any definitive statements be made about their kaiju status? Intriguingly, multi-headed monsters of literature and legend harken back to ancient times.

At first glance, *Jurassic Park's* universe of spectacular cgi-enlivened dino-monsters (e.g. *T. rex,* 'raptors,' *Spinosaurus*, 'compys,' etc.) would appear to have few connections to Ray Harryhausen's heart-stopping, stop-motion animated mythological monsters featured in films, such as *Jason and the Argonauts* with its bat-winged harpies and seven-headed Hydra, or *The 7th Voyage of Sinbad* with its Cyclops, two-headed winged Roc and Dragon. But there are connections, metaphorically linking search for fossil vertebrates from anqituity to present day. Intriguingly, 'monsters' as known to the ancients and men of medieval times in a sense, centuries later, spawned a protean host of pop-cultural movieland dino-monsters and other reconstructed prehistoria.

But (my emphasis here), *multi-headed* forms ... like Ghidorah? Yes, and we can address such oddities too, that is, in the context of kaiju-conjuring.

Perhaps it's unsurprising that several multi-headed monsters were brought to life through Ray Harryhausen's talents. After all, a variety of frightening monsters including multi-headed (and multi-limbed) forms abound throughout Greek mythology. Researcher Adrienne Mayor noted that "According to Homeric-Hesiodic lore, strongmen of myth were often said to have multiple heads or limbs."[2] Is this because men noted abundant vertebrate fossil deposits scattered throughout the Mediterranean region, associating gigantic bones with their deified mythological heroes and their foes who walked or plundered Earth so long before? If such bone beds represented a condition of mass mortality (more than, say, one co-deposited skeleton of two or more elephantine Mastodon individuals represented therein), then in their early efforts to rationalize and reconstruct remains of the giant or 'dragon,' a composite (chimeric), multi-headed beast resulted.

Thus, in Edith Hamilton's *Mythology* (1942), besides Cyclops, evil Centaurs, Griffins, winged Harpies, Spinx, the Chimera, Laestrygons, Minotaur, the Monster of Troy and an assortment of ogres and sea-serpents defeated by the Argonauts, Hercules, Perseus, Odysseus and heroic Theseus, for example, we *also* read of multi-headed creatures: the nine-headed Hydra, scaly-winged Gorgons so hideous with snaky, serpentine 'hair' that a mere glance could turn a mortal to stone, the three-headed giant Geryon, and the Underworld god Pluto's three-headed dog Cerberus. Most imposing of all, however, was Typhon, "A flaming monster with a hundred heads, who rose up against the gods. Death whistled from his fearful jaws. His eyes flashed glaring fire." (Hamilton, p.67) Only Zeus could destroy this terrible monster with his lightning bolts (perhaps not unlike the lightning-like gravity beams emanating from Ghidorah's three-electrified heads).

Certainly, a hundred-headed titanic beast inspired by Greek monster Typhon ("Typhoeus"), memorialized by a story written by Grandmaster science fiction writer Robert Silverberg, titled "Call me Titan,"[3] would qualify as kaiju per Lees's definition.[4] The front cover of the *Asimov's Science Fiction* issue in which Silverberg's story appeared, illustrated by "Jael," conjures images of a multi-headed monster composited with a human torso. Freed after 2,500 years during a spectacular eruption from within Mt. Etna's bowels, Typhoeus—who describes 'himself' accordingly, has an old score to settle with Zeus. "I am ... something of a monster ... four hundred feet high, let us say, with all those heads, dragon heads at that, and eyes that spurt flame, and thick black bristles everywhere on my body and swarms of coiling vipers sprouting from my thighs.

The gods themselves have been known to turn and run at the mere sight of me." Such a monster as Typhoeus would even give Godzilla the willies. By comparison, Ghidorah seems rather tame.

Like many of the Greek or Roman gods and monsters of antiquity, Silverberg's Typhoeus is anthropomorphized, which is how ancients tended to conceptualize big, stony bones (i.e.—as human-like skeletons). And being human-like forms no basis for exclusion from kaiju status. Giant walrus "Magma" from Toho's *Gorath* (1962) and Harryhausen's 'Walrus giganteus' from his *Sinbad and the Eye of the Tiger* (1977) would both pass Lees's first two kaiju criteria, yet would clearly fail the third, invoking mental capacities being beyond animalistic range.[5] Walruses aren't human-like, but, in contrast, Frankenstein's Monster certainly is. While Mary Shelley's (or Universal Studios') prototypical Frankensteinian Monster would not be sufficiently large as those two giant walruses to topple even smaller building structures, Toho's version of the Frankenstein Monster could. Isn't he smart enough to pass Lees's third criterion? Toho's *Frankenstein Conquers the World* (1965) introduced a menacing human-like kaiju monster, uncannily akin to Godzilla (or perhaps Silverberg's Typhoeus), instilled with special powers of indestructibility and cellular regeneration.

Now back to Ghidorah ... although this time not Toho's, but the *Russian* 'Ghidorah.' Many of you probably didn't realize that a decade *before* release of *Ghidrah, the Three-headed Monster*, another (Russian/Soviet) *three-headed*, fire-breathing, winged dragon had already blazed its way across the silver screen in *Sword and the Dragon* (U.S. release, 1960).

Unlike many of Harryhausen's movies, *Sword and the Dragon* (or in Russian *Ilya Muromets*, pronounced 'moor-mitz', released originally in 1956) isn't staged in classical antiquity, but during medieval times, in the days of armored knights, a period when many scholars (for example Aldrovandus, Kircher, Gesner) believed in existence of dragons! By the Middle Ages, we even find a number of stalwart Christians earning their saintly credentials after slaying 'dragons.' And both in China and elsewhere, fossil remains of prehistoric vertebrates were "... supposed to be the genuine remains of either dragons or giants, according to the bent of the mind of the individual who stumbled on them."[6]

In the Russian release, the dragon is named "Zmey Gorynych," but in the American version it is named "Zuma." So, to what extent is Zuma a kaiju? First, understand this isn't really a 'kaiju film,' per se, but more of a typical knights-in-shining-armor folkloric tale. In fact, the plot setting doesn't at all mirror Ghidorah's circumstances following its extraterrestrial voyage to Earth. The titular 'Ilya' becomes a hero after claiming a sword relinquished by a great godlike knight, once defender of the state. After defeating an awful robber gremlin who exhales hurricane wind gusts, Ilya eventually leads the Russians into battle against an invading horde of Tugars who also command monstrous Zuma, whose origins are unknown. Dousing the dragon's fiery emanations with buckets of water, Ilya and his knights sever each of its three heads, one-by-one. Then, victorious Ilya's son accepts the sword and rides off into the sunset seeking other adventures.

Be wary of purchasing the film simply for the dragon action, because beforehand the action was, well, 'a-draggin.' Zuma is merely a climactic monster prop (built both full-size and miniaturized) whose quite impressive scenes last but three minutes. And the film's English translation is atrocious.

Not all dragons would be regarded as kaiju, and I would hesitate to categorize Zuma as such.[7] Zuma is merely a fearsome animal summoned into battle not unlike, say, elephants ridden into battle by Hannibal's cavalry warriors during the Punic Wars. Zuma lacks purposeful intelligence. However, at times Ghidorah, who rarely 'lost its head' in battle, does display purposeful intelligence above and beyond that of many dragons of film and lore (including Zuma). So Zuma may be regarded as the 'Russian Ghidorah' only for its prototypical appearance and similarity of powers, but not for its overall behavior, or kaiju-ness.

A premier dinosaur paleoartist from Russia—then the Soviet Union, Konstantin Konstantinovich Flyorov (whose work remained largely 'invisible' to the western world until recently)—was also "… a sought-after consultant for Soviet films featuring prehistoric animals. Presumably he adhered more strictly to fossil evidence as an advisor than he did as an artist."[8] And yet a recent, clearly incomplete, Wikipedia entry, "List of films featuring dinosaurs"—the term used quite loosely, and in a pop-cultural sense, shows only two Russian titles. (The U.S., Japan and the United Kingdom lead the way in entries on that list. And there are no references to "Zuma's" film.)

A possible reason for the apparent dearth of Russian sci-fi movies between the 1930s and early 1960s featuring prehistoric themes is outlined in a recent *Filmfax Plus* article.[9] In 1932, Joseph Stalin inaugurated a "social realism" policy that permitted censorship of any artistic endeavor that strayed from "real life social conditions," or that inspired the populace with "traditional values." Accordingly, "This policy ruled out much of science fiction…"[10] However, it remains plausible that while in "… many countries and historical contexts, paleoart and kindred forms of illustrations are second-class citizens, relegated to a rung far below fine art … In the Soviet Union, this separation worked to the paleoartists' advantage. They dodged the scrutiny directed at professional fine artists, and were able to create less didactic images favoring open interpretation over instruction."[11]

Clearly any Soviet movie involving a 'Russian Ghidorah' wouldn't have seemed so stricken with 'social realism,' but that movie might have inspired traditional communist values among the people. And Flyorov who studied zoology and paleontology—earning science degrees, couldn't have had much to do with designing a three-headed 'prehistorian' monster.

And thus the monster Zuma somehow survived, procreating into Toho's Ghidrah a decade later!

Notes to Chapter Ten: (1) See J. D. Lees's *The Official Godzilla Compendium* (1998) pp.132-133 for a concise history of Ghidorah's major appearances in film. Lees discusses Ghidorah's *two* origins as suggested by (A) the monster's 1965 film debut in the Americanized *Ghidrah, the Three-headed Monster*, and also in (B) a 1991 time travel sequel—*Godzilla vs. King Ghidorah*. Jeff Rovin includes an entry for 'Ghidrah' in his *The Encyclopedia of Monsters* (1989), pp.112-113; (2) Adrienne Mayor, *The First Fossil Hunters* (2000), Princeton University Press, p.149. Among other remarkable successes, Mayor has identified a 3,000 year old painted vase now held in the collection of Boston's Museum of Fine Arts depicting Hercules destroying the legendary Monster of Troy, strikingly associating real occurrences of vertebrate fossils found *in-situ* with the ancients' interpretation of mythological monsters; (3) Robert Silverberg, "Call Me Titan," *Asimov's Science Fiction*, vol. 21, Feb. 1997, pp.18-36; (4) In *G-Fan* no. 78, Lees stated that for a creature to be classified as kaiju, "It's size must be beyond the upper limit of natural and enable the entity to destroy buildings." Next, Lees claimed, "It must present a menacing aspect." The last part of Lees's test is that "It must show some degree of purposeful intelligence beyond that of any comparable animal." Furthermore, "... other traits (special powers, a mysterious origin, etc.) may be present, but are not essential."(5) However, arguably, Toho's Frankenstein Monster fails Lees's third criterion when we note how the Monster's intelligence, while greater than that of the immense non-kaiju walruses of filmdom, doesn't represent anything beyond an ordinary range for humankind. Also, while Universal's Frankenstein Monster isn't gigantic like Toho's, its great strength might allow it to destroy buildings of modest size. For more on Toho's *Frankenstein Conquers the World* see Bob Statzer's "Frankenstein: The Toho Terrors," in *Castle of Frankenstein* no.30, 2001, pp.41-53. Mystic connections between Godzilla and Frankenstein's Monster are quite readily apparent when we consider, respectively, the two creatures' uncanny abilities to regenerate tissues after being wounded, and the fact that both originated following exposure to 'rays' of incredible intensity—radioactive in Godzilla's case, and a "ray beyond the ultraviolet captured through the lightning, the one that gives life" in the case of Frankenstein's filmic lore. Their association is understood through Marc Cerasini's 1996 novel, *Godzilla Returns* (pp.137-

138), and Stefan Petrucha's 2006 novel *The Shadow of Frankenstein* (pp.165-168, 177, 240, 244-246); (6) Charles Gould, *Mythological Monsters* 1886, p.199; (7) Here I'm not pronouncing whether or which) dragons—which were introduced into film in 1924—are indeed 'kaiju.' A more authoritative review of dragons is first needed. Clearly both dragons and dino-monster design has been influenced somewhat inter-relatedly. For example, just look at Gorgo's ears which reappear in dragons of the silver screen a year later in the winged dragon-like 'harpy' animated for *Jack the Giant Killer*, and a two-headed dragon featured in Bert I. Gordon's *The Magic Sword* (1962), which was actually an 11-foot high prop that emitted fire. According to Tom Triman, "Rumor has it that when the dragon's operator activated the wrong switch the mechanical monster inhaled the flames instead of exhaling them and blew itself up." Incidentally, Zuma also resembles a three-headed version of Reptilicus with wings; (8) See Chapter VIII in *Paleoart: Visions of the Prehistoric Past* (2017), by Zoe Lescaze, quote from p.234; (9) Jack Hagerty, "Cosmic Journey," in *Filmfax Plus*, no.151, March-May 2018, pp.66-67; (10) Quotes from Hagerty, pp.66- 67; (11) Quote from Lescaze's *Paleoart*, p.223.

Chapter Eleven — *Many Faces of Gfantis*

And so I finally decided to sculpt a "Gfantis." This is the monster inaugurated as mascot by J. D. Lees for G-Fest 1999. Due to numerous priorities and obligations, sculpting took nearly a year from bare bones beginnings to finish. To my knowledge, while there have been many drawings and paintings of this dino-monster appearing in *G-Fan* magazine, as of 2012 there have been no miniature, full-bodied, 3-dimensional representations made … possibly until now. Okay – I do recall seeing a well-executed Gfantis costume worn at G-Fest. Well, I'm all thumbs when it comes to stitching and sewing cloth material, but I've occasionally been known to creatively enliven cold lumps of clay. So here goes.

Internet searches reveal that Hiroshi Sagae sculpted a highly impressive head bust of Gfantis, but it seems he didn't complete the animal. Besides the fact that Gfantis was never a "real" Japanese movieland creature, another possible reason for why Gfantis sculpts are rare may be that most sculptors usually don't begin projects unless they intend to mold the prototype and cast resin replicas for sale and profit. With all its majestic spikiness, not only is attempting a Gfantis rather challenging, but would also be a costly venture when considering materials and labor involved in molding and casting many individual spikes. From design stages, I never intended for my Gfantis sculpt to be replicated. Accordingly, the armature was constructed as a one-of-a-kind item. Take a look (see back cover).

This model, made of baked Super Sculpey, is about 14.5 centimeters tall and approximately 18 centimeters long (with its swerving tail, which by the way, is elevated above and not dragging on the base.[1]) It was inspired by a Gfantis image appearing on the cover of the G-Fest XVIII "Official Program Book" (2011). Also, Bob Eggleton's painted head of Gfantis appearing at the *G-Fan* website was a great reference. Originally, I had intended to pose my Gfantis in a stance and with a model base not unlike that of the 1963 Aurora Godzilla plastic model, with a name plaque "Gfantis" emblazoned below a row of crumbling, smashed buildings and skyscrapers. But my efforts to recreate demolished structures might have looked too cheesy and so that part of the project was abandoned.

Also, I had intended to use this project as a hands-on example that would appear in my book, co-authored with dinosaur sculptor Bob Morales and Diane Debus, *Dinosaur Sculpting*, 2nd edition.[2] McFarland published this expanded edition in the summer of 2013. Gfantis was literally the first sculpture I'd attempted in a decade, and so I was anxious as to whether I could still do anything halfway reasonable! And so I began in January 2012. But as the weeks and months marched on I found myself at long last enjoying the practice of sculpting once more, and even completing yet another three miniature sculpts as well involving a science fictional dinosaur theme (not shown here but appearing in *Dinosaur Sculpting*), that for a time took priority over my Gfantis.

I consider this model to be a decent working rough – not worthy of molding (i.e. even if the armature was suitable for such a purpose, which it isn't), but serving as a good draft prototype, test ""visual" or study piece for the second, more detailed and vivid sculpt that instead could later be replicated for sale. (Although, I wasn't planning to enter the Gfantis figurine market.)

Here's an outline of how the creature was made, including a few in-progress photos. Personally, I don't consider myself a modeler or much of a diorama builder; resin castings can be so darn expensive! But I do like to make my own creations when the spirit urges. (The term usually reserved for this practice is "scratch-built.") This Gfantis was made using only 1½ one-pound boxes of Super Sculpey. A very inexpensive proposition with respect to materials overall! Given patience, time and ingenuity, yes—you too can make your own daikaiju as well! Here are the basics of how.

Sculptures always begin with one of your own sketches (for better or worse). In this case I drew a crude rough of how I wanted this sculpt to look on a napkin during a lunch break. I anticipated adding crumpled city buildings and a plaque, mimicking ambiance in the Aurora Godzilla model. With much trepidation, and after visiting a crafts store, and raiding our closet for the choicest, sturdiest coat hanger wire (never leave your wife's clothes laying on the floor!!), I began the project in earnest several weeks later. The following is a mere overview of the process. However, for all the inspirational step-by-step details on how to sculpt prehistoric animals, necessary tools, and a variety of reliable techniques which may also be applied to the sculpting of science fictional prehistoric daikaiju, read our *Dinosaur Sculpting*.

Yeomen sculptors reading this already know the drill. Rely on sufficiently sturdy wire for the task at hand, stiffer wire for the backbone and tail, but use softer wire (such as thinner gauge copper or aluminum wire for places where the model may need to be severed, in case you want to mold it in separate sections). Bend the wire into the desired shape. Legs can be attached at the hip area with additional wire, which may be tied using thin copper wire that is wrapped and twisted around the main vertebral wire (i.e. 'backbone'). Then to hold those legs in place so they don't slip, you can squoosh a small 'blob' of polymer clay (i.e. Super Sculpey) into the hip socket 'joint.' Depending on how you've designed your creation, you may need to consider adding a little extra, measured length of wire at the end of one of the legs that may be inserted into a block of wood to support the armature while you are sculpting onto and around it. In this case, my Gfantis' right leg was anchored temporarily into the temporary base shown here using a bit of extra wire protruding into the wood.

Then bake this intermediate assembly in a kitchen oven according to manufacturer's instructions. The clay blob will harden after cooling and the legs will be rigidly joined to the backbone. Light filler material such as aluminum foil (my preference) should be placed around the backbone/leg assembly. Pack this in tightly conforming to your model sketch and then secure the filler in place tying thin gauge wire around it. Apply a thin coat of polymer clay around this package, scratching it lightly with a sharp-pointed sculpting tool so that later unbaked layers of polymer clay will tightly adhere to the inner layer; bake according to manufacturer's instructions. Never allow a single layering of clay that is ready to be baking) to exceed half an inch thickness.

Then, after this cools, proceed to sculpt the exterior of the body. Normally, since I sculpt dinosaurs that are narrow-bodied with respect to width, it became an "awakening," realizing how much extra polymer clay was needed to fatten up Gfantis' legs, torso and tail. (For the record, yes, Gfantis was my introduction to daikaiju sculpting.) Arms of bipedal creatures can be sculpted around wire armatures that have been pushed into the abdomen. Spikes, claws, teeth and eyeballs can be "pre-baked" separately from the main body of your creature and then (after proper cooling) simply embedded or pushed as hardened pieces into proper positions on the still soft (unbaked) sculpt. (Rationale for doing certain things the proper way is described at length in the book.) Super Sculpey can be baked over and over several times if needed. In this manner, you can add on additional layers of clay as you proceed, making your creature even fatter, baking the sculpt sequentially. Super Sculpey may also be carved even after it has been oven-baked to hardness. There are other polymer clays besides Super Sculpey that one may use.

Representations of our greatest monsters, like Frankenstein's Monster, Godzilla and even famed *Tyrannosaurus rex* of paleontological lore do not remain static in time. Their appearances and personas 'evolve.' Gfantis has also evolved, as witnessed through the pages of *G-Fan*. Without elaborating, examine Gfantis as depicted, for example, in issue no. 64 – where, in "Gfantis Lives!" Timothy A. Gagnon's portrayal looks especially pointy and spiky, displaying quadrupedal motion. In issue no. 72, after-the-fact conceived powers of both Gfantis and "Mecha-Gfantis" are described on pp. 36-37, featuring Jeff Rebner's art. (I won't recount Gfantis' mysterious, daikaiju powers here though.) Rebner's and Frank Parr's (see issue no. 67, p. 20) depictions of Gfantis have become more or less accepted as standardized visuals. Matt Frank's Gfantis, resplendently seen in issues nos. 86, 87 and 90 ("Gfantis, Heart of the Beast"), is adorned with elaborately decorated head spines, with even tufts of fur (?) protruding from behind the jaw, exemplifying mammalian/near-humanoid traits. Musing on these few examples, doubtlessly, inventing Gfantis as a mascot was a very good idea. For this conjured dino-monster offers many artistic (e.g. visual as well as literary) opportunities.

Yes, I've applied perhaps more than a tad of artistic license in my version of Gfantis, like other artists who have tackled this particular dino-shaped monster. Yet I have not striven to create the canonical image of G-Fantis; that responsibility lies with Gfantis' creators. This is simply my impression[3] of a peculiar beast tracing its origins to some three-dozen *G-Fan* issues ago, based on others' artistry. It's merely a one-of-a-kind piece to pose, perhaps, on a shelf adjacent to my Aurora Rodan model.

Notes to Chapter Eleven: (1) Unfortunately to many referred-to images readers must track down and see no.101 of *G-Fan*; (2) *Dinosaur Sculpting: A Complete Guide*, 2nd ed., (2013), p.67; (3) Diane Debus made the cool, oval wooden base. I entered this Gfantis model sculpt in the Model-Building contest at G-Fest in July 2013 and won 1st Place in the "Scratch-built" category.

Chapter Twelve — *The Flying Death: An American 'Rodan' Progenitor from 1903?*

Over the decades, precious few published prehistoric monster sci-fi stories (or movies) have emphasized pterodactyls. Usually, winged pterosaurs are just another shape in the paleo-entourage encountered when dinosaurs emerge from an erupting volcano, or while cryptozoologists explore surviving denizens of a 'lost world,' (e.g. Kong's Skull Island) where time has 'stood still.' But only once, to my knowledge has a complete work of printed or written fiction been framed upon the anachronistic occurrence of a winged pterosaur—save for Toho's giant monster 'Rodan' (with screenplay based on Takeshi Kuronoma's story). Toho's widely known, *Rodan, the Flying Monster* (1956/57) may have been the first monster film to feature a pterodactyl as *the* central menace out of prehistory, but it was not the earliest paleo-monster work of fiction featuring a pterodactyl as sci-fi culprit, or primary subject of horror. Could another older entry, predating Toho's by half a century, resurrected here from the dusty library shelves of yesteryear be our Rodan progenitor? Here I speak of Samuel Hopkins Adams's (1871 – 1958) now obscure 1903 tale, *The Flying Death."* Species of *Pteranodon* figured plot-wise in both cases.

Admittedly, Adams's old story—more of a detective mystery than a solid science fiction tale doesn't hold up so well today. But for kaiju fans and lovers of prehistoric monster stories, it does have its moments, inaugurating means of suspending disbelief, to be emphasized here, that would seem wholly familiar today—cliché, (even predating Arthur Conan Doyle's *The Lost World* masterful novel). Adams's story is about a group of vacationers on the Montauk coast of Long Island, when tragedy ensues—a shipwreck, from which several individuals are rescued, and a victim who was apparently murdered by a long sharp pointed object. A doctor, journalist and an entomology professor compile clues and observations, at first presuming the murderer was one of those rescued, a trick knife-thrower named 'Whalley' who quickly escapes. More victims (human and animal) pile up and the perplexed trio, with several lovey-dovey lady friends who are also present, eventually find themselves entertaining an utterly fantastic explanation—that maybe a prehistoric flying reptile could be responsible for the strange deaths.

This kind of semi-paleontological mystery tale would have been unusual back in 1903. And yet if it is to succeed then it would require masterful suspension of disbelief in the face of a seemingly impossible conclusion. And so we eventually encounter that by now traditional 'science speak' section in the middle of the book, where the learned professor appeals to our logical minds, by announcing a theory soaring mightily on the wings of fancy, literally out of the blue not unlike the flapping wings of the anachronistic pterosaur itself. Here is how Adams went about that aspect of his writing task.

Most convincingly, there are strange tracks to consider surrounding victims on the beach sand that are similar to five-clawed fossil tracks found in slabs on a stone wall, which the professor later identifies as pterosaurian footprints. (However, genuine fossil pterosaur tracks hadn't been positively discovered until 1952.) And death-wounds on each of the victims were evidently made by a "broad-bladed, sharp instrument with power behind it." There are also queer noises in the sky and a set of high-altitude kites launched for aeronautic measurements has been disturbed. The odd possibility of a murderous "wandering ostrich" doesn't fit the bill.

So, based on similarity of the fossil footprint impression and the recent sand tracks, the professor excitedly jumps to a conclusion in this scene, dramatically announcing that the footprint on the stone slab is dated from the Cretaceous Period, possibly 100-million years old (but "at least 10-million")! Therefore, the killing creature, "…was not a bird. It was a reptile. Science knows it as the *Pteranodon*…a creature of prey. … This monster was the undisputed bird of the air, as dreadful as his cousins of the earth, the dinosaurs, whose very name carries the significance of terror." Its long beak could have produced the stabbing puncture wounds found on victims.

So why is it alive and where did it come from? The professor continues, luring us in: "Science has assumed they were extinct … But … we have prehistoric survivals. The gar of our rivers is unchanged from its ancestors of fifteen million years ago. The creature of the water has endured; why not the creature of the air?" The professor speculates it may have survived hidden away at either the North or South Poles, or "in the depths of unexplored islands; or possibly inside the globe. …there still remains the (Earth's) interior, as unknown and mysterious as the planets. In its vast caverns there well may be reproduced the conditions in which the *Pteranodon* and its terrific contemporaries found their suitable environment on the earth's surface, ages ago." Furthermore, since the professor seems to favor the latter possibility, "recent violent volcanic disturbances might have opened an exit." This may have been a reference to Krakatoa's 1883 eruption.

The professor doubles down on his theory citing legends and lore of giant birds, like the "roc," and thunder birds of the Native Americans. He also mentions reports from 1844, alleged sightings in England where "reputable witnesses" found tracks after a night's snow fall of a "creature with a pendant tail (note – not present in genus *Pteranodon*), which made flights over houses…" Overall, a most "epoch-making discovery." Certainly it would seem the professor has gone bonkers, right? Although how much different are these passages from others in similar tales that came afterward invoking the prehistorical?

After a couple more strange deaths, there's a 'let down' moment when it appears as if the professor's theory is wrong, and Whalley is the murderer (instead of a pterosaur). But Whalley's actions do not account for all the evidence and eventually *Pteranodon* makes its appearance, just as the book is ending. It is Whalley who (spoiler alert!) stabs the attacking pterosaur with one of his knives, and the "…lone survival that had come out of the mysterious void, bearing on its wings across the uncounted eons, joy and sorrow, love and death" first rises from the beach, only to sink into the sea.

Before paleontologist O.C. Marsh's 1876 description and naming of the 20-foot wing-span *Pteranodon*, known from the Cretaceous chalk of Kansas, Jules Verne had incorporated references to pterodactyls in 1860s editions of his novel, *Journey to the Center of the Earth*—appearing in Axel's waking dream where he imagines Earth's geological history, from the Recent to the Primordial. Among more contemporary stories published during Adams's time, a handful of others also recruited monstrous pterodactyls, but not as *the* central, *singular* menaces of prehistoric provenance. For example, none of the following entries (an April 1, 1906 *Chicago Tribune* spoof wherein a horde of dinosaurs invade Chicago; Sir Arthur Conan Doyle's *The Lost World* – serialized in 1912; Jules Lermina's 1910 novel *Panic in Paris*; Edgar Rice Burroughs' 1914 novel *At the Earth's Core*) predate Adams's 1903 story. Pterosaurs in these yarns usually are relegated to cameo roles and are often quickly dispatched or circumvented, with exception of Burroughs's insidious, intelligent rhamphorhynchoid "Mahars" race.

Doyle's 1912 story is rather interesting in context of Adams's. *Lost World's* fictional Professor Challenger, back in London yet seeking to convince reporter Mr. Malone that 'curupuri' paleo-monsters still live on the 'lost' Amazonian plateau, exhibits a photograph bearing an image of a large pterosaur resting on a tree, and a two foot length of preserved pterosaur wing—distinctive from a modern bat's wing in its skeletal structure. A suggestion is made that this pterosaur may be of genus *Dimorphodon*. Later, atop the lost plateau, the Challenger expedition has three separate encounters with living pterodactyls, none of which is stated to be genus *Pteranodon*, although one that buzzes over a campfire has a 20-foot wingspan. Then at the novel's climax, Challenger releases a pterodactyl with a 10-foot wingspan from a packing case during a scientific meeting at the Zoological Institute. This creature, which finally persuades Challenger's former detractors, flies around the room before disappearing through an open window into the night sky.

I like to think that *Flying Death's* pterosaur may also be that which flew the coop in Doyle's subsequent story after crossing the pond, although chronology of Adams's and Doyle's stories remains incompatible with such a notion. It is widely accepted that paleontological elements in Doyle's story were written using E. Ray Lankester's 1905 popular text, *Extinct Animals*—several passages (pp. 230-235) of which discussed pterodactyls and specifically *Pteranodon*. In the 1925 silent film, *The Lost World*—based on the 1912 novel— before surmounting the lost plateau, a large *Pteranodon* is seen by the Challenger expedition circling a rock pinnacle, substantiating Challenger's outlandish theory. (Then a few years later another magnificently stop-motion animated *Pteranodon* battled Kong atop Skull Mountain, to its death.)

It is interesting to consider how writers of the past dealt with suspension of disbelief in their pterodactyl-peril monster stories. In Toho's *Rodan* scientists rely on a blurry photograph of a wing in motion (instead of an 'actual' wing sample as in Doyle's story), a small section of eggshell and ultimately Shigeru's suppressed horrific memory of seeing a gigantic hatchling in the mine cave. Measurements are made in a laboratory: a state-of-the-art "electronic computer" is used to calculate the actual dimension (100,000 cubic feet in volume) of the original egg, and (comically) a "carbon-14" age date report suggests the Late Cretaceous. Then comes their conclusion as a drawing-restoration of the *Pteranodon* is produced, thus sealing the deal—the flying monsters are clearly genus *Pteranodon*, species *rodan*. Next, a 'flight of fancy' theory is offered: the eggs were hermetically sealed after being buried in a volcanic eruption thus "perpetuating the germ of life" for millions of years—a suspended animation. Atomic explosions fractured Earth's crust admitting air and warm water, permitting the egg to hatch. Voila—just add water and stir!

Remarkably, there really isn't too much more evidence facilitating scientists' consideration of a giant flying monster in *Rodan* than what the group of vacationers had gathered in *Flying Death*. Photographs are key in both Doyle's better-known tale and in *Rodan*. It might be said that Adams pioneered the (American) literary technique for documenting the 'reality' of living prehistoric monsters from circumstantial evidence, relying on now familiar and cliché means of suspending disbelief in their anachronistic occurrence, even before Doyle's Challenger. Regardless, Adams's precociously written 'science speak' section—cited above—of his mystery tale, so typical of what came later, probably left little impact on writers more commonly renowned for redefining the genre during the early and middle decades of the 20th century.

In such examples, efforts were made to use evidential science to convince readers or audiences in the utterly impossible, or the 'what-if'. Key differences among the three cases, however, are: in Adams's novel we aren't certain there really is a pterosaur woven into the plot until the very end after all the clues are laid out; at the outset in Doyle's novel we suspect that Challenger's assertions could be correct because there are those drawings of allegedly living prehistoric animals in Maple White's notebook, the section of wing—plus don't we simply want to believe? And in Toho's *Rodan,* before paleo-scientists have convinced themselves there are *living,* giant pterosaurs founded upon—let's face it—relatively flimsy evidence, audiences know they're roving about because we've already seen a gigantic hatchling in a cavern and an adult specimen emerging from a volcano.

The Flying Death novel was once offhandedly proposed for filming, around the same time as was *The Lost World.* And yet today, hardly anyone has heard of Adams's first foray into mystery writing—also his only venture into 'paleo-fiction.' The story of how this went down is recounted by Stephen A. Czerkas in his illuminating 2016 book, *Major Herbert M. Dawley: An Artist's Life,* (p.62). In a discussion with Dawley's friend, writer Chester F. Bray, Willis O'Brien, who was participating with Dawley in completing the 1919 silent film, *The Ghost of Slumber Mountain,* learned that Dawley was planning to do another film stocked with stop-motion animated dinosaurs—Doyle's *The Lost World.* O'Brien, who had previously done several short films incorporating prehistoric animals before teaming with Dawley, asked Bray to write him an original story that could be used as a screenplay for such another movie. Bray declined, instead suggesting that O'Brien build his film around two other published works, either *The Flying Death* or *The Lost World.* O'Brien knew little of them, but Bray brought him a copy of the *latter,* and the rest is history. One wonders though, what if Bray had handed O'Brien a copy of *The Flying Death* instead?

Subsequently, Dawley had a severe falling out with O'Brien, who was charged with stealing ideas from Dawley while misrepresenting/inflating his (lesser) apprentice-type role in production of *Ghost of Slumber Mountain.* While O'Brien and Dawley both independently invented stop-motion puppet animation, Dawley patented the process.

As outlined on Wikipedia, Samuel Hopkins Adams was a prolific writer, working for many years on scam patent medicines, research which significantly paved the way toward America's Pure Food and Drug Act of 1906. *The Flying Death* was his first novel, and other stories of his such as "Night Bus" made it to Hollywood, transformed into films—in this case 1934's *It Happened One Night,* starring Clark Gable and Claudette Colbert. *The Flying Death* was serialized by *McClure's Magazine* in 1903, and republished in book format in 1908, also by McClure's.

Still …where did Adams get the idea to write one of the earliest works of paleo-fiction involving a pterosaur, no less?

Chapter Thirteen — *Zilla's Neglected Novelized Nemesis of '98—GARGANTUA*

Mankind faced additional ecological challenges embodied within the time-honored "man vs. nature" theme in 1998's *Gargantua*, a Tor Books novelization by "K. Robert Andreassi," otherwise known as the prolific author Keith Candido, (based on Ronald Parker's screenplay). Think of Candido as a major *Star Trek* novelist, although he's done much more than that! Indeed, *Gargantua* is only a very minor facet of his authorial output. Quick perusal of his website (Decandido.net) reveals not only a host of geeky *Star Trek* universe writings, but also "Buffy," "Battle Tech," "Dr. Who," "Firefly," "Resident Evil," "Supernatural," "Xena" and "Hercules" short stories and novelizations and, additionally, numerous comic book story contributions, essays, articles, reviews, voiceovers, with new books scheduled for release (as of May 2014 and presumably beyond) to boot.

Largely forgotten today, *Gargantua* was a movie televised on the Fox Network. *Gargantua's* premier broadcast was synchronized with the 1998 *Godzilla* release, obviously intended to capitalize on furor over TriStar's feature. *Godzilla* was released on May 20th –*that* titular monster often referred to now as 'Zilla'—while *Gargantua* struck first one day earlier for television viewers on May 19th. More recent IMDB ratings, however, reveal which film prevailed over time. TriStar's maligned entry has a 5.3 stars rating (out of 10)—as of December 27, 2018, whereas *Gargantua*'s rating is only 2.8 stars. Another striking contrast is that while nearly 168,000 reviewers thought kindly enough of the former to spend time rating it, only 1,065 viewers of *Gargantua* voted. *Gargantua* was relatively ignored and not even remembered today by most dino-monster fans. (Full disclosure—I didn't even watch this movie until several years later after I purchased a VHS tape.)

I distinctly remember in May 1998 walking into downtown Chicago bookstores seeing shelves lined with paperback copies of Andreassi's novelization, (but absolutely no copies of Stephen Molstad's *Godzilla* in sight).

At the time of its initial broadcast (and to best of my knowledge), *Gargantua* received virtually no coverage in mass market monster fanzines such as *Fangoria* or *Starlog*. It's worthwhile summarizing this tale here. Also, by focusing on the novelization here, it's perhaps easier to glean insights about the story's message or idea, as opposed to relying solely on the video, where key passages may have been omitted. Therefore, I'll defer to the movie too, particularly catchiest scenes, although please remain mindful that, plot-wise, film follows novel quite faithfully.

Besides having a cool-sounding name, Gargantua was one of the relatively few American kaiju to storm through a televised film and promotional novel. Although during the 1930s, Gargantua was the name of a real gorilla billed by the Ringling Brothers & Barnum & Bailey Circus as the "world's most terrifying, living creature," the titular Gargantua of 1998 was neither ape nor reptile. And while adult male Gargantuas resemble horned ceratosaurs crossed with a *Jurassic Park Velociraptor*, they're neither. Instead the filmic Gargantuas are amphibious salamanders—mutated *newts*. The *Gargantua* story borrows mainly from *Gorgo*.

The titular monster "Gargantua" is perhaps the largest of the brood (shown on the paperback cover – artwork by Gabriel Galluccio), although not quite as 'scaled up' (a pun!) as an Ogra (i.e. *Gorgo*). The novelization is another tale about family values, borrowing on this theme from *Gorgo* (or *Gappa: The Triphibian Monster*). But *Gargantua* isn't by any means simply a rehash of those. Lovelace's *King Kong* takes place during the pre-WWII, Depression years; "Carson Bingham's" (pseud. for Bruce Cassiday (1920-2005) *Gorgo* novelization reflects early post-nuclear Cold War anxiety, while *Gargantua* is framed during the reactionary environmental movement when use of personal computers became widespread. And the televised movie was first broadcast during *The X-Files'* heyday. But the novel was also written at a time when Candido could avail himself of all the preferred elements and ambiance of more recently published monster tales – such as *Jaws* and *Jurassic Park*.

Gartantua's premise rests on three circumstances: 1960s dumping of pesticide wastes polluting the Iozima Ridge marine environment off the coast of a south Pacific Island named Malau; undersea earthquakes permitting an oceanic migration route for a mutated species of salamander from the Iozima Ridge to the Malau coast; and, of course, the theme of paternal love. Here, Man is the predator, and manufactured products like Cheese Puffs, referred to as "vile" food in the novel, are Nature's ill. Furthermore, even ecological abominations like the Gargantua species have a place in nature and shouldn't be destroyed; their respect for family relationship would make them appear more honorable than humans who would profit from their capture or covet them as trophies or specimens.

The story is set within the luxurious Pacific isle of Malau. Here, azure, tranquil waters unleash a terrifying new incipient species of amphibian, in adulthood—an ecological horror, product of mankind's oceanic pollution. (To a certain extent, *Gargantua* belongs to the same intellectual "clade" as *Horror of Party Beach*.) Following an earthquake (mindful of that in *Gorgo*, although far less cataclysmic), and an introductory *Jaws*-type expiration for two swimsuit babes, we are introduced to the main characters. Widower Jack Ellway and his young son Brandon are on Malau to study local marine biology. (Brandon is rather the counterpart of "Sean" in *Gorgo*, as he befriends a juvenile 'gargantua' creature.) The pretty female lead, physician Alyson Hart, more than eye-candy yet a mere story 'prop.' Paul Bateman (a distant analog of journalist "Steve Martin" in *Gojira*) is the island's sole reporter and newspaper man. Dr. Ralph Hale is an Australian geologist who is conducting field investigations on Malau. After another death ensues, the scientists naturally begin putting things together.

Tension builds when Jack senses that some form of marine life was responsible for these deaths. Next, Brandon discovers a dog-sized "salamander" living in a swamp that he names "Casey."Then on the beach, amazingly, a 10-foot tall deadly, horned "lizard" sub-adult creature (with a short row of horns along its dorsal side) emerges from the surf, injuring several people. This is utterly crazy, thinks Paul Bateman, "because big scaly monsters don't exist outside of movies." Here is one of many blurred references to describing the creature(s) as lizards – even though Jack believes they're amphibian (i.e. salamanders) in nature. This second creature, nicknamed "Ghidrah," is subdued with a dart fired from a tranquilizer gun and captured.

So Malau President "Manny" calls a halt to boating until this mystery is resolved. There's also an assortment of stereotypical bad guys; the henchman is a New Zealander named "Derek" Lawson. Of course, Derek, whose livelihood is based on fishing and offshore tourism is outraged! As you might have guessed, the scientists would rather study the animal properly, but Derek eventually realizes how much money can be made through exploiting it, turning it into a major gate attraction. Meanwhile the President grows concerned about protecting his people in case there are more of these monsters out there lurking in the waves. They might even have to be killed for reasons of public safety, a possibility that disturbs young Brandon who by now has grown quite attached to "Casey." But it's Paul who mutters one of the book's big ideas, "…whenever humanity has power over nature, Nature's usually the big loser." (However, in the long run—think geologically, this philosophy is shortsighted and highly debatable.)

With scientists temporarily in charge of the giant salamander operation, Ralph and Jack conclude that a population of creatures might be breeding somewhere along a 2,000-mile long submarine mountain chain named the Iozima Ridge, explaining why nobody's noticed them before. But, conveniently for this story, a fault line resulting from recent seismic activity leads to the nearby shallow waters off Malau. This is a circumstance, "…kinda like an expressway that's suddenly developed an off-ramp." Incidentally, the Iozima, a veritable undersea 'lost world' setting, is not unlike the subterranean canyons offshore from New York City haunted by the Rhedosaurus in *The Beast From 20,000 Fathoms*. Thus, the two 'mad scientists,' Jack and Ralph, need to check it all out in person in a submersible craft named the *Scorpion Fish*. The vessel gets knocked around considerably during the dive, prompting Jack's considerations that there indeed may be another, larger (unseen) specimen out there after all. Or was it merely another quake, (or both happening simultaneously)?

Meanwhile, neither Brandon, nor Jack in the cases of Casey and Ghidrah have been able to determine what kinds of foods their respective critters like to eat. But Paul learns from the Internet that during the 1960s, the Iozima Ridge was used as a dumping ground for thousands of containers of waste DDT pesticide. As he states, "Concrete can leak, you know." Now the scientists begin putting things together fast. Jack recounts the case of mutated frogs which had been in the news back then, resulting from their residing in contaminated ponds. Frogs and salamanders are both amphibians, so could the strange 10-foot tall monstrosity they've captured be a mutation, resultant of exposure to pesticide? When DDT is collected in a water sample collected during the submarine dive, Jack finally utters that fabled conventional phrase, "… we created a monster… and it's probably not the only one of its kind." Yes—a foreshadowing of what is to come, and segue to following chapters in the novelization when two more gigantic "gargantuas" finally appear. So it's perhaps no wonder anthropomorphized "Casey," later likened to a teddy bear, finds Cheese Puffs to be the only food Brandon can find to feet him with. Artificial food ("vile foodstuff… laced with more chemicals and preservatives than the average pesticide") must prove tastiest to mutated organisms.

With respect to "dinosaurians," so far in the novel we've met only cutesy Casey and the 10-foot, horned, sub-adult critter. What has been responsible for the several human deaths and 'Nessie'-like sightings at sea? Eventually – given the book's title, something truly 'gargantuan' oughta show up, right? And *they* do. A blurry photograph of a large swimming "superlizard" becomes the next bit of evidence. From this photo taken by a passenger capsized in a motorboat, Jack notices that the interfering beast was horn-*less* and approximately 35-feet tall. He therefore conjectures that this third specimen is a female, quite possibly the mother of the 10-foot creature held in captivity. If this larger one comes ashore, there could be some serious consequences. The President then calls in the military, led in the guise of a U. S. Marine, Colonel Wayne, who isn't a stereotype cigar-chomping cuss. In fact, he does his amiable best to cooperate with the scientists. Together, they strategize how to best handle the third, largest creature so far, should it arrive on shore. Later Wayne's impression of the huge mother is that of a tyrannosaur, except with considerably larger arms that could possibly function as forelegs.

When Big Mama Lizard does emerge from the ocean, the military retaliates with a barrage of firepower, and the mother dino- (which actually only turns out to be 25-feet tall and 40-feet long) is repelled back into the water after killing two soldiers. Mother lizard makes a second appearance just as Jack and Brandon attempt to use Casey as a bargaining chip in coaxing the creatures away from Malau. But this time Mama is felled by a rocket launcher, prompting eco-minded Jack to mourn, "We discover three members of a brand-new species and immediately reduce it to two. We're the human predator at its worst." Meanwhile, Derek and his crew, in attempting to steal the 10-footer, accidentally allow it to escape into the sea. Well, never fear, Jack, for there's still at least one more—a 60-foot long horned male—which makes its single startling onshore assault when it comes in the night to reclaim Big Mama Lizard's corpse.

Using Casey's voice amplified through underwater, high frequency speakers as a lure, they attract the two surviving males (i.e. 10-foot tall and 60-foot long ones), tracked using sonar, who demolish Derek's trawler. Jack is hypnotized by majesty of the large male, as it combines "plumage of horns" seen in the 10-foot male, with the mother's magnificence in grandeur and size. Now reunited, the creatures swim downward toward their Iozima Ridge home. After this happy ending, Derek is finally arrested, and an explanation is offered as to the "reptiles'" occurrences on Malau, with the usual "I don't think we'll be seeing them again," concluding phrase added. This is partly because, on cue, soldiers have sealed the Ridge "off-ramp" using explosive depth charges. (Or could it be because producers declined any need for a "Gargantua II" sequel?)

Jack Ellway believes that "...Junior and the baby went exploring out of the fault line and found themselves on Malau. Mom came after them and ran afoul of the human predator. Dad came to retrieve Mom's corpse and finish her job of bringing the kids home." Family loyalty at work, just like in the human world where Brandon who now appreciates his father Jack more than ever, and Alyson Hart's pretty face seem to all fit together now as a picture-perfect future family!

There are no more of the gargantuan dino-monsters out there ... at least, so far, there have been no sequels. For reasons not unlike why TriStar's 1998 *Godzilla* generally seemed to fail among the diehard fan base, so too would *Gargantua* seem rather off-putting. Whereas Lovelace's 1932 *King Kong* and Carson Bingham's *Gorgo* novelizations strike readers as original for their times—both holding me on the edge of my seat during several harrowing passages, Candido's *Gargantua* is just too, well, 'nice'. I probably got a cavity or two from re-reading it. It was perhaps intended for younger audiences. This 'flaw' may be attributed to the scriptwriter, however, not the experienced writer, Candido. Curiously, Lovelace's "Kong" has attained classic status in a recently published edition, while Bingham's *Gorgo* remains virtually unknown. However, Candido's *Gargantua*, rife with clichés, seems destined for perpetual obscurity, not unlike Stephen Molstad's 1998 *Godzilla* novelization then competing for bookstore shelf space. (See Chapter Five.)

Despite their saurian appearance, the Gargantuas aren't misconstrued as 'prehistoric' *per se*, as they're instead viewed as a recent product of man's pollution of the marine environment. However, the Gargantua 'species' is rather whimsically compared to dinosaurian monsters throughout several passages in the novel ... *Tyrannosaurus*, Ghidrah, *Jurassic Park* raptors, etc. Eventually they're identified as modern salamanders, mutated into dinosaurian form. Few kaiju-fans may realize the significance of salamanders in the annals of European paleobiology and paleo-themed speculative fiction, and their importance warrants a short digression.

Swiss physician and naturalist Jacob Johann Scheuchzer (1672-1733) who became fascinated by accounts of the biblical Deluge and the fossil relics evidently remaining from that catastrophe, indoctrinated one of the earliest known fossil vertebrate specimens that become of significance in the developing field of paleontology. This was his specimen obtained, in 1715 and described in 1726 as *Homo Diluvii testis*, known as a preserved "human witness to the Flood." Although Scheuchzer excitedly declared this fossil human had drowned in 2306 B.C., two centuries later, revered French anatomist Baron Georges Cuvier (1769-1832) noted the fossil vertebrate was not human. Instead, Cuvier proved that this "witness of the Flood" was a large form of fossil salamander from the Miocene epoch, closely related to the modern Japanese giant salamander. Later, however, during the 19th century considerable confusion reigned over the nature of 'Cheirotherium'—fossil 'hand tracks' resembling those of humans found in Triassic strata. Eventually the fossil imprints were assigned to an extinct 'frog-amphibian' known as the *Labyrinthodon*.

Next, Austrian biologist Paul Kammerer (1880-1920) committed suicide after it was divulged that data he'd gathered in support of the 'Lamarckian' theory of inheritance through acquired characteristics had been falsified, possibly by co-workers. Kammerer's research, performed on salamanders, was disproved after doubtful (Darwinian) British scientists couldn't reproduce his experimental results. A 'hoax' was unveiled when his preserved salamander specimens were shown to have been tampered using india ink. (For details on this intriguing affair, read Arthur Koestler's 1971 account, *The Case of the Midwife Toad*.)

Then in 1936, Czech writer karel Capek (who also coined the word 'robot') published a novel, *War With the Newts* (English translation), in which a marine variety of large salamanders are aided technologically by an unscrupulous, unwitting (human) sea captain. Soon the intelligent and wildly successful newts overpopulate the globe, and their efforts to flood the continental land masses threaten mankind's extinction. While amphibious newts and salamanders aren't classified as reptiles, the fact that *War With the Newts* merited discussion by University of Chicago Professor W. J. T. Mitchell in his scholarly volume, *The Last Dinosaur Book* (1998), gives pause. Intriguingly, Capek described the 'newts' as an "evolutionary resurrection" of Miocene salamanders, meaning they're conceptual descendants of Scheuchzer's 'Witness to the Deluge.' Also, the newts' erect, bipedal saurian nature resembles 20th Century-Fox's televised Gargantuas.

In the typical 'alternative kaiju' tale, authors avail themselves of symbols, tropes and metaphors used as literary devices … such as diving bells and submersibles (i.e. thus penetrating the lair of the beast), earthquakes and volcanoes (signaling alarm to mankind—Nature's fury), lighthouses or towers (symbols which, when destroyed or 'conquered' represent mankind's emasculation), retaliatory nuclear or other weapons of mass destruction (symbolizing degradation of our environment—our pending extinction), the sea (a monster's emergence from the sea, where life originated represents Nature's retaliation), regeneration (the kaiju cannot really be destroyed because it is *within us* and won't go away until our inevitable extinction), and prehistoric ambiance (an anachronistic revisiting of the prehistoric suggests extinction).

While Gargantua monsters probably are, at best, marginal kaiju, their story as told in a fantastic novelization reveals more about humankind than we're willing to accept.

Chapter Fourteen — *Skullduggery: A Piltdown 'Elementary'*

Many already know the lowdown on Piltdown, so I won't digress too deeply into history of this copper-fooling caper, save for mentioning the essentials. Between 1912 and 1915, several fossils were found in Sussex, England that were attributed to Great Britain's supposedly oldest human form, a "missing link" named *Eoanthropus dawsoni*, or Piltdown Man. Quite possibly, according to British scientists, these fossils dated from the Late Pliocene. Of course, as we now know, Piltdown Man was a carefully contrived fraud, consisting of a human skull and the jaw of an ape. In the early 1950s, British scientists unraveled circumstances of the forgery using methods unavailable forty years before.

Over the next four decades several individuals having known connections to the Piltdown case were accused—notably Charles Dawson, an amateur archaeologist, antiquarian, and paleontologist who identified the original fossils allegedly 'discovered' by laborers working in a gravel pit in 1908. These were brought subsequently to Sir Arthur Smith Woodward's attention in February 1912. Smith Woodward, Keeper of Geology at the British Museum, is decidedly not the culprit of the outlandish crime; he is merely one of the many victims. Under suspicion are French philosopher/paleoanthropologist and French Jesuit Priest, Pierre Teilhard de Chardin, thought by Stephen J. Gould and a handful of others to have assisted Dawson; British anatomist Sir Arthur Keith, a recently-accused suspect; former British Museum Curator of Zoology Martin A. C. Hinton; and the focus of this presentation – writer (and medical doctor) Sir Arthur Conan Doyle—yes creator of Sherlock Holmes and intriguingly, author of *The Lost World*, a famous novel, which according to accusers, offers clues to Doyle's knowledge of, or involvement in, the dastardly deed.

During my first, 1979 reading of Doyle's extremely entertaining and classic paleo-novel, I recall one curious passage causing me to do a double-take. Here one of Doyle's characters states, "If you are clever and know your business you can fake a bone as easily as you can a photograph." Hmmm. Yes, I was fully aware that Piltdown Man was discovered around the same time that Doyle's novel was published (first printed as a serial in *The Strand Magazine* beginning in April 1912, with the novel published in book form that fall). But until four years later, I had no idea how closely tied was Doyle to the East Sussex countryside where the merriest paleo-prankster(s) ever in Jolly Old England had some devilish, if not downright devious, fun.

In this fossil-foolery excursion, about to unfold, I claim little originality. Nor shall I blame anyone thus far not already implicated as perpetrator. Numerous papers and books have been published on the subject of the hurtful Piltdown Man forgery. (Moreover and sadly enough, some 500 doctoral dissertations were published on the supposed evolutionary significance of Piltdown Man, prior to the fraud's revelation in 1953.) One of the best recent publications to delve into this interesting topic is John Evangelist Walsh's *Unraveling Piltdown: The Science Fraud of the Century and its Solution* (1996). I am borrowing freely from this source, as well as a critical *Science 83*, September article, "The Perpetrator at Piltdown."

Do I really think Doyle instigated Piltdown, or was Doyle otherwise somehow involved? 'Elementary,' my dear readers! By Jove, let's just say it would take a smashing bit of detective work indeed (on par with Sherlock Holmes' abilities, perhaps?) to prove Doyle was the mastermind who dun-it! So, light up your pipes, settle into your easy chairs, warm yourselves by the fireplace from the foggy chill, and be prepared to weigh the evidence.

As mentioned previously, Charles Dawson claimed that the first remains attributed to Piltdown Man were discovered in 1908, even though, mysteriously and inexplicably, over three years passed before the potentially important fossils were brought to Woodward's attention. (Teilhard had been an acquaintance of Dawson since the summer of 1909; they had gone on fossil collecting trips together, especially to a quarry in Hastings.) One wonders why any sensible person would hesitate so long before drawing attention to the specimens, if indeed their scientific significance were comprehended. According to Dawson, (unidentified) workmen informed him of a 'coconut' (i.e. the alleged intact Piltdown Man skull) they had inadvertently smashed while digging in a gravel pit in 1908. Later, on his own, Dawson collected a few more skull fragments, 'eoliths'—crude tools made of flint – and fossils of extinct mammals, the latter aiding researchers' attempts to date the deposit.

Thereafter, enlisting Woodward and Teilhard, the trio discovered additional fossils in the pit on the Barkham Manor grounds in East Sussex during the summer of 1912, including a 'fossil' jaw, strikingly 'chimpanzee-like' in aspect, bearing two intact molars. Unfortunately, the jaw was fragmented in the chin region, and also precisely where it would have been possible to determine whether and how the hinge truly articulated with the associated human cranial bones. The molars were worn down (i.e. filed by the forger) to appear characteristic of humans, but not of apes.

The discovery proved an absolute sensation! For Piltdown Man evidently even predated the recently described (1907) Plio-Pleistocene paleo-human, Heidelberg Man, and was apparently far more evolutionarily advanced. Those who were enamored with the prospect of the world's oldest, advanced humans having evolved in the British Isles instead of on the European continent pooh-poohed a plausible alternative, a coincidental possibility that the 'chimp' jaw came from an entirely separate animal than represented by the cranial (human) remains, having been simply co-deposited in the gravel pit. It all seemed too good to be true (and sadly, it was).

Commenting on Woodward's reconstruction of the skull at the 1913 meeting of the International Congress of Medicine, held at the British Museum on August 11th, Arthur Keith acerbically stated, "If a student had presented a skull like this one he would have been rejected for a couple of years." Keith claimed prophetically that, as he anticipated, if Piltdown had possessed long ape-like canines, such teeth would have prohibited human sideways jaw movements, making the molar wear observed on molars in the 'chimp' jaw an impossibility! Then, triumphantly, only three weeks later, on August 30th, 1913, Teilhard 'found' one 'missing' canine tooth attributed to the 'chimp' jaw, curiously shaped like an ape canine although worn as if in a human jaw. Such great timing and luck. And how uncanny that Teilhard made such a wonderful discovery on his final Piltdown dig! Hmmm. Stranger still was the unearthing of delicate nasal bones and turbinate remains also attributed to Piltdown Man, a 'discovery' which American Museum paleontologist Henry Fairfield Osborn later claimed was "almost a miracle."

Not everyone was completely fooled by Piltdown Man, though – yet. In particular, French paleo-anthropologist Marcellin Boule claimed Piltdown was no "dawn man," re-designating it as "Homo dawsoni." Boule and several American and German scientists protested, claiming that the jaw didn't belong with the cranial remains. However, Piltdown's detractors were in for stunning surprises during the summers of 1914 and 1915.

For in 1914, Dawson excavated a fragment of fossil elephant bone, strangely if not suspiciously-shaped like a British cricket bat, which was theorized to have been carved by tool-using Piltdown Man. A year later, Dawson confounded detractors with his discovery of yet another set of human cranial fragments found at a site in Sheffield Park (sometimes referred to as "Piltdown Two"). Sheffield Park is only two miles distant from the original site – Barkham Manor. Here, uncannily, cranial remains were found associated with an apish, fossilized canine. This was too much for mere coincidence, and it was sufficient to silence the critics.

American Museum paleontologist Henry Fairfield Osborn, formerly a skeptic (who instead had theorized that Man's evolutionary ancestry stemmed from Asia), finally relented, "Paradoxical as it may appear ... it is nevertheless true." Regarding the Piltdown Two 'fossils,' Osborn beamed, "If there is a Providence hanging over the affairs of prehistoric men, it certainly manifested itself in this case, because the three minute fragments of this second Piltdown Man found by Dawson are exactly which we should have selected to confirm the comparison of the original type...."

After Dawson died in 1916, Woodward examined further remains allegedly belonging to a third Piltdown individual that Dawson had found at yet another site in nearby Barcombe Mills. According to Dawson, these fossils (skull fragments, a preserved brow ridge and a right lower second molar) had been found in 1913, although they had been strangely neglected until after his death. Details surrounding the discovery and excavation of the Barbombe Mills specimens has been largely lost, as Dawson mysteriously—yet characteristically—recorded little about their recovery. Following Dawson's death, Smith Woodward made frequent solo fossil hunting excursions to Barkham Manor, in vain, for no fresh discoveries were attributed to Piltdown Man.

Artists restored Piltdown Man in-the-flesh during the early 20th century. Notably, Belgian scientist Louis Rutot sculpturally portrayed the tool-shaping "Man of Sussex," and zoologist and artist Dr. J. Howard McGregor sculpted a striking head bust for a 1920s American Museum of Natural History display. Smith Woodward's 1913 controversial skull reconstruction had been completed prior to Teilhard's discovery of the straight canine tooth, and didn't incorporate the 'fossilized' nasal bones. In his preliminary, 1914 reconstructions, McGregor placed the canine tooth in the upper jaw. By 1922, however, Osborn and McGregor decided the canine belonged in the lower jaw instead.

So what is Doyle's documented relation to Piltdown and *Eoanthropus'* celebrated discoverer – Charles Dawson?

Doyle lived in East Sussex, in Crowborough, a town just eight miles north of Barkham Manor, also situated near Dawson's Lewes domicile. Both Dawson and Doyle were members of the Sussex Archaeological Society and they did socialize and correspond. Furthermore, Doyle had relatively easy access to the Piltdown locality because he owned a motorcar. Evidently they met in 1909, after Woodward suggested that Dawson, a recognized expert on the Wealden dinosaur fauna found in the region, should pay a visit to Doyle, who claimed to have found a "huge lizard's tracks" in a nearby quarry. In November 1911, Dawson had consulted Doyle at his home on the nature of alleged dinosaur bones which Doyle thought he'd identified. (Dawson's conclusion was that Doyle's specimens weren't dinosaurian after all, but an iron/sand concretion instead.)

While walking the grounds with Doyle, however, Dawson happened to find a "beautiful flint arrowhead."

It is possible that during this occasion, Doyle may have spoken to Dawson about his literary forthcoming publication, *The Lost World*, the first installment of which appeared in April 1912. For Dawson wrote to Woodward following the visit that Doyle had written "A sort of Jules Verne book ... on some wonderful plateau in S. America with a lake, which somehow got isolated from Oolitic times ... and where ancient animals and vegetation still survived." Although Dawson most certainly had knowledge of Piltdown Man by then, there is little documentation of any discussion between himself and Doyle concerning the tribe of 'missing links,' which incidentally made their way into Doyle's fictional tale. On December 16, 1912, over a year later, and following publication of *The Lost World*, Doyle visited the (by then) famous Pilltdown pit. Surviving records do not indicate Doyle knew anything of Piltdown Man's existence until the fall of 1912, long after Dawson's discoveries had become publicized. (An interesting, fictional account of Dawson's first meeting with Doyle appears in Peter Marks' imaginative novel, *Skullduggery*. (1996), pp.80-85.)

Next, let's outline the evidence pointing to Doyle as a possible culprit and recount the plot of *The Lost World*.

Had it been scripted today, *The Lost World* could have been adapted into an exemplary *The X-Files* kind of episode. Arthur Conan Doyle's fast-paced action novel is a can't-put-it-down page-turner that still works on many levels. It is one of our finest examples of paleo-fiction literature, even though few may have read it. More would prefer instead to enjoy the 1925 silent film based on the novel. Themes explored throughout are whether a man's heroic nature is gained by rite of passage through proving one's courage in the midst of insurmountable danger vs. innate bravado, and the (ironic) fickleness of woman's desire in choosing a lover. But was Doyle's novel written to serve another, darker purpose?

One notices parallels to Jules Verne's *Journey to the Center of the Earth*, published half a century earlier. For example, in the latter novel it is fictional alchemist Arne Saknussemm's written records guiding explorers into the bowels of the Earth where they find traces of a 'prehistoric' fauna. In Doyle's novel it is Chicagoan Maple White's scrapbook which leads Professor Challenger and company to a 'lost' plateau inhabited by prehistoric animals. In both Verne's and Doyle's novels, prehistoric ape-like creatures are discovered living alongside other varieties of extinct animals then known to contemporary science. Instead of a Central Sea, as in English translations of Verne's novel, Doyle incorporated a central lake – Lake Gladys—named in honor of the fickle woman to whom the narrator, Ned Malone, is trying to prove his valor by joining the expedition.

In *The Lost World's* opening, intrepid and hot-tempered Prof. Challenger has already made a reconnaissance of the base of the South American plateau, first accessed by Maple White. However, Challenger's claims of a living prehistoric fauna are ridiculed by colleagues. He manages to convince a reporter, (protagonist) Mr. Malone, of his utter sincerity in the matter and, following a raucous London meeting, an expedition (including, principally, Malone, Challenger's scientific colleague—Dr. Summerlee, professor of anatomy, and a curious sporting champion—Lord Roxton) is launched to judge veracity of Challenger's contestable conclusions. The year is, presumably, 1908.

Arriving at their Brazilian destination, Malone, Challenger and company scale the plateau where they encounter a fauna enlivened from Wealden Cretaceous strata setting where Doyle and Dawson lived, in Sussex. Before an iguanodont herd is spied by the party, (immediately convincing Summerlee that Challenger is not a madman after all), Summerlee finds a fresh set of huge tracks. Challenger as pompously as ever, proclaims "Wealden! I've seen them in the Wealden clay. It is a creature walking erect upon three-toed feet, and occasionally putting its five-fingered forepaws upon the ground. Not a bird, my dear Roxton—not a bird...No; a reptile—a dinosaur." Is this, perhaps, an allusion to Doyle's meeting with Dawson? (Challenger professes his theory of Maple White Land's native fauna occurrences and evolutionary forces at play there on p. 242 of Doyle's novel, 1925 ed..)

The adventurers find a living *Stegosaurus*, which Maple White had illustrated in the notebook recovered by Challenger, and a *Phororhacos* too. When they disturb a pterodactyl rookery, they are enveloped in a pterodactyl swarm (the Professors can't decide if the offending genus is *Dimorphodon* or *Pterodactylus*). But the most hair-raising scenes happen when Malone unwisely takes a walk in the dark down to the central lake only to be pursued by a carnivorous dinosaur hopping relentlessly after him in kangaroo-fashion.

Their chief menace though is the tribe of warring, red-haired tribe of 'missing links' (debated to be either *Dryopithecus* or *Pithecanthropus*—and notably *not* Piltdown, by Challenger and Summerlee) which hurl their unfortunate victims off the edge of the plateau cliff. One ferocious, tree-climbing ape-man is described by Malone as having:

> "...a human face—or at least it was far more human than any monkey's that I have ever seen. It was long, whitish, and blotched with pimples, the nose flattened, and the lower jaw projecting, with a bristle of coarse whiskers round the chin. The eyes, which were under thick and heavy brows, were bestial and ferocious, and as it opened its mouth to snarl what sounded like a curse at me I observed that it had curved sharp canine teeth."

These creatures are arguably more orangutan-like than Piltdown men as envisioned by British scientists.

Unconscious, Malone falls—avoiding impalement—into a pit manufactured by ape-men for catching wild animals, thereby escaping cruel jaws of that menacing kangaroo-like carnivorous dinosaur. While Malone recovers, his comrades are captured by the 'missing links.' Roxton escapes, finds Malone and they rescue the professors. Then, reunited, they team with another more recently evolved tribe of Indians, which also inhabit the plateau, to battle the apes. Aided by their blazing rifles, the company of Indians and Europeans cause near-extinction of the 'missing links.' Expressing a tone decidedly politically incorrect under current standards, Challenger justifies their decisive military maneuver. "Now upon this plateau the future must ever be for man."

In my opinion, Doyle's 'missing link' ape-men seem derivative of another fictional tribe of ape-humans described by Jack London in his imaginative novel, *Before Adam*, (1906, published in 1907 by The MacMillan Company. MacMillan had publishing houses both in New York and London.) Also, Doyle's indians versus 'links' tribal war seems reminiscent of climax scenes in London's 1907 novel. Furthermore, conflict between co-existing ape-men and native indians seems analogous to a distant future envisioned by British author H. G. Wells who populated the landscape of circa 800,000 A. D. with our evolutionary descendants—hideous Morlocks and their contemporaries, the gentle, human-like Eloi.

The party descends the plateau and triumphantly returns to London, where they reveal their findings at a well-attended public meeting. Challenger steals the show when he produces a living pterodactyl, captured by Roxton, from a crate shipped from the plateau. The monster flies around the auditorium, shocking the attendees, before flying out an open window into the night sky. Meanwhile, Malone discovers that his girl, Gladys has married during his absence. Her new husband, having no adventuring inclinations whatsoever, remarkably has an absolute opposite persona of the hero Malone has become during his journey to the lost world. Gladys' husband's profession is given as a mere 'solicitor's clerk.' Of note, although chiefly remembered today for his (avocational) paleontological excursions, Charles Dawson's vocation had been as a solicitor in Uckfield.

Beyond Doyle's written descriptions, today, "Lost World" imagery is not often encountered, but can still be found. Artists Harry Rountree and Joseph Clement Coll had each illustrated published serializations of *The Lost World* for, respectively, *The Strand Magazine* and *The Sunday Magazine*. Their wonderful visuals showing dramatic scenes from the story were recently reproduced in a charming book, *The Annotated Lost World—The Classic Adventure Novel* by Sir Arthur Doyle, Annotated, with an Introduction by Roy Pilot and Alvin Rodin," (Wessex Press, 1996). Pilot and Rodin were not supportive of Doyle's possible involvement in the Piltdown affair, claiming he has been 'exonerated of any complicity.'

And in 1925, five years prior to Doyle's death, First National produced a definitive silent film adaptation of *The Lost World*. This movie is still revered for its marvelous special effects which are comprehensively discussed in *The Stop-Motion Filmography* (1999), by Neil Pettigrew (McFarland, pp. 425-438). Some liberties were taken with sequences in the novel to produce a more visually dramatic movie, featuring famed dinosaur 'puppets' created and animated by Willis O'Brien and Marcel Delgado. Most interestingly, in light of the charges made on his good name, Doyle revealed his impish nature when, in 1922, he presented test footage from the forthcoming movie as sort of a promotional prank on the magician Houdini and other astounded guests who were stymied as to how the dinosaurs had been filmed. (Hmmmm—a most interesting character trait, eh, Watson?)

Belief in Piltdown Man's authenticity began unraveling in 1949 after Dr. Kenneth Oakley's initial chemical tests (fluorine analyses), intended to quantify Piltdown Man's age, produced puzzling results. According to the first round of fluorine test data, Piltdown Man may *not* have been older than 50,000 years, instead of half a million years old (or older), as originally suspected. Rather astoundingly, Oakley had knocked Piltdown off its Dawn Man pedestal, relegating the apparently primitive species to an all too young geological stage, an age shared by Cro-Magnon humans of modern anatomical form. Therefore, an atavistic Piltdown Man could not qualify as a 'missing link' to modern humans. But this was only the beginning. For, by the Fall of 1953, anthropologists Oakley and Joseph Weiner had determined that "distinguished Piltdown scientists had been the unwitting victims of "a most elaborate and carefully prepared hoax." (Quotes from Walsh, 1996, p. 70.)

Fearing the worst, Weiner and Oakley teased out details of how a cleverly designed fossil caper had been engineered. Subsequent age dating efforts established that Piltdown I's cranium 'fossils' were a mere 620 years old. These cranial fragments had probably been unearthed from a local British grave site. The 'chimp-like' jaw proved instead to be a (modern) orangutan's. Instead of being half a million years old, the orangutan jaw was only half a millennium old (i.e. 500 years). Furthermore, Weiner determined that abrasion marks noted on the jaw's molars, and wear patterns evident on the canine discovered by Teilhard, had been fabricated by the perpetrator(s) using hand held tools. Weiner even made a fake 'fossil' tooth himself to prove how easily this could be done. Then they discovered traces of chemicals on the alleged fossils, proving they had indeed been artificially stained using oxidizers to simulate the appearance of great, naturally iron-stained, age. Because most researchers were only permitted to study plaster casts of the reconstructed skull instead of the original 'fossil,' the staining was not obvious to others who might have detected it decades earlier. Having debunked key Piltdown evidence, Weiner and Oakley scrutinized other specimens attributed to Piltdown Man.

Next, Piltdown 'fossils' found at Sheffield Park (Piltdown Two site) also proved to be modern. The 'cricket bat' elephant –carved implement supposedly utilized for hide-dressing turned out to be a genuine fossil, although it was disproved that any paleolithic human could have carved the once fresh bone using a flint 'eolith' knife. Furthermore, the forger had gone to the added trouble of iron-staining the bone 'bat' to make it appear even older than it was. Other mammalian vertebrate fossils found at Piltdown were obviously planted in the ground alongside other Piltdown remains. In several cases Weiner was able to determine their probable geographical origin, typically outside of Britain.

Four 'eolith' tools were fakes too, for the 'natural' iron-staining could be dissolved in hydrochloric acid solution. ('Eoliths' were described as crudely formed implements, being 'rolled or abraded flints,' vaguely distinguishable from other 'freshly worked flints' Dawson had supposedly discovered.) Weiner even uncovered a signed note from one of Dawson's contemporaries, a local flint collector—Fred Wood, who had died in 1940. Wood alleged that several flints in his private collection (not comprising part of the Piltdown collection) had been "stained with of permanganate of potash and exchanged by D. for my most valued specimen!" Wood also warned prophetically, "Watch C. Dawson...." (Quotes from Walsh, 1996, p. 86.)

While Weiner implicated Dawson as Piltdown's perpetrator, he considered other suspects too. However, to me, the most intriguingly original possibility to present itself in recent years was, as John Hathaway Winslow and Alfred Meyer suggested in 1983, that Arthur Conan Doyle had masterminded the Piltdown fraud. In Winslow's and Meyer's September, *Science 83* article, a number of previously overlooked details pointed to Doyle. As the authors stated, "To be on Doyle's trail is to be on the trail of the world's greatest fictional detective himself," a reference to the famous Sherlock Holmes of Scotland Yard.

The case against Doyle rests on four lines of evidence. First, as discussed previously, Doyle and Dawson were acquaintances around the time of Piltdown Man's discovery and Doyle had ready access to the excavation site. Secondly, Doyle possessed technical skills, knowledge and whereabouts to engineer such an intricately designed fraud. (For example, his fictional character, Sherlock Holmes, often relied on well-crafted elucidation of forensic and anatomical evidence to solve perplexing crimes.) Thirdly, careful reading of Doyle's *The Lost World* story, offers striking congruities between fictional plot and the fictional 'dawn man' of Sussex. But did Doyle have a motive? Well, finally, and possibly the clincher, Doyle may have used Piltdown to demonstrate the gullibility of British paleontologists on theoretical evolutionary matters, although the whimsical prank may have simply gone too far.

How circumstantial is the evidence against Doyle?

Sir Arthur Conan Doyle was a trained medical doctor, having knowledge of chemistry and anthropology. It's a cinch he would have known how or been able to stain bones 'authentically,' had he wanted to, and he also would have known which bones to stain in order to confound evolutionary theorists with a convincing fabrication. Winslow and Meyer believe that breaking an orangutan jawbone skillfully, so as to suspend disbelief sufficiently that it should be composited with cranial remains of a 'fossil' Piltdown Man, would have been easy for someone with Doyle's expertise.

Furthermore, Doyle had traveled to the Mediterranean region in 1909, where he could have acquired several fossil elephant and hippo teeth, chompers known from collecting localities there, which were subsequently planted into the pit at Barkham Manor. During his visit to Tunisia, Doyle could have also obtained several flints found in the Piltdown gravel as well. Several Piltdown flints seem characteristic of those known from a paleolithic flint 'factory' situated at a Tunisian town named Gafsa. Even if Doyle didn't acquire these flints himself, he might have gotten them from an acquaintance, Norman Douglas—who had a reputation as a 'flint maniac.'

Weiner suspected that Doyle had become inspired by Dawson's discovery of Piltdown Man, although Walsh suggests Doyle didn't know about Piltdown Man (even though he and Dawson already were acquaintances) until long after *The Strand's* first installment of *The Lost World* had already been published. (Woodward was first informed by Dawson of the 1908 'discovery' on February 15, 1912, four months before the first installment of *The Lost World* appeared in *The Strand Magazine*.) Winslow and Meyer stated in their *Science 83* article that "Weiner did not consider the possibility that instead of Doyle's book being inspired by the Piltdown excavation, the Piltdown hoax was inspired by, or developed hand-in-hand with, the plot of *The Lost World*."

Besides that incriminating quote about faking a bone as easily as one can fake a photograph, Winslow and Meyer noted several citations in *The Lost World* which could be construed as instances of Doyle offering clues to what he had mischievously done. For instance, a character states that a practical joke (of such or analogous nature to fakery) "...would be one of the most elementary developments of man." Also, the blended admixture of prehistoric fauna in *The Lost World* setting, analogous to the occurrences of non-contemporary remains planted at Piltdown sites, seems peculiar in light of Charles Dawson's 1911 note to Woodward, where he wrote concerning Doyle's forthcoming novel, "....I hope someone has sorted out his fossils for him." The red-haired, nest-building, 'nondescript' 'missing links' of Doyle's story may be compared to modern orangutans, a jaw of which was actually doctored into the reconstructed Piltdown Man's jaw.

The 'lost' plateau in the novel is similar geographically to the layout of Sussex, where Piltdown Man was discovered. For instance, it seems as if the Sussex region has been uplifted and the ancient remains interred in Wealden and younger deposits have become revivified in Doyle's fictional account. In fact, one character in *The Lost World* muses that the 'lost world' is an area "as large perhaps as Sussex, (which) has been lifted en bloc with all its living contents." A map published in the novel reveals other general geographical resemblances between the Weald and the fictional 'lost world' setting.

Characters in Doyle's novel are thought to have been based on real historical figures. Professor Summerlee, for example, may have been inspired by Woodward. Lord Roxton was based on Doyle's friend, Roger Casement, who fought for native's rights in the Congo, Peru, Brazil and Columbia. Professor Challenger is a composite of several British scientists, principally the 'formidable' paleontologist Sir Edwin Ray Lankester (1847-1929). Lankester was a staunch Darwinian evolutionist. Lankester's popular book, *Extinct Animals* (1905) proved a major inspiration behind Doyle's descriptions of animals living on the 'lost plateau,' and, in fact, Lankester read Doyle's 1912 novel with enthusiasm. Like Lankester, Doyle also was an evolutionist. 'So, where's this leading?' you ask?

Importantly, Lankester ridiculed the decidedly phony spiritualist movement (e.g. often involving seances to communicate with the dead) then prevalent in Britain, which Doyle advocated. Winslow and Meyer suggest that Doyle's (well documented) defense of spiritualism, stoked Lankester's ire. Lankester had speculated in popular articles published between 1906 and 1909, and subsequently lectured in 1911, about the possibility of Pliocene Man—something akin to that later found at Piltdown. So a very early non-primitive human became an expectation of what scientists could possibly realize someday in the primate fossil record. Perhaps Doyle found this opportunity to good to pass up. Lankester simply may have "set himself up." For, much of what was discovered at the Piltdown sites validated Lankester's predictions. In another of Doyle's stories, The Land of Mist (1926), an ape-man, *Pithecanthropus*, is summoned during a seance. Wouldn't the fictional medium have summoned a Piltdown Man instead, if Doyle really was the perpetrator offering hints?

Winslow and Meyer conclude, "Unwittingly, Sir Ray (Lankester) ... provided a list of objects to be discovered or verified as being man-made, and the hoaxer obliged him on every count....Piltdown Man, as he saw it, represented a large step in the direction of his hypothetical Lower Miocene apeman. It was a case of self-fulfilling prophecy, with the hoaxer providing the wherewithal." When Doyle wrote *The Lost World*, a novel he suspected Lankester would read, perhaps he thought Lankester would realize that he, Smith Woodward and other

Piltdown Man investigators had been duped. But if that was Doyle's intention, nobody caught on to the storyline clues, that is, if Winslow and Meyer are right, until 70 years later. "Such gullibility must have exasperated Doyle, or made him howl with laughter." (Quote from Winslow and Meyer, p. 43.)

After Winslow's and Meyer's article was published, many howled, but not with laughter! For instance, paleontologist Stephen J. Gould, who in three meticulously researched *Natural History* magazine articles had made a compelling case for Teilhard's duplicity—as Dawson's 'accomplice' at Piltdown, led the attack. In a letter to *Science 83*, Gould called their Doyle-as-conspirator hypothesis, "evidence-free based on speculations about motive." Winslow and Meyer replied, thoughtfully challenging Gould's implication of Teilhard, "Could it be that history is repeating itself? That the deep-seated animosity held by leading evolutionist Ray Lankester toward the Spiritualists finds a parallel in the case of Gould, the no-nonsense evolutionist, vs. Teilhard, the evolutionary mystic?"

Authors of the two most comprehensive and recently published treatments of the Piltdown affair, (i.e. Walsh's previously cited book and *Piltdown: A Scientific Forgery*, (1990) by Frank Spencer), couldn't support possibility of a Doyle-led Piltdown conspiracy either. Walsh devotes an entire chapter to "The Writer Accused," carefully evaluating the Winslow/Meyer theory (pp. 107-127). Significantly, Walsh noted:

> "....perhaps the most telling observation to be made against any possible relevance of *The Lost World* to Piltdown relates to Doyle's picture of his imagined ape-men (Note—these were illustrated by Rountree in *The Strand Magazine's* serialization of *The Lost World*.) ... the heads of Doyle's ape-men are depicted as quite unlike that of Piltdown Man.... Doyle's ape-man ...is a veritable beast, an unevolved animal. It's lower jaw is 'projecting,' and those 'bestial and ferocious eyes' peer from under brows that hang 'thick and heavy' like those of any simian. This portrait of a creature supposedly half-man and half-ape has nothing markedly human about it ... Neither, it seems, did he know anything of the crucial straight canine found by Teilhard in that spoil heap. If he had, he would scarcely have specified a 'curved' canine for his own creation."

Doyle meticulously guided Harry Rountree who illustrated scenes from *The Lost World* for *The Stand Magazine* serialization. Here, Rountree depicted Doyle's imagined ape-men with curved (and not straight), upper canines.

And Spencer questions why Doyle would have written excitedly to Dawson in November 1912 about the latter's discovery of Piltdown Man, if he had already known about the events leading to its fabrication. Spencer also wonders, "... why did (Doyle) not spring the trap he is supposed to have so carefully set for Lankester? It is submitted that the reason was because the trap was not set in the first place." (p. 168)

Taking all into consideration, it seems unlikely that Doyle perpetrated the Piltdown fossil fraud. Instead, the real perpetrator may have tried to dupe Doyle, perhaps upon that 1911 occasion when Dawson happened to find an interesting flint arrowhead at Crowborough. Referring to this incident as a "very close call," Walsh states, "What it was that saved Conan Doyle.... from becoming captives of Dawson and Piltdown, in the fashion of Teilhard and Woodward, is not on record. With Doyle, perhaps it was simply a heavy burden of work, both his own writing and his constant public involvement." (Quote from Walsh, p. 198.) (In Peter Marks' novel, *Skullduggery*, Doyle actually solves the caper in Holmesian fashion, proving Dawson's duplicity.)

Who-dunit then?

I think Walsh makes a very convincing case for Charles Dawson. Without recounting all his evidence (and here I do highly recommend his 1996 book *Unraveling Piltdown*), even before Piltdown Man's 'discovery,' Dawson was a known forger of numerous archaeological artifacts and was notorious for plagiarizing others. Evidently, Dawson's motive was to win acceptance in professional scientific circles, possibly in order to keep pace with the accomplishments of his two younger brothers who had become very successful in their respective careers. (See Chapter 11, titled "The Wizard of Lewes" in Walsh's book for the evidence.) It may be that Dawson was suffering mentally and just couldn't help himself from harming others in this regard.

In 1996, charges were 'unequivocally' brought against a former British Museum curator of zoology, Martin A. C. Hinton. During his investigation of bones stained like those found at the Piltdown sites which were found in a trunk bearing Hinton's initials, King's College professor of paleontology Brian Gardiner concluded that Hinton (and not the "incompetent" geologist Dawson, who Gardiner believes was duped by Hinton) was sole perpetrator. In Gardiner's (tepidly convincing) scenario, Hinton used Dawson vengefully to fool Woodward, who was the real intended target. (*Nature*, vol. 381, May 23, 1996, pp. 261-62)

What may have served as Piltdown Man's inspiration? In 1983, Winslow and Meyer suggested that Doyle may have been influenced by an anthropological fake, a "taxidermic frolic" known as "Nondescript" he was familiar with, created by Charles Waterton in 1825. Waterton had attended the same preparatory school as Doyle did, many years later. In Waterton's 1825 book, *Wanderings in South America* he claimed to have killed an ape-man in South America. Fossil foolery of this sort had been played before, notably half a century earlier in France. One celebrated incident involved a faked human fossil jawbone and associated hand axes allegedly found at Moulin Quignon by workmen that was presented to Boucher de Perthes. (See Chapter Three in *The Neandertals: Changing the Image of Mankind* (1993), by Erik Trinkaus and Pat Shipman.)

In any event, it cannot be presently proven that either Dawson or, maybe, Hinton drew inspiration from predecessors' pranks. I highly doubt that Doyle's *Lost World* novel was intended to suggest that its author was a perpetrator, as there were sufficient reasons and inspirations for Doyle to write his novel independently of publicity concerning Piltdown. (See Pilot's and Rodin's introduction to *The Annotated Lost World*, for example.) I also think that Doyle may have been interested in creating an original 'take' on the evolutionary themes previously explored by Jules Verne, H. G. Wells and also by Jack London (i.e. *Before Adam*, 1907). According to paleontologist Jose Luis Sanz, Doyle's purpose in writing *The Lost World* was to prove that it was entirely possible to create new types of adventure novels, other than the old standbys, typically involving pirates and quests for treasure. Although in this case, the 'treasure' sought by Doyle's protagonists was scientific knowledge. (*Starring T. rex*, 2002, p. 16)

So why did I write this? Until I completed this research (in May 2002) I was admittedly wedded to the prospect of Doyle's implication at Piltdown, nearly as much as were those innocent yet entirely duped British scientists who desperately (and with prejudice) wanted to believe in the existence of Great Britain's distinguished Dawn Man. Perhaps even though I ended up falsifying my belief, I initially wanted to believe there could be a connection between Doyle and Piltdown Man that could be deciphered through careful reading of his classic 1912 novel, *The Lost World*.

I can't lay out everything for you here though, and I can't definitively make the case against any possible perpetrators. So, channeling the great Doyle himself, my work is done if I have entertained you or, even better, inspired any of you to delve further.

Chapter Fifteen — *Rise of the Gigantis Imposters*

(With apologies and regret, I am unable to reproduce movie posters and comic book covers referred to here. Seek the original printing in G-Fan no.103 for the visuals.)

Amidst crumbling city buildings stands a colossal bipedal monster that really doesn't resemble Godzilla (even in caricature). Godzilla fan have surely seen these peculiar fellows before, as shown in this vintage poster for *Gigantis the Fire Monster*, released in the USA in May 1959.[1] The artist's caricature of the "Anguirus" monster seen at right is fairly identifiable, but have you ever wondered who, or what, the monster standing on the left really was … or whatever 'happened' to him?

In the background are two volcanoes – one of them erupting, possibly a vestige exemplifying the movie's abandoned American film title and concept, as conceived by Ib Melchior, "The Volcano Monsters." But the bipedal dino-monster as shown is also erupting flames, not only from its mouth but *also* issuing laser-like from its nostrils and eyeballs (i.e. like the 'Anguirus' caricature). Huh? And its warty skin texture is not exactly "godzillean" either. Most noticeably, however, those "fins" are entirely wrong. They're spiky, resembling those of an iguana, or more to the point much like those seen on older dinosaur restorations and (at least) in three predecessor dino-monster films. It's difficult to discern whether fins seen on its tail end are double-rowed, or if only a shadow of a single row is projected onto background smoke.

So, maybe this is "Gigantis" perhaps, (but certainly not Godzilla). Let's see where this curious bipedal creature came from, and where it went.

In *G-Fan* no.98, in an article titled "Triumphant Triumvirate: Godzilla's Dinosaurian 'Progenitors'," I outlined how certain paleoartists' famous dinosaur restorations proved pivotal in designing the original Godzilla suit for 1954's *Gojira*.[2] In particular, as noted, there were several inspirations supporting Toho's decision for the ostentatious osteoderms (i.e. "fins") adorning Godzilla's backbone, including Rudolph Zallinger's famous *Tyrannosaurus* painting from the *Age of Reptiles* mural, a small version (or "cartoon") of which was reproduced in *Life Magazine* (Sept. 7, 1953). Godzilla's suit was also influenced by contemporary restorations of *Stegosaurus* and *Iguanodon*. These real dinosaurs were melded into an artistic conception that synergistically 'worked' on screen like no dino-monster suit or model design had ever before (or since). But what about the curious pseudo-godzilla shown in the aforementioned 1959 movie poster (for which Warner Bros. extended copyright for newspaper and magazine reproduction)? What's his story? Well – in a sense, it goes back, way back.

While of course artists added (reptilian) spiky vertebral scales in many of the earliest known dinosaur restorations, the practice became more of a cliché during the turn of the 20th century. For instance, in an 1886 publication by astronomer Camille Flammarion, an artist named "Motty" engraved a picture of a gigantic *Iguanodon* with triangular scales adorning its vertebral column, standing in the corner of a city block, peering into a 7th –floor window. No flames or erupting volcanoes indicated there though. Then by 1897, Charles R. Knight began restoring dinosaurs resplendent with a short arrangement of triangular backbone scales for articles in *Century Magazine*, including his famous pair of "leaping" *Laelaps*. We see this appearance carried over into restorations of the North American dinosaur *Ceratosaurus* as well, such as in a 1901 restoration produced by Knight and Gleeson for McClure, Phillips & Company. *Ceratosaurus* is seen again with its short array of requisite vertebral scales in W. D. Griffiths' early silent film, *Brute Force* (1913). This monster was crafted not as a stop-motion figure, but rather as a "full-sized dinosaur mock-up."[3]

Then two carnivorous dino-monsters were introduced in popular culture, both of which sported the spiky backbone trait. The first of these was bipedal in stature, and the second (appearing in 1953) was a quadruped. Both of these monsters are suggested to have possibly inspired *Gojira*, although clearly the latter did more so than the former. What are these monsters? Respectively, they're the mutant *Tyrannosaurus* menacing Metropolis in Max Fleischer's *Superman* cartoon "The Arctic Giant" (1942) and Ray Harryhausen's "Rhedosaurus"—true stop-motion animated star in *The Beast From 20,000 Fathoms*. Examine the vertebral spines on the 1942 creation. Doesn't Max Fleischer's prehistoric creature rather resemble the monster seen in the 1959 "Gigantis" poster (minus the flames and smoke)?

Oddly enough, we see another similar design as published in Donald F. Glut's *The Dinosaur Scrapbook* (1980, p.148), where we spy "Godzilla as he appeared in an early preproduction drawing, showing a somewhat different version of the Japanese monster." Here Godzilla, shown in a death pose, sports a single slightly recurved (e.g. like Gfantis) row of triangular dorsal scales adorning his backbone.[4] The creature in this pre-production drawing more closely aligns with another early design based on the Zallinger tyrannosaur-esque 'prototype' (described in *G-Fan* no. 98). Fans of the first Godzilla movie (both

Japanese and US versions) also would recall that a re-drawing of Zallinger's *Life Magazine* tyrannosaur, replete with short triangular dorsal scales, appears in the brief slide show presented to Japanese officials in the Diet Building following Dr. Yamane's visit to Odo Island. Alas—we may never exactly know the full story of how the short triangular "fins" became translated into the tall, magnificently rounded, sculptured and highly contoured triple-row of 'stegosaurian' osteoderms evident in final suit design. A stroke of genius!

Another Godzilla movie poster was created by Ren Brown for USA release of 1964's cryptically titled *Godzilla vs. The Thing*. While in this particular poster, Godzilla is named as one of the film's major combatants, the strange dinosaurian really doesn't resemble any 'godzilla' we know and love. Notice fins on the monster. To be fair, the largest, most distinctive dorsal fins aren't visible in this poster. But those which are depicted do rather resemble dorsal scales evident in the 1959 "Gigantis" poster, as opposed to any self-respecting Godzilla suit. The creature shown in the 1964 movie poster also vaguely resembles the Godzilla seen in the alleged pre-production drawing published in Glut's book. Of course, artists take liberties and utilize artistic license when creating movie poster art, but as Godzilla fans everywhere realize, such *abstract* visuals purportedly representing our favorite dino-monster would no longer pass muster.

The first Godzilla film I saw at a movie theater was *King Kong vs. Godzilla* during its June 1963 release. Given that the first Godzilla perished at the end of 1954's *Gojira*, and that the second Godzilla was "Gigantis," then arguably the reptilian which King Kong battled was Gigantis—not Godzilla. This was one of the most exciting episodes of my early life …. which seems strange to admit now. "Kong vs. Godzilla" perpetuated the theme of ape pitted against reptilian, which was conceived three decades earlier in RKO's *King Kong*.[5] Kong represents humanity, and as W. J. T. Mitchell claims such conflict "evokes the ancient antagonism between the cold and warm, reptile and mammal, dry bones and living creatures."[6] That insight may suffice as a 1933, Depression era, filmic interpretation, but why, 30 years later, would Toho feature giant wrestling, tussling prehistoric monsters battling on screen? Well, besides the drama of witnessing all that fun-filled carnage and horsing around, those giant monsters supposedly mirrored two super powers sparring for nuclear and geopolitical supremacy during a high-tensional, angst-filled early Cold War period.

Whatever—the theme of theropodous dinosaur vs. giant ape continued throughout the 1960s although in other stories and guise, and as we shall shortly see while featuring dino-monsters quite similar to the "Gigantis" creature visible in the cited 1959 poster. Examine the cover of Charlton comic, *Konga* no.23, issued in November 1965. Doesn't the dino-monster named the "Uuang-ni," closely resemble the Gigantis monster seen in the 1959 poster! About the only major difference is that the Uuang-ni sports a ceratosaurian-like horn on its snout. In this tale, (art by "Montes – Bache"), we see a Kong-like theme merged into sort of a "volcano monsters" story. Giant ape Konga had battled giant prehistoric creatures before in "lost world" island settings (as in *Konga* nos. 2 and 3, August and Oct. 1961, respectively), but this new story was clearly influenced by Toho's blockbuster, *King Kong vs. Godzilla*. And that dino-monster evident in " Masulli + Rocke's" startling cover surely was inspired by that 1959 "Gigantis" volcano monsters movie poster. Interestingly, there are *two* fiery Uuang-ni monsters residing within the active volcano that emerge from within to terrorize the natives, which heroic Konga ultimately dispatches. Notice how the Uuang-ni seen on the book cover even snorts fire from its enlarged nostrils, much like the Gigantis visible in the 1959 poster (but not from its eyeballs – which of course would be impossible, right?). Furthermore, examine how similar are those pointy, prominent dorsal spines to those seen in the poster. And so pseudo-'Gigantis' rose again, only to be defeated by a 'Kong' imposter!

1965 was certainly a banner year for a 'return' of the 1959 Gigantis movie poster monster, although as seen in comic books. For besides its appearance in *Konga*, a variant of that stylized dino-monster was also seen both in an issue of *The War That Time Forgot* (Nov. 1965) and *Superman's Pal – Jimmy Olsen* (no. 84 April 1965). In the "War" series published by DC comics, US military soldiers continually encounter lost world islands populated by scores of humongous dino-monsters and other prehistoria, usually in the Pacific. One character they befriend is an enormous white-furred gorilla who appears in several issues. But in one story, titled "Terror in a Bottle," (*Star Spangled War Stories* no.123), the white ape returns to fight yet another scaly-backed, 'gigantis-like' dinosaur (which however does not breathe flames).

In "Jimmy Olsen," a giant gorilla from Krypton named Titano is transported using a "time-and-space-warper" invention to a volcanic Pacific island where the cub reporter endeavors to make his own monster movie. But the only decent footage is obtained after another giant creature from Krypton, the "Flame Dragon," also materializes from space-time to combat Titano. Thus, on the cover of the issue readers spy a highly suggestive, colossal scene reminiscent of *King Kong vs. Godzilla*, featuring a flame-breathing dino-monster highly resembling the 'Gigantis' creature seen in the 1959 movie poster. However, as it turns out, artists failed to show the Flame Dragon's wing appendages on the comic book cover. Within pages of this issue it is apparent that the flying Flame Dragon from Krypton resembles neither Gigantis nor Godzilla.

Toho movie aficionados may also recall King Kong's mighty battle with Gorosaurus in *King Kong Escapes* (USA release 1968). Close inspection of the Gorosaur suit reveals a series of short dorsal fins situated along the spine and tail.[7] Charles Knight, famous for his 1897 leaping "Laelaps" duo painting, would probably have appreciated Gorosaurus' nimble combat theatrics.

Perhaps I'm just over-sensitized to noticing vertebral spines on dino-monsters of late because of all those spines I sculpted on a 14-centimeter tall Gfantis model in 2012, which seemingly took half of forever to complete. (See *G-Fan* no.101, and Chapter Eleven.) It would certainly have been easier to fashion a single row of such spines on a dino-monster model! While it may seem that that signature Gigantis volcano/fire monster mien has faded into obscurity, peripheral searching reveals this is not the case. For I own a plastic 1980s dino-monster toy whose appearance strangely echoes the creature in that old 1959 poster. Evidently photographers have even incorporated the toy into odd forms of 'paleoart,' as exemplified in a figure printed in Mitchell's *The Last Dinosaur Book* (1998), as you can in the figure shown.[8]

Perhaps further distinction should be made between variations in the nature of dorsal appendages brandished by dino-monsters. We have seen short triangular "scales" in some (as in Zallinger's *T. rex*) versus more elongated, stiffened scales (longer than a school bus) as in the Gigantis creature featured in the 1959 poster. And then there are outright "bony spikes" situated over the dorsal area such as on the creature named "Gaw" created by artist Joe DeVito for *Kong: King of Skull Island* (2004), written by Brad Strickland with John Michlig, or even the infamous Zilla (also known as "GINO"). The bony spike ambiance takes us back to the early 1850s, with Benjamin Waterhouse Hawkins' Victorian visions of the armored dinosaur *Hylaeosaurus*, which as I mentioned in a *G-Fan* letter of long ago, in general outline rather resembles Toho's kaiju monster Varan.[9] There may be other interesting examples as well.

As mentioned, the 'Gigantis' dino-monster was most recently spied lurking about in Mitchell's book …. should we fear its inevitable return?

Notes to Chapter Fifteen: (1) In the 1959 poster the spelling of the artist's name, signed at lower left, is hardly visible in the poster, yet seems to be something like "stenerger." What was "Stenerger" (sp. ???) thinking when he made this dramatic portrayal? And does anybody know anything about him? (2.) Reprinted in Chapter Ten of my 2017 *Dinosaur Memories II* book; (3) See p. 118 in Donald F. Glut's *The Dinosaur Scrapbook*, 1980; (4) In personal communication dated March 2, 2013, Glut stated that this picture was received from writer and fan Saki Hijiri. Now Glut is not certain it's a true pre-production drawing; (5) Arguably, the essence of *KK vs. Godzilla's* climactic battle scene had been played out already in *Unknown Island* (1948). But you may read my thoughts on this 'ceratosaur-tyrannosaur' vs. giant furry sloth scene in *G-Fan* # 76;.(6) Quote is taken from W. J. T. Mitchell, *The Last Dinosaur Book: The Life and Times of a Cultural Icon* (1998), p. 252; (7) See, for example, photo of Gorosaurus in Glut (1980), p. 157; (8) See Mitchell (1998), p. 204 and title page; (9) See my letter to *G-Fan* # 61, p.28. Also besides dorsal spines, scales and spikes, there's the ancillary matter of dinosaurian humps and sails, commented upon further in my 2013 *Prehistoric Times* no.104 article, reprinted as Chapter Twenty-two of *Dinosaur Memories II* (2017).

THREE

Front cover of Konga no. 23 (November 1965). Artwork exudes the primeval battle between apes and dinosaurians Copyright Charlton—author's collection. See Chapter Fifteen for more.

Apes & Godzilla (redux)

I say "redux" here because I've addressed this topic several times elsewhere. It's a theme dear to my heart possibly because the very first Godzilla movie I saw in a theater was *King Kong vs. Godzilla* back in 1963—more on that experience later in Chapter Twenty, though. By 'apes' I mainly mean 'gorillas' of paleontological sci-fi, spurred by Paul B. DuChaillu's (1831?-1903) 'discovery' of the Gorilla in 1859, the same year that the 1st edition of Charles Darwin's *The Origin of Species* was published. Thereafter, the gorilla became a central topic of late 19th century evolutionary conversation and debate, perhaps culminating decades later with the most famous paleo-ape/'human' of them all—Kong.

In particular, *King Kong vs. Godzilla* mirrors more than the then still-recent World War conflict between the U.S. and Japan. This movie is also symbolical of the ideological clash between (and historical movement from) a pre-nuclear age and a dawning 'atomic' period. A clash between a time when mankind methodically exploited and 'conquered' once 'discovered,' pristine, 'lost' horizons of the planet—and a time of a dawning 'atomic world'—when man's secretive laboratory discoveries reached a crescendo, taking us to the brink, threatening to annihilate civilization in a single day.

This section encompasses several articles reorganized into book chapters that I've sprinkled through several fanzines over the years. These articles acknowledge a perceived, metaphorical 'battle' for eco-planetary supremacy staged between our close evolutionary cousins (apes—representing our own dim paleo-history), and the fierce, savage reptilian horde whose unlucky demise 66-million years ago cleared a path for our kind. For the "Part 1" referred to in Chapter Nineteen's title, readers must refer to my *Dinosaur Memories II* book (2017).

Just as there was a tie-in novelization, written by Greg Keyes, for Legendary's *Godzilla King of the Monsters* movie printed in 2019, indeed, may we perhaps anticipate the first official *King Kong vs. Godzilla* novelization to be published in 2020—hopefully coinciding with release of the titular blockbuster film?

So, hang on to your seats; here we go again!

These chapters originally appeared in the following publications:

Chapter Sixteen—previously not published.
Chapter Seventeen—*G-Fan* no.88, Summer 2009, pp.26-30.
Chapter Eighteen—*Mad Scientist* nos. 29-30, Summer 2014 and Winter 2015, (respectively, pp.25-39 and 35-43).
Chapter Nineteen—*G-Fan* no.121, Fall 2018, pp.16-19.
Chapter Twenty—*G-Fan* (in press as of 2019).

Chapter Sixteen — *Yet Another Tale of 'Lost World' Paleo-Monsters*

By now the fantastic idea of prehistoric monsters that are anachronistically prehistoric, in the sense that they're alive in our present when they shouldn't be here because they should be extinct, has a long and vivid past, pre-dating 1950s filmic phenoms Godzilla or the Rhedosaurus, or Superman's awesomely huge 1942 foe—the dinosaurian Arctic Giant. Dinosaurs, dino-monsters or 'kaijusaurs' are usually imagined terrors attacking mankind in our urban centers, but occasionally we see other varieties such as marine plesiosaurs, Devonian gillmen, winged pterosaurs and even assortments of paleo-mammals. In movies, one of the most famous denizens of the savage past to attack a city-scape was a stop-motion animated *Brontosaurus* in 1925's *The Lost World*. Not long after, Kong—viewed as a *prehistoric* ape in Delos Lovelace's 1932 novelization—ravaged New York City in RKO's *King Kong* (1933).

Over a century's worth of published stories (and artwork) have also seduced readers into suspending disbelief when it comes to prehistoric monster fare. One long forgotten (until recently!) example was a 1906 *Chicago Tribune* farce printed on April Fool's Day. Another is a newly discovered curiosity, "Menace of the Monsters," written by John Hunter for 1933/34 issues of a British zine titled *Boys' Magazine*. This 'lost' tale (not the only such Kong-inspired fiction story then), was salvaged in 2017 from the abyss of time by giant- and crypto-monster scholar, Justin Mullis to whom I am indebted for apprising me of this story. However, the first post-world war 'modern' to reflect upon this story may have been E. S. Turner who wrote of it in his 1948 book, *Boys Will be Boys* (reprinted in 2012).

Boys' Magazine was intended for adolescent young men who fancied an absorbing yarn or two. Hunter's dino-monster "Menace of the Monsters" tale was serialized into a dozen weekly installments, (i.e. beginning in November 1933, continuing into early 1934). The story kicked off urging "Chums" to enjoy an "Awesome! Thrilling! Stupendous! Even greater than King Kong … breathless spectacle tale (that) will hold you spellbound." Here I'll mainly riff mainly on installments I found independently of Mullis in 2018, as well as accompanying artwork. This will offer a reasonable gist of the overall story. Admittedly, I would not have known of this title were it not for Justin.

While not the most outrageously well written example of paleo-fiction, Menace of the Monsters is not lacking in charm. And so we are bombastically introduced to the story of four adventurers—Professor Laban Twick (so absent-minded he doesn't know exactly when to fear), "Bigshot Bruce" the intrepid big game hunter, and (of course) two young men, Harry Langham and Pongo, his (not so funny, really) "comical pal." They have somehow rather miraculously captured an astounding menagerie of prehistoric specimens from the Place of Mists, a mysterious place where "life had stood still," never geographically identified but which seems to have been situated somewhere in the southern hemisphere. Eventually, during a storm of the century, their ship "913" freighting the "most terrible cargo that ever was loaded under hatches," smashes along the British coast, freeing the monsters from their cages for an attack upon civilization. As Turner interestingly opines, dinosaurs have been symbolically substituted for a German army invasion of the island. Such a metaphorical geopolitical conflict is pursued, as (presumed) English scientist "Twick" pooh poohs theories of his chief German scientific rival, Professor Waldenheim.

It isn't readily apparent who drew the imaginative illustrations throughout the series, or the cover image for the initial November 1933 installment. (See Frontispiece.) This cover image showing a scaly carnivore—probably the story's Tyrannosaur—shown gigantically much larger than in life with a low column of jagged vertebral spines, smashing through Sydenham's Crystal Palace—recalls advertising art for 1925's *The Lost World*, in which another attacking theropod dino-monster is portrayed. Interestingly, the Crystal Palace grounds are sort of a 'ground zero' for the dinosaur craze because it is where the very first attempts at sculpting life-size dinosaur statues were made; those antiquated, Victorian dinosaurs, made by Benjamin Waterhouse Hawkins during the 1850s, remain on outdoor display there. Other illustrations in Part 1 show a number of dinosaurs—copied from Charles R. Knight's magnificent late 19th century paintings—wallowing in their swampy misty land, prior to capture. In Part 1, the giant gorilla is also illustrated in full twice—once caged in the ship's hold amidst snarling reptilians each in their cramped, separate electrified cages, and in a later scene after it escapes, shown destroying the ship's turbine.

Other drawings show a plated stegosaur charging the adventurers before leaping onto another smaller ship, and then drowning at sea. After the ship grounds along the British coast near Dorset, members of the reptilian horde are shown in separate installments as the heroes battle each, now widely separated, dino-monster… to the death. So eventually we see a huge dinosaur-attacking snake, and a horned dinosaur based on Knight's 1897 painting of the

"Agathaumas" (also a dinosaur animated for *The Lost World*). In one installment this overly gigantic horned dinosaur is depicted ravaging a soccer field. (Those Agathaumases and stegosaurs keep turning up in subsequent segments, however.) We also witness a caveman menace who quickly becomes an ally, a brontosaur swimming at sea, and a croc with a gigantic gaping maw, called a "pelagosaur." Quaint illustrations such as these clearly bearing the mark of early fantastic Kong and Lost World-fueled dino-monster lore and fascination must have certainly seemed exciting to readers of this strange tale!

It's the *Stegosaurus* that breaks loose first while the ship is at sea—an early death. In fact, the living cargo is likened to having "Death himself as a grim passenger." When, during the height of the storm, concern that specimens might suffer or die as they sustain electrical shocks from their cages, a decision is made to shut off the electricity—an ill-fated mistake. For as the Professor implores (as science professors are wont to do since time immemorial), if the animals die their scientific value would be diminished. And so, with the electrical barrier neutralized, our giant ape tears itself free, killing several men and then with its fists wrecking the ship's gyroscope. At the end of Part 1, when one of the dinosaur escapees is sighted edging toward land through the pounding surf, the Professor comically declares, "That's a dinosaur. I think we'd better run." Yes—that's what men have been doing ever since on such occasions! But there's also an entire herd of Mammoths roving about to flee from too. One wonders, how four people managed to capture all these animals in the wild, let alone squeeze them into the hold of one ship!

The main "menaces" in Part 2 are a giant snake, and a trio of those horned Agathaumases. Here the Professor calmly decides to go forth and simply "have a look" as the latter raid a soccer field. Meanwhile one of the horned monsters overturns a steam locomotive even though it's soon dispatched by "TNT-charged bullets" issued from Bigshot Bruce's rifle. Then in Part 3, it's divulged that Bigshot Bruce wants utmost to kill the … "great tyrannosaur," so far untracked—on the lam. For the tyrannosaur was "The king demon of all those monstrous creatures that the wreck had let loose on the land," and later, "a thing more hideous than the most evil dream that ever man dreamed."

Another stegosaur demolishes a country inn, but then disappears. But the main battle staged in this particular segment is between a shambling, axe-wielding "Man Thing"—prehistoric man versus an enraged *Triceratops*. Next, a swimming brontosaur is torpedoed at sea by a navy destroyer, and the cave man takes off into the cold night. Then we learn that the great lizards are attracted to London's 'pea soup' fog, likened to the monsters' natural environment in their enveloping Place of Mists, yet also alluding to the famous opening paragraph in Charles Dickens' 1850s novel *Bleak House*, in which one can almost imagine *Megalosaurus* trodding up Holburn Hill.

After escaping the 120-foot long Pelagosaurus's "gaping jaws of death," the heroic band surmises there still remain a *Dryptosaurus* (made widely known via Knight's 1897 painting) a *Diplodocus* and the tyrannosaur (also depicted in paintings by Knight now famous in pop-culture) to reckon with. These three veer toward London's famous foggy atmosphere. While long-necked *Diplodocus* is hemmed in by "cliffs" of tall buildings along city streets with the hungry *Dryptosaurus* lunging at its heels we read, "Away north, another shape was coming down into the fog from beyond the heights of Hampstead. The Master Slayer was moving down on London, coming into the fog he loved. The great tyrannosaur was descending on the Metropolis." And as the editor advertises "Death and destruction sweeping on London! Look out for the crashing climax of this mighty monster yarn next week." And I would be remiss if not divulging how it all ends.

So with dryptosaur, *Diplodocus* and tyrannosaur all converging on London, the two former menaces collide on Westminster Bridge, "They fought a fight which the Place of Mists had seen scores of times, and which no human being saw that night," (i.e. because of enveloping fog). As a metaphorical storm gusts through the city dispersing the fog, *Dryptosaurus* emerges from that battle to fight a climactic, statues-wrecking battle with the tyrannosaur in Trafalgar Square! An army fusillade helps to extinguish the pair.

Notice we haven't read of the mighty ape since Part 1? So where is that alleged yet oh so elusive 'giant ape vs. T. rex' segment—presumably inspired by that dramatic scene in 1933's *King Kong*? Yes, Justin planted this provocative seed in my mind in 2017, and I like to believe their mythical battle did occur. There are *two* apes in the tale, and they tear up Tower Bridge in London before meeting an evident demise.

Another clue, however, comes from an older obscure book pried from my bookshelf—James Van Hise's *Hot-Blooded Dinosaur Movies* (1993). Therein, on page 54 one spies the cover of *Boys' Magazine* (Oct. 28, 1933) advertising a *King Kong* feature story, printed just weeks prior to Menace of the Monsters. Then on page 65 of Van Hise's book, there's an image of a Kong-like ape battling a possible tyrannosaur (or is it the dryptosaur) amidst wreckage of London's Tower Bridge. (See back cover.) Artwork on this latter image seems quite similar to that appearing in the Menace of the Monsters. (See Back Cover.)

According to Mullis who had independently discovered these images beforehand, the latter 'ape vs. theropod' image was originally printed at the end of the Kong story, but as an advertisement for the forthcoming Menace of the Monsters tale, which is probably where and how Van Hise originally found it too.

This summary is rather cryptic and fragmentary, but scholar Justin will likely complete the picture for us.

Despite its arresting title, Menace of the Monsters is light on suspense—not even, say, to a 'Hardy Boys' baseline. And there isn't much character development: no lead female characters. Plot-wise though, it carries interest. No—despite the incomplete nature of the story, this paleo-monster tale is not exemplary. Clearly not as good as Delos Lovelace's masterful 1932 *King Kong* novelization. But I wouldn't dismiss it so readily if I were half a century younger today.

Chapter Seventeen — *Komodo: Ancestral Daikaiju*

The greatest giant movie monsters are creatures like Godzilla, Kong, the Rhedosaurus, and hordes of daikaiju, right? But what's their common denominator of inspiration? Dragon myths? Fossil bones of extinct elephants found by the ancients, transformed into legend? *Tyrannosaurus rex*? Well, close and all very relevant, perhaps, but that's not quite the whole story. Here I suggest the stem basis and documented source of inspiration for many of the giant *movie* monsters was the extant monitor lizard—the "Komodo Dragon," *Varanus komodiensis*. These fascinating 'prehistoric' creatures link dragon lore, dinosaurs and daikaiju, with "Lost World," cryptozoological elements. Savvy readers already realize that *King Kong* (RKO, 1933), for example, was directly inspired by Komodo dragons displayed in New York, an anecdotal story which we'll come to shortly. And to a certain degree, *Gojira* (Toho, 1954) was, in turn, founded upon *King Kong's* influential box office successes.

Few have considered how Komodo Dragons are in a sense 'ancestral' to the greatest of classic, giant movie monsters. So has 'Komodo' been entirely neglected? Hardly! Furthermore, fittingly, yet perhaps surprisingly, Komodo Dragons claim traditional monster status too, that is, in fantasy literature and film, which we shall also explore. And so, finally, it is time to pay homage to that often overlooked, ancestral (pseudo-) 'daikaiju' itself—the Komodo Dragon!

Zoologist, explorer-naturalist, W. Douglas Burden, documented his publicized discovery of Komodo Dragons in a number of publications, most noteworthy being his 1927 book, *Dragon Lizards of Komodo: An Expedition to the Lost World of the Dutch East Indies.* Impressed by China's national draconian emblem, Burden suggested that, "it is just possible the Komodo lizards are the very ones which have given rise to all the dragon mythology." Actually, western scientists had known about the dragons of Komodo since 1910 from the rumors of Malay pearl-divers startled to see them roaming about; the first scientific description was written by P. A. Ouwens in 1912, founded on a single dead (shot) specimen shipped to Buitenzorg, Java. But Burden yearned to see them alive in their native habitat, and also to bring back examples to New York

City. And so, in very roundabout fashion—on the way, linking up for a time with Roy Chapman Andrews in China—he eventually sailed to the volcanic island chain, 'where there be dragons,' arriving in June 1926. He exclaimed, "With its sharp, serrated skyline, its gnarled mountains, its mellow sun-washed valleys and the giant pinnacles that bared themselves like fangs to the sky, (Komodo) looked as fantastic as the mountains of the moon ... We seemed to be entering a lost world." *That* movie, based on Arthur Conan Doyle's 1912 novel, had been released the year before.

Setting out bait (rotting boar corpses) to attract *Varanus*, the huge lizards began to arrive. Burden described his first sighting with wonder ..."It was ... perfectly marvelous ... a primeval monster in a primeval setting ... Had he only stood up on his hind legs, as I now know they can, the dinosaurian picture would have been complete." Burden caught (and shot) several dozen specimens—several intended for an American Museum display diorama. Two were shipped live to the Bronx Zoo. But the live specimens perished in their confines. As Burden lamented in 1927, "Perhaps ... it is just as well, for any reptile that has been caged is but a poor shadow of his former self. After watching these great carnivores in the wilderness of romantic Komodo, it was painful to see the broken-spirited beasts that barely had strength to drag themselves from one end of their cage to the other." Then somewhat prophetically, he added, "Perhaps as in the case of many mammals, *Varanus komodoensis*, in order to survive, demands the freedom of his rugged mountains." Sounds like an epitaph for Kong.

I lack sufficient space to recapitlate Burden's expedition, a spirited adventure which caught fervor of the western world. Remember, this was a time during *National Geographic's* heyday when adventurers traveled intrepidly to exotic, 'lost' places of the globe searching for crypto-species, at times with astounding success. Nor may I outline the Komodo Dragons' natural history. Here, it will suffice to state that *Varanus komodiensis* (typically 10 feet long and weighing 150 pounds) is now a protected species, representing vestiges of a 40-million-year-old lineage originating in Asia, with older marine ancestors, Mosasaurs, that swam in Mesozoic seas alongside extinct plesiosaurs and ichthyosaurs.

Fortuitously, Merian C. Cooper, *King Kong*'s producer-director, lived in New York City, near Burden. Cooper's creativity was influenced by several factors surrounding the Komodo story, which sent fantastic notions swirling through his mind. While Cooper, an adventurer - explorer in his own right, had been intrigued since childhood by Africa's jungle apes, and later by Charles R. Knight's prehistoric life murals in New York City's American Museum, Burden's "King of Komodo" story proved *most* inspirational. Captured wild Dragons, later languishing and eventually expiring within the confines of civilization, melded with the idea of a woman (Burden's wife, Babs) joining the expedition, exposed to sudden peril. Indeed, Burden's Komodo story transformed into Kong's incarnate!

In a letter addressed to Burden, Cooper stated:

"When you told me that the two Komodo Dragons you brought back to the Bronx Zoo, where they drew great crowds, were eventually killed by civilization, I immediately thought of doing the same thing with my Giant Gorilla. I had already established him in my mind on a prehistoric island with prehistoric monsters, and I now thought of having him destroyed by the most sophisticated thing I could think of in civilization, and in the most fantastic way ... to place him on top of the Empire State Building and have him killed by airplanes."

Yes, Kong's "whole basic story" *is* that of the fated Komodo Dragons.

Thematically, Cooper's Kong tale is draconian. The idea of a bound female (Ann Darrow, played by Fay Wray) 'presented' in sacrifice to the huge monster borrows from the Greek myth of Andromeda, a beautiful woman chained as sacrifice to a Dragon but rescued by the heroic Perseus. (Perseus also slew the Gorgon.) Essentially then, Kong—rather than prehistoric reptiles he defeats on Skull Island—has become a metaphorical 'dragon,' not unlike those from the lost 'prehistoric' island of Komodo.

Accordingly, perhaps the first 'komodo-esque' monster to appear in film was in *King Kong* (RKO). Do you recall that scene where knife-wielding Jack Driscoll (played by Bruce Cabot), hiding in a cave, slices a thick vine while Kong probes for him from above the gorge? (Also see Chapter Seven.) This happens just after sailors are shaken from the log-bridge and prior to the great *Tyrannosaurus* battle scene. Driscoll cuts that vine because a huge, spiny reptile is incongruously climbing up toward him from the ravine. Allegedly, this reptile sequence was spliced into the film after it was decided that the giant spider sequence documented in Ruth Rose's script and Delos Lovelace's 1932 novelization of the script, was either too gruesome for audiences, or that it slowed down the movie pace. And so instead of a spider, Cooper authorized substitution of that strange reptile lacking hind limbs, with only two claws on each hand. Is that weird reptile 'substitute,' which few have understood the basis of before, until now—a 'nod' to Burden's Komodo Dragons, homage memorializing an inspirational factor underlying Cooper's film? Were its hind limbs missing because, in deleting the spider sequence, last minute sculpting alterations were necessitated; therefore a hastily prepared unfinished 'dragon' was all they had time for?

Or quite possibly, Kong's weird reptile wasn't the first cinematic Komodo-like 'dragon' after all. Readers may refer to a photo of a draconian illustrating my article in *G-Fan* no.84, (Summer 2008; "Prehistorical Daikaiju Evolution," p. 46). This was a 'still' of a mechanical special effects dragon, defeated by Norse hero Siegfried in Fritz Lang's 1924 silent movie classic, *Siegfried* (Part 1 of the *Die Nibelungen Saga*). Siegfried's impressive, spiny quadrupedal dragon foe rather resembles an enlarged variety of extinct monitor lizard known as *Megalania* (a genus related to *Varanus komodiensis* which we'll come to shortly). Although this film predates Burden's Komodo expedition by two years, in time, far greater movie dragons would come, including those of the magnificent 'drago-dinosauroid' variety, (i.e. Godzilla).

There were a number of significant factors leading to Godzilla's invention, RKO's *King Kong* being one of them. Tomoyuki Tanaka (Godzilla's creator) was most likely highly motivated by the big ape's persona. We read in *The Official Godzilla Compendium* (1998, pp. 11-12) that:

> "Why not make a film about a giant monster? There were several reasons why Tanaka's imagination went in this direction. In 1952, the grandfather of all giant monster films, Willis O'Brien's *King Kong* (1933), was re-released in Japanese theaters. The box-office earnings were spectacular, and the film's profits convinced Warner Brothers to release *their* new monster feature, *The Beast From 20,000 Fathoms*, in Japan the following year—with equally spectacular results."

One wonders whether, if Tanaka had never seen Kong (or Ray Harryhausen's Rhedosaurus, whose quadrupedal stance curiously mimics that of the Komodo Dragon) 'live' on film, would there have been a Godzilla? Special effects master Eiji Tsuburaya also held special fascination for *King Kong*.

Two other early 1950s curiosities link Godzilla to Cooper's "prehistoric gorilla" or "beast-god" Kong, and hence Burden's Komodo. In Toho's *Gojira* (1954), we learn that Odo islanders "identify (Godzilla) with a legendary monster of the past that could be appeased only by virgin sacrifices, which puts him squarely in the tradition of the fire-breathing dragons of folklore." (John J. Pierce quoted from *The Official Godzilla Compendium*, Random House, 1998, p.17) Such Kong-ish scenes, with Godzilla claiming a virginal prize, were never filmed, however. Also, according to Toho folklore, the name "Gojira," denoting "gorilla-whale" (blending 'gorira' - gorilla, with 'kujira' - whale), later anglicized to "Godzilla," was borrowed from a burly Toho press agent's (or stagehand's) nickname. Sci-fi novelist, Shigeru Kayama—who authored the original "Gojira" story—also recalled being told that the monster was supposed to be marine, "a cross between a whale and gorilla." So in name, Godzilla/Gojira melds modernity's greatest predecessor terrestrial and marine monsters—Kong himself and Melville's archetypical leviathan, "Moby-Dick." (Thematic connections can also be made associating Moby-Dick with dragons, and therefore in derivative fashion linking Leviathan to both Kong and Godzilla.) As Steve Ryfle suggests in *Japan's Favorite Mon-Star* (1999, p. 23), it is quite possible that the Toho employee allegedly lending the "Gorilla-Whale" moniker never existed.

Let's examine several recent science fictional/horror examples reflecting the Komodo Dragon's monstrous, yet unheralded legacy.

Noted science fiction writer Robert Reed's raw-edged "The Dragons of Springplace" (1996) merges modern Godzillean angst, "corrosive effluvia of the Nuclear Age—old reactor cores, dirty plutonium, dismantled bombs," with Arthur Conan Doyle's myth of the "lost world" plateau. Here, in Springplace, there be Dragons! Giant mutated Komodo Dragons, that is, which grow to sixty-feet long, bioengineered for survival in a superheated, radioactive, faux-primeval ecosystem. Springplace was constructed in our future (although in the story's distant, two-hundred thousand year old, past), as a repository for nuclear wastes. It is a "sarcophagus of concrete and glass" of geological proportions devised to seal in contaminants. Through the ages, more vaults were built overlying the older ..."each new sarcophagus thicker and thicker; the Earth's crust eventually bowing under the weight of so much glass and dirty plutonium."

If civilization should fail, then future generations of humans would be thwarted from scaling the polluted plateau, or mining its awful substances, through a number of engineered hazards including sculptural warning signs (i.e. in case people could no longer translate written 'ancient' languages), the presence of "crocodiles bigger than a bus" and of course, mutated Komodo Dragons. Due to their cold-blooded metabolisms, a population of hundreds of giant Dragons can be supported on Springplace, guarding mankind's "poisons." A ruffian named Daniel, who has assimilated a 'dragon' persona, rescues humanity from evil-minded terrorists who threaten to blow up the plateau, thwarting their aim to scatter plutonium dust throughout the atmosphere with high tech nuclear weapons.

Komodo's shade figures in TriStar's *Godzilla* (1998) as well. Near the beginning of that film, audiences spy several reptiles 'watermarked' with imagery of maps, in particular one showing French Polynesia. While most of the reptile footage features iguanas (and an egg nest), for about a second we also see a strolling Komodo Dragon. A hydrogen bomb test is shown, suggesting that one of the reptilian eggs exposed to radiation in French Polynesia hatched a baby Zilla. For the record, French Polynesia and the island of Komodo are thousands of miles apart in the Pacific Ocean. And while Komodo Dragons were indeed briefly shown in the film, it is more likely that the producers were intending for their monster to be a mutated iguana rather than a 'Dragon.' This ambiguity persists in H. B. Gilmour's 1998 novelization of Dean Devlin's and Roland Emmerich's screenplay, intended for younger readers; on page 1, Gilmour refers to "the six foot-long reptile" whose nest becomes irradiated.

A "normal (varanid) monitor lizard" was also progenitor to a loquacious, 500 foot-tall, 50 ton "radioactive cancer" named Gojiro in Mark Jacobson's witty novel, *Gojiro* (1991). An atomic bomb test transformed his isle into "Radioactive Island." Since then the very sentient and sometimes 'monstrous' B-movie star Gojiro, teamed with his pal "Komodo, the world-famous Coma Boy," stride across the Pacific into Hollywood and the New Mexico desert in search of sobering truth and the meaning of existence and identity" in a world in which neither belongs." Analogously, Todd Tennant's and Michael Bogue's eye-catching "King Komodo," featured online and within pages of *G-Fan*, presumably resulted as a mutation of an ordinary Komodo Dragon.

Not every 'Dragon' of horror and science fiction is of the extant Komodo variety; far more frightening was the very real, yet extinct, Australian *Megalania*, which grew up to 30-feet long—nearly four-times as large as the Komodo Dragon! *Megalania*, whose fossils were described in 1889, became extinct only 25,000 years ago. While we may speculate whether it hunted early humans possibly existing then on that continent, this Dragon most certainly ambushed large contemporary marsupial animals. And, yes, *Megalania* headlined in Grandmaster science fiction writer L. Sprague de Camp's 1993 short story, "The Honeymoon Dragon." Here, De Camp's expert time safari guide, Reginald Rivers and his wife decide to take a time journey to the Pleistocene in a colleague's newly constructed time transition chamber, located in Australia. Unbeknownst to Rivers, however, this rival (who plans to eliminate Rivers) had already made a previous trip whereupon he encountered a cave infested by *Megalania*. So this time, Rivers and his wife are insidiously lured into *Megalania* terrain, that is, after his rival covertly removes shells from Rivers' gun: premeditated, attempted murder! Fortunately, others notice their dilemma in time, shooting the charging Dragon before Rivers and his wife can be devoured.

Komodo (Komodo Film Productions, 1999), directed by Michael Lantieri, seems somewhat inspired by *Megalania*. Here's the spine-tingling synopsis from the DVD packaging... "On an island ... something was hidden, waiting for the chance to strike. One family is missing and the lone survivor is so shocked he is unable to describe the terrible events. Now psychologist Victoria Juno (played by Jill Hennessy) will uncover the truth by going back to the nightmare. But the last thing she ever expected was that the nightmare was real." Or in other words, borrowing the tag line, "Welcome to the bottom of the food chain."

It seems that Pontiff Petroleum has encountered a rare, endangered species on Emerald Isle off the coast of North Caroline where they're drilling for oil, in the process poisoning the environment. With natural food sources exterminated, a starving population of near *Megalania*-sized Komodo Dragons devours humans for nourishment. How did the Dragons arrive in the United States? Nineteen years earlier, a smuggler of rare animals left a box of smelly Komodo Dragon eggs by the roadside; they hatched, survived and now they're hungry. Pontiff executives hired a biologist handy with a gun to covertly deal with the "problem," thus (illegally) avoiding the "environmental crusade" resulting if people discovered Pontiff was slaughtering an endangered species. Ultimately, a youth whose parents were eaten by Dragons, adopts a vengeful, caveman-like, 'dragon-slayer' persona to rescue bloodied survivors. The cgi-animation was particularly well done for this captivating, yet rather predictable film.

The next pair of Komodo monster movies are disconnected to *Komodo*. They're about an experiment gone awfully wrong, creating a "Jurassic Park" meets Wellsian "Food of the Gods" sort of affair, merging classic science fiction with horror. While *Megalania* was certainly impressive and largest of its kind, it never reached giant Japanese monster proportions. However, this 'short'-coming has been alleviated by Hollywood producers of the current millennium.

In *Curse of the Komodo* (2003), giant, cgi-animated Komodo Dragons run amuck on an isolated Pacific monster island, Isla Damas, where they were exposed to a chemical 'serum.' An experiment known as Project Catalyst was, according to scientists monitoring the tests, intended to produce exponential increases in food production. But the U.S. Navy was *also* interested in the results. Could the new chemicals be converted into potential weaponry, "Playing god with nature just to make some generals happy?" Yet playing with Nature only creates mutations, and here the Dragons seem most effected. In the end, the experiment is a bust, as 150-foot long Komodo Dragons, their digestive processes amplified, chase puny humans around, eventually penetrating their "Forbidden Planet-like" defensive force field. Now the Navy wants all the evidence to be "cleaned up," so that no one can trace what happened on Isla Dumas. Meanwhile, the Komodo Dragons seem invulnerable to high-powered rifles and handguns (that magically never run out of bullets). In a scene, borrowed from *Jurassic Park*, a Dragon chases a truck moving at 40 miles per hour. This prompts one revelatory exchange... "I thought they were extinct." The professor replies, "That's not a dinosaur." Exposure to Dragon saliva creates a septic condition, causing infected individuals to stumble around in crazed, zombie-like fashion. Despite the occasional wooden acting and several plot inconsistencies, the idea is relevant to our times.

A sequel, *Komodo versus Cobra* (2005), re-explored Isla Dumas's strange circumstances. This time, reporters associated with an environmentalist group, "One Planet," intend to divulge the truth about the military's project, exposing how carnivores were abused by "ruthless capitalists." Predictably, members of the news team are soon dispatched by a giant Dragon, appearing rather different than in the previous film. And, of course, a doomed resident scientist manages to pass along key theoretical concepts—how the military ultimately intended to use "biocatalyst" to breed superior soldiers. But this time another species has been effected, a cool-looking, slithering cobra snake, over a hundred feet long! Yes, this is the cgi- battle of the new century, Komodo vs. Cobra, which takes place in the film's waning moments with napalm bombs exploding all around. All too late, one victim opines that "we shouldn't have messed with Nature ... too high a price." In the end, we see how biocatalyst indirectly impacts human DNA, spawning a fork-tongued, hybridized human, suggesting yet another sequel.

Occasionally, as in early 2009, people are killed by Komodo Dragons (which today live on four islands in eastern Indonesia). Dragon attacks on humans are rare, yet their frequency has increased in recent years. A sign of Nature's imbalance? To ignore the Komodo's fate is tantamount to forsaking our future. We are as one: man and beast must live in harmony ... or perish.

Oh—and let's vow to not mess with Nature!

Chapter Eighteen — *Kong versus Godzilla – A Battle of Novelizations*

We've enjoyed them, yet many do find monster and sci-fi movie "novelizations" to be of lesser interest or value than the respective movies they're intended to publicize – so they tend to languish or be forgotten. What is meant by a movie novelization anyway? For our purposes, it's a published novel or extended story written from or based on a movie script or screen play. *The Lost World* novel written by Arthur Conan Doyle (serialized in *The Strand* 1912) isn't a novelization because First National's filming came long after (in 1925) – even though later/post-filmic editions of Conan-Doyle's classic tale have appeared. Ditto and straight forward with other literature such as Mary Shelley's *Frankenstein (or the Modern Prometheus),* Bram Stoker's *Dracula*, H. G. Wells's *The Time Machine*, or even Michael Crichton's *Jurassic Park*. So when the hot movie idea and script-writing offers distinct opportunities, and business people sense promotional advantages in pumping out a book telling the story in words, primarily to further advertise the movie, we can look forward to reading the story (more or less as later witnessed on the silver screen) sometimes weeks in advance of the much vaunted film in book format. Intriguingly, novelizations may offer interesting plot variations and details that were excised or difficult to follow in the cinematic version.

Promoters commission a talented writer to do the job admirably, ideally someone who can write to a deadline and won't blow the budget. Some of the better results (extending from way "olden" times of my youth) might include Vargo Statten's *Creature From the Black Lagoon,* W. J. Stuart's (pseud. for Philip MacDonald) *Forbidden Planet,* Alan Dean Foster's *Alien*, Donald F. Glut's *The Empire Strikes Back* and, of course, Arthur C. Clarke's *2001: A Space Odyssey*. And don't forget those early *Star Trek* movie novelizations, or one of the very first – Thea von Harbou's novel *Metropolis* (1927). But in the arena of giant dino-monster literature, several compelling novelizations of the past (and present day) invite further consideration. Readers may know of Delos W. Lovelace's *King Kong* (1932 – based on a story by Edgar Wallace and Merian C. Cooper), Carson Bingham's (pseud. for Bruce Cassidy) *Gorgo* (1960 – based on an original story by Eugene Lourie), Dean Owen's *Reptilicus* (1961), K. Robert Andreassi's (pseudo. for Keith Robert Andreasi Decandido) *Gargantua* (1998 – based on the teleplay by Ronald Parker), Christopher Golden's *King Kong* (2005 – from the screenplay by Fran Walsh, Philippa Boyens & Peter Jackson – based on a story by Merian C. Cooper and Edgar Wallace), and Greg Cox's *Godzilla* (2014).

Novelizations aren't published quite as often today as they used to – I believe, because fewer people read these days as opposed to during olden times. (Computer games and texting have taken over valuable reading time.) Novelizations really didn't begin to appear until the movie industry was attaining a state of maturity, that is, near the dawn of the era of "talkies." Nevertheless, let's examine novelizations leading to a King Kong and Godzilla giant-sized, back to back 'battle of the books.'

In the case of the Delos Wheeler Lovelace's (Dec. 2, 1894- Jan. 17, 1967) novelization of *King Kong*, for which he was paid a princely sum (then) of $600, originally there was no paperback pulp edition (as usually would be the case today). The book was proudly published by Grosset & Dunlap in hard cover with a vividly illustrated book jacket showing some of the movie's exciting scenes (making use of an artist's paintings rather than photographic stills). There were also abridged serializations by a "Walter F. Ripperger" in 1933 Feb. to March issues of *Mystery Magazine*, and another entry published in 1933 issues of *Boys' Magazine*. Other than his famed monster book, Lovelace mostly labored as a newspaper and freelance writer, but also authored other books including a historical novel in 1953, *Journey to Bethlehem* and also a wartime biography, *General Ike Eisenhower*. Delos Lovelace was married to Maud Hart Lovelace an author of "Betsy-Tacy" novels written for girls. She and her husband also collaborated on many stories and three other books, one of which featured Indian legends.

Delos Lovelace obviously worked from early screenplay versions, as the novel contains a full writeup of the "spider pit" sequence, and, for instance Captain Englehorn's vessel is named the *Wanderer* (not the *Venture*). Also we have a scene with dinosaurs not featured in the film – namely the striking 'Kong vs. *Triceratops*' scene. Overall, sci-fi author Greg Bear observes that Lovelace's "prose style is quick and colorful, not always polished but seldom getting in the way, and his language is often brusquely poetic. He captures much of the film's personality and adds a few details of action and character we can't see on the screen – all good stuff for a novelization." [1]

The jacket of the 1932 edition rather confusingly states that the novel was "conceived by Edgar Wallace and Merian C. Cooper, screen play by James A. Creelman and Ruth Rose, novelized from the Radio Pictures by Delos W. Lovelace." So Lovelace didn't exactly receive top billing for his efforts. Decades later, when the novel was reissued (for at least a third time), although by The Modern Library (New York, 2005), attesting to its quality, Ruth Rose's name had been chopped from the title page. Ruth Rose was the wife of Ernest Schoedsack, Cooper's dedicated partner, who edited Creelman's screenplay, resulting in its "authentic quality... the dialogue sparkled and kept the story moving ... 90 percent of the final dialogue was hers."[2]

Edgar Wallace (1875-1932) was a famous and prolific fiction writer, who authored 167 novels within a 14-year span, several of which later became movies. He even penned some science fiction including a novel concerning a comet hellbent on a collision course with Earth, causing the unification of humanity, titled *The Day of Uniting* (1926), and another in 1929 (which would be of interest to Leonard Nimoy's "Mr. Spock" persona), *Planetoid 127*, about "Vulcan" a once hypothesized planet on the Sun's far side. Cooper's 'treatment' of the Kong story was handed off to the prolific Wallace who at David O. Selznick's beckoning arrived from London on December 5, 1931. Wallace's initial screenplay version was completed on January 23, by which time the working title "King Kong" had been accepted. Wallace then promptly died of pneumonia in late February but his manuscript, which Cooper found disappointing, survives.

Wallace's version had a 50-year old "Danby G. Denham" sailing with his entourage to "Vapour Island" with "John Lanson" and "Shirley Redman." There they'd encounter a band of convicts before doughty Capt. Englehorn subdued Kong with gas bombs. A crucial plot element of the final version – Ann Darrow's status to the islanders as a sacrificial offering during the ceremony to Kong was not present. Also, Kong was to be electrocuted by lightning atop the Empire State Building, rather than the familiar, famous ending, overcome from strafing army planes. This, despite the fact that Cooper had suggested in an RKO memo to Wallace that he "Please see if you consider it practical to work out theme that John attempts single handed rescue on top of Empire State Building if police will let off shooting for a minute. ... Then when he fails, airplane attack, or something along this line." Not that there's anything wrong with these ideas, but ultimately Wallace's approach was redone into the masterpiece we know and love today.

Why then if Cooper wasn't satisfied with Wallace's draft, did he ultimately share story credit with him not only for the film but also the novel? As Mark Cotta Vaz explains, " ... Cooper, who lived by a code of honor, had promised story co-billing to Wallace and felt duty bound to extend the credit. There was also more than largesse involved – Cooper and Selznick both felt Wallaces's famous name would add prestige and bring attention to the picture."[3]

An RKO attorney indicated that his contract with Wallace conveyed joint publication rights to the story "in book and/or serial form." This meant that Cooper reserved right to publish a novelization under his name, but that credit be shared as "By Merian C. Cooper Based upon (or transcribed from or adapted from) original scenario by Edwar Wallace." As Mark Cotta Vaz states, "it was a sticky issue, sharing story credit, since Cooper was disappointed with Wallace's work." Cooper wasn't bashful about noting how "The present script of Kong, as far as I can remember hasn't one single idea suggested by Edgar Wallace. If there are any, they are of the slightest." (After all, Cooper himself had considerable writing experiences—having worked as a newspaperman—concerning his explorations.) So instead, as conveyed in a July 20, 1932 memo, Cooper decided to market the forthcoming novelization as, "based on a story by Edgar Wallace and Merian C. Cooper, if you want to use Wallace's name. I don't think it fair to say this is based on a story by Wallace alone, when he did not write it, though I recognize the value of his name, and want to use it."[4]

Kong, of course, was absolutely conceived by Merian Coldwell Cooper (1893 -1973)! Cooper amazingly barnstormed through a life of wanderlust and derring-do, as chronicled by admirers.[5] Guiding influences inexorably led to his enduring filmic masterpiece *King Kong* released in March 1933. He *conceived* the idea for Kong, yet was credited neither as screen play, nor novelization writer. So without Cooper there would have been no immortalization of his "prehistoric gorilla" – Kong – "… fifty times as strong as man – a creature of nightmare horror and drama." Carl Denham (played by Robert Armstrong) became essentially Cooper's persona instilled within the movie. Finally, few might realize that in 2005 Cooper's idea was re-embellished by Joe DeVito and Brad Strickland in a new novel titled *Merian C. Cooper's King Kong* (St. Martin's Griffin, New York), although this wouldn't qualify as a "novelization" under terms of this discussion. (I label it as a seamlessly rewritten "re-novelization" of the 1932 novel.)[6]

The 1932 Lovelace novel itself is very well paced (like the movie), presenting a literary realm within the science fictional tradition of Conan Doyle's *Lost World* and Burroughs' *Land That Time Forgot* "Caspak" trilogy. For those whom had not yet seen the film, but read the book prior, a great sense of mystery, fear and sufficient suspension of disbelief is conjured in the first third of the novel as the *Wanderer* carries its crew toward the fateful island. Denham, full of bravado, is a "lucky" man, who always was lucky and gets what he wants. Early on he's contemplating the 'Beauty and the Beast' theme for his movie; Ann Darrow is his flower, as yet unbloomed, plucked from the New York city streets who tries on dresses for rehearsals that make her appear like some kind of strange "bride." But Denham alone knows the ship's destination, in search of the Malay Kong legend. Jack Driscoll falls quickly for Ann, in spite of Denham who warns, prophetically, that when a male goes soft for the girl, he'll lose his edge. Better stay away.

Meanwhile, shipmate "Lumpy's" pet monkey Ignatz befriends Ann in a scene inspiring Denham's Eureka moment – a Beauty and Beast angle, as he says to Driscoll:

"The Beast was a tough guy, tougher than you or anybody ever written down. He could lick the world. But when Beauty came along she got him. When he saw her, he went soft. He forgot his code. And the little ham-and-egg fighters slapped him down. Think that over, young fella."

In another scene while still onboard, Jack, Ann and Ignatz are in the ship's crow's nest, foreshadowing what will happen eventually atop the Empire State Building (substituting Kong for Ignatz). Here, not unlike Kong, Ann realizes that simian Ignatz is jealous of Driscoll's advances toward her.

The island with the Skull Mountain, barricaded by an enormous wall, is concealed in misty fog. But they arrive safely, thanks to the map Denham obtained from a Norwegian sailor (who isn't a 'visible' character in the story). Lovelace cleverly suspends our disbelief as to the island's myth. Everything about this voyage is precipitated because of the Norwegian's map, which was drawn from a dying native's description – who was the only survivor in a canoe blown off course from Kong's island and picked up in the Norwegian's ship. (Conveniently, the islander died before they reached port.) So when Capt. Englehorn asks "Does he believe the native's story, this Norwegian?" Denham (via Lovelace) answers, "Who cares? I do!" (And therefore so should we, gentle readers.) Denham continues, "Do you think a picture as detailed as that (i.e. map) could grow entirely out of the imagination?" Nah – of course not! You can't make this stuff up. We *want* to believe!

The landing party is astonished by the colossal wall, reminiscent of ancient Egyptian structures. They inadvertently (and inadvisably) observe the Bride of Kong sacrificial ceremony, which has been desecrated by their intrusion. Now usually in such stories, the resident paleontologist becomes the "explainer" of the local dino-monstrous fauna. But there is no one with such credentials aboard the ship, so interestingly Ann becomes spokesman for paleo-science. She's not only a scream queen, but also not just a "dumb" blond. (Sorry for that pejorative.) Ann establishes the prehistoric provenance of the indigenous fauna soon to be encountered, beginning with the most significant creature – Kong, even before she witnesses the mighty "beast-god" up very close, in the flesh. Denham speculates as to whether Kong (*if* he exists) may be a gorilla based on ceremonial garb worn by the natives, albeit a vey large specimen.

Ann adds, "But there never was such a beast … At least not since prehistoric times." Driscoll and Englehorn readily disagree. But Denham is completely convinced by Ann, further suspending disbelief as to what will soon happen, stating "Remember? Both of you said we wouldn't find any Skull

Mountain Island. But here it is. Why shouldn't such an out-of-the-way spot be just the place to find a solitary, surviving prehistoric freak?"

Well, now the natives need a substitute "bride" for Kong, and their sighting of Ann's blond hair remedies the desecration; Ann's offering will appease the beast god. So, later, the natives kidnap her from the *Wanderer* and proceed with their ceremony, while Englehorn, Driscoll and Denham scramble to launch a rescue party... but only too late, for Kong snatches Ann from the sacrificial pillar. Here, readers are treated to one of monster-dom's first cries of terror when scream queen Ann spies the huge beast approaching from the dark jungle. "It did not look *up* at Ann upon her pedestal. It looked *down*. Moving closer it stared down between the two pillars. ... And Ann's scream sped piercingly into a dead silence."

Meanwhile, Driscoll, Denham and Englehorn are rustling up a search party. With a dozen other men they go barreling through the opened gate into Kong's dark jungle. As in the movie, they soon encounter an assortment of prehistoric monsters. But there are key differences in the lineup. Following the massively sized footprints, a *Stegosaurus* soon stands in their way. It isn't named as such, and Lovelace does not mention its signature characteristic – the bony plates adorning its spinal column and tail. But from the other distinguishing features, especially its "huge spiked tail" and tremendous hind legs, it clearly must be that genus. Now "Denham and Driscoll grew cold from their full realization of the true scope of the island's mystery." True—soon they'll be pondering their "... utter helplessness. Of what use was such guile and wit as theirs against the huge fantastic beasts of this nightmare island?"

After two rifle bullets fail to kill the charging *Stegosaurus,* it is immobilized by one of the gas bombs reserved for Kong, flung by Denham. Then Denham fires a round into the scaly reptilian head, thus dispatching it. Now Ann's hypothesis has been proven. "Prehistoric life!" Denham exclaims. "She was right... Ann, I mean. But she only had the start of it. She guessed the beast-god was some primitive survival. But if this thing we've killed means anything, the plateau is alive with all sorts of creatures that have survived along with Kong." Notice that "plateau" is referred to here; a possible slip by Lovelace thinking of Conan Doyle's novel, perhaps? Now the mist is thickening around them as the jungle grows denser. There's a stream to cross and a flimsy raft to build.

Yes, and another dinosaur to flee – resulting in Denham's first use of the term "dinosaur." This time it's a *Brontosaurus* disturbed from its submerged slumber. The aquatic monstrosity overturns the raft, dislodging rifles and the rest of the gas bombs carried into the jungle for their pursuit of Kong. "No weapons remained." The brontosaur chases one unlucky soul, but the rest manage to reach the opposite shore. Although a staunch herbivore, *Brontosaurus* eventually eats the unfortunate man who had attempted to find refuge in a tree branch, as from across the valley, a "thin, distant scream drifted to their ears." A pitiful scene as described by Lovelace, causing the hardy men to cringe and even hurl. But now they're confronted by another prehistoric landscape, an "asphalt field," most likely inspired by southern California's Rancho La Brea Tar Pits of Pleistocene age. This "… asphalt morass … was there before the first beast came into being. A sink of hell! There'll be thousands of carcasses buried rods deep in it. If we try to cross, Jack, we'll have to be careful we don't stick and sink too."

But cautious as they must be, they spy their quarry lumbering toward them across the asphalt; Kong carrying an unconscious Ann as if she were a ragdoll. Kong's "… primitive brain valued this strange possession for reasons it could not understand. Much as a prehistoric woman might have cradled her baby, he carried the girl's inert form in the crook of one arm." But Kong was being pursued by other prehistoric beasts, ones which were not featured in the RKO film. Denham summons his book knowledge and identifies them as *Triceratops*. When Driscoll asks what they are, Denham – assuming Ann's *in absentia* role as paleo-expert – declares, "Just another of Nature's mistakes, Jack. Something like a dinosaur. … They got their names from the three horns on their heads." Fair enough – for evolution, like Time apparently stands still on Skull Island.

Now things tighten as casualties quickly mount. Avoiding sticky asphalt, Kong tenderly places Ann down and commands a dry mound facing the pursuing *Triceratops* herd, one of whose feet had already become mired in tar. The remaining pair amble along drier paths toward Kong, while Kong heaves huge boulders of asphalt toward their armored heads, wielding devastating blows. Kong pounds his breast, roaring triumphantly. The men are amazed by his mighty strength, but soon realize they must clear the area, back through the bushes toward a ravine. The only way they can keep up with Kong's pace will be to cross the ravine over a fallen log, bridging the chasm. The sole surviving *Triceratops*, wounded yet dangerous, gores one of the men, forcing the rest onto the relative safety of the mossy log bridge.

Kong angrily shakes the log and men begin to tumble off into the misty chasm. Yeah – we know what's down there. A can of Raid pesticide won't do! Driscoll makes it to the other side of the ravine, while the other men (minus Denham, who knows when to retreat for arms and reinforcements), meet horrible fates. There's been a lot of intrigue as to whether or not there was even such a "spider pit sequence" ever filmed or added into the RKO movie, but recent information indicates that this may have been the case.[7] But what isn't generally described or discussed in this context is that (to me) the giant bugs lying in wait at the bottom of the ravine represent not a radioactively mutated breed of arthropods, but instead another throwback to Paleozoic times – yes, the Carboniferous or Coal Age when insects, and centipedes truly grew to enormous proportions due to increased concentrations of atmospheric oxygen. And that giant lizard which climbs up the vine toward Driscoll while he's trapped on the ledge .... Well, in the book, it's a giant creeping spider instead. So, like the "asphalt" which is intended to be Pleistocene, the ravine bugs are wallowing in their muggy, buggy "Coal Age" environ. Lovelace is giving us little 'living' vignettes from deep time ... as now we step back into Cretaceous times.

A "grotesque" *Tyrannosaurus*-like monster with a snake-like, toothy head enters the picture, although Lovelace's description of this beast borrows more from earlier popular culture (such as Conan Doyle's description of a meat-eating dinosaur in *Lost World*) than from later incarnations, such as those witnessed in Charles R. Knight's paintings, down through *Jurassic Park*. The key difference is that this tyranno-monster incongruously *hops* toward its prey rather than runs or strides.[8] As we now know, no self-respecting tyrannosaur would hop, not unlike a kangaroo, toward its adversaries (such as *Triceratops*). In fact why are there both *Triceratops* and *Tyrannosaurus* present in the novel? Probably thanks to Charles R. Knight who had immortalized the creatures facing off against one another in at least two world famous paintings by then (i.e. for the American Museum of Natural History and Chicago Natural History Museum, reprinted in numerous books and magazines). So it is significant that Kong proves mightier than either of these two monstrous brutes from the Cretaceous. Kong dispatches tyrannosaur after a tumultuous battle and carries on his way through the jungle. Driscoll, observing the fight, muses that "Any critical observer would have realized that Kong had met enemies of the meat-eater breed before and had worked out a technique of battle which served well when he was not too enraged to use it." But now the hungry bugs, a giant lizard and other awful things, sensing there is fresh carrion left in Kong's wake, come scurrying up from below.

Driscoll continues tracking Kong who is carrying Ann tenderly toward his cave lair high atop the island. Yes—ape-like Kong has become 'elevated' (literally, due to his ascent to Skull Mountain's pinnacle) to a brutish cave-"man" king of the island. Meanwhile, Denham, who has made it back to the wall, explaining what happened to the first rescue party, offers insights as to Kong's nature. "Kong ... is the one thing outside the wall that is something more than beast. He's one of nature's errors, like all the others, but he was nearly not an error. And in that huge head of his is a spark. Ann means something to him. ... He sensed that Ann was different. ... It was Beauty and the Beast over again." High atop Skull Mountain, it remains Nature red in tooth and claw – all the way, all the time!

For no longer has Kong made it back "home," when a slippery aquatic reptile emerges from the stream pool source, which attacks him – coiling about him like a snake. (This is an *Elasmosaurus* at the time also made famous by Knight's museum paleoart.) Kong's persona is becoming more human-like, as we read, "He shivered, so full he was of the loathing his species has had of reptilian things since the dawn of time...." And after a brief tussle with the "water beast" he begins to solicitously disrobe Ann. Driscoll understands that Kong was "… plainly trying to find some connection between the frail tissue and the whiteness he had exposed." Before drifting into a racy porno here, we must declare this behavior as highly improper, (especially for the 1930s). Then Kong's attentions are diverted to a menacing pterodactyl, which is readily dispatched. Yet chaste, virginal Anne appeals to Driscoll, "Jack! Don't let him touch me again." Ann is spurning her massive protector, the beast-god who has saved her from several horrible fates from reptilians within the jungle already and yet who is now violating her. What more must a lustful "cave-man" do to show valor toward his "bride." But demure Ann craves human solace, and so she and Jack jump into the stream, which carries them back toward the sacrificial altar and the wall. The oddest love triangle ever!

After Kong is subdued with gas bombs after breaching the wall in his mad dash pursuit of Ann, now Denham wants to do more than simply make a film. He's stoked to capture Kong, only to tragically exploit him back to Broadway, billed as "King Kong! The Eighth Wonder!" Now Kong is caged, drugged and forced to view puny humans who have come to gawk at him. Denham even admits that they've "knocked a lot more of the fight" out of him. After a photographer's camera flashes thrice, and when Jack puts a guarding arm around Ann's shoulder to protect her from the agitated beast god, Kong rips free of his bonds. The implication is that Kong is striving to protect Ann (Beauty) once more, even in this wretched place called "civilization."[9]

Denham was right all along. For after the planes' machine guns mortally wound Kong, "For a breath then, high above the civilization which had destroyed him, he hung in the same regal loneliness that had been his upon Skull Mountain Island. Then he plunged down in wreckage at the feet of his conquerors." And of course, as in the movie, Denham muses with Kong lying at the base of the Empire State Building. "The aviators didn't get him…. It was Beauty. As always, Beauty killed the Beast."[10]

So where is this all leading? Only to the greatest giant monster "battle" of them all. King Kong versus Godzilla!

Despite what's implied in the chapter title, there was never an English novelization for the 1962 Toho film *King Kong vs. Godzilla*. Here, I simply refer to two recent novelizations, one corresponding to Peter Jackson's Universal Pictures 2005 cgi-effects laden movie *King Kong*, and the second a just-released novelization conforming to the script of Legendary's 2014 film *Godzilla*. Now let's examine these two novelizations authorized from movie scripts, the first being Christopher Golden's *King Kong* (2005), based on a screenplay by Fran Walsh, Philippa Boyens and Peter Jackson, and based on the story by Merian C. Cooper and Edgar Wallace. And the second novel, Greg Cox's *Godzilla* (2014), based upon the screenplay by Max Borenstein; story by David Callahan. These mythic monsters represent the East's and West's most famous, celebrated giant prehistoric monsters.

Both novels were written during a time of heightened recognizance of what harm and plunder the latest 'geological force'—namely mankind—could render upon a fragile Planet Earth equipped with meager (albeit punishing) means of exerting strategic self-defense. Whereas in their respective novelizations, Carson Bingham's *Gorgo* and Delos Lovelace's *King Kong* project a more traditional 'man vs. nature' theme, Cox's Godzilla (2014) reflects that Nature seeks a balance independent of man although an implication is that if man seeks to live in harmony with planetary processes, he can become a part of the natural order. We've seen the movies, but now let's indulge and see how the recent Kong and Godzilla stories stack up in written form.

The Delos Lovelace 1932 *King Kong* novel was written during Americas' great Depression, whereas decades later Christopher Golden strove to recapture the essence of that dreadful time during the continuation of a bubble period in our economy, soon to enter a Great Recession. To me, in many ways Golden is a more pleasing, satisfying writer for our time than Lovelace, even though his 2005 novel will never be remembered as the 'official' literary version (because it is based upon a movie 'remake'). Golden's novel was written at a time when in the popular vein, America had gone 'dinosaur crazy,' and also when ranks of the "99%-ers" were beginning to look askance at corporate greed and Wall Street. Could those guys in charge (or any ranking official for that matter) be trusted on moral principles, or were they merely scoundrels with ulterior motives? Like Carl Denham?

Like K. Robert Andreassi's 1998 *Gargantua* novel, Golden's *King Kong* explores ecological, environmental themes, specifically man's capacity to endanger the environment through questionable or unwise undertakings, and our inability to live alongside resulting gigantic crypto-prehistoric species—once they're discovered, especially if we're somehow responsible for invading their terrain and then cruelly subjecting them to civilization's confines. Only this time, one man's greed is at the root of the problem, resulting in a dreadful path of destruction.

Golden's novel offers more character development than in Lovelace's, and of course, several of the key character names have differing roles. Most significantly, Jack Driscoll isn't a salty sailor sort, but instead an emotionally conflicted playwright who falls in love with down-on-her-luck vaudeville comedienne Ann Darrow. He's writing the script for Carl Denham's adventure film in the hold of a tramp steamer, the *Venture*, when they finally set sail. Carl Denham isn't quite cold-hearted, yet he's wincingly willing to sacrifice anything or *anyone* to attain his maniacal goals Denham also has another *serious* problem, namely paying his employees; Jack Driscoll for his writing, or Captain Englehorn for the voyage he's chartered. It's very hard to like this rascal. Mainly though, what Denham himself doesn't understand is that by filming something mysterious, primitive, hidden from the eyes of civilized man, thus filling in the blank spaces on the world map, he begins to destroy, "making it no longer an enigma"! (This is something author Joseph Conrad understood, but Denham does not.)

We also read more about the crew, particularly first mate Ben Hayes, a courageous former soldier who draws inspiration from Joseph Conrad's novel *Heart of Darkness*. There is also considerable, illuminating 'backstory' concerning several sailors, the native primitives, as well as Kong and his Island itself. Because we've endeared ourselves to the crew, we're a bit sorrier when many of them get eaten, speared or trampled later on. Also, there are more instances of comic relief spliced in amidst moments of sheer terror in Golden's novel (than in Lovelace's). Golden pulls everything off expertly.

Some of the humor stems from in-jokes, for example, as Denham flees the studio in a taxi, desperately looking for a new pretty actress to play in his movie alongside "Bruce Baxter" the high-priced veteran film hero hired to be the star. Denham pleads to his personal assistant 'Preston,' what about Myrna Loy, Joan Blondell, Clara Bow or Mae West? Won't work, Preston replies, but then Denham exclaims enthusiastically, "Fay's a size four," meaning she will fit into the wardrobe they already happen to have onboard for their actress. Of course, this is a reference to Fay Wray, of 1933 *King Kong* fame and lore. But she's already doing some (other) movie for RKO. Hah—nicely done Mr. Golden! Baxter's best line in the novel comes when he's trapped in the bottom of the ravine surrounded by giant carnivorous bugs. He lets loose with a volley of words (and bullets), yelling, "Nobody get in my way! I'm an actor with a gun and I haven't been paid!" (Why did he have me thinking of Ted Baxter, from the *Mary Tyler Moore* show?)

When Denham spies Ann Darrow, he instantly realizes she's the missing piece in his puzzle; only she can fill the bill. Always considering how his actors will appear before the camera, he falls in love with her glimmering looks. "Her clothes were a bit tatty, but her face was *luminous*, her eyes bright even in the gathering gloom of evening, her golden hair falling around his face as though she was some kind of angel." There's more than an element of destiny as Denham proclaims, "'You're the one, Ann'…. as though it was some kind of epiphany, as though he really believed it. 'It was always going to be you.'"

The map! Denham isn't the only one onboard who knows of Skull Island – the uncharted island perpetually cloaked in fog, its mysterious 100-foot high wall, or the name "Kong." Hayes and 'Lumpy,' (ship's cook and physician), heard of the legend from a shipwrecked Norwegian who later died. They have more than an inkling of just how treacherous the place must be, striving to steer Denham from his mad course with a tale of terror, but Denham nonchalantly replies, "You've got to do better than that. Monsters belong in B-movies."

Meanwhile, Preston has discovered Denham's crumpled map showing the location of Skull Island in a trunk, confronting Denham with it. Denham is resolute, belittling Preston, "Are you afraid …. or are you a *filmmaker*?" Denham will sacrifice nearly anything for mere *belief* in what's revealed on the crudely drawn map! At one point, as few survivors emerge from near certain death in the ravine, Englehorn likens Denham to a cockroach. By the end of their horrific stay on Skull Island, Denham is despicably using Jack and Ann as bait to lure Kong into his trap.

The scream! And yet, despite the seemingly insurmountable odds, they do arrive at the Island of the Skull in the middle of the Indian Ocean. When Denham leads an unauthorized film reconnaissance party to shore in a rowboat, they rehearse a scene in which Ann screams for all her life, glancing upward at some nameless, unimaginable horror. Although it's "only acting," she lets loose. But then something amazing happens. For, "… then as the last echo of her scream disappeared, there came another sound. An unearthly, bestial cry that was part scream itself and part roar from somewhere deep within the jungle's interior. The sound was inhuman and yet she had no doubt it was a reply to her own primal scream."

The 'reply' is so loud that even Englehorn's crew can hear it far away aboard the *Venture*. Bruce Baxter, there on shore with Ann and the others, tries to steady his nerves, suggesting hopefully that they had only heard a bear. But now Denham charges up through and over the gigantic, ancient wall to see what really bellowed that terrible jungle cry. Later, when a native spears Mike (Denham's sound man), Ann screams for dear life and in answer (again) a blood curdling roar reverberates from within the lush jungle canopy. The natives fear that Ann is 'summoning' their beast with her anguished cry.

The Wall and the Tomb! One of the most intriguing aspects of the 2005 *King Kong* novel is Golden's enriched description of the mysterious wall, neighboring stone ruins embossed with idols and menacing statues, hinting of an powerful ancient civilization since lost to an immensity of time and degradation, swallowed by the jungle. In fact the wall is a "feat of construction no less extraordinary than the great pyramids, towering hundreds of feet in the air. It began in the water and ran up onto the shore, creating a barrier that effectively cut this part of the island away from the rest. It ran inland from here, and disappeared into the jungle…" Closer to shore, there's also a dark tunnel with stairs hewn from rock within a natural cavern, leading upward through a cliff, to a tomb above. This invigorated yet macabre archaeological, anthropological dimension and implications is quite fascinating. Soon after surmounting the cliff tunnel, observing the extensive necropolis of mummified corpses and human skeletons entombed throughout and beyond, Denham realizes that he's "traveling on a journey through time from present to past." Moreover, the "island itself was a kind of phantom remnant from another age." But of course there are extant villagers present there too, although these may not be of the same mighty race who constructed the great wall visible beyond the tunnel, or who built the now dilapidated ruins scattered about the plateau in between.[11]

The fauna! Denham isn't the only member of the rescue party who senses the prehistoric essence of Skull Island. Jack Driscoll summons a literary reference. "Dinosaurs …Somehow this place had survived in its primordial state from prehistoric times. Arthur Conan Doyle had once written about a Lost World, and here it was in all its glory. Skull Island." Ann, who by now has endeared herself to every man onboard, is abducted from the *Venture*, prompting launch of an intrepid rescue party through the wall's gate, into the jungle. Ann screams again and again when Kong claims her, but among the rescue party, only Denham catches a glimpse of the mighty beast – a 25-foot tall gorilla—after he's snatched her from the sacrificial altar. "In all his life – in his entire career as a filmmaker – he'd never seen anything so fantastic. It was as though he'd gotten a glimpse of the remnant of another world, of an age when the impossible strode the earth." They're collectively about to encounter many other "impossible" things, besides Kong. Curiously, when they're attacked by their first dinosaur, it partially owes its description to the outmoded "kangaroo" pre-1930s paradigm. A very strange dino indeed, sporting a beaklike snout, shoulder spikes, horn-covered plates "all over its green, mottled hide," and a fin along its head – *Stegosaurus*. Riddled with gunfire, the monster dies, suffering extinction, like its kind was "supposed" to be. But, of course, that's not all. Kong also has an epic, 'nature red in tooth and claw' battle versus not one, but *three* tyrannosaurs, dubbed "V. rex" by Ann. The "V" stands for 'voracious' or 'vicious' partly because V. rex is even larger than the bones of its extinct cousin she'd seen in a museum, T. rex! Kong manages to dispatch them all. But there's also a "raptor" dinosaur species, as well as giant leathery bats which swarm over Kong, plus all those nasty carnivorous bugs and spiders waiting at the bottom of the infamous ravine for the men to tumble off a log shaken by Kong.

The camera! After arriving on the island, listening to Kong's roar, Denham senses that his camera has a date with destiny, no matter what the price; he will be famous, his name will become legendary – perhaps more so than the animal he seeks. Yet Denham has an uncanny gift, a 20th century curse—an "unfailing ability to destroy the things he loves." Deep in a jungle canyon, twelve brontosaurs stampede, triggered by arrival of a pack of flesh-eating dinosaurs. In their feeding frenzy the flesh-eaters also claim several of the rescue party. In escaping with his precious Bell & Howell film camera undamaged, Denham sacrifices one of his assistants, who is devoured. After all, by this time, Denham is rather disingenuous about following Kong into the jungle. For Ann's safety has become a secondary concern to his getting a lot of great footage for his film. "Whatever what happened to Ann, he would still have his movie." Denham even manages to hang onto his accursed camera during their aborted raft river crossing while they're attacked by a deadly "Piranhodon" (a gigantic piranha form).

Kong's throne! Ann notices a gigantic, severely weathered stone statue of a gorilla deep in the jungle, "a powerful figure that might have been some ancestor to Kong…. another remnant left by that ancient civilization." When she tries to make him understand that it is a likeness of himself, Kong realizes how "monstrous" he must appear. After being protected by him on several occasions, and after delivering her safely to his cave lair high atop Skull Island, Ann has lost her fear of Kong. But it is here that Ann notices the skull and scattered bones of another giant gorilla. To her, Kong is a more reassuring, reliable friend or protector than Denham ever could be, even though both have rescued her from disagreeable circumstances, respectively, from within their own islands (Manhattan vs. Skull Island). Kong's ancestors are gone, but beauty and beast have achieved a special connection, mutual understanding, a nearly unfathomable relationship. They are as a "family." Kong is alone, not unlike herself in this primeval red in tooth and claw world, or including that world beyond the visible coastline from which she has escaped, one that Kong cannot possibly comprehend.

Their roles eventually reverse following Kong's capture, in New York City, where Ann does her utmost to protect Kong (first from Denham, then) from 'faunal horrors' a sprawling Metropolis has to offer. Majestic Kong – the last of his kind—does not triumph here, as he falls from civilization's tallest peak, an ending that tugs at your heartstrings. And those final fateful words, "It was beauty killed the beast," ring hollow this time, in this novel (i.e. as opposed to Lovelace's). For, to me, it wasn't so much Ann's beauty as the driving 'fault' or force at all, but rather Denham himself who should be largely blamed for Kong's death, for greedily exploiting and raping nature, furthermore, fatefully extinguishing the Kong species forever. All for sake of a movie; Denham's name should forever live in infamy.

Christopher Golden vs. Greg Cox! Both Golden's (born in 1967) and Cox's (born in 1959) achievements are artfully done. They're both prolific novelists. Since the mid-1990s Golden, an award winning author, has written numerous original novels, and dozens of television show tie-ins, as in the "Buffy," *X-men*, and *Battlestar Galactica* universes. Well over 8 million copies of his books are in print! Meanwhile, Cox also is a best-selling author, with numerous stories dating back to 1983. Among his genre writings include novelizations of *Underworld: Rise of the Lycans, Man of Steel,* and *Dark Knight Rises*. He's penned over half a dozen *Star Trek* novels and short stories. Cox also has several "Buffy" tales up his sleeve. Both writers skillfully infused themselves into, or rather *beyond* their respective scripted templates.

Perhaps unsurprisingly, Cox's isn't the first attempt at a Godzilla-based *novel*.[12] But his entry claims status as the first (non-"GINO") Godzilla movie *novelization* published in English intended for a grownup, as opposed to targeting a younger reading audience. Yet still, as many of you may suspect, Cox's *Godzilla* isn't the first *novelization* because popular Japanese sci-fi/fantasy author, Shigeru Kayama actually got to that hallowed ground first (to date, published only in a Japanese mid-1950s edition). We learn from Donald Glut's *Classic Movie Monsters* (1978), for example (p.403), that "An authorized Godzilla novel based on the first two movies and written by Shigeru Kayama was published in Japan." Thus, both Cox and Kayama tackled the same idea; 'godzi-novelizations' featuring giant monster battles (e.g. presumably an 'angilas' type adversary in Kayama's case, and a M.U.T.O – massive unknown terrestrial organism—in Cox's book). Cox's novelized tale also represents the 4th 'origin' story for the ever popular monster based on films, quite satisfyingly different plot-wise.

Once again emphasis is placed on man's senseless destruction of the environment, tampering with nature, but this time in a more unbridled fashion. In the first chapter, a nuclear device is detonated in 1954, spewing forth radiation and contaminated particles into the ocean-atmosphere system, in order to destroy a "leviathan." In the next chapter we learn Dr. Serizawa (a biologist) is visiting a waste-producing strip mine in the Philippines to investigate a strange yet enormous fossil skeleton found within the mineral enriched deposits. Later a parasitical larva makes its way to a nuclear power plant in Japan where it feeds on radiation, growing, metamorphosing, mutating into a gigantic praying mantis-like monster. Then a hydrogen bomb is set to explode near San Francisco. There is the ever present dread of radiation exposure.

And so, with this latest tale, we're exposed to a nuclear reactor accident in Japan, and a tsunami (mindful of the 2011 Fukushima tragedy). The MUTO organism, regarded by scientists as a "living fuel cell," is considered valuable (and therefore furtively studied) because of its implications for "solving the world's energy crisis." Appearance of the monsters is even regarded by some gamblers in Las Vegas as a "'…hoax, 'like global warming.'" Bad bet to ignore an 'alleged' attack (let alone a planetary greenhouse forged in reality) that is potentially apocalyptical, for the MUTO is symbolically "laying waste to the entire world. It felt like a prophecy!"[13] Yes, in the apocalyptical war against giant monsters (and ourselves), "Gaia" (i.e. the 'living' planet Earth) and our flimsy, interglacial civilization will not be left unscathed. After all, monster movies and their science fictional stories build on and capture the current wave of public fear. But even so, Cox's story can't be categorized under the current (heat) wave known as climate science-fiction, (or "cli-fi").[14]

Cox's characters are more righteous and concerned about taking a correct course of action, with Godzilla's unwitting help. There is no Denham-like scoundrel scrambling about for personal gain. Accordingly, at a time in our history when many lauded the anti-establishment whistle-blowers, much of the novel centers on an *X-Files*-like conspiracy. Joe Brody is a nuclear engineer upon whom weighty responsibility has been placed; he seeks truth behind a devastating reactor meltdown covered-up for the Press as an earthquake incident that killed his wife back in 1999. He sounds crazy to his son, Ford, until the parasitic MUTO emerges from its cocoon, but not before emitting an electromagnetic pulse (EMP) knockout punch. Compared to the MUTO's hugeness, ironically, man is but an insect!

From there, much of the novel centers on Ford striving to reunite his own family, while preventing the MUTOs –a 200-foot tall winged male and 300-foot tall wingless female (the latter emerging from a radioactive cocoon disposed of at Yucca Mountain in 1999)—from spawning their monster 'family' offspring which would result in humanity's doom. Irradiation of the eggs by the Bomb in San Francisco will only "catalyze" the larval bugs, resulting in an onslaught of unimaginable magnitude. (Perhaps a dangerous 'world overpopulation' subtext is intended here; everyone knows how quickly nasty bugs can proliferate.) Ford, who eventually thwarts their ability to reproduce, patrols with trepidation through a train tunnel leading toward his wife and son in San Francisco. The train's rail route, dubbed by Cox as a "heart of darkness" winds past a nuke-seeking and devouring MUTO. Eventually, during the final battle against the MUTOs, Ford personifies with Godzilla. In a sense he *becomes* the great mega-saurian, Godzilla (who, reciprocally, has become rather anthropomorphized).[15] If they haven't become 'family,' after what they've been through they're at least a monster-fighting 'team.'

But two of the most interesting if not appealing aspects of Cox's novel are the conspiracy angle, coupled with perspective offered by contemplative Dr. Serizawa on the balance of Nature. Furthermore, there is the not quite so unique monster-radiation 'origin' theory to ponder, as well as Godzilla's colossal struggles versus the MUTO. Serizawa's persona interestingly melds together bits of Melville's Captain Ahab, *Gojira's* Professor Yamane, and Mary Shelley's mad Dr. Frankenstein. Serizawa carries his father's pocket watch, which ceased working at the moment (and place) of the 1945 Hiroshima atomic explosion, yes—his very own symbolic "clock of doom." In a somewhat tensional moment, Admiral Stenz of the USS *Saratoga* reflecting US historical interests in ending that war can only awkwardly reply, "Quite the collector's piece." This brief exchange brilliantly underscores Serizawa's keen perspective. It also captures Japan's vs. America's clashing positions concerning use of nuclear weaponry, while also hinting at how the meanings of a "Godzilla"-type monster must fundamentally differ between our cultures.

So, no wonder Serizawa prefers allowing the symbolic forces of nature to duke it out ("Let them fight."), hoping that 'his' charge, Godzilla, will restore the balance threatened by the MUTOs. "Nature has an order, Mr. Brody," Serizawa states early on to Ford, "A power to rebalance." Yes, but our special problem is that Nature also has no special need or concern for man's trivial existence. Serizawa eloquently states the story's theme. "The arrogance of Man is thinking Nature is in our control, and not the other way around." This is a more pervasive Japanese perspective, one which the western world is only coming to terms with.

Serizawa's project is aptly and ironically known as Project Monarch (as in the butterfly). The gigantic finned, semi-reptilian creature he and Dr. Graham have been tracking the whereabouts of for so long (yet merely seen glimpses of) isn't a metamorphosing mega-bug (like the repulsive MUTOs). But the project name certainly is apt for the massive cocoon-weaving parasite that once ravaged alpha-predators (animals we as fellow vertebrates would more readily bond with or feel connected to) such as the Godzilla species in prehistory. Why such creatures would nourish themselves on radiation doses is intriguingly explained. "The (Godzilla) animal – and others like it – *consumed* that radiation … But as radiation levels on the (Earth's) surface naturally subsided, these creatures adapted to live deeper in the oceans, farther underground, absorbing radiation from the planet's core." This declining source of adaptive radiation 'theory,' suspending disbelief, is not an entirely new science fictional idea, but one which was introduced in part and at least as recently as in D. G. Valdron's short story, "Fossils."[16] Project Monarch has been managed in a clandestine manner for decades, adding to the *X-Files*-ish nature of the tale, and the well played conspiracy angle. A key, driving question for Serizawa and his Monarch team is, "did they ever actually destroy Godzilla with a nuclear bomb back in 1954 …. or not?"

Fortunately for us, *not*. For Godzilla emerges into our world not merely as a giant monster but as a *natural force*, first in midst of a tsunami, but later likened to a towering volcano – his fiery breath is described as "volcanic blue fire." Our monster is ready to deal with his primordial prey, assuming the role of restoring natural order by attacking the MUTOs, which appear like "giant alien insects" (not unlike, in form and function, those featured in the 1979 film, *Alien*). Serizawa believes only Godzilla, as Nature's balancing force, can contain the MUTO. "I believe he is that power."

When the monsters wage their climactic battle, "They charged at each other like divine beasts out of myth and legend." But this is not merely myth proven reality, it is also fundamental planetary biology. "It was survival of the fittest – on a grandiose scale," driven by "powerful biological instinct." An "invincible leviathan" versus "alien insects" encased in "obsidian shells." Godzilla is as foreboding and harrowing a creature as was Carson Bingham's Gorgo of 54 years earlier.[17] But whereas Gorgo's (and Ogra's) appearance served a stern "warning," Godzilla and the MUTOs fully embody that vengeance and wrath, incarnate. In Cox's capable hands, Judgment Day unfolds!

The battle rages through San Francisco, with the MUTOs nearly reigning triumphant, until Ford's torching of the egg sacs in the "infernal vista" that could have easily passed for the lower pits of Hell" underlying China Town distracts the female bug, allowing a badly injured yet vengeful Godzilla precious time to regroup and subdue the male in a fierce tussle, a Darwinian battle of tooth and claw. Godzilla's incinerating ray blazes through the female, vaporizing steely chiton and ichor. As Ford's son, Sam, knows from playing with his toys, in a bit of foreshadowing, "The dinosaur always wins."[18]

Whereas in foreign giant monster films, figures in authority are hesitant to pull the trigger on nuclear weapons wielded in order to defeat the approaching menace (preferring alternate, less panicky or disagreeable, fictional strategies such as 'oxygen destroyers'), this isn't quite the 'rule' in American films (and their novelizations when they're written). Ultimately the nuclear warhead discharges near San Francisco, even as MUTOs are already dying, presumably (ironically?) to rid the world of its solitary 'savior,' Godzilla.

Ultimately, Serizawa reflects, contemplating Godzilla's role, "Godzilla destroyed the MUTOs ... as Nature intended." Furthermore, "Godzilla was Nature incarnate, eternally resilient and unstoppable," truly the creature assigned to restoring balance in the natural world. "Nature, red in tooth and claw, had created him to be the ultimate predator and he had claimed that title beyond any doubt. Where humanity and all its technology had failed, he alone had saved the planet from being overrun by a plague of giant parasites."

So why all the fuss about giant dino-monster novelizations, especially since they aren't as routinely done as in yesteryear (although movie and television 'universe' tie-ins remain prevalent)? Novelizations generally garner only exceedingly short shrift attention in comparison to the parent films. Well, in case I haven't already conveyed this above, consider – where else can you enjoy those Cool Monster descriptions, but in print! (e.g. like MUTO's "obsidian shell)," and those blow-by-blow battle descriptions...... Hey, it's not like reading Shakespeare but it's all wonderful commuter train or lunch break reading.

The human brain is ~ 90% devoted to the sense of sight, which is why so many of us would rather see the movie or play a video game based on its "universe" than read a tie-in book. Too few of us read (much) anymore and it's a shame that these often neglected, on the mark novelizations aren't mined for the added information, explanations and dimensions they provide concerning the story behind underlying respective movies. I know a picture is said to be worth a thousand words, but the several thousand words and descriptions in movie novelizations fill in a great deal that may be omitted or left unembellished, or otherwise confusingly portrayed in the finished film.

Furthermore, without preaching (probably to a 'choir' here), if you're a fan of fantasy films and giant movie monsters, it's worth it to explore our founding material, often entrenched in the older classics: Mary Shelley, Edgar Allan Poe, Jules Verne, H. G. Wells, Bram Stoker, Arthur Conan Doyle, Edgar Rice Burroughs, H. P. Lovecraft, John Taine, John W. Campbell, Ray Bradbury, Isaac Asimov, Arthur C. Clarke, etc. Indulge! However indirectly, from their sinister inklings issued in a long dark, labyrinthine cascade came forth nightmarish, superhuman creatures leading to those such we cherish as Kong and Godzilla, as well as those stalwart human characters who would conquer them to quell our deepest (draconian) fears.

Notes to Chapter Eighteen: (1) From Greg Bear's Introduction to *King Kong* (The Modern Library, 2005 ed.), p.xxv. (2)Mark Cotta Vaz commenting in the Preface to *King Kong* (The Modern Library, 2005 ed.), p.xii; (3) Mark Cotta Vaz, p.xi; (4) Quotes in this paragraph, as recounted by Mark Cotta Vaz, pp.x-xi; (5) James V. D'Arc, Foreword to *Merian C. Cooper's King Kong* (2005), by Joe DeVito and Brand Strickland (pp.ix-xv); Mark Cotta Vaz, *Living Dangerously: The Adventures of Merian C. Cooper, Creator of King Kong* (2005); Pierre Du Chaillu, *Explorations and Adventures in Equatorland Africa* ; W. Douglas Burden, *Dragon Lizards of Komodo: An Expedition to the Lost World of the Dutch East Indies* (1927). Burden, an American Museum zoologist who studied Komodo Dragons on their Indonesian island habitat, inspired Cooper with visions of a battle between "his" giant gorilla and Burden's Komodo dragons. A term "King of Komodo" was a term uttered by one of Burden's expedition members, big-game hunter F. J. Defosse.;(6) Brad Strickland, John Michlig along with DeVito also produced a remarkably well written and fully illustrated novel "sequel" to the Kong myth titled, *Kong: King of Skull Island* (2004). I have discussed their book in my 2006 book, *Dinosaurs in Fantastic Fiction*, as well as in *Dinosaur Memories II* (2017); (7) Paul Niemiec, "Back to the Spider Pit! The Lost King Kong Footage did Exist and Now We have a Witness!" *Filmfax Plus*, Summer 1013, no.134, pp.56-64, 107; (8) I've discussed the background on Tyrannosaur locomotion, including the incongruous hopping trait in popular culture sci-fi, at length in my 2009 book, *Prehistoric Monsters: The Real and Imagined Creatures of the Past That We Love to Fear*, (McFarland); (9) In the 1933 movie, so many of the living, reptilian menaces (and sanctuary) encountered on his home island are replayed in New York City, although this time analogously and metaphorically being steel & concrete structures & war machine adversaries. Even Ann notices this as she foreshadows (i.e. in the novel) from a supposedly secure hotel room across the street where Kong has just broken loose, "It's like a horrible dream like – being back there – on the island." To me, this strange revisiting of events represents the true art of

Cooper's story, which is conveyed a little stronger in the 1933 film, rather than in the 1932 novel. The famous movie scene where Kong dispatches the snake-like elevated train is not written into the novel. And there are no filmic 'analogous' human machinations for Lovelace's Triceratops herd. (10) The first version of "Beauty and the Beast" published in 1740, was by Gabrielle-Suzanne Barbotde Villeneuve. See Wikipedia entry online. (11) Although not a "novelization," the year before release of Peter Jackson's *King Kong*, in 2004 a fascinating sequel to the classic tale was published by DH Press. This visually captivating novel, *Kong: King of Skull Island,* (created and illustrated by Joe DeVito; written by Brad Strickland with John Michlig), offers considerably more about Skull Island's fauna, the Kong lineage, and the native islanders. Rather than simply redoing another King Kong film someday, perhaps producers might consider this intriguing story instead; (12) During the late 1990s Godzilla craze, several Godzilla novels were written by Marc Cerasini and Scott Ciencin. None of these were "novelizations,' however. Cerasini's were tagged as "young adult novels," favoring readers aged thirteen and up, while Ciencin's were "juvenile novels," for readers 8 to 12. H. B. Gilmour did write a novelization (based on the screenplay written by Dean Devlin and Roland Emmerich) of the 1998 TriStar *Godzilla* film, although this was published by Scholastic Inc. and was intended for junior readers. Meanwhile, Stephen Molstad wrote TriStar Pictures' novelization for more mature readers—*Godzilla* (1998). (13) This characterization is not unlike 'Ogra's' and Moira's roles in Carson Bingham's 1960 novelization of *Gorgo*. (14) Lily Rothman, "Nature Bites Back," *Time*, May19, 2014, pp.50-52; (15) Two other examples of highly anthropomorphized giant dino-monsters include titular creatures featured in novels, Mark Jacobson's *Gojiro* (1991) and William Schoell's *Saurian* (1988); (16) Valdron's story was published in *Daikaiju!: Giant Monster Tales* (Agog! Press, 2005), edited by Robert Hood and Robin Pen, pp.181-195; (17) For more on Bingham's 1960 *gorgo* novelization, see Chapter Twelve in my *Dinosaurs Ever Evolving* (McFarland, 2016); (18) Well, maybe *not* always. For example in Alex Irvine's 2013 novelization, *Pacific Rim*, ("from Director Guillermo Del Toro; story by Travis Beacham; screenplay by Travis Beacham and Guillermo Del Toro"), we learn that giant kaiju attacking through a breach set into the ocean floor are distant evolutionary descendants of dinosaurs (which themselves were cruder, experimental kaiju versions) See Chapter Twenty-five in my *Dinosaur Memories II*.

Chapter Nineteen — *Modes of Survival: Godzilla 'versus' Kong (Part 2)*

While a parade of Kongs have appeared in films and stories since 1933, perhaps the main dichotomy stems from impossibly super-giant versions versus the ('merely') moderately large kind. The former may be daikaiju, while the latter arguably do not overreach the bounds of biophysics. Therefore, non-kaiju, moderately-sized kongs are likely more naturalistic beings, subject to laws of evolution, having a prehistoric past. How did such creatures originate … and *survive* extinction into modernity? Let's delve into this by contrasting two recent examples—Legendary's filmic King Kong from 2017's *Kong: Skull Island*, and Kong as portrayed in the similarly titled *King Kong of Skull Island* (2017), a book created and illustrated by Joe DeVito, co-written by Brad Strickland.[1]

Certainly Toho Studios deserved credit for introducing super-gigantic Kong—twice![2] The most famous and successful of these films was 1963's *King Kong vs. Godzilla*, where the idea of a possibly 'kaiju Kong' originated. Godzilla's foe in that classic film strained credulity, due to its gi-normous size. (I'll refrain from commenting on quality of costume's appearance.) Even children watching this battle of the century knew that no gorilla or ape could have attained such a huge size. And, for us kids who had already seen Kong's first appearance on television by then, Toho's Kong then also seemed entirely removed from the prehistoric fauna it lived alongside in the 1933 RKO classic—through 'red in tooth and claw' survival. (This lack of any prehistoric aspect was alleviated somewhat in Toho's second Kong movie—1967's *King Kong Escapes*, when that 3rd Kong variant battles another prehistoric denizen, Gorosaurus.) True—battling larger-than-in-life, anachronistically 'prehistoric' tyrannosaurs and pterodactyls on a misty island that Time mysteriously forgot (a hallmark tradition of dino-monster tales ever since), as in the 1933 RKO film precedes Kong's persona as a creature somehow out of prehistory.

Of course, in order to fight a towering menace such as radiation-fueled Godzilla, said opposing monster must also be gigantic—no matter what, Kong included, even if apes could never grow to such sizes due to biophysical constraints. But in Legendary's 2017 film, *Kong: Skull Island*, a prelude to his eventual, forthcoming 2020 blockbuster battle with Godzilla, this new ape has also attained gigantic sizes. The cycle repeats. So how 'naturalistic' is Legendary's Kong? While historical and prehistorical background on Toho's half-century old Kong entries seems highly obscure (merely borrowing or 'built upon' taken for granted information viewers may have recalled from watching the 1933 film), we may glean a bit more info on Legendary's Kong in this new millennial presentation.

"Monsters Exist!"

That's cryptozoologist Bill Randa's (John Goodman's) early utterance in *Kong: Skull Island*, as Project Monarch prepares to embark on a 1973 post-Vietnam War expedition to Skull Island—"the land where God did not finish creation." Among numerous, nearly unimaginable fauna encountered, most of which are non-dinosaurian, they collect seismic evidence supporting a "hollow Earth" theory, suggesting an underground lair inhabited by "ancient species" such as the Skull-crawlers (having two forelimbs, but no 'legs') that emerge to ravage terrestrial ecology. These, according to Randa, exemplify how "The planet doesn't belong to us." Such ancient species and others, like Kong, "owned the Earth long before mankind." And they threaten to "take it back" from us. Two kinds of fauna, however, would appear to be 'Mesozoic-esque,' as we witness vicious winged pterosaur-like creatures on the attack, and in the Kong cemetery we also see a fresh bone-white skull of three-horned *Triceratops*, indicating this animal must have died recently during Kong's long lifetime.

But here, in this place "where myth and science meet," Kong is enormous, terrifyingly towering perhaps 250 to 300-feet in height! Kong, summoning kaiju strength, swats, bashes and flings military helicopters out of the sky. He is the last of his kind though, having birth parents (witnessed as skeletons[3]) as well as a history recorded in pictographs by tribal natives. Kong does seem 'naturalistic,' to the point where he is represented as an ecology protector, an environmentalist monster. Kong restores eco-balance, preventing Skull-crawlers from running amuck on the island; he opposes man's destructive tendencies, recognizing men of warfare as 'bad'; he defends his territory; he even saves a giant 'water buffalo' from suffering under a downed 'copter. This island doesn't seem quite as 'prehistoric' as the 1933 Skull Island (or Peter Jackson's 2005 Universal version). Certainly though, here "Man is not king: Kong is"—not worshipped as a god, but regarded as a purposeful, defiant king.

The ecological theme is elaborated further in a tie-in graphic novel "Skull Island: The Birth of Kong" (2017), by Arvid Nelson and 'Zid.' Following conclusion of the graphic novel story, readers may read appended field notes 'written' by Bill Randa in 1973. Presenting science notes Skull Island's ecology has become sort of a tradition since publication of *The World of Kong: A Natural History of Skull Island* (2006) with a Foreword by Peter Jackson. (An interesting article concerning contrasting views on said ecologies that have been published about Skull Island's 'parallel universe' ecologies would make an interesting 'read.')

In the graphic novel's instance (and in complete contrast to the 2006 volume), however, there are few surviving Mesozoic-aspect creatures—or dinosaurs, not even a 'gorosaurus'—abounding there, with which to dramatically link Legendary's last living Kong to prehistoric provenance (as in both the 1933 and 2005 films). The Kong species has survived for millions of years on Skull Island, according to Nelson and 'Zid,' making Kong more or less a 'living fossil.'

DeVito's (Pre-)historical, Naturalistic Kongs:

No lineage of kongs could contrast more fully with Toho's than those created and illustrated by Joe DeVito for his 2004 and 2017 books, respectively, *Kong: King of Skull Island*, and *King Kong of Skull Island* (although my focus here will be on his 2017 'opus'). In their latter volume, DeVito and Brad Strickland offer a complete natural history and evolutionary glimpse on kongs of Asia that immigrated with an ancient civilization of people to Skull Island in the Indian Ocean, fleeing a major geological-volcanic, eruptive event 80- to 100,000 years ago ('conveniently' destroying much of the terrestrial fossil record in that region). The humans carried several dozen kongs aboard their massive ships from the mainland to Skull Island at that time.

Although *the* kong that later became known as 'King Kong' in the early 20th century was an aberrant 30-feet tall, normal height for an adult of the species was 'only' 18 to 20 feet. Somehow it is not entirely beyond the bounds of disbelief to accept that apes of such a size could have possibly evolved.[4] And yet these kongs—thought to have evolved 15 to 20-million years ago on the Asian continent—are thought most likely *not* closely related to the large Asian genus, *Gigantopithecus,* according to (fictional) paleontologist Dr. Vincent Denham (Carl Denham's 'son' in the new tale). To date, fossils belonging to the kong species have only been discovered on Skull Island. Part of being in the natural world order is having an evolutionary past—that, to me, makes a species perhaps more interesting, although somehow less strange or "mysterious" in any kaiju 'beast' sense. (Conversely, 'origins' due to extreme radiation exposure, or 'flight' through outer space would seem far more 'mysterious.')

It is their behavior and intelligence that makes kongs so distinctive, yet again not so unimaginable. Attributed with opposable thumbs made kongs downright 'handy' as tamed co-workers of the Zantu, the warrior tribe that befriended and tamed them so long ago. DeVito's kongs are club-wielding, highly intelligent and communicative using a "secret language" with their Zantu cohorts. They even helped build Skull Island's 700-foot long, reinforced Wall and, wearing leather armor breast plates for protection, triumphed over the indigenous dinosaurian 'deathrunners'[5] on military excursions. But in primeval times, as the tribal "Tagatu" civilization declined, the kongs reverted to a feral and savage state—King Kong himself being the most illustrious of the clan. (Like Legendary's 'legendary' monster, King Kong also had parents, as mentioned in DeVito's writings.)

Still, DeVito states that Kong is "certainly more than ape" (p.228), possibly due to his more complex brain. Kong may be the evolutionary apotheosis of his lineage. But then if he's somehow 'special' in the cranial department, deviating beyond others of his species, then is he sufficiently kaiju, despite his more modest bodily dimensions (i.e. relative to Legendary's or Toho's version of Kong)?

So is Kong Kaiju?

Ah, yes—finally that burning question, or dare I say the 100,000-pound 'gorilla' in the room! Everybody thinks they know exactly what a "kaiju" is, right?[6] "Kaiju" is a term in popular lexicon so overused it has lost meaning. And why should we care whether or not Kong is a kaiju? Well, for example, in a recent G-Fest (semi-objective) 2018 panel and polling of experts (see Chapter Thirty-nine), Kong tied with Godzilla as the greatest prehistoric monster of science fiction. Plus—he's the most interesting 'test case.'

Some of us are infatuated with "giant monsters," others prefer those that appear prehistoric but are still science fictional, and to many of these creatures the term "kaiju" is often loosely applied. In *G-Fan* no. 78 (Fall 2006, p.72), J. D. Lees questioned whether King Kong is a kaiju. He opined definitively:

> "The Toho Kong certainly qualifies, but the American version(s) are iffy, on two counts. First at 22 feet or so, Kong is large, but it doesn't require rewriting the laws of physics to imagine a mammal of such proportions. The prehistoric mammal 'Baluchitherium' topped out at 25 feet. So Kong's size, though impressive, may not be beyond the upper limit of what could occur naturally. Secondly, Kong's creators have taken pains to retain animal-like behavior in the big ape, giving doubt to whether he possesses any human-level reasoning ability. Unarguably, Kong does exhibit some humanlike behaviors and expressions, but not necessarily any beyond what most lay-people would consider possible for a gorilla. If push came to shove ... I would no longer include Kong as a kaiju, even though I have considered him one since my childhood ... in his American incarnation, Kong doesn't make the cut."

Shrewd reasoning. My opinions?

Both Legendary's and DeVito's Kongs are naturalistic. Both have an alluded prehistory ancestry. Comparing Legendary's and DeVito's, the main significant distinction, respectively is the relative sizes of each of their 'last' kongs. So on the surface, isn't Legendary's Kong more kaiju-like than DeVito's simply because of relative size? Legendary's version defies the force of gravity and biophysical scaling rules. Is greatly exaggerated size *the most* defining parameter of any kaiju contender?

We may suspect that DeVito's Kong lineage leading to King Kong most likely is non-kaiju, per Lees's definition. However, does DeVito's suggestion that King Kong's superior intelligence makes him anomalously 'more than ape' outweigh significance of the clear evolutionary lines, underscoring his 'naturalistic' element—connections to deep prehistory? In King Kong's case, which of Lees' two 'no. 3' criterions (see Note 6, below), should be given priority when assessing kaiju status? 'Inexplicability of origins' versus heightened intelligence! *In the case of DeVito's Kong*, the latter criterion favors kaiju-ness, while the former criterion (further mentioned in Note 6) does not.

Although, to me, not kaiju in terms of its 'naturalistic' element, Legendary's King Kong has (at least) two inescapable kaiju-qualities (great size and super-strength). But is Legendary Kong's instinctive 'eco-mindedness' demonstrative of yet another kaiju quality too—heightened intelligence, perhaps even on a profound 'Gaian' (planetary) plane?[7]

It may be that the individual monsters most of us fans like best are kaiju, gigantic and perhaps 'prehistoric' as well. Godzilla scores on all three points—Kong is a definite 'maybe'! In 2020, however, no doubt we will witness two towering behemoths collide again for their first grudge match in nearly six decades—Godzilla versus King Kong.

I can't wait!

Notes to Chapter Nineteen: (1) DeVito's 2017 take on Kong is an "expansion" containing considerably more information and historical background on the famous monster's ancestry and life, as published earlier in his (with Brad Strickland and John Michlig) 2004 DH Press book, *Kong: King of Skull Island*. I've discussed this in my *Dinosaur Memories II* (2017) book. And in Part 1 of "Modes of Survival," (also included in *DM II*), I introduced the basic premise for why the 1933 RKO classic Kong is to be regarded, anachronistically, as a 'prehistoric ape.' (2) John LeMay discusses other little known Kong movies that had some traction in Japan in his *The Big Book of Japanese Giant Monster Movies: The Lost Films* (2017). Presumably most of these Kongs would have been super-sized too. (3) Likewise, Legendary further hinted at Godzilla's prehistoric past through the use of a fossil skeleton belonging to the godzilla genus found in a mine shaft in 2014's *Godzilla*. (4) That kongs evolved on the mainland (perhaps during the Late Pliocene or Early Pleistocene epochs?) would not conceptually deter their trend toward larger growth. But large dinosaurians inhabiting Skull Island would seem counter-intuitive, when considering that on insular areas most species tend to evolve into 'pygmy' varieties via 'dwarfism.' This despite the remark appearing on p.185, "Everything has become gigantic!" Dinosaur remains recently found in the Transylvanian region of Europe are a fitting example of dwarfism, or the pygmy mammoths of Wrangell Island—in other words reduction in overall size and mass. But then over 100 millennia, shouldn't the kong species have also attained/evolved smaller sizes after arriving on Skull Island? (5) DeVito offers interesting speculative notions on the island's feathered "deathrunners," Gaw, and kongs (extrapolating from his 2004 volume). Gaw and the deathrunners, are sentient (rather anthropomorphised), "greatly evolved descendants"—dinosaurian survivors: they have opposable thumb-like digits, and have left archaeological traces of 'cultural' traits. It is speculated that Skull Island was formed "because of the cataclysm that destroyed the dinosaurs everywhere else, enabling any final vestiges of the Mesozoic era that somehow made it there to survive in that one special spot…" (p.222) Also, on Asia there once roamed another ape-like genus related to the kongs, the more primitive, subcutaneously armored "Thakus kongs," now extinct. See pp. 222-228 in DeVito's 2017 volume. (6) J. D. Lees defined kaiju as creatures having a triad of characteristics. Briefly, in his first estimation, "Its size must beyond the upper limit of natural and enable the entity to destroy buildings. It must present a menacing aspect. There must be something inexplicable in its origin or nature." However, Lees decided that the 3rd criterion was "superfluous," and so he replaced it with "It must show some degree of purposeful; intelligence beyond that of any comparable animal." My opinion, as readers may have surmised from reading this article, is that the 'inexplicability of origin and nature' term could still be retained as a useful 4th 'decider' criterion to settle certain cases. (Lees, "What is a Kaiju?" *G-Fan* no. 78, Fall 2006, pp.68-72) Meanwhile, Lees is now considering the possibility of a more elaborate kaiju "classification" system … (5/12/18, pers. comm.) Must kaiju factors be numerically 'weighted'? I'll leave such questions for greater minds to ponder.(7) Within a chain of personal e-mail communications dating from May 2017, Author Mike Bogue stated, "…the definition of kaiju can be somewhat elastic, but there is the danger of stretching it to the breaking point. … a past Supreme Court justice … said that he couldn't define pornography, but he knew it when he saw it. Perhaps the same thing is true of kaiju." Although I wonder … is

Legendary's Kong merely a faux-kaiju? Part 1 of this essay appeared in my *Dinosaur Memories II*, (2017).

Chapter Twenty — *King Kong Days*

So what's your favorite Godzilla movie? Not 'best', but favorite. Today, there are so many to choose from, But back in the day—my 'day,' there were only a handful. Two of which I'd seen televised (*Godzilla, King of the Monsters* and *Gigantis the Fire Monster*), and of the other three (*King Kong vs. Godzilla, Godzilla vs. the Thing,* and *Ghidrah the Three-headed Monster*), I saw two at a movie theater in Chicago Heights. It was hard then deciding which of these Japanese films was truly 'best,' but even more difficult leaving RKO classic Kong out of the equation. Why? Because Kong and Godzilla just seemed to 'complete' one another, even if they'd only battled in a single film—a non-decisive score perhaps to be settled in 2020!

As I mentioned in *Dinosaur Memories II* (2017, p.276) in "Planet of Apes versus Dino-Monsters," the primal theme of mammal-versus-reptile seems ingrained, "… universal. Does it have anything to do with the Cretaceous-Paleogene mass extinction event 66-million years ago, when—as the story is usually told, dinosaurs lost in the great evolutionary sweepstakes, whereas we, or rather our mammalian ancestors, 'triumphed'?" Therefore, shouldn't we naturally root for Kong (instead of Godzilla)?

Well first, to answer my lead question, I'll need to slip back in reverie to … my King Kong days.

By the time we moved to Park Forest, Illinois in August 1961, where I began 2nd grade at the Illinois Elementary School, I'd already seen *King Kong* with my dad at least twice, televised. What an epical thrill ride that movie was, and I confess that I became eternally grateful to my dad for introducing me to this film! He, in turn had seen 'Kong' at the movies when he was my age, escorted by his parents—the same year they saw animatronic prehistoric animal displays at the Chicago World's Fair—1933. And so if you asked me during the summer of 1961 which was my all-time favorite movie, the answer, in a heartbeat, would have been *King Kong,* featuring its heroic, yet tragic dinosaur-thumping brute.

But that fall, things changed, thanks to local Chicagoland television networks, when I became exposed to Japan's radioactive colossus—Godzilla!

By 1961 the world had changed considerably, since 1933. A second world war had been fought and won. Nuclear weapons had been invented and used both in war and repeatedly in test operations. The Soviet Union had quickly followed in pursuit of these bombs thanks to spy tactics. Korea had come and gone, but still not quite. While America's Kong was a relatively 'safe' monster, an endangered crypto-creature beast confined to his insular lost world—not posing dire threats to the planet, Japan's towering Godzilla was another matter entirely—reflecting what Man was beginning to fear most, an end to civilization—utter annihilation from the Bomb!

It's difficult to recall so many years later when I first saw *Godzilla, King of the Monsters* on a Chicago network, but Ted Okuda's and Mark Yurkiw's 2007 book, *Chicago TV Horror Movie Shows: From Shock Theatre to Svengoolie* offers clues. My experiences with Godzilla (and Gigantis) surely must have come from broadcasts on programs like *Thrillerama, The Big Show,* or *The Early Show,* all airing in Chicagoland during the early 1960s. These are the programs I avidly watched weekly while fortifying my knowledge base on the whole of monster-dom. When it came to monster movies then I was an equal opportunity TV-watcher. Godzilla and Gigantis were rare yet recurrent features during this period, and so I became enraptured with this formidable, fire-breathing, seemingly invulnerable 'dino-dragon' plunging the planet in peril.
These movies were dark, somber—unlike Kong's debut. While 1933's Kong seemed heroic and tragic, Godzilla was angst-incarnate, a dreadful omen. Yet, as I realized even then at so tender an age the Toho films reflected something sinister and disturbing, ineffably precarious and threatening about modern times. And yet how different were *those* times from 21st century modernity!

I can't be sure of the exact date but sometime in early June 1963 sight of a small ad in the Saturday movie section of *The Park Forest Star* shifted my world. (And with some swagger I might state that I was no stranger to the *Star*, having had my 2nd grade school poem "On Halloween Night" published in a kiddie section of the paper.) Anyway, that movie ad was more recently reproduced in Don Glut's *The Dinosaur Scrapbook* (1980, p.154). OMG—my mind was in a complete whirl. It was as if *everything* I'd been learning about monsters and dinosaurs led to this VERY moment! There on the newspaper page

I saw a mesmerizing drawing showing "The Two Mightiest Monsters of All Time ... in the most colossal conflict the screen has ever known! KING KONG VS. GODZILLA (In color)." Air force jets and helicopters were zooming in, train cars were falling from Godzilla's clutches, buildings were toppling and ships were torn asunder in frothy seas below their towering figures. In this ad Godzilla looked absolutely formidable, dinosaurian. But Kong ... well ... he didn't look right, not quite like I remembered him from his prior film appearance when he defeated the tyrannosaur and *Pteranodon*. But maybe that was because Kong had only been ineffectually portrayed in a drawing, instead of a photo. Regardless, I absolutely did not care because as I stated my terms to my dad right then & there in the living room, I simply HAD to see this movie ... ASAP!

And so he—being an old-timey Kong lover ... *he* took my brother Rick and I to see the movie that VERY afternoon. I still remember the thrill of standing in line outside the theater as the crowd was slowly escorted inside, seeing those black and white movie stills posted outside the ticket booth, showing photos of scenes we were about to solemnly witness. Now Kong's look seemed even further 'off' but I couldn't wait to feast my eyes on Godzilla—*in color*, no less! I should know, right—because I had become *the* 'expert' on such giant monsters (probably unlike all those other kids filing in line behind me). Nevertheless, I'm sure my dad must have been snoozing in his seat by the time Kong also was being lulled to sleep with soma berry mist while sitting atop the Diet Building.

Well, as we know the two monsters fought to a draw, but to me Godzilla 'won' overall because Kong's suit just didn't seem kong-ish or gorilla-like enough, while Godzilla had never looked better. Throughout the movie I kept forcing myself to foolishly believe that this particular King Kong version was the same as seen before (in the 1933 film) ... but convincing myself of this 'fact' was as difficult as trying to understand how or why Kong had gotten so huge. (Remember I was only 8 ½ years old then, and quite willing to be gullible whenever necessary to suspend my own disbelief.)

In an odd way, a scene from *King Kong vs. Godzilla* imprinted itself on my psyche. It stems from that part of the movie where Godzilla has destroyed the train and is pounding relentlessly toward the frightened girl. Well, I used to have a recurrent harrowing nightmare where I was playing in the (suburban) grassy field behind our house; suddenly air raid sirens would begin blaring! Then out of a rapidly expanding tornado storm cloud approaching from the north, Godzilla would materialize—heading, stomping directly toward, (gulp!) me ... *me*—left with no place to hide or duck and cover. But remember, it was still Cold War times.

Satisfyingly, former 'themes' of pending nuclear threat and also striving to explain Godzilla as a dinosaurian remained in the movie. There's radioactivity detected when the nuclear submarine is destroyed in an early scene while

Godzilla emerges from the glacier—a nice tie-in to the prior film's icy conclusion in *Gigantis the Fire Monster*. And until the end, when it is decided that the best way to vanquish both monsters is to let them destroy each other on Mt. Fuji (e.g. because they're "instinctive rivals"), use of the atomic bomb remains a possible, "last resort" alternative to relying on conventional weaponry. The atomic bomb is specifically referred to a handful of times throughout the movie. At one point it is mentioned that ten million people might perish if Godzilla reaches Tokyo—hence the necessity of pouring millions of Caltex gasoline into a deep trench (which will later result in a major groundwater pollution problem), and amping up those electrical high tension wires (that Kong later uses to floss his teeth) around the city.

The prehistoric provenance for both monsters certainly seemed so exciting to my dinosaur-minded self then. Following Godzilla's dramatic emergence, American news commentator Eric Carter states how the world is stunned to know that prehistoric creatures are alive in the 20th century! A scientist then states that Godzilla is akin to a "cross" between a tyrannosaur and plated *Stegosaurus*, having thrived 97 to 125-million years ago, founded on Japanese fossils resembling the Godzilla species. Viewers are supposed to accept that its resurgence in modernity owes something to "suspended animation." Regardless, I was enchanted that the scientists founded his argument on basis of a children's book shown in the movie that we actually had at home—Darlene Geis's *Dinosaurs and Other Prehistoric Animals* (Grosset & Dunlap, Inc., New York, 1959), with illustrations by R. F. Peterson. Meanwhile, Kong's prehistoric status was taken more for granted, since audiences were more familiar with his 1933 exertions on Skull Island. Nonetheless, despite their brain size disparity, this pairing of giant monsters were considered 'hated enemies,' further suggesting prehistorical conflict between them.

I came away absolutely enthralled with what I'd witnessed on the big screen that afternoon. A year later, our maternal grandmother Justina purchased those 1964 Aurora model kits of both Godzilla and King Kong—whose size and appearance had evidently been based on the 1933 movie, fortunately *not* the '63 film. I immediately latched on to the Godzilla model, while my younger brother Rick kept Kong. Our toy train set would soon be ravaged by these creatures out of the prehistoric realm. Now, approaching six decades years later, only my original Godzilla remains, (partially intact—i.e. minus its skyscraper stand). Months later the Godzilla Board Game arrived, (although artwork on the box cover proved far more enchanting than inside contents). When the Academy Awards were televised, I naively believed that *King Kong vs. Godzilla* might win for special effects! I remember being so disappointed when the movie wasn't even mentioned (or nominated).

I had always thought that Kong was 'supposed' to be the good guy in this movie, and Godzilla the scarier 'bad' monster. Partly this is because Kong, as an anthropoid, was more akin to our species, while Godzilla was a saurian dino-monster, emitter of radiation. But surprisingly a 2019 *G-Fan* article by Ted Johnson ("Godzilla: King of the Good Guys," no. 122, pp.56-60) altered my view on Godzilla's motives during that mid-1960s period. Maybe Godzilla was just minding his own business all along—or as Curly Howard of the Three Stooges might say, "I'm just a victim of circumstance."

A few years later, from my then 11-year old perspective, with the Cuban Missile Crisis safely behind us, and given that the Soviets hadn't started a nuclear war (as some feared) in the wake of JFK's assassination … for a little while the angst of living in the nuclear age diminished slightly. Godzilla no longer seemed to be the looming metaphorical radioactive threat it had once posed to the world as projected, say, on Chicago's Thrillerama show. For a discrete interval of time it didn't seem as if self-extinction of the human race was imminent or plausible, even though Toho conjured up another world-threat for us to face—Ghidrah (later known to American audiences and fans as King Ghidorah)!

Yes, the *Park Forest Star* seduced me once again, transmitting with such urgency and with evidently so much at stake … "Ghidrah! The Three Headed Monster Battles Godzilla, Mothra and Rodan For the World! All new sights! Never to be Forgotten!" I'll say. It all looked so cool!

Meanwhile my parents were becoming discouraged with my love of sci-fi shows and monster movies. And so thanks to news appearing in the *Park Forest Star*, despite much greater reluctance than before, in the fall of 1965 I was 'approved' to see *Ghidrah the Three-headed Monster,* at the same theater. No Kong this time, but Rodan was a welcome sight—a television favorite by then and from newspaper advertising this new monster Ghidrah looked eminently cool. (I just couldn't wrap my head around how fuzzy-looking Mothra could fare well against these bigger and more formidable brutes—but then that's why I needed to see the movie.) Now Ghidrah represented an all-consuming world threat, supplanting Godzilla's prior 'role' and meaning—now if *only* Earth's terrible monsters could cooperate to defeat the even more menacing outer space destroyer!

Crashing in from space not unlike that asteroid which doomed the dinosaurs 66-million years prior, Ghidrah took fiery form in a dramatic scene! Meanwhile Godzilla and Rodan were already bickering, before in the end teaming with a sensible Mothra larva to fight highly impressive Ghidrah, lest our planet be utterly destroyed by this common enemy. Yes—there was that geopolitical lesson therein. Stemming from an inexplicable heat wave gripping the island, a sense of utter urgency that 'something terrible or cataclysmic was going to happen' pervaded throughout the first half of the film—thanks to the 'Martian' Prophetess. She warns of the arrival of giant doomsday monsters and the ensuing destruction to come. Humans would suffer extinction when Ghidrah passes, an outcome which already happened to the civilization on Mars—now a "dead world." She even prays to the almighty universe to save our planet from annihilation! One might say she's… uh … 'my favorite Martian.' Yet when scientists hypnotize her using that weird machinery, probing for the truth during that 'You're in a deep sleep' scene, probably my dad was already snoozing.

Meanwhile Japan's authorities (again) claim they cannot authorize use of atomic weapons … especially when Mothra—the smartest and bravest of the monsters—can convince Rodan and Godzilla to collectively fight Ghidrah because with everything in the balance, Earth still belongs to them as much as it does to us.

To me, although not Oscar-worthy like *King Kong vs. Godzilla*, "Ghidrah" was great—until that jolting 'monster talk' scene. Yikes! While the buildup to the climactic battle was quite good, the monsters' comical antics ultimately seemed undignified, undermining the gravity of the circumstances Planet Earth was supposedly facing according to the Prophetess. Thereafter, Godzilla's persona transformed into a saurian super-hero, while Kong's diminished (until the new millennium).

Back in those VCR-less times, a full decade went by before I saw *King Kong vs. Godzilla* again. By then I was in college and left unimpressed by those embarrassing scientific explanations offered at the beginning of the movie concerning Godzilla's paleontological provenance. Why had I ever thought this movie merited an Oscar? My 'King Kong' days seemed sadly over.

But my dad had made a triumphant last stand in his quest to see a 'Kong' sequel that lived up to the original he'd seen as a young lad in 1933. That excursion led to a family drive-in movie in 1967 to see *King Kong Escapes*. Nuff said on that. I don't think he saw the 1976 remake.

In his *Japan's Favorite Monstar* (ECW Press, 1998), Steve Ryfle rated both *King Kong vs. Godzilla* and *Ghidrah the Three-headed Monster* 3.5 out of 5 stars—not too shabby. For comparison the parent 1954 film *Gojira* rated 5 out of 5, (while for purposes of comparison the awful *Godzilla vs. Megalon* only rated 1 star).

Something I've reminisced about is a strange correlation, indicating just how these movies affected me then. After I saw *King Kong vs. Godzilla* during the summer when I was a pending 4th grade student, my grades were very blasé— C+ overall at best. But during the fall of 1964, when *Godzilla vs. the Thing* came out my parents refused to let me see it—and lo & behold, I became a 5th grade star pupil. Then in the fall of 1965, when *Ghidrah* arrived, to my shame what an indifferent sort of student I became in 6th grade … 'earning' a miserable D in effort and an F in conduct by the spring. So you can gauge just how and to what indelible extent these monster movies got in my head. (Without making up excuses though, my paternal grandmother Edna suffered a stroke in early 1966, leading to her death in October, and for many months afterward, for me, that was a terribly depressing, stressful period.)

And so now we have a new entry, Legendary's *Godzilla, King of the Monsters*, its title recalling that of the very first Godzilla movie I ever saw, although apparently founded on the giant monster battle I knew in 1965 as *Ghidrah the Three-headed Monster*. A lead character in the blockbuster 2019 film states: "The mass extinction we have feared has already begun, and we are the cause … we are the infection. But like all living organisms the Earth unleashed a fever to fight this infection…" represented by the mythological-sounding "Titans" which will "guarantee that life will carry on." This theme and perspective is rather "Gaian" (e.g. treating Earth in totality as a living organism, as theorized by James E. Lovelock, as for example outlined in his popular 1979 publication, *Gaia: A New Look at Life on Earth*). But for the view of humans as infection, or perhaps a virus, I recollect Loren Eiseley's views, expressed in his 1970 book, *The Invisible Pyramid*.

Clearly the concept of extinction as conveyed 21st century style seems more spectacularly dire, as opposed to how us kids received analogous messages six decades ago. I fear this must mean planetary circumstances truly are worse now than during days of my youth.

How Kong's reemergence in 2020—as tragic hero or Godzilla's subversive foil, might soften Mankind's ultimate fate remains to be seen. Which monster will likely be Earth's mightiest savior, Godzilla or King Kong?

FOUR

1970 cartoon by Joseph Farris, Oct. 1970, copyright The Chicago Tribune. See Chapter Twenty-four for more.

Paleo-Books

I was raised in a house of wall-to-wall books—or at least well-stacked bookshelves in every room, including my own. Our dad had an amazing collection of old scientific books in his personal library forming a basis for his published researches. He once wrote an article on his 'dinosaur book' collection that was published in *Dinosaur World* (no.2, June 1997, pp.10-14, 42). Dad's collection of rare scientific volumes was later donated by my mother in 2016 to the Science History Institute—previously known as the Chemical Heritage Foundation—in Philadelphia. Included in his collection were many old tomes on the geological sciences, which I would often peruse. And so, more or less, his love of books was passed along to his three sons. And yet—referring to Chapter Twenty-one—my parents couldn't stand comic books! I suppose they didn't even think they were worthy of the word 'book,' which I first heard used in that context (i.e. individual comic books considered as 'books') during a 1990s conversation I had with Don Glut who worked in that industry for many years.

Entries in this section all have something to do with 'books'—that is, books with ties to paleontology in one form or another—from 'highbrow' to perhaps lesser-exalted works of literature. Chapters here run the gamut from comic books and sci-fi 'pulp-ish' dino-monster stories, to 'chapter' books that have caught the imaginations of both young and old. Only a mere sampling of what's 'out there'!

A few notes of possible interest.

Referring to Chapter Twenty-two, I found it satisfying when, through a mid-2010's reading of Alexander Winchell's *The Sketches of Creation* (1870), I came across the term, "paleo-artist." This term was perhaps then used for the first time to describe a growing profession that's become a mainstream avenue of dinosaur popular culture only in recent years. A little backstory: in December 1999, as a result of copious readings, I informed artist Mark Hallett of my finding that (as I then believed) he had coined the word "paleoartist" in referring to modern artists who reconstruct images of prehistoric life, as published in a 1986 volume. But as I later discovered sixteen years later, I was wrong; instead it was Winchell who had used the term over a century earlier.

I strongly encourage readers to read Jared Diamond's 2006 book, *Collapse: How Societies Choose to Fail or Succeed,* in context of Chapter

Twenty-four (here) to see where we're headed. Diamond discusses twelve eco-factors—or, to say the least, "problems"—that seem intractable to this reader, in terms of resolving our social plight on Planet Earth. Also, consult Diamond's illuminating *Upheaval: Turning Points for Nations in Crisis* (2019).

And finally, Chapter Twenty-five merits a bit further preliminary discussion. As John LeMay has commented, once you finish one book, you finally discover all the things you omitted from it. That was a difficulty I dealt with on the heels of my 2006 book, *Dinosaurs in Fantastic Fiction*, prompting a short series of additional articles printed subsequently in *Prehistoric Times* magazine. So the marine paleo-monster menagerie subject in this chapter addressed another set of stories that escaped recognition in my 2006 "*DIFF*." Furthermore, there've been so many newly printed dino-sci-fi stories published since the mid-2000s—thanks in part to universal availability of the Createspace platform—that it's rather difficult to keep up with those titles (so I've given up the ghost). Regardless the aim of *DIFF* was, as indicated in its subtitle, to function as a "thematic survey." Arguably none of the new-millennial publications fall outside any of the fundamental themes outlined in *DIFF*.

Of course, one of the most time-honored, repetitive themes of fantastic 'dinosaur,' and primeval monster (or perhaps, more broadly, "cryptozoological") fiction remains the "lost world" setting. So, for example, those marine crypto—relic paleo-monsters lurking beneath the waves—represent one more kind of 'lost world,' relatively unexplored depths! From their undersea realm *Megalodon* and *Kronosaurus* have surfaced, coming back to haunt us! But the *real* question and most disturbing always remains '*why*'?

Interestingly, cryptozoology monster scholar Justin Mullis has come across another lost world tale, originally printed in 1908, Henry Francis's "The Last Haunt of the Dinosaur" (*English Illustrated Magazine*). The tale involves two professors' safari expedition to cruelly obtain a trophy of an African sauropod-ish dinosaurian. This very short story predates Arthur Conan Doyle's *Lost World* novel by four years. Mullis also suggests that Francis's paleo-monster could have represented a 'first' for the fabled (although undiscovered) 'mokele mbembe,' of more recent cryptozoological intrigue. His discussion of this topic, "Science fiction and Rise of Cryptozoology," is presented in a 2019 volume—*The Paranormal and Popular Culture: A Postmodern Religious Landscape* (Caterne and Morehead, eds.).

These chapters originally appeared in the following publications:

Chapter Twenty-one—*Prehistoric Times*. Nos. 59-60, April/May 2003, pp.18-19; June/July 2003, pp.18-19.

Chapter Twenty-two—*Prehistoric Times* nos. 117 Spring 2016, pp. 42-43; 118 Summer 2016, pp.41-43; 121 Spring 2017, p.6.

Chapter Twenty-three—*Prehistoric Times* no.126. Summer 2018, pp.48-51.

Chapter Twenty-four—*G-Fan* nos. 119 Spring 2018, pp.26-33; 120 Summer 2018, pp.22-30.

Chapter Twenty-five—*Prehistoric Times* no.83, Fall 2007, pp.22-23, 27.

Chapter Twenty-one — *Traggnificent! A Comic Book of Prehistoric Bearing*

My parents despised comic books, but grandparents 'corrupted' me anyway. So during the early 1960s, when I tried to sneak comic books from my grandparents house over to our first home in Park Forest, IL, once my Mom found them, out they went into the trash! Superman, Batman, Spiderman, the Fantastic Four, Hawk Man, Metamorpho (remember that guy?) and Green Lantern may have used their intellect, death-defying invincibility and might to thwart the most dreaded villains imaginable, yet, "POW!" "BAM!", "OOF!", my superhero pals were powerless in presence of vile "Garbage Man," proving deadlier than Kryptonite in the eradication of my comics! I even created my own superhero character—Electroid—for an original comic book series that I drew with such woefully lacking skill, but I only mentioned that so I could finally get the title into real print after all these years. See how sneaky us comic book 'creators' can be? Just like Don Glut—as you'll soon read!

    I didn't realize there was another breed of comic book heroes, intrepid dinosaur-battling champions of a 'prehistoric' era, stalwart characters such as Turok—Son of Stone, Tor of a Million Years Ago, and Conan the Barbarian. Even Tarzan confronted dinosaurs and other prehistoric animals on numerous occasions in the comic strips. V. T. Hamlin's long-standing Alley Oop was a more comically-centered 'toon' figure. In Alley Oop, Hamlin often explored social and satirical themes instead of featuring dramatic tales of adventurous proportions. Yeah sure, Superman fought dinosaurian menaces in the comics too, but your chances of survival vastly improved when paleolithic 'experts' like Turok confronted the restless dinosaur horde! Sigh! If I hadn't been so sheltered, I would have devoured these tales faster than a tyrannosaur gulped down gobbets of 'Tops' flesh.

    Not until September 1982, when Donald F. Glut's *The Dinosaur Scrapbook* (1980) fell into my welcoming hands, did I realize just how prevalent dinosaur figures were in comics. Chapter 10 of Glut's classic collectible volume outlined all that I had missed about the history of dinosaurs in the comics. Glut wasn't just writing as an 'outsider' to the field either, as he had conceived and written comic book series of prehistoric, scaly brood-encountering heroes, notably the intrepid warriors Dagar and Tragg, and the world's first liberated cavewoman—Lona (a.k.a. - 'Devil Woman'). Curiously however, although Glut has written so much about dinosaurs-in-popular-culture, he has modestly published little concerning his own 'book' creation, 'Tragg.' (The trials and tribulations of Glut's aborted "Man-Lizard, Stone-Age Avenger," yet a fourth, original persona, a 'prehistoric-age' superhero, were recounted in his *Jurassic Classics* (McFarland, 2001, pp. 109-118. Man-Lizard was inspired from Joe Kubert's Tor as well as the influential film, *One Million B.C.,* 1940.) So, in early June 2002, I decided to get the full scoop on Glut's dinosaur-taming heroes, face-to-face, from the master himself, intending to 'demystify' Don's comic book character, Tragg.

Don's first attempt at an original comic hero 'Man-Lizard,' conceived during the mid-1960s, never reached fruition. But this 'Stone-Age Avenger,' who would have been outfitted with a caveman superhero garb—a Hawkman-like mask made of a ceratosaur skull, a suit made of dinosaur hide and mammoth hair trunks, paved the way to his commercially successful creations—Devil Woman, Dagar and Tragg. Dagar the Invincible, a sword-wielding hero who confronted remnant mammals from prehistoric times and huge saurians, made his Gold Key comics debut in October 1972. (For more on Dagar's 'sword & sorcery' adventures spanning 18 issues, see "The Trials of Dagar, Prehistoric Warrior," in *Jurassic Classics*, by Donald F. Glut, 2001, pp. 119-125.) Alternatively, Tragg's 'evolution' took a convoluted and 'alien' course through its run of 9 issues.

Don's first published 'Tragg' tale, titled "Cry of the Dire Wolf" (May 1975, *Mystery Comics Digest*) concerned prehistoric origins of the werewolf myth. Ages ago, a flying saucer crash-landed in a lake, releasing "alien chemicals" which react with the blood of a dire wolf. When a caveman drinks the fluid by light of the full moon, he is transformed into a were-(dire) wolf. Tragg kills the hairy demon with his axe, cast from silver ore. Cry of the Dire Wolf was intended as a one-time 'spinoff' from another Glut story "Scaly Death," (June 1970, *Vampirella Magazine*), however, Tragg ventured forth once more in "Cult of the Cave Bear," (*Mystery Comics Digest* no. 9), predating Tragg's own book. Here, a caveman zombie-master inadvertently resurrects a dead cave bear worshiped by the clan.

Next, Don's editor permitted Tragg his own unique publication, that is, on a conditional basis. First, each Tragg story had to stand on its own merit, because missed issues might throw readers off-track. Secondly, Tragg had to have experienced 'close encounters' with an alien race.

For those of you who may not recall, during the early 1970s America was in the throes of a post-UFO era, *2001: A Space Odyssey*-inspired wave of alien madness. In February 1970, Putnam had published writer and researcher Erich von Daniken's immensely popular book, *Chariots of the Gods?—Unsolved Mysteries of the Past*. It was easy to get caught up in the fervor. Von Daniken claimed to have obtained "astonishing proof," addressing questions "... unanswerable until our own space age," such as "Did astronauts visit the Earth 40,000 years Ago?" (Assuredly, he claimed.) In fact, he concluded, many of the famous (as well as obscure) archaeological structures and artifacts had an alien origin, having been being built, formed or created by ancient astronauts, cosmic "Gods," who visited Earth in prehistory. Yes, it certainly sounds like a 'stretch' today, like an *X-Files* television series episode. But at the time it all proved enormously fun and influential. A motion picture documentary by Sun International fueled Von Daniken's fire.

Seven years and three books later, with release of *Von Daniken's Proof: Further Astonishing Evidence of Man's Extraterrestrial Origins* paperback, Von Daniken, master of pseudo-science, still had supporters. In his idiosyncratic "Proof" (1977) Von Daniken claimed, "....we did not originate by ourselves and by chance." (p. 251) Yes, the 'Gods' from outer space had even molded the course of our evolution. (Can you almost hear that eerie-alien, danger-instilling *Day the Earth Stood Still* Gort-in-motion theremin sound playing in the background?)

Anyway, Don's editor at Gold Key gave Don the nod to write a "Sky Gods" comic, even though Don initially had aspirations to turn the Tragg book into a straight caveman/dinosaur feature patterned after Kubert's Tor instead. Don sneakily 'compromised' by weaving Tragg into the plot, and so by June 1975 issue no. 1 of *Tragg and the Sky Gods* ("T&SG") finally made it into print!

The basic storyline was that the alien sky gods, known as 'Yargon' who appeared as "magnificent specimens of humanity in skimpy outfits with gold skin and green hair," had taken aside several Neanderthals and tinkered with their evolution to give humankind a much-needed evolutionary assist. Then, over two decades later, the Yargon returned to review the results of their experiment,' encountering the characters of Tragg and his mate Lorn, who appeared as Cro-Magnon. Don recalls that the name Yargon was an 'in-joke' as his then wife's maiden name was Gray, letters of which were incorporated into the alien-sounding name.

Because of their physical and intellectual differences, Lorn and Tragg become shunned by their tribe. After discovering the truth of what the aliens had done to them, the pair became more intimately tied to the Yargon race. (And, in case you're wondering where the planet Yargon was, Don simply says, "... somewhere in space, maybe in another galaxy.") A female alien who is supposed to be the designated mate of the Yargon commander is attracted to Tragg, making

Lorn jealous. This turn of events causes the commander to become a "raving lunatic." In fact, on one Tragg cover, Lorn and the Yargon female are shown in a catfight, wielding spears and pulling each other's hair.

Besides being forced to "survive the myriad terrors of their own world," eventually, Tragg and Lorn must also thwart a Yargon military invasion. However, in 1977, the series ended before Don could complete the story. In one planned issue, the Sky God commander would have gone on a rampage because the woman he loved had fallen for a savage, (i.e. Tragg).

Don related one interesting 'flashback' issue of *Dagar the Invincible* wherein the titular hero was mystically zapped further back into prehistory, sharing a panel with Tragg. This occurrence may be the most convoluted aspect of Tragg's history and evolution. Don had wanted to write a crossover scene, where Tragg met Dagar, hence the latter's projection back through time. But an editor, who feared readers wouldn't be able to distinguish the two similar-looking heroes, decided that the Tragg figure should look more primitive, like a Neanderthal. "So, suddenly Tragg wasn't 'Tragg' anymore," Don recollected, "..and was renamed Jarn." But Don surreptitiously still had his way, "For if you look closely at one panel in the comic, the real Tragg look-alike figure made a 'guest appearance,' unnoticed by the editor."

Caveman Jarn, who confronted dinosaurs and pterosaurs with Dagar, was later revealed to be Tragg's brother! As it turned out, Jarn only looked more primitive than Tragg because the Sky Gods hadn't tampered with his genetics while Tragg, a product of Yargon bioengineering, turned out to be Dagar's direct ancestor. Sheesh, that's quite a 'yarn,' huh? Maybe Don and his editor should have conspired on some soap opera scripts too.

Tragg was illustrated by comic book artist veterans Jesse Santos and Dan Spiegle. According to Don, compared to other comic publishers, Gold Key was rather 'anal' with the scripts. "Every panel of the 25-page book had to be ruled off and descriptions were written in the panels to direct the artists." However, even though Don would meticulously attach xeroxes of the animals scripted into various scenes to help artists improve the accuracy of their illustrations, the artists would never fail to draw 3- or 4-fingered *T. rexes*.

The comic book industry has transformed remarkably since the mid-1970s, during Tragg's heyday. But even more so have abilities of movie animators—just recall the box office success of 2002's blockbuster release *Spiderman*. (Nuff said). So, just 'thinking out loud,' interwoven into the Tragg story, you've got dinosaurs, pterodactyls, prehistoric mammals, huge aquatic reptiles, with Neanderthals, spear-wielding heroes and scantily-clad cave-beauties living alongside aliens, all at large on a primeval, unspoiled earth! What a delightful, inspirational combination of ingredients, potentially, for a new movie taking advantage of the latest advances in computer animation.

And so, I don't wish to 'bragg,' but that's the 'rag' on Tragg!

Now let's survey the tales of Tragg, beginning with the debut issue, a story titled "Spawn of Yargon" (T&SG, no. 1, June 1975), written by Don Glut and illustrated by Jesse Santos.

"Spawn of Yargon" opens dramatically on a cliff-setting with stalwart, axe-wielding Tragg (resembling a muscular, long-haired 1970s rock star, sans beard) defending his mate, the buxom and beautiful red-haired Lorn, from a *Pteranodon*. With deftly aimed spear Tragg rescues Lorn from *Pteranodon's* clutches. The pair then seek refuge in a nearby cave from a band of uniformed troopers bearing weaponry—the Yargon! Although as human in appearance as Tragg and Lorn, the green haired and golden skinned intruders are....alien.

Next, a flashback explains the odd juxtaposition of armed humanoids with Pleistocene age cave-people. Twenty-five years before, an altruistically-minded team of Yargon scientists completed their expedition to Earth at a time when Neanderthals represented the 'zenith' of humankind, in a strange place where relic populations of Mesozoic saurians still thrived, intermingling with the contemporary sabertooth cats and woolly mammoths. The Yargon scientists, named Sky Gods by the local tribe, experimented with two pregnant Neanderthal females using their 'evolvo-ray' apparatus, evidently causing their developing fetuses to acquire 'advanced' traits, as possessed by the Yargon. When their offspring, young Tragg and Lorn, reach one year in age, the Yargon scientists return to their planet in their spherical spacecraft.

Of all the tribal members, Tragg's older brother—a Neandertal named Jarn—is the most accepting of Tragg's and Lorn's physical differences. Desperately fighting for their lives when they reach maturity, Tragg and Lorn are ostracized from their tribe, exiled in their primordial world amid untold dangers.

Meanwhile, we learn of a cultural revolution on the planet Yargon, wherein the scientifically-minded rulers are overthrown by those "...craving the excitement of battle and conquest." It is these Yargon warriors, not the scientific expedition, who invade Earth bent on destruction and conquest.

Pariahs Tragg and Lorn recognize the newly arrived Sky Gods wielding their weaponry against Earth's mightiest saurian, *T. rex* and *Triceratops*. The Yargon fire upon our protagonists before Tragg kills a Yargon warrior in self-defense, this bringing realization that the Sky Gods are not invincible after all.

In T&SG no. 2, (September 1975), Tragg and Lorn are captured by a disbelieving band of tribal members they tried desperately to warn of the Yargon's menace. Tragg and a Neandertal-like brute, Gorth, trek towards the Yargon's landing site to test the validity of Tragg's claim, but Gorth instead proves deceitful by attempting to kill Tragg en route. While Tragg battles for his life with an enraged sabertooth cat, Gorth wanders into the Yargon camp, only to be captured and brainwashed by the enslaving Yargon.

When Tragg survives his ordeal, the lead female Yargon officer, Keera, (named after two of Don's former cats), takes a shine to him, provoking bitter resentment from the Yargon commander, Zorek. The bond between Keera and Tragg only deepens after he rescues her from a bloodthirsty *T. rex* felled by Tragg's axe.

The Yargon capture Tragg, whom *they* refer to as 'mutant,' Zorek revealing his plan for conquest. Zorek schemes to unleash the great dinosaurs, woolly mammoths and sabertooth cats of this territory to crush and devour the tribe. While the Yargon trigger a stampede of behemoths directed to kill Neanderthals, Tragg escapes and assisted by Keera, repels the dinosaur charge. Yet, afterward, when the tribal leaders question whether it was Tragg or the Yargon who triggered the stampeding animals, Keera—now jealous of Lorn, lies, claiming Tragg was responsible. By shifting blame, Keera ensures continual worship of her people, the Sky Gods, while the tribe casts Tragg and Lorn into the wilderness once more.

In T&SG no. 3 (December 1975), titled "Slaves of Fire Mountain," the Sky Gods conspire against Tragg's former tribe, forcing their Neanderthal 'slaves' to build a fortress within a volcanic crater. Following a brief pterodactyl attack, outcasts Tragg and Lorn, joined by Tragg's brother Jarn, are captured by the Yargon and forced to do their bidding. Keera professes her love for Tragg, although he virtuously upholds his loyalty to Lorn. Zorek, now jealous for Keera's shifting affections, threatens Tragg with his 'devolvo beam' which would "... promptly revert (Tragg) to the form of an even more primitive (anthropoid) ancestor."

But Keera comes again to Tragg's aid, sabotaging the devolvo-machine and releasing Tragg. Now sealed within the volcano's subterranean depths, a monstrous croc *Phobosuchus* corners the small heroic band on the rim of a pool of fiery lava. In destroying the ridge underneath the crocodylian (causing *Phobosuchus*) to topple into the lava, "...a scalding, melting agony that lasts but briefly," the pursuing Yargon block their means of catching the escapees.

By the time of T&SG's no. 4 (February 1976) release, "Project Sabre-fang," the Yargon have become thoroughly nested inside their volcanic lair, their Neanderthal 'slaves' in total servitude. Tragg and Lorn come to the aid of their tribal members, those who had scorned them for their physical and anatomical differences and advanced mental attributes, features resulting from Yargon experimentation a quarter century before. But the scheming Zorek is intent on destroying the elusive Tragg, because his intended mate Keera is attracted to hunky Tragg's "...noble qualities..ALIEN to the warriors of Yargon...." So what's next?

This time Zorek uses his devolvo ray on two subjects simultaneously, a captive sabertooth cat and also the unscrupulous Neandertal 'slave' Gorth, to produce a hideous hybrid 'were-lion,' sabre-fanged monster. Sabre-fang pursues Tragg and Lorn relentlessly through the jungle and nearly kills Tragg, until Keera interferes. Keera spares Tragg's life but orders the fearsome creation to kill Lorn, who now flees for her life, swimming across a sauropod-infested river to safety. The were-lion meets its sticky, gruesome end in a tar pit surrounded by carrion-eating scavengers.

Following Sabre-Fang's sticky demise, next, in T&SG no. 5, "Attack of the Man Apes" (May 1976), Tragg and Lorn, seeking other tribes to form an alliance with against the Yargon menace, encounter a colony of primitive, hairy hominids who revere fire. They have deified flames emanating from a subterranean oil pool as their precious 'god.' When the fire is extinguished following a brush with a *Centrosaurus*, the natives grow restless because they don't understand how to make fire. So they become vengeful, blaming Tragg and Lorn for killing their Flame God. In the ensuing scuffle, Tragg and Lorn fall through a fissure into a cavern where so many large beasts have already met their cruel fate in the underworld. Only after a very close encounter with an *Acrocanthosaurus* does a flaming torch rekindle the oily residue, at the same time dispatching the dangerous predator.

Female warriors Lorn and Keera clash in T&SG no. 6 (September 1976), "Death-Duel." Teaming with another Neanderthal tribe of styracosaur-riding warriors, hero Tragg with Lorn launch an assault against the Yargon fortification in the volcanic Fire Mountain. Keera, having previously aided Tragg, is now Zorek's prisoner, but is sprung from her cell by an admiring Yargon officer. Keera's arrival prompts a 'cat-fight' between her and Lorn, when she claims her red hair is satanic, (while Keera's own green hair is that of a 'goddess'). Keera loses her struggle, proving that Keera is not a 'goddess' after all. Then Keera confesses Zorek's evil plan to attack the cave people. Enlisting the *Styracosaurus* 'cavalry' brigade, Tragg soon engages the Yargon within their lair, to free his fellow tribesmen.

T&SG no.7, "Battle For a World," (November 1976) chronicles an attack upon the Yargon's Fire Mountain. Slipping into hidden passages within the volcano, Tragg and Jarn sneak into the Yargon hive, only to be captured. Meanwhile, outside the Yargon warriors launch an aerial attack, causing the styracosaur 'tanks' to be buried in an avalanche of rock, a fate they readily extricate themselves from. Even Yargon's blazing stun-rays and blasters can't repel the styracosaur advance, as the inner lair is penetrated. Next, Yargonian machinery and devices fall into the lava, igniting mysterious chain reactions and a stirring within unstable earth. While Yargon forces fly to safety aboard their jet packs, the styracosaur team flees with freed comrades just in the nick of time—before Fire Mountain explodes is a massive "BA-RRROOOOOM."

Next, in "Master of the Living Bones," T&SG no.8, perhaps the series' strangest entry, an alien spiritual entity, 'Ostellon,' is restored by a trio sorcerers intending to change the future course of history. Using Ostellon's mystical powers, they hope to eliminate Tragg's descendants (i.e. Glut's creations Dr. Spektor and Dagar) from the shadowy time line leading to modernity. That way, Evil will rule for eternity on planet earth instead of Good..... Ahahahahahahahah! Ostellon is granted immortality and the ability to restore dead bones to life, a fiendish combination of powers. For how can Tragg fight and kill skeletons that are already dead?

Keera overhears the scheme and warns Tragg after he defends Lorn from an attacking *Deinonychus*. Keera warns that "...this Ostellon is as insane as Zorek..." Soon 'living' Neanderthal *skeletons* attack Tragg and his tribesmen, but Keera dispatches them with her blaster ray. When a 'living' *Tyrannosaurus* skeleton menaces Tragg, it is sent hurtling into the sea. Fortunately the bony *T. rex* is not buoyant and cannot swim. And when the sorcerers see how Ostellon has failed his mission, he suffers, "... the same power, becoming likened to the very bones (he) could once command..." In the final frame, Lorn is shown recording the history of their encounters with the Sky Gods on their cave wall.

For T&SG no. 9, "Where Prowls the Devil Shark," (September 1976), Don introduced a monstrously-sized, 120 foot long 'Meg' shark, (*Carcharodon*) which Tragg must ultimately grapple with underwater. By now, Keera is exiled from the remnants of the Yargon camp and living with Tragg's tribe. When Keera, unsuccessfully, attempts to subdue a rampaging ceratosaur, she is spotted by flying Yargon troopers. Keera refuses Zorek's affections while in the adjacent waters the hungry shark prepares to feed. Zorek orders the execution of Keera; but the squad's blasters only stun her, perhaps deliberately. She falls into the pounding waves below. Tragg and Lorn rush to her rescue, diving into the frothing waters...... only to confront the huge 'devil shark,' which, thanks to Tragg's spear, becomes bait for other marine saurians.

A second issue no. 9 was released in 1982, although this was principally a 60 cent reprint of T&SG no. 1 to which a shorter, new segment was added titled "Race Against Death." Here Lorn must battle a *T. rex* with her spear during her quest to obtain plants which the tribal shaman will use to cure Tragg of a venomous snake. A tenth (unpublished) issue was scripted.

After September 1976, no more Tragg issues were published. Perhaps by the early 1980s, the erroneous yet fantastical concept of dinosaurs coexisting with cave-people had become outmoded, especially given the popularity of novels such as Jean Auel's *Clan of the Cave Bear* (and sequels), and paleontologist Bjorn Kurten's *Dance of the Tiger,* staged in more realistic, late Pleistocene settings. As we know, however, pop-cultural icons are often recycled, and the Tragg 'universe' is certainly amenable to a retelling, perhaps in some form on the silver screen.

Chapter Twenty-two — *America's First Popular 'Dinosaur Book'?*

Ordinarily, consideration of classic paleo-books and tomes zeroes in on the old heroes of 19th century geology in their discovery of the rock record and expanse of geological time, emphasizing Scottish, British and French participants. American science was then considered a backwater – (i.e. 'us' striving to catch up to European overseas academic and professional standards). But as we know, American paleontology *would* close the gap, and beyond behind-the-doors scientific investigations and (then) abstruse monograph publications, eventually what was gleaned would also be, in turn, disseminated to the public through a variety of means, such as museum displays, film documentaries and of course – those old *popular books*. This nurturing, edifying practice is lovingly continued today.

An inherent difficulty for the casual researcher with identifying historical 'firsts' is that shortly thereafter, another still earlier example surfaces. And so here, I'm merely suggesting what *may* be considered America's first popular 'dinosaur book,' after discussing several earlier 'near misses' or, in hindsight, pretenders to a then as yet un-invented throne. Furthermore, despite the nationalistic flavor herein, it is not intended to be disruptive. For we should bear in mind that for their times, immensely popular books which today would be loosely considered as 'dinosaur books,' had already been published in England and France, long before American writers prospered in this time-honored arena. For example, the names—Gideon Mantell, Georges Cuvier, William Buckland, Louis Figuier, or Henry N. Hutchinson, Camille Flammarion, from England and France, may be especially familiar to readers.

So as a preliminary, we must qualify what is meant by "popular" book. For our purposes, rather simply, college textbooks and scientific monographs written by American scientists are excluded. Conversely, books incorporating published texts of a lecture series for edifying the public, or books otherwise intended for laymen – often those lavishly illustrated with restorations of primeval times (often printed on glossy paper)—are in. Generally, although not a hard set 'rule,' one notices that popular dinosaur books tend to show more life restorations (e.g. reproductions of museum paintings and sculptures) of prehistoric vertebrates within – thus riskily introducing more speculation, while in text books, drawings or photos of actual fossils and skeletal reconstructions are usually favored.

And so accordingly, perhaps it won't be too surprising when we eventually learn that the author of America's first popular dinosaur book was also a pioneering advocate of the science and art of paleo-restoration! But we'll come to that. Along the way too, herein, we'll learn who 'first' used the hallowed term "paleoartist" in its modern parlance. (Interestingly, it wasn't Mark Hallett after all.)

And so, given the above, consider two peculiar early examples as possibilities. Edward W. D. Cuming's (1862-1941) *Wonders In Monsterland* (1902) – involving two children on a time travel journey in which prehistoric life is encountered—was published in America; the publisher's parent firm was situated in London, and presumably the prolific author/journalist Cuming was not American but British. However, children's books perhaps merit their own special category, albeit a subset of 'popular' books, which I will not dwell further upon here. Francis Rolt-Wheeler's 1916 full-fledged novel *The Monster Hunters*, regarding a boy who learns about Earth's paleontological history from his Uncle during the course of two scientific expeditions was published in Boston. This is a charming book, incorporating many of Charles R. Knight's paintings. Its author, Rolt-Wheeler, was British. Despite their popularity, novels, including sci-fi books such as (Chicagoan) Edgar Rice Burroughs' 1918 trilogy beginning with *The Land That Time Forgot* with its distinctive phylo-evolutionary theme, are excluded too.

So naturally, in searching for America's first genuine and popular 'dinosaur book' per my definition, one will encounter several questionable cases. For instance, consider H. Alleyne Nicholson's *The Ancient Life History of the Earth* (1st ed. 1877), later published in New York in 1897. In his Preface, Nicholson (1844-1899) stated his hope that readers would find the book sufficiently "untechnical" such that general readers might also find it useful. However, this *is* a college textbook and, anyway, Nicholson was British, not American. Then, William F. Denton's *Our Planet: Past and Future – Lectures in Geology* (Wellesley, MA: 1868), although striving for more popular readership would appear to be a bit more scholarly in tone, or text-bookish in nature than a true 'popular' entry. And Denton (1823-1883) was also British. When it comes to objective categorization, there are few sharp lines.

Certainly one would anticipate that earliest instances of popular dinosaur book writings would be disinterred from among the ranks of popular geology or natural history, including "evolutionary epic" (of the non-textbook variety) themed books dating from the mid -19th century or before, now usually found upon dusty, rare book library shelves. This is because traditional means of writing about paleo-history for enthusiasts and laymen alike (or for younger readers) became pronounced within the time-honored theme of 'exploring' or describing life – fauna and flora—as it was thought to have existed, leading *sequentially* 'upward', through the dim corridors of geological time to the present.

Many life through geological time-themed examples exist – indeed, over two centuries' worth, even beyond the realm of popular books extending into other venues, such as museum displays! We also see this theme conveyed in film documentaries, perhaps the most recent discovery being a recently restored 1925 film titled, "Fifty Million Years Ago," a rare, 7-minute trade film (Service Film. Company), first released in July 1925. This explored paleontological life through time, utilizing stop-motion animation and two-dimensional graphical techniques. This film begins with the nebular hypothesis of planetary formation, and then briefly presents visual dioramas of life from the Cambrian Period through the Recent Ice Age. Dinosaurs, including a *Stegosaurus* fancifully wriggling its (rubbery-looking, flexible) plate-osteoderms, were emphasized as several of Charles R. Knight's 1897 painted and sculptural restorations were brought to life in realistic looking dioramas. The film was preserved in 2015 by Colorlab Corp, for the National Film Preservation Foundation. And, as another example, there were series of (7.5" X 10.5" size) collector promotional cards printed in color, titled "Tiere der Urwelt," by The Kakao Company of Hamburg-Wandsbek, Germany, issued in (circa) 1905. The artist of the particular 30 card set I own was "F. John."[1] But this latter example came from overseas in Germany; it's not American in origin and wasn't published as a book.

There were other early examples in the ever-popular life through geological time vein. But now, cognizant that other such examples may yet fall out of the woodwork, let's get back to other early, now obscure, or otherwise neglected books leading to my 'straw' suggestion for America's first dinosaur book.

Beginning in the mid-1800s, there were, of course, textbooks published on paleontology and geological history in America. But the topic only became more popularized by American writers here during the latter half of the 19th century, with extensive works such as by Alexander Winchell (1824-1891) and William D. Gunning (1828-1888), appearing. At the time, compared to today, while bones of large prehistoric vertebrates generally proved most fascinating to the public, dinosaurs were often less appealing to the public than perhaps the large extinct fossilized mammalian species. In fact, dinosaurs really didn't become much of a 'fad,' until the early 20th century (i.e. not considerably long after more of their remains had been discovered in America, analyzed and displayed). So during those early decades (rather ironically considering my use of the term), what we typically might refer to as 'dinosaur books' in modern parlance—which we may take to include popular 'life through geological time' works on paleontology including discussion of the Mesozoic Era and its predominant life forms—*wouldn't* necessarily have emphasized dinosaurs in proportion to certain fascinating fossil organisms existing in other Eras.

A number of recent scholarly works have shown how prevalent during the mid to late 19th century, was the theme of writing popularly about Earth's life through geological time. Many of these (pioneering) writers weren't 'scientists' *per se*, yet they found the theme profound and compelling, as did their loyal readership. However, this recent scholarly focus has been on Victorian era (British) popular writers, while ignoring contemporary American entries.

To me, America's first 'dinosaur book' (however obscure it may be today) should truly emphasize dinosaurs to some degree, implying that there must be sufficient scientific backing for stating something more concretely about them. However, said dinosaur discussion may be embedded within a pleasing life through geological time 'tour.' And to qualify (in this admittedly subjective analysis), naturally the item should be 'book length' (neither a magazine or newspaper article, nor a short museum pamphlet), yet published in the United States, and its author should be American.

Okay – we're approaching the moment you've been waiting for – which early book may have been America's first popular "dinosaur book"? …. I've pared down a list of possible contenders to three. The first two authors in question are Alexander D. Winchell and William D. Gunning. Let's consider their early entries first. How peculiar that very few readers may have ever heard of either!

Winchell's biography and accomplishments were nicely summarized within George P. Merrill's massive *Contributions to the History of American Geology* (Report on U.S. National Museum, 1904: Washington, 1906, pp.189-733). Born in New York, his early career focused on lecturing and teaching about science, especially geology – beginning in 1849, and was appointed a professor at the University of Michigan in 1853. A biographer stated of Winchell that, "His largest educational platform … was the public platform. Here he was under no constraint by reason of youthful auditors. No limits were set to his rhapsodic scientific eloquence, no courteous regard for the amenities of possible professorial etiquette hampered the free flow of his criticism, or the exultant prophecy of the betterments of the future… Himself a working geologist in the field, he was well acquainted with geological methods. A teaching geologist in the university, he was skillful in imparting his own knowledge and in training others…" Merrill added, "Professor Winchell was one of the best known of popular writers and lecturers in geology." A "Robert Bakker" or 'Edwin H. Colbert' of his time perhaps?

By 1869 Winchell was commissioned as Michigan State Geologist in 1859 and directed the State Survey. Although he also held other prominent positions throughout his career, he became more widely known as a writer of popular geology books. Besides his first, i.e. *The Sketches of Creation* (1870) published in New York, he authored *Doctrine of Evolution* (1874), *Pre-Adamites* (1880), *Sparks From a Geologist's Hammer* (1881), *World Life* (1883), and a textbook, *Elements of Geology* (1886). The subtitle of Winchell's clearly written *Sketches of Creation* (to be outlined below) is lofty, grandiloquent, yet also revealing: "*A Popular View of Some of the Grand Conclusions of the Sciences in Reference to the History of Matter and Life, Together With a Statement of the Intimations of Science Respecting the Primordial Condition and the Ultimate Destiny of the Earth and the Solar System.*" Not exactly a 'theory of everything,' but definitely taking readers well on the way there. Besides the geology outlined therein, Winchell's entry also is an older school excursion into 'natural theology.'

Aptly referring to paleontologists as "astronomers of time-worlds," Winchell's *Sketches of Creation* covers the totality of geological time sequentially, introducing each new cast of extinct species onto the primeval 'stage' as the metaphorical 'curtain' unfolds – a long since abandoned style of transitioning through geological history. Eventually he lapses into speculations concerning futuristic planetary visions. *Sketches of Creation* is eloquently written and amply illustrated (borrowing several prehistoric restorations by Edouard Riou from Figuier's *World Before the Deluge*).

Over a century before the word became in vogue, Winchell uses the term "paleo-artist" in modern parlance on p. 203, and writes – possibly appearing for the first time in a popular book – about the then relatively new dinosaurs, *Laelaps* and *Hadrosaurus*, titanic reptiles which he thought may have lived some 100,000 years ago. His focus is on North American geological events, only occasionally delving into European discoveries. Winchell views geological development on a steadily cooling Earth, however, as being anthropocentric/providential for mankind, preordained, leading to our ascension. Interestingly, he dismisses Darwin in several passages. I was delighted to learn of a "Col. Wood's Museum" formerly existed in downtown Chicago, a few blocks from where I worked, that housed a 70-foot long *Zeuglodon* skeleton, no doubt destroyed in the great Chicago fire of 1871. In concluding his book, Winchell swiftly summarizes the whole of geological time, written in the style of the "waking dream," popularized by Jules Verne in a famous *Journey to the Center of the Earth* passage.

So, although "popular," Winchell's intriguing entry – an attractive, engagingly written publication—shall be dismissed here from further consideration as America's first popular dinosaur book. Despite its life through geological time flow, there is too much emphasis on economic geology and futuristic planetary speculation to properly qualify. Next, what about Gunning's book?

According to William Henry Larrabee's article in *Popular Science Monthly* (vol. 50, Feb. 1897) William D. Gunning delivered countless popular lectures nationwide on "life, evolution, American antiquities, and social theories," but mainly concerning geology. Born in Ohio, this degreed, spiritual individual may also be clearly counted among the early 'Darwinians' in America. Larrabee mentions that Gunning was often attacked by those opposed to his evolutionist views, in which he preferred to express the "facts." Arguably, his most famous book was *Life History of Our Planet* published in Chicago in 1876, illustrated by his second wife, Mary Gunning. (It was rare then for a woman to receive acknowledgment on the title page of any kind of scientific work, as was the case here.) Today, Mary's curious frontispiece showing "Wyoming in the Later Eocene" is perhaps known slightly better than the volume in which it originally appeared.

Overall, Gunning's *Life History* is another kind of life through geological time tour. He takes what would be considered a possibly 'Gaian' view of planetary biogeochemistry (p.30), and ponders immensities of time (e.g. hundreds of millions of years) for occurrence of certain geological events. In one interesting passage he takes a bird's-eye view of life history, roving backwards through time for readers' perspective. Given Mary's frontispiece, there is expanded discussion of the *Uintatherium* (pp.67-69, 249) and the Eocene age: he uses the metaphor of a palimpsest for the rewriting life's history within the Archaean rock record. (p.90) Gunning adeptly discusses similarities between avian and reptilian hind limb anatomy (commenting on *Archaeopteryx's* significance), during the book's outline of Mesozoic vertebrates. "Birds came from reptiles, and Dinosaurians represent nature in the transition." (p.140)

Perhaps unsurprisingly, his digressions on the Carboniferous, recent Ice Ages, then recent embryology studies and interpretations, evolution of the horse, and origins of man are considerably longer than those on dinosaurs. While recognizing that the latter group disappeared mysteriously, Gunning has relatively little to say throughout about extinction. In in contrast to Winchell, Gunning is a disciple of Haeckel, Darwin and Huxley, favoring processes of natural selection, adaptation, descent and modification for the development of life through the ages, up through the origin of races of man. Gunning is also familiar with Sir John Lubbock's works including his influential 1865 book, *Prehistoric Times*.

So did Gunning publish our first popular dinosaur book? No. It trends too far into anthropological matters to qualify. And so on to the final entry. Ta da!

Frederic Augustus Lucas (1852-1929) was born in Plymouth, Mass. He had little formal education and was not an 'academic.' He learned taxidermy techniques during an 11-year long stay at Ward's Natural Science Establishment in Rochester, New York, gaining expertise in the anatomical mounting of birds.

This led to a position of Curator at the National Museum of Natural History in Washington, in 1882. He eventually was appointed Director at the American Museum in 1911. Lucas is credited for his study of the *Stegosaurus*, resulting in collaborations both with G. E. Roberts and Charles R. Knight who each produced painted life restorations in 1901, and also subsequently a 1903 sculpture by Knight. Lucas' novel idea was that this dinosaur has a *double* row of plates, including the staggered pair geometric arrangement.

During this time, Lucas also authored his guideline on the techniques of paleorestoration, titled "The Restoration of Extinct Animals," (Smithsonian, 1901), perhaps the first published instruction manual of its kind. Much of this material was incorporated into and expanded upon in his 1901 book, *Animals of the Past: An Account of Some of the Creatures of the Ancient World* (American Museum of Natural History, 1901), intended for general audiences. While there is general life through geological time 'movement' throughout chapters of his book, along the way Lucas also emphasizes what may be gleaned from study of fossils and how restorations of extinct animals are made. Dino-aficionados will note that Lucas pointed to the lack of fossil evidence for tail-dragging in dinosaurs, considered that birds and dinosaurs are not so "far apart" skeletally, and even suggested that dinosaurs' blood may have been warmer than that of reptiles. After delving into evidence for evolution of horses and elephants in later, separate chapters, in a distinctively modern 'take,' Lucas also pondered possible causes of biological extinction ... while already warning of dire, devastating consequences of mankind's eradication tendencies. In short, Lucas' book has all the hallmarks and flavor of a popular modern 'dinosaur book' ... (although, guilty as charged, I've unfortunately explored this topic via 'whig history' – that is, employing modern standards of categorization to circumstances of the past.) In any case, Lucas' 1901 book records when distinctively 'modern' (American) approaches in outline emerged in regard to publishing facts about dinosaurs (and those 'other' prehistoric animals) for lay readers.[2]

Finally, all three books (i.e. by Winchell, Gunning and Lucas) do delve into primeval man, although each with different 'spin.' (Charles Lyell's book, *The Geological Evidence of the Antiquity of Man* had been out since 1863, with various authors taking up the theme to some degree.) Gunning maintains a pronounced evolutionary perspective (in one instance even mentioning Neanderthal), while Lucas soft-pedals the matter, instead discussing other contemporaneous Ice Age mammals such as the mammoth and mastodon. And that is another indication, for modern popular dinosaur books usually do not also feature much information on the origin and evolution of mankind, which is the field known as paleoanthropology.

(Thanks to my father, Allen G. Debus, for making several of these old volumes known to me, obtaining editions of them decades ago.)

Note to Chapter Twenty-two: (1) However, I misspelled the artist's - i.e. "F. John" - name 4 out of 5 times erroneously as "F. Long" in several of my prior books; a regrettable case of "Long-John(s)" dyslexia, perhaps? (2) Of course, by Lucas's time, far more—scientifically, was known of dinosaurs than during decades when Winchell and Gunning were writing.

Chapter Twenty-three — *Two Most 'Significant' Dinosaur Books of the Dinosaur Renaissance*

During the mid to late 1960s, a peculiar phenomenon began in such subdued fashion that few would have been able to trace or predict the course of its direction or extent in half a century's time. Even worse for the quietly burgeoning 'movement,' as it were—using slang of the 1960s, was the fact that it flowered ever so slowly in a field of science long considered (literally and figuratively) dead, then dominated by fuddy duddys often derided as 'stamp collectors.' That field of endeavor was paleontology, the movement therein began with dinosaur science, and within a decade it had its name—the Dinosaur Renaissance.

Now nearly five decades since its beginning, lasting longer already than the ages of many who will read this, there has been such an outpouring of books published on dinosaurs during this interval, for various streams of audiences addressing such a wide variety of venues and themes—ranging from fiction to continually updated facts. It is impossible to remember, let alone own them all. Surely there have been hundreds (thousands?). And yet we supposedly still remain within an evolving, ever-branching dinosaur renaissance ("DR") state (or instead is it a dinosaur "revolution"?). But of all these noble and well-intentioned publications, arguably, two stand out as 'most significant' during this DR episode in the history of paleontology.

Here I'm not referring to favored children's books from the past, lavish 'coffee table' books brimming with magnificent paleoart, or those concerning mass popular culture of dinosaurs in everyday life (parks, toys, museums, trading cards and comic books, movies, etc.). Instead, let's identify those two popular books that really instilled the newly minted message for intellectually minded souls, inaugurating heightened phases of understanding, convincing laymen and enthusiasts alike that we were/are in midst of an irrevocable quantum shift in our perception of dinosaurs. The two books? Let's hit the highlights! They are Adrian J. Desmond's *The Hot-Blooded Dinosaurs: A Revolution in*

Palaeontology (1975 – 1st ed. UK, published by Blond & Briggs; 1st US ed. printed in 1976 by The Dial Press, w/US paperback ed. by Warner Books in 1977), and Michael Crichton's *Jurassic Park* (Alfred A. Knopf, 1990). These, more than the rest, fostered new ideas from the old, and became iconic—each distinctively.

Although origin of our "renaissance" in dinosaurs is usually referred to discovery (1964) and scientific description of North American genus—*Deinonychus* (*Osteology of Deinonychus antirrhopus, an Unusual Theropod from the Lower Cretaceous of Montana*, by John H. Ostrom, Peabody Museum of Natural History—Yale University, Bulletin 30, July 1969), for me personally, I was alerted to the fact that something was significantly 'in motion' regarding the study of dinosaurs during the summer of 1971. Fortuitously, I came across an editorial in *Nature* magazine (vol. 229, Jan. 18, 1971, "Changing Dinosaurs—but not in Mid-stream"), discussing Robert Bakker's astonishing view that sauropods were terrestrial, not semi-amphibious, swamp-wallowing creatures! Incredibly, this idea had even been proposed three years before, unknown to me (well, I was only 14 in 1968). This eye-opening article also briefly introduced ideas on tail-lashing sauropods and dry land-lubbing hadrosaurs. But what really riveted my attention, was a reproduction of Bakker's 1968 restoration showing a pair of parading *Barosaurus* with necks held high and tails outstretched, elevated from the ground in 'natural' position—looking so different, so alien when compared to convention, or—what was 'supposed' to be correct. Although Bakker's *Barosaurus* painting left an indelible impression, the new, fresh ideas seemed to fade thereafter in the public sector, that is, until 1975. And just who was this upstart "Bakker" who dared to challenge convention anyway.

Before Bakker became an iconic paleo-persona, there was Edwin H. Colbert (1905-2001), another great popularizer of dinosaur science. While the dinosaur renaissance—as it would eventually become heralded—remained in its infancy (c. 1964-73), Colbert published three sumptuously illustrated, popular books on dinosaurs, targeting savvy, non-juvenile audiences. I learned my dinosaur science and the history of this peculiar profession from Colbert's outstanding, 'trilogy': *Dinosaurs: Their Discovery and Their World* (1962, but I owned a copy from the 4th printing—1964); *Men and Dinosaurs: The Search in Field and Laboratory* (1968); and *Wandering Lands and Animals: The Story of Continental Drift and Animal Populations* (1973).

During my youth, Edwin H. Colbert became a 'scientific hero' of mine. While, for his time, Colbert might have seemed 'progressive' (and diplomatic), he would later be considered 'old school.' He espoused generally 'cold-blooded' views on dinosaurian physiology and a conventional 'thecodontian' evolutionary origin for the saurischian dinosaurs (while that of ornithischian genera was "more obscure" and as yet unresolved). Colbert, also an expert in paleo-mammals, was a geological gradualist—terms like mass extinctions had yet to be widely accepted in context of modern paleontology. Colbert made no definitive evolutionary ties between birds and dinosaurs. But Colbert contributed to acceptance of plate tectonics theory through his co-discovery of *Lystrosaurus* bones on Antarctica, as outlined in his *Wandering Lands* volume. His wife, Margaret was a noted paleoartist whom I interviewed during the late 1990s for an issue of *Dinosaur World*.

Surely there were other, more generalized, informative paleo-books newly circulating then. Two stand-out titles were Herbert Wendt's *Before the Deluge* (1968) and Willy Ley's *Worlds of the Past* (1971), the latter of which seemed adapted for a more adolescent readership than the former. But when it came to mainstream learning about the nature of dinosaurs, say, preparatory for college training needed to become a real dinosaur paleontologist—Colbert was my 'go to' guy!

Yes—Colbert's was the world of dinosaurs I'd grown into, but change was afoot. Besides plate tectonics, Stephen J. Gould and Niles Eldredge were hammering out tenets of an evolutionary tempo corresponding to "punctuated equilibrium." Meanwhile, John Ostrom and Bakker were re-exploring not only more direct, possible evolutionary connections between certain kinds of dinosaurs and modern birds, but reevaluating the entire basis of dinosaurian biology and metabolism—finding that, like birds, many were also warm-blooded creatures. Astrophysicists pondered whether radiation from a 'nearby' supernova extinguished the last of the dinosaurs circa 70-million years ago. (In time the Late Cretaceous 'sliver' of the geological time scale would also be further refined.) And a paleoartist (Sarah Landry) had the audacity to adorn a (non-avian-) dinosaur (*Syntarsus*) with feathers! While all this was brewing in scientific literature, with some sensationalized ideas dribbling into the popular press, an author was compiling a truly wonderful book capturing the essence of our vastly shifting vista on dinosaurian world order.

At first Yale University became *the* locus for America's shift in dinosaur philosophy—modernized views extending to every aspect of dinosaurs. Besides Ostrom's aforementioned July 1969 landmark monograph on *Deinonychus*, the new ideas were also memorialized in two Yale Peabody Museum *Discovery* publications, (vol. 3, no.2, Spring 1968, and vol.5, no.1, Fall 1969). In the former, Bakker rhapsodized on "The Superiority of Dinosaurs," launching an all-frontal attack on convention, while in the latter, Ostrom in more subdued, concisely worded fashion reconfigured our image of theropod dinosaurs based on skeletal anatomy found in *Deinonychus*. And although these publications were not widely circulated therein, for the first time the public was able to see Bakker's outstanding and convincingly 'real' paleoart restorations. Colbert's influence waned by the mid-1980s while other scientists attracted readership, publishing in-depth, insightful, serious-minded yet popular accounts of paleontological history, perspective and methodology. Stephen J. Gould (*Ever*

Since Darwin; The Panda's Thumb) and Martin J. S. Rudwick (*The Meaning of Fossils*) were in this fold, although of this new wave, only Adrian J. Desmond wrote a true 'dinosaur book' like none ever before.

So which key/core features made *The Hot-Blooded Dinosaurs* ("HBD") so startlingly different from what came before? Relying on Bakker's and Ostrom's prior studies and opinions (with nods to P. M. Galton and Loris Russell), Desmond presented views on the 'hot-blooded' question that not only seemed, but were, radical even 'blasphemous' for their time—albeit *logical* ideas stirring controversy. His points (initially disseminated in the 1975 UK edition) were fortified with several eye-riveting paleoart examples depicting dinosaurs and pterosaurs in poses that departed significantly from imagery dating from the cold-blooded theoretical 'era.' In April 1975, *Scientific American* (via Bakker) declared we were in midst of a "dinosaur renaissance," (a term not coined by Desmond). Over a decade later, Bakker would refer to such views as dinosaur "heresies."

True—there were other, later publications of merit and popularity, documenting further advances and nuggets of knowledge in dinosaur science, but essential groundwork running the gamut of popular terrestrial and aerial Mesozoic saurians had already been compiled by 1975 through Desmond's revelation of the (*HBD*-subtitled) "*revolution* in palaeontology." What is generally forgotten today—given that so many ideas outlined in *HBD* are now accepted both in popular and scientific view—is what a storm of opposition and controversy ideas within Desmond's book met with from the outset. (Not that my neophyte opinion mattered, however, but I did write one favorably-minded college paper on *HBD* then, for which I was awarded an "A" grade.[1]) Surprisingly, Desmond also reported that dromaeosaurs, "ostrich-dinosaurs" and pterosaurs were nearly 'bird-brained'—intelligent!

Interpretive harmony was achieved by conciliating (deceivingly) *paradoxical* views, in light of new evidence concerning dinosaurian anatomy and physiology. Channeling Bakker's and Ostrom's logical conclusions well founded in biology and phylogeny, Desmond had audacity to document not only how dinosaurs and pterosaurs played active "endothermic" roles in prehistory, but furthermore that birds were derived from warm-blooded (dinosaurian) ancestors. Likenesses in form between coelurosaurian dinosaurs and birds were less likely a result of evolutionary convergence as many had thought. Instead they established a much closer, direct genetic relationship. It was the biological "superiority" of such creatures (and *their* "pseudosuchian" thecodontian ancestors) that held paleo-mammals in check for a staggering ~ 200 million years! Thus, in this revitalized "unified theory" of Earth's evolutionary history, ascendance of we mammals was not somehow preordained. We rose to prominence only following a cosmic accident enduring for a "catastrophically short time": thereafter, only the 'small' inherited Earth. Who knew?

Colbert never was so bold (or believing). Yet the evidence and convincing rationale was all there to behold; if only not beholden to convention. A few scientists (e.g. Hitchcock, Huxley, Cope, Seeley, Beecher, Brown, Heilmann, Holland and even, oddly, Owen) had even precociously glimpsed and/or grasped key essentials of the argument decades earlier, although by the 1930s the paleontological field generally slipped back into a 'saurian consensus,' with many reconciling to the old reptilian paradigm, as commonly portrayed and reinforced through contemporary paleoart. So, by 1975, impact of *HBD's* theory and philosophy was keenly felt.

Desmond, now an Honorary Research Fellow in the Biology Department at University College London, went on to write about Victorian paleontology—*Archetypes and Ancestors* (1982), and a major biography of Thomas H. Huxley. His background and resume' couldn't differ more from novelist Michael Crichton's—a physician trained at Harvard Medical School, who also wrote the second most significant book of the dinosaur renaissance—*Jurassic Park*. During the 1980s, a bevy of popular paleo-books materialized upholding the new ideas and views on dinosaurs, which must have facilitated writing of Crichton's masterpiece. One pivotal (non-paleo) book in this 'stew' however, was James Gleick's *Chaos: Making a New Science* (1987).

Overall sales volume isn't a factor here in my paired choices; *Jurassic Park* far outsold *HBD*, for example. Whereas *HBD* was a scientific entry, *Jurassic Park* ("JP") was a major, bestselling blend of fiction, artfully founded in fact, expertly suspending disbelief. In an odd sense, Crichton's *Jurassic Park* may be considered as a vivid, natural outcome 'manifestation' or illustration come to life of Desmond's *The Hot-Blooded Dinosaurs*, analogously not unlike how Jules Verne's 1864 novel *Journey to the Center of the Earth* was a 'living' embodiment of Louis Figuier's 1863 text *Earth Before the Deluge*. *Jurassic Park* is where the rubber meets the road for us dinophiles in our dinosaur renaissance. Toward his creative successes in *JP*, Crichton acknowledged a number of prominent paleo-people—researchers of enlivened dinosaurian views, including Gregory S. Paul, John Horner, Douglas Henderson, William Stout and of course Bakker and Ostrom. (Interestingly, Margaret Colbert's artistic contributions were also credited.)

Isla Nublar's bioengineered dinosaurs are a Frankensteinian merging of paleo-chemistry, undue faith in supercomputers and, ultimately, fateful hubris in man's misguided effort to control natural systems. Here we see greedy, power-mad characters coaxing and re-wiring Nature into a tainted, unnatural and perhaps forbidden state. DNA power is considered more potent and menacing than nuclear power. Vicious *Velociraptors* breeding (despite built-in controls to thwart mating) are a metaphorical warning of man's irresponsible, projected use of (genetic) science. Nature supersedes man's sense of order on every level in idealist John Hammond's unfinished Jurassic Park. But of course, in *JP*, a primary 'take away' is the characterization and behavior of the dinosaurs and pterosaurs, especially those tiger-striped "raptors," which are frequently likened to highly intelligent, modern birds; they build nests for their young and hunt in packs. Throughout his novel, Crichton outlines varied types of Mesozoic saurians discussed at length in *HBD*, animating each vibrantly.

But (in my humble opinion), allied to *JP's* dinosaurian mayhem, Crichton also harnessed essence of an idea—another relatively early stage 'dinosaur renaissance' outcome—formulated in the kiln of chaos theory and Gould's 'contingency.' Contingency catalyzed as a concept in paleontology following 1980s scientific intrigue over the unsettling 'randomness' of a large asteroid strike on Earth that happened 66-million years ago. By definition, chaos is unpredictable in the same manner that evolutionary 'contingency' is founded upon sheer randomness. (Entropy eventually re-equilibrates unstable natural systems.) Step by step, contingency proceeds chaotically over time, resulting in albeit unpredictable patterns in spurts of speciation and mass extinctions episodes that (macroscopically) may seem similarly shaped (graphically) and recognizable. Just like we can't 'wind back the tape' from 4.5 billion years ago and restart, hoping to produce identical results in evolutionary history, any 'butterfly effect' produces system instability that over time vastly increases chances of unpredictable results in discernible ('chaotic') patterns.

Thus, *Jurassic Park* dinosaurs are chaos-incarnate, fabricated upon chance preservation and discovery of certain biting insect fossils in amber, exacerbated by headstrong human ingenuity and incautious, best-guess choice in genetics manipulation (e.g. use of amphibian DNA—a crux of the resulting problem—to fill in gaps). They're therefore 'unnatural' organisms per one definition, yet uncontrollable as living, 'natural' beings. And as we know, despite chaos, Life always "finds a way."

JP wasn't the first sci-fi story to address the dilemma of the bioengineered 'hot-blooded' dinosaur, although it set the standard for all those to come thereafter. Meanwhile, Desmond's landmark *HBD* remains a mine of information on that heyday of the dawning dinosaur renaissance stage. Together, they're a dualistic interplay of once 'heretical' ideas.

I recall one of my college geology professors cautioning back in 1973, not to be too eager to "jump on the continental drift bandwagon"—which, of course, had become the prominent paradigm shift in the Earth Sciences of the time. Now I wonder what he thought (then) about Bakker's and Ostrom's theories, which were steadily gaining traction. Near the end of the *JP* novel, in a fit of delirium (leading to his death), mathematical 'chaos' researcher Ian Malcolm mutters "Not ... paradigm ... beyond." Thanks to studies documented in Desmond's *HBD*, later brought to life through *JP*—15 years later, it became clear that, witnessed through the marvels unrestrained within Crichton's figurative *Jurassic Park*, we were then already well 'beyond' *onset* of the paradigm shift in dinosaur science.

Additional Reference: Allen A. Debus and Diane E. Debus, Chapter 36, "Renaissance Dinosaurs," (pp.234-242) in *Paleoimagery: The Evolution of Dinosaurs in Art* (McFarland Publishers, 2002).

Note to Chapter Twenty-three: (1) At risk of annoying readers, here is text of that early paper on the *HBD*. I include this because while most readers are familiar with Crichton's *JP* contents, many aren't as well versed with details of Desmond's *HBD*—hence this 'hey, what's in there?' summary may be useful:

> The decade of the 1970s became a focal point in our understanding of vertebrate paleontology. Perhaps this was made most evident to the layman following the publishing of Adrian Desmond's acclaimed, but highly controversial book, *The Hot-Blooded Dinosaurs* (1975). As this is examined it becomes evident that although certain individuals sometimes may temporarily deter its progression, the quest for truth in paleontological understanding ensues. As has been pointed out in previous chapters, these patterns of thought have greatly influenced public conceptions about how prehistoric animals existed.
>
> *The Hot-Blooded Dinosaurs: A Revolution in Palaeontology* should be welcomed as a meticulously prepared recapitulation of the debate over saurian endothermy and phylogeny written by a devoted scholar. In this book author Desmond successfully bridged the gap between the professional paleontologist and layman who may desire less technical detail by frequent use of verbal and pictorial restorations or reconstructions.
>
> Long abandoned theories have been treated in proper historical fashion by aptly demonstrating how and why these ideas were significant in an earlier period of scientific inquiry. In this matter, Desmond gained readers' confidence by placing scientific concepts and figures in historical perspective.
>
> One prominent figure in 19th century natural history was Sir Richard Owen, the adamant, anti-evolutionist whose conception of dinosaurs conformed to then popular theories of Creationism and Catastrophism. Sir Owen bequeathed the name 'Dinosauria' to three known species of fearfully great or 'terrible lizards.' The name held. Unfortunately for the rapid maturation of the science, so did the theological implications for dinosaurs until eventual 'triumph'

of Darwinism in the 1860s led publically by the "bishop-eater" Thomas Henry Huxley.

Sir Owen was cast more as a stubborn chap who refused to consider opposing viewpoints even when his own arguments failed than as a scientific villain who callously used facts to substantiate his beliefs. Owen was often referred to as the British Cuvier for his achievements in anatomical studies. Nevertheless it was largely because of Sir Owen's influences that Mesozoic vertebrate paleontology became ensnared in the great 'theology versus evolution' debate for nearly fifty years. It is remarkable to note that such emotional intensity could have been inspired from so few fossil remains. As a result, detailed studies of dinosaurian physiological characteristics, phylogeny and paleoecology were held in abeyance until this central issue became more or less resolved.

As the European evolutionary debates waned, American paleontologists were assembling complete dinosaur skeletons such as Joseph Leidy's *Hadrosaurus* and numerous treasures unearthed in the western United States during the Cope-Marsh vendetta. The resulting restorations did not at all resemble those completed at London's Crystal Palace by Benjamin Waterhouse Hawkins under Owen's guidance. Although the basic reptilian (cold-blooded) reconstruction prevailed throughout this period it is evident that dinosaur mobilities, eating habits, reproductive characteristics, appearances and probable ranges of habitat (niches) were being seriously reconsidered.

W. J. Holland's 1905 restoration of Andrew Carnegie's *Diplodocus* stimulated a classic confrontation concerning the animal's mode of locomotion. Oliver P. Hay insisted on basis of its enormous weight and its reptilian physiology that *Diplodocus* must have maintained a creeping stance. Holland was vindicated following reconsiderations of the relative lengths of the ribcage and the thigh bones. A creeping *Diplodocus* would have incongruously required a channel carved into the ground for its belly in order to move. Discovery of the famous Glen Rose tracks in Texas in 1938 fortified Holland's position, but never fully explained how this immense creature's alleged 'reptilian' metabolism could have powered its motion. In dealing with many problems of this nature, paleontologists were discovering that the functional morphology of *reptilian* dinosaurs was becoming very paradoxical.

It was not until 1969 that the conventional model of the cold-blooded dinosaur finally disintegrated. John Ostrom of the Peabody Museum at Yale University was able to account for many of the apparent discrepancies that had been plaguing Mesozoic vertebrate paleontologists for years on the notion of dinosaurian endothermy. Ostrom's position was fortified following Armand

Ricqles' 1969 study of similarities between mammalian and dinosaurian internal bone structure, and Robert Bakker's demonstration in 1972 that dinosaur predator/prey ratios closely resembled those extant in mammalian populations.

Many paradoxical paleontological difficulties were reconciled by application of the endothermy concept. For example, *Tyrannosaurus* could run, fight and stalk prey rather than acting as a lazy scavenger because it possessed a warm-blooded metabolism. Marsh's popular concept of the amphibious, snorkeling *Brontosaurus* persisted for decades despite the hydrostatic, lung-crushing problems of breathing air while the body bulk was deeply submerged. Using the endothermic model this and many other morphological discrepancies could be resolved by depicting sauropods as giraffe-like animals.

Deinonychus, an eight-foot long Cretaceous carnivore, possessed five-inch sickle-shaped talons on each foot because in life it was energetically capable of standing on one leg while slashing its prey with its other leg. *Struthiomimus* looked and appeared as a rapid sprinter because on the basis of its thigh/tibia bones ratio it *was* a sprinter.

According to one theory, dinosaurs apparently inherited their endothermy (warm-bloodedness) from ancestral, bipedal thecodonts. This was inferred from studies indicating that bipedal pseudosuchians such as *Euparkeria* had already developed physiological means for sustained, fast running by the Early Triassic. Presumably, development of an erect posture is one important prerequisite for the great dinosaurian diversification that ensued.

The most controversial topic—that of avian descent—was carefully broached by noting that morphological similarities between coelurosaurian dinosaurs and modern birds had been realized since Edward Hitchcock's observations of the famous Connecticut Valley fossil footprints in 1848. Whether or not birds can be considered as dinosaurian descendants is largely a classification formality. If birds branched at the pseudosuchian level before emergence of dinosaurs, they cannot be classified as descendants of dinosaurs. However, birds could be considered as surviving dinosaurian stock if they branched from coelurosaurian ancestors. Much of the controversy focused over the central issue that *Ornithosuchus*, a probable pseudosuchian ancestor, and *Archaeopteryx* both possessed bird-like clavicles (collar bones), whereas coelurosaurs did not. Any similarities between the latter two were therefore considered to be resultant of evolutionary convergence.

Recent analysis shows that *Archaeopteryx* highly resembled the coelurosaurian dinosaurs and that coelurosaurs did possess collar bones. We are left with the unresolved impression that coelurosaurian dinosaurs were ancestral to proavians, and that feathers evolved on these smaller dinosaurs as a heat conservation mechanism.

It is customary for any book about dinosaurs to address the subject of the great Cretaceous extinction. Several deterministic models were discussed in *Hot-Blooded Dinosaurs* including the 'supernova theory' which received great notoriety several years ago. Conversely, Desmond gave no consideration to David M. Raup's statistical computer models treating Cretaceous extinction patterns as natural consequences of species longevity. This final chapter was interesting, but given the present state of knowledge there is little opportunity to incorporate the endothermic model conclusively with the dinosaur extinction problem.

Desmond's presentation of the 'modern dinosaur' as a warm-blooded creature possessing established affinities to avians provided readers with a much clearer understanding of the problems which vertebrate paleontologists encounter when dealing with the question of how these organisms may have existed in life. Illustrations by Robert Bakker and Gregory S. Paul were included showing galloping *Triceratops* herds, darting hadrosaurs and streamlined sauropods contrasting sharply with our fond remembrances of the Chicago Field Museum murals designed and painted by Charles R. Knight over fifty years ago, showing classical, lumbering reptilian poses.

An even greater contrast in scientific ideology over the matter of hot-bloodedness versus cold-bloodedness in Mesozoic saurian is illustrated in the restorations of pterodactyls prepared by E. Newman in 1843, and more recently by Jessica Gwynne. Newman's restoration outrageously depicted carnivorous, bat-like marsupials engaging in aerial combat. Nonetheless Newman was among the first to propose that pterodactyls may have been fur-bearing creatures. In doing so he was forced to also suggest that they must have been marsupials because reptiles could not have had hair. He even challenged the great anatomists Cuvier and William Buckland (1784-1856), who objected to Newman's proposal.

Desmond noted that we have progressed in our thinking about pterosaurs since these early times. Accordingly, Gwynne's restoration represents the latest in paleontological understanding about these animals. Her illustrations show a soaring, graceful animal, clothed in white fur, which was probably very similar in manner to the modern seagull. The suggestion is also made on basis of convincing fossil evidence that these animals were quite intelligent and warm-blooded. Gwynne's restoration is quite unlike traditional ones we've been accustomed to, portraying leathery-looking, 'bat-like' monsters.

The book's historical perspective must be considered once more. While these illustrations indicated that as *new* fossil evidence is discovered, restorations of dinosaurs will be modified accordingly, an incident occurring in 1981 suggested how scientific interpretation can adversely bias our concept of dinosaurs. Vertebrate paleontologist Othniel C. Marsh was responsible for providing the wrong skull for a *Brontosaurus* skeleton discovered headless in Como Bluff, Wyoming in 1879. Although Marsh's choice of skull had been challenged several times later on, he capped the headless skeleton with the stubby-shaped fossil skull of a *Camarasaurus* found elsewhere, the matter was not completely resolved until a century later. Now most *Brontosaurus* skeletons throughout the world have been reunited with their proper, sleeker-looking skulls or skull casts. The incident is significant because it exemplifies how new information may be obtained or fresh ideas may have to be considered concerning the probable *Brontosaurus* diet and habitat from reexamination of its teeth.

The endothermic model, as presented in *Hot-Blooded Dinosaurs* appears highly convincing today. However, because new fossil, geological, biochemical, paleoclimatological, biogeographical, or even sociobiological evidence someday may be discovered that will alter this view, one must be equally cautious in accepting new ideas as in avoiding tendencies to adhere to traditional concepts (as did Owen). I recall being cautioned by a geology professor in 1973 not to be too eager to "jump on the Continental Drift Band wagon." Particularly in this discipline where a great potential exists for a single discovery to shift scientific climate, the reader is similarly encouraged not to blindly accept the theory of 'hot-blooded' dinosaurs without due consideration.

In the short time since publication of *Hot-Blooded Dinosaurs* a new highly controversial 'asteroid collision' extinction theory has received much publicity. A more philosophically satisfying (to those who prefer non-catastrophic solutions) extinction model proposed by Bakker in 1977, related *several* major extinctions of the Mesozoic to correlated intervals of sea regression and tectonic quiescence. Fossil remains of a large carnivorous dinosaur whose size must have dwarfed even *Tyrannosaurus* were recently discovered in London. A new fossil jawbone was discovered in Arizona which may represent an important (once) 'missing link' between mammals and reptiles. As we regard these few examples and the enigmatic circumstance of Roland T. Bird's 'swimming dinosaur' as discussed in *Hot-Blooded Dinosaurs* we understand that the endothermy concept has not yet been successfully incorporated into every dinosaurian model, and that discovery of new startling evidence will modify current perceptions of these fascinating extinct creatures.

Scientific development as a human machination has often become obscured by scientific bias toward certain ideas, personal conflict, or theological barriers. Therefore we see another purpose to Desmond's historical perspective. Just as the Recent diversity of life is better understood on basis of evolutionary narrative, scientific expression and its effect on the public sector is best understood in terms of its historical development.

(College book review paper by Allen A. Debus, later modified in 1983.)

Chapter Twenty-four — Coming Dark Ages: Four Metaphorical 'Horsemen' of the Paleo-pocalypse

Physicist extraordinaire—Stephen Hawking—believes that humans have a mere century to escape Earth, before it's too late. True, there *are* finite probabilities of asteroid impact, perhaps one even as massive as that which doomed the dinosaurs which even during our lifetimes could extinguish a fragile, teetering civilization. Eruption of a super-volcano or irradiation by gamma ray bursts from distant colliding neutron stars could also deliver a knockout punch to mankind. But such cataclysms could threaten any planet to which we might flee in the cosmos.

Perhaps what should be most concerning to us now though isn't what the galaxy could do to us, but instead what we might inadvertently end up doing to *ourselves*, right here. That nightmarish, ill-fated 'unnatural,' self-inflicted disaster caused by Man! Yes, the old Pandora's Box theme, that Apple in the Garden of Eden, the Genie freed from the bottle, or the Frankenstein mythos—Mankind the menace, unheeding Mother Nature's call for steady voices of reason. Meanwhile we recklessly tinker with the atom, or (worse!) the gene, at large—things so small we can't even see. And it's all there to bear witness and warn, spelled out metaphorically for us by beloved paleo-monsters in science fiction lore! Don't believe me, read on. Or are you too scared.

By the way, while I'm not formally 'reviewing' or 'critiquing' any of the books outlined below, I do highly recommend them all. They held my attention or at times instilled abject fear.

Nuclear:

We live in a peculiar facet of the so-called 'nuclear age.' While conventional Cold War tensions have relaxed, somehow atomic war seems more imminent or inevitable than ever. Activists such as Nobel Peace laureate Beatrice Fihn do what they can to achieve de-escalation of the global 'nukes' stockpile (see *Time* 12/25/17 – 1/1/18, pp.12-13), but are their efforts merely an illusion of progress? Meanwhile, interest in "disaster tourism" emphasizing "post-apocalyptical" sites where nuclear accidents happened such as Chernobyl grows. Tourists thrill vicariously in nuclear tourism spotlighting significant destinations such as Hanford, Washington where the Manhattan Project National Historical Park opened in 2014. (See *Distillations Magazine,* Fall 2017, "Greetings from Isotopia," by Rebecca Boyle, pp.26-35) Given such signs of the times, readers would highly enjoy Mike Bogue's new 2017 book, *Apocalypse Then: American and Japanese Atomic Cinema, 1951-1967*. Feeling safe?

The most frequently recognized 'horseman' (i.e. by fannish devotees, myself included)—harbinger of doom in this regard is the Nuclear threat. Less commonly examined by scholars is a distinguishable paleontological metaphor leaning on atomic annihilation or self-extinction via global, thermonuclear war. Thus we see recent provocatively titled fiction such as Matt Dennion's *Atomic Rex* and are reminded of a host of those creature feature films of the 1950s concerning emergence of gigantic dino-monsters from the depths of time hell-bent on vengefully destroying our civilization in the aftermath of nuclear bomb testing. A coupling of recent books delves deeply into the nature of a pending Paleo-pocalypse, cautioning that dark path best not taken, mirrored in filmic fare. These titles are Mike Bogue's *Apocalypse Then*, and my own *Dinosaurs Ever Evolving: The Changing Face of Prehistoric Animals in Popular Culture* (2016). Here, I'd like to briefly meld some of the ideas woven into Bogue's scholarly tome, representing how the nuclear threat's unnatural specter warns by getting those 'glow-in-the-dark' dino-monsters directly in our faces, screaming for us to abaodon this juggernaut before it's too late.

Despite all the nuclear bomb testing conducted during the late 1940s and early 1950s, the true dangerous and destructive nature of *radiation fallout* wasn't fully realized until after 1954's Bikini Atoll Hydrogen Bomb-*Lucky Dragon* vessel incident. Most of us think of Godzilla, as originally featured as the incarnate symbol of nuclear destruction in two films, 1954's *Gojira* and its Americanized 1956 version, *Godzilla: King of the Monsters*. But prior to 1954, there was the Warner Brothers' *The Beast From 20,000 Fathoms* (1953), also with its foreboding anti-nuclear messaging. (And for the first time nuclear and paleo- themes merged, but only like ships passing in the night, in 1951's *The Lost Continent* released by Lippert Pictures.) Of course, as Bogue points out, the nature of Japanese messaging concerning nuclear threats in science fiction films of the day differs remarkably (yet understandably so) from contemporary U.S. sci-fi films. Very briefly, unlike those of us in the States, Japanese actually suffered the horrors of nuclear attack (twice at the conclusion of World War II), and then in 1954 families were stricken by radioactive fallout in what became known as the "Lucky Dragon Incident" in the aftermath of Hydrogen Bomb test Operation Bravo in the Pacific Ocean.

So while Japanese sci-fi themes are of the 'it would be best to suspend proliferation and all testing of nuclear warheads,' lest, ultimately, an unmitigated atomic war might break out persuasion, American audiences were instead mollified by inherent values and ultimate good of nuclear science, which usually 'saved the day' by dispatching those oft-mutated metaphorical monsters. Faith in the military was unwavering as well in 1950s films. Never fear—chaos can be tamed and controlled by American wherewithal and scientific know-how, thus securing that inevitable happy Hollywood ending. Oddly, words spoken by actor Raymond Burr at conclusion of *Godzilla: King of the Monsters* ,"But the whole world could wake up and live again," having been set free of the menace, are perhaps ironically reminiscent of the title of H. G. Wells's 1914 novel *The World Set Free*, in which development and use of nuclear weapons are presaged. Indeed Wells coined the term "atomic bomb," "… and detailed with some accuracy the apocalyptic power of chain reaction weapons in his phrase, 'continuing explosives.'" (*The Science Fiction of H. G. Wells,* Frank McConnell, 1981, p.4)

Of fourteen prominent titles (give or take several 'Americanized' versions) invoking anachronistic 'prehistoric' monsters with a parallel radiation theme in Bogue's book, one (i.e. *The Giant Behemoth*) was a British production and not discussed at length, ten were Japanese—typically all featuring daikaiju. The only two *bona fide* western (American) films introducing dino-monsters covered in Bogue's book (and here I must affirm that Bogue discusses *many other* films besides just those involving prehistoric monsters!) were 1961's *Gorgo* (i.e. only tangentially tied to nukes), and 1953's *The Beast From 20,000 Fathoms* (of which more shall be stated shortly). However, 1957's "almost atomic" *The Deadly Mantis* (featuring a 'prehistoric' bug freed from Arctic ice), and *The Monster That Challenged the World* (1957), although presenting impressive non-dinosaurian monsters, also featured creatures of prehistoric ilk.

And so we are left with seven Godzilla movies (in which the nuclear theme echoes), *Rodan, the Flying Monster*, and two Gamera movies. Unlike American entries mentioned above, the Japanese daikaiju films listed by Bogue unwaveringly hit the theme dead-on.

The Toho series begins with *Gojira* (1954) in which the antinuclear theme rings loud and clear. So much has already been written of this foundational, A-class film, in which fears of radiation exposure play prominently, that it isn't worth glossing over its stark, depressing message and symbolism here. Thereafter, in the early Godzilla sequels, radioactivity and prospect of nuclear devastation caused by war and/or nukes testing was no longer central to the plot, was more causally mentioned as a disbelief-suspending 'device,' and therefore was overall less concerning to viewers. In 1955's *Godzilla Raids Again*, references to H-Bomb testing remain, however. As Bogue suggests, when Dr. Yamane states that killing Godzilla is "hopeless," this is an allusion to the persistence of the nuclear threat, which likewise seemed 'hopeless' to the Japanese. Furthermore, Godzilla "becomes angry when he sees bright lights," because they brings back "memories of the hydrogen bomb testing." And as Bogue notes, "Just as such memories brings back Godzilla's brush with nuclear testing, so did the sight of Godzilla (restore) memories of Hiroshima and Nagasaki." And in one dismal scene, we see Osaka ablaze with flames erupting above the city, reminiscent of a mushroom cloud. In the Americanized 1959 version, *Godzilla Raids Again* (formerly titled *Gigantis the Fire Monster*) Bikini H-bomb test footage has been spliced into the introduction.

Toho's 1950s flirtation with radioactively awakened giant monsters concludes with 1956's *Rodan,* with its pivotal scene where Dr. Kashiwaga speculates whether the winged pterodactyl "is the result of nuclear bomb testing. It not only has contaminated the ocean and the air, it's also had great effect on the Earth. This tremendous energy may have awakened Rodan from his 200-million-year-old sleep." The 1957 Americanized version even begins with a spliced shot of a hydrogen bomb test!

By the 1960s, association between occurrences of giant Japanese monsters and radiation, although detectable, had diminished. For instance, in 1964's *Mothra vs. Godzilla* (Americanized as *Godzilla vs. the Thing*), Mothra's Infant Island is identified as the place where atomic tests were conducted resulting in its peculiar, wasted ecology. Here Bogue notes that the Japanese version "comes closer to identifying America than any Toho film" as the perpetrator during the period under his examination (i.e.1951 to 1967). However, in the Americanized film, "Japan is likened to be guilty of nuclear testing by association with the United States, even though the latter country remains unmentioned." (pp.188-89) Thereafter Godzilla's persona softened as—from audience perspectives, his symbolic radiation levels seemed to diminish. Then in *Godzilla vs. the Sea Monster* (1967) the Red Bamboo operating from another Pacific Island use 'slaves' from Infant Island to manufacture heavy water, As Bogue interprets, "the heavy water factory is a stand-in for the nuclear threat." (p.197) And in 1969's *Son of Godzilla*, scientists on yet another Pacific island tamper with the atom again, striving to control weather systems using radioactive isotopes. While their aim is noble—for the benefit of mankind, perpetrators realize that if their technology fell into the wrong hands that "The results would be the same as a nuclear holocaust." Yes, if terrorists obtained their scientific knowledge, a global "nuclear winter" could ensue. (pp.199-200) By 1964/65, introduction of King Ghidorah (seen as a global symbolic menace) into the mythos (i.e. *Ghidrah the Three-Headed Monster*) drastically reduced the nuclear threat's scope. In the 1965 American version, a minister of defense limply states, "We cannot assume responsibility for authorizing the deployment and use of atomic weapons." (p.201)

Beginning sequences of 1965's *Gamera, the Giant Monster* "brim with Cold War tensions … If you removed the monster from the first few minutes, this introduction could be the beginning of a nuclear war movie." (Bogue, p.194) Furthermore, as Russia and the U.S. brace for pending World War III, some worry about a possible geotectonic consequences of nuclear war—a shift in Earth's axial tilt. *Gamera* doesn't stand up to the Toho features listed above, but I concur with Bogue's assessment of the Cold War build-up. Meanwhile, western films—*Gorgo, The Monster That Challenged the World, The Deadly Mantis*—don't rely on a nuclear theme quite as much, although viewers will note allusions to consideration of deploying nuclear warheads, radiation as a catalyst for hatching of monster eggs, and an outline of the U.S. early warning missile system spanning the Arctic, respectively.

One significant reason for why the gloomy radiation/nukes theme expressed in the first Godzilla movie appeals to me is because I first watched this film several times during the early 1960s Cold War stage, during the Cuban Missile Crisis period, when it was aired on local Chicagoland television stations. Even then, at age 8, I wasn't 'protected from' or oblivious to news reporting of the dire circumstances. I 'knew' what nuclear weapons could do. Many other movies and television shows of the time played up the nuclear theme mightily too, but none left such an impact upon my young impressionable self then as did the somber *Godzilla, King of the Monsters*. But hey—Godzilla was also a cool-looking pseudo-dinosaurian, and I was a burgeoning dinosaur geek, enjoying dinosaurs of every possible persuasion—fictional, statuesque, toy-land types and of course the foundational scientific/evolutionary kind.

Most daikaiju and prehistoric 'irradi-creatures' featured in films of Bogue's study period are Japanese. While American movie dino-monsters tied to the radiation theme fizzled by the 1960s, such connections waned more gradually in Japan. Bogue notes, "Godzilla was born in the hell of an H-bomb, but by the mid-1960s, his ties to the nuclear threat had become frayed ... from 1961 on, Japanese giant monsters began to shed their atomic attachments, gradually at first, and then completely by the late '60s." Furthermore, in the case of "…graphic depictions of nuclear war ... American films minimize the effects of the nuclear genie, hoping to rob Pandora's Box specter of some of the power…". (p.257) However, "The Japanese weren't afraid to 'show it like it is.'" (p.275)

By the time of the scientific dinosaur renaissance in America, from 1969 onward, appeal of dinosaurs and other 'monstrous' creatures of the past secured a more scholarly, research and reappraisal role. Dinosaur extinctions continued to be debated, taking center stage in the 1980s when "nuclear winter" could become philosophically equated with the comet impact ash-lifting cloud extinguishing sunlight for years, thus dooming dinosaurs and other species 66-million years ago. During that episode, dinosaurs increasingly became stand-ins for us—mankind, via the conceptual "Dinosauroid."

Feeling safe? In 2018, the U.S. Center for Disease Control (CDC) posted guidance online—safety tips—to follow in event of a nuclear attack! How would you react when the words "Incoming missile inbound. Duck & cover! This is not a drill" flash on your tablet screen? On Jan. 13, 2018 what turned out to be a false alarm was transmitted using those scary terms announcing an incoming ballistic missile headed for Hawaii. Still feeling safe, or just … 'fail-safe'? Many of us who were alive on Sept. 26, 1983 didn't know of a Russian heroic savior—Stanislov Petrov, who prevented launching of a 'retaliatory' nuclear strike resultant of a satellite systems glitch, when the chances seemed "50-50" to the Soviets that an *apparent* American attack was real. (*Time* 10/2/17, p.17). Whew—only a false alarm. World War III was averted.

Still feeling safe?

Viral:

Both frigid Arctic and Antarctica regions are not without their science fictional dino-monsters. Polar species of both regions are of relevance here—that is, to consideration of the viral vector as we shall now see, focusing principally on high latitude dino-monsters.

Bill Gates opined in 2018 that "Of all the bad things that could happen—a nuclear war, an asteroid, a gigantic earthquake—the one that's the most scary is a big epidemic, like a flu epidemic sweeping the world as it did in 1918." (*Time* 1/15/2018, p.42) Due to overuse of antibiotics, we've created resistant "superbugs" (ironically) thriving in hospital settings. (My wife survived one such 'bug' that nearly killed her in 2011.) The constitution of those annual flu shots is at best educated guesswork and often misses the mark as to which strains will proliferate and spread mostly during "flu season." Despite scientific and substantial medical gains since the germ theory of disease was fully accepted, illnesses still and will continue to plague mankind.

Outbreaks of dreaded Ebola occasionally happen, instances which someday could remain unwisely unchecked if key nations chose to ignore the problem. Recall how the AIDS epidemic flared seemingly out of nowhere, killing millions during the 1980s and beyond. And, annually, medical practitioners dread rise of another 1918-type influenza pandemic that killed 21 million worldwide. Bird flu, swine flu … there are other deadly viruses out there too, or on verge of mutating, that must be controlled such as the Marburg virus, as author Richard Preston sensationalized in his book, *The Hot Zone* (1995). Preston also cautions there are "demons in the freezer"—viruses such as smallpox retained for scientific usage that could be illegally procured and *weaponized* by certain countries or terrorist groups hell-bent on evil-doing. In this "jet age," we're all so quickly connected geographically permitting germs to spread faster than ever before through airports and megacity-scapes. So how have dino-monsters (of all things!) metaphorically represented this horrific, feared yet possible fringe of our pending Paleo-pocalypse? Consider…

In 1953's *The Beast From 20,000 Fathoms* a strange four-legged dinosaurian creature is released from Arctic snow and ice following test detonation of a large nuclear device. Most of this film's thrills and chills stem from Ray Harryhausen's stop-motion animated "Rhedosaurus" stomping through New York City. This film's attributes have been summarized and described countless times before by film scholars and sci-fi movie fans, but not (until now) quite in the context outlined below. Here's a slightly different spin from usual perspectives.

First, a little further background. By the early 1920s, in the wake of horrors of the Great War—World War I, public fears centered on the burgeoning military industry's ability to manufacture increasingly sophisticated war planes and weapons of mass destruction. Many feared, what if a hostile nation spread poison gases or (worse) "… perhaps even biological agents…" from *above*, using military aircraft! Sci-fi writers, including H. G. Wells, had written of the presumed devastating capabilities of futuristic aircraft: pre-atomic societies dreaded a day when such warfare would come to pass. By 1932 Winston Churchill for one predicted that "…disease germs would also be used as weapons." According to historian of science Peter J. Bowler, "The terrors (then) imagined … were fully as comprehensive as those depicted in fictional accounts of a nuclear holocaust … during the (later) Cold War." (*A History of the Future*, 2017, pp.151-52) Under contemporary terms 1930s public reaction to prospect of *disease germs* raining from the sky was like, to citizens of the late 1950s, a *radioactive* ash cloud equivalent.

When Rhedosaurus is wounded by police bullets and bleeds on Manhattan city streets, people are suddenly exposed to a virulent bacterial strain held dormant in deep freeze for hundreds of millennia. Now people are getting sick from paleo-germs that should have been left frozen in Arctic ice. (i.e. Thanks to Operation Experiment which freed the colossus and its pathogen.) So it seems that one representative means of destroying civilization—as contemplated by Winston Churchill—leads to a second option. And now only a strategy for killing the dino-monster without scattering and splashing its flesh and blood everywhere, thus avoiding use of conventional weapons, can be employed. (In the end a symbolic 'good' isotope is fired from a dart gun into an open wound, thus killing Beast in a palatable happy Hollywood ending.)

Don't such insecurities represented in 1953's *Beast* more or less a symbolically 'bridge' or link that earlier form of raining deadly terror (disease germs) which the public once feared could be used in air warfare, to a second rapidly dawning angst posed by invention of the Hydrogen Bomb, given the latter's propensity for destruction on an unprecedented level? Months later, Cold War fears as projected in film had escalated from prospect of biological warfare toward the fiery hazards of nuclear arms with concomitant spewing of deadly radiation (not just germs) as represented in 1954's *Gojira*—insinuating how horrific would futuristic warfare be if conducted using hydrogen bombs. Yes—by then American and Japanese citizens were becoming increasingly cognizant of risks posed by radiation exposures (not just the fiery initial blast around 'ground zero'). *Beast* coming before *Gojira* uniquely offers a more 'even balance' in regard to the two forms of terror—older, prior fears of exposures to disease germs transiting (historically) toward nukes—the current concern. Thus 1953's *Beast* offers a very distinctive blend for its time, messaging arguably since 'lost' on modern viewers.

Merging of *Beast's* (primary) opening nuclear theme with another introduced later in the movie during the actual City attack, entrenched in biological plague, was probably *not intentionally* executed to record a 'transition' of the latter (biological) kind of threat evolving into the former more modern angst—marking even as how science progressed, everything stays the same in terms of social consequences (i.e. terror). *Beast's* bio-terror theme was instead probably conceived as just a cool sci-fi, natural consequence (paleo-germs to which modern man has no immunity) of digging into a forbidden past, ironically facilitated with (then) futuristic nuclear weaponry via "Operation Experiment."

But, although *Beast* is now well over half a century old, bio-terror on a global level hasn't lost its ... well, ugh... shall we say, 'charm' in the minds of certain sci-fi writers who want to mess with our minds, as we shall now see. Here are two such, rather exemplary literary examples.

Frank M. Robinson's outstanding 1993 short story, "The Greatest Dying," (published in *Dinosaur Fantastic*, 1993, Mike Resnick and Martin H. Greenberg, eds.), goes far, far beyond *Beast*. Whereas *Beast* offers that conventional Hollywood ending, as American sci-fi movies usually do, novels and printed stories may tread much farther into dark, sinister territory.

In Robinson's tale, a museum scientist technician discovers a flesh and blood claw embedded in a sample of amber (from Peru), dating from the Late Cretaceous, when dinosaurs were extinguished from Earth. In process of extracting the immaculately, hermetically sealed fossil from amber matrix, the sharp claw tip accidentally slices into the technician's skin. You guessed it—there are toxic microbes preserved within the fossil and he soon spreads an infection, which then wildly proliferates uncontrollably around the globe causing Mankind's extinction—another 'Greatest Dying'—the same way dinosaurs died out 66 million years ago, via plague (i.e. not, in this scenario, an asteroid impact).

Best not tamper with the Past ... which leads us to Lake Vostok in Antarctica and its 'Jurassic Dead.' Let me introduce a horror-stricken world in which catastrophic germ warfare is waged using cadaverous zombie dinosaurs as the vector.

Many are the kaiju and dino-monster invasion novels lately published by the Severed Press. But, to date, my favorite entry is the "Jurassic Dead" trilogy authored by Rick Chesler and David Sakmyster. The series gets off to an intriguing start in the first installment, *Jurassic Dead* (2014), and with many plot-wise embellishments following in sequels. Incidentally, you might be interested to know that there is a real "Lake Vostok" on that harsh, bitterly cold southern continent. This water body is 'subglacial,' capped by 3,640 meters of exceedingly old ice. There, ancient microbes and bacteria comprising a prehistoric gene pool, warmed by geothermal activity— isolated in a 'sealed-off' environment for as much as 25-million years, were recently discovered.. Also— yes, real dinosaur fossils and remains of other prehistoric life have been found in Antarctic sediments! Coupling these ideas and ingredients for suspension of disbelief … (with added popularity of a certain AMC zombie program) and— voila!—we have submitted for your approval … the *Jurassic Dead*!

The main story line begins with discovery of an incredibly well preserved *Tyrannosaurus* corpse found via deep drilling through the ice. But when it thaws (e. g. not unlike Superman's 'Arctic Giant' decades earlier) on an Antarctic research station to utter dismay of onsite U.S. and Russian human investigation teams, it's restored to life … or rather a state of undead animation. Other dinosaurs are recovered in similar states of (partial) decomposition from the Lake. Yes—they're 'zombified' too. When a biochemist analyzes *T. rex's* cellular tissues, revealing the destructive antagonistic viral nature of the specimen's cells toward normal healthy tissue, evil-doers in the story realize they've got ground-floor essentials of an amazing "terror weapon." Driven by hunger, when the dinosaurs infect humans through biting (e.g. saliva injection into blood stream) or scratching, their victims don't simply get sick as in Robinson's story; in turn they're turned into "reptilian zombie things" rapidly via virus transmission! In time the hideous zombie horde—humans plus dinosaurs— quickly escalates in numbers.

Eventually, pseudo-science behind the madness is revealed, as a power-driven madman named "DeKirk" (who also shows up in each of the two sequels) professes provenance of the zombie dinosaurs. "I've had a little theory for quite some time now. I never did buy into the whole asteroid-killed-the-dinosaurs explanation. … and now I know exactly what drove them to (extreme cannibalism) A parasite. One glorious, devilish little bug that can bring down the mightiest of species. Including our own." (pp. 144-45) It turns out that this particular parasite is a "prion" likened to Mad Cow Disease. So DeKirk's team is insanely working on a means of modifying yet weaponizing, the prion such that "All the benefits of being a zombie—the near indestructibility—without the drawbacks—the mindlessness and the rotting flesh" can be realized. "Yes," DeKirk hellishly divulges, via bio-terrorism he intends to become a "God." (p.145) A "new world order" is in the making.

One may wonder why such a strange and wide assortment of dinosaur genera that didn't live together congregated to Antarctica's subglacial lake. This is explained in the first sequel, 2015's *Jurassic Dead 2: Z-volution*, as the result of primordial instinct during the Late Cretaceous extinction when "the prions that had infected the specimens had driven their instinctual tendencies, urging them south like migrating butterflies until they had converged on this spot (i.e. Antarctica), the most likely place to afford protection amidst the drastically changing climate system. The one place they could literally disappear and yet remain frozen, waiting for a future thaw ... Waiting for their reign to come again." (p.10) Like "migrating butterflies"? While well written, I couldn't help from snickering while reading that curious passage.

Following a nuke-triggered volcanic explosion on Adranos Island, the crazed zombie outbreak is seemingly contained to Antarctica, and the two heroic survivors (Veronica and Alex) believe they're safe—but of course they aren't. For DeKirk also escapes after discovering another incredible cache of other dinosaur specimens in the ice! (Later, resurrected paleo-mammals invade from Siberia.) In the first sequel, DeKirk invades Washington D.C. with a zombie army of men and dinosaurs. DeKirk now (neurally) hive-mind-controls the "prehistoric microbe" and thus behavior of his superior forces—ultimately, their all-consuming hunger—thanks to use of microchips and modified protein sequences in the zombie virus permitting him to direct the attack remotely. DeKirk, who has even transformed himself into an "Alpha" kind of human/zombie-saur hybrid, is taking over ...everything, and soon he'll control America's nukes! Meanwhile, "the prehistoric creatures (were) attacking the city like Godzilla on steroids." (*Jurassic Dead 2: Z-volution,* 2015, p.93) Eventually, in *Jurassic Dead 3,* (2015) development of a cure is pursued.

Note how each of the first two unnatural (i.e. Nuclear and Viral) catastrophes can still be metaphorically linked to paleo-monsters and daikaiju resembling famous dino-monsters—further underscoring how pervasive 'dinosaurs,' a term used loosely here, have become in western popular culture. While certain individuals could conceivably destroy us wielding bio-terror weapons or launching nuclear strikes, however, diseases would follow us through the stars, although maybe not in the form of maleficent bio-terror.

Sooner or later a large asteroid *will* hit us, or if any of those other errant natural or cosmic forces also manifest during the Anthropic age of modern man, per Hawking's reasoning maybe all of this helpless hand-wringing over human-caused bio-terrorism and nukes will be moot.

(Other Reference: Laurie Garrett, *The Coming Plague: Newly Emerging Diseases in a World Out of Balance* (1994).

Genetic:

In the 1990 novel *Jurassic Park*, Michael Crichton's fictional chaos theoretician Ian Malcolm professes:

> "…things are going very fast now. Fifty years ago, everyone was gaga over the atomic bomb. That was power. No one could imagine anything more. Yet a bare decade after the bomb, we began to have genetic power. And genetic power is far more potent than atomic power. And it will be in everyone's hands." (p.313)

Some three decades hence, mankind is already wielding early fruit of this relatively new, uncharted genetic territory, where there be 'dragons.' Yes—dragons and monstrous mutations, perhaps especially in the mind of Jeremy Robinson, author of the "Project Nemesis" series, which we shall soon come to.

Use of scientific surgical/medical methods to transform or modify biota into other forms, whether to suit our own purposes better or simply to perform basic research goes back millennia. For instance, through trial and error experimentation mankind learned how to breed the family dog and many species of vegetation now harvested as crops, which greatly contributed to a global human population 'bomb.' By the late 19th through early 20th centuries—yet decades before discovery and description of DNA[1], the intriguing notion of biological transformation became more of a 'thing' in the minds of stalwart sci-fi writers such as H. G. Wells (e.g. *The Island of Dr. Moreau*, 1896) and Aldous Huxley (*Brave New World*, 1932).

Now, however, in the early 21st century, we take for granted concepts like cloning, DNA-forensics and "genetically modified organisms." The human genome has been established. How many of you have gotten your own DNA analyzed through '23andMe' or Ancestry.com? Meanwhile, experts warn what might happen if scientists were to clandestinely experiment with the aim of genetically 'perfecting' the human species (or *uber*-species). *National Geographic* magazine featured an article on "The Next Human," for instance (April 2017)—product of biological engineering. Wouldn't the rest of us then become 'obsolete' or considered inferior—and you just *know* where that would lead. As always, there are pros and cons (although the cons usually don't surface right away).

And along the way, post-Wells and Huxley, other creative authors have stood on the shoulders of 'giants' who see and forge ahead toward where the new technology is heading, or where it would wend if wielded by unscrupulous hands, giving us their 'what-if' sci-fi scenarios to shock and thrill. *Jurassic Park* is merely one of those (both) literary and filmic (1993) metaphorical (but perhaps not quite so implausible now) examples. Although certainly not the first of its kind, *Jurassic Park* was also something of a catalyst in the imagining of many other unclean genetics-terror driven stories and movies, a pattern that goes beyond my purpose here. Beyond the pages of sci-fi novels, genetic experimentation isn't just a science fictional authorial 'device' (like radiation often was employed during the 1950s and '60s in producing strange mutations); it also has a very real dark and slippery slope.

Why did I find myself appreciating the first book in Jeremy Robinson's "Nemesis" series? I suppose it's because I've so enjoyed television programs like *The X-Files* and *Fringe* through the years, wherein federal agents investigate strange cases, discerning peculiar if not uncanny or irreconcilable phenomena. And that's how Robinson's *Project Nemesis* (2012) struck me in its early chapters: I couldn't put it down! But is Robinson's primary monster "Nemesis" a *dino*-daikaiju—another grim manifestation of the Paleo-pocalypse? Based on Cheung Chung Tat's eye-riveting front cover art for the book, it kinda' looks like it could be, but let's consider.

While in Toho's efforts to suspend disbelief, it becomes possible to equate origins of dino-daikaiju such as Godzilla and Anguirus, to dinosaurian stock[2], "Nemesis," however, would not appear to be genetically 'saurian.' Just because a fantastic monster vaguely *resembles* a dinosaur does not mean it absolutely is 'dinosaurian.' That is unless the author or script-writer also makes a concerted effort to associate or otherwise imbue said kaiju with dinosaurian evolution in some way—then we can romp freely through the bounds of utter speculation.

So Nemesis is clearly daikaiju, but is it also a *dino*-daikaiju, like Godzilla? Well, *which* Godzilla do you mean? After all, it remains conceivably possible, using genetic engineering, to produce a daikaiju that in form vaguely resembles a dinosaurian daikaiju, even though it isn't *genetically* tied to real dinosaurs known to science by way of fictional origins, or artistic costume and graphic design.[3] In fact, Nemesis is 'alien' in aspect, not terrestrial, and its murky origins render it ultimately 'not of this Earth.' But because artwork associated with the *Project Nemesis* novel portrays Nemesis as *so* suggestively 'dinosaurian' or at least *dino*-daikaiju in appearance, for our purposes let's regard this fantastic, thoroughly entertaining novel as a new-millennial, post-*Jurassic Park* 'paleo-apocalyptical' extrapolative metaphor.

Before delving further into *Project Nemesis*, and only as an aside, John LeMay offers a tempting 'what could've been' morsel in his 2017 book, *The Big Book of Japanese Giant Monster Movies: The Lost Films*. Therein on p. 57, under the subheading "Godzilla and the Chariots of the Gods?" he refers to a 1979 film concept with a working title, "Godzilla: God's Angry Messenger." In this movie script, "… aliens arrived in the days of the dinosaurs and experimented on one until it became Godzilla. The monster is then put into suspended animation and hatches millions of years later from an egg on the ocean floor and proceeds to besiege the world as a 'last judgment.'" Although never made, given its cosmic connections, the overall plot of this mostly forgotten 'lost' Japanese movie idea is rather aligned thematically to the gist of Robinson's *Project Nemesis* (and sequels). The resulting Godzilla in this unmade film would have seemed not unlike the original (albeit *non*-dinosaurian) "Nemesis Prime' in Robinson's story.

LeMay also reported on that "one that got away" film, 1994's aborted Jan De Bont American Godzilla attempt, scripted by Terry Rossio and Ted Elliot. Here, Godzilla was envisioned as "… a bio-engineered monster made from dinosaur genes" by aliens (presumably during the Mesozoic Era), whose purpose is to defend Earth from a rival species of aliens that send probes to various planets which use the local life to create a doomsday beast." (p.101) In the original script, (online link kindly provided to this author by LeMay), a surgeon explains, "… so they created a virus that would reprogram DNA and then transfer RNA." Another character (i.e. "Marty" channeling an alien persona) replies, "We left it in stasis … created from dinosaur genetic template. Alien probe would awaken it." Also 'Marty' in entranced, broken speech, describes Godzilla then as a "… doomsday beast… it reproduces. Destroys. Wipes out dominant life forms. Later, the aliens come. To a planet already … conquered." Again—simply another case of great minds thinking alike! And certainly after *Jurassic Park's* record breaking successes—plus the proven capabilities of cgi effects, by the mid-1990s DNA and bioengineered monstrosities were all the rage in science fiction.

Both in the 1979 film script and in De Bont's, Godzilla was cast as a (deliberately) mutated *dinosaurian*, therefore 'dino-daikaiju-like.' Although still creating another wave in the beckoning Paleo-pocalypse, Jeremy Robinson's Nemesis—a DNA-enhanced genetic mutation, pseudo-dinosaurian only in appearance—and the rapidly evolving, inadvertently-mutated monster in *Shin Godzilla*, are both not *dino*-daikaiju, although definitely daikaiju. But hey, even if they only appear dinosaurian—still a valid form of paleoimagery, then such concocted beasts can still effectively conjure or recollect fearful denizens out of Earth's Mesozoic past, signaling thoughts of pending extinction (if only subliminally).

So *Project Nemesis*, building on prior events taking place in a prequel—Robinson's *Island 731* (2013), a tale reminiscent of Wells's *Island of Dr. Moreau*, concerns bio-engineering of a daikaiju from alien monster DNA (possibly in ancient *historical* times, during the time of an ancient Atlantis civilization), that steadily takes on gigantic pseudo-dinosaurian features. The 'alien' DNA is fused within two humans, madman "General Gordon" and remains of a murdered female child named "Maigo," who eventually becomes the invulnerable, titular Nemesis monster. Of course this is 'Frankensteinian' science run amuck. Perhaps analogous to the Godzilla seen in 2016's *Shin*

Godzilla (i.e. mutated from a non-dinosaurian progenitor), Nemesis also undergoes rapid evolution as the alien DNA, derived from a "Nemesis Prime" fossil, takes hold morphologically in Maigo's tissues, transforming 'her' into a towering, city-smashing daikaiju, possessing special mysterious powers besides brute force.

Meanwhile, two agents Ashley Collins and Jon Hudson quickly come to blows with both of the transforming monsters. (Hudson is connected to a section known as the FC-P, which investigates crypto-zoological and paranormal phenomena.) When the hybrid DNA-altered Maigo awakens, she's already developing monstrous features and still growing by the minute, molting through four stages, each time transforming into a new deadly form. General Gordon, although now a genetically enhanced monster too, grows more slowly, retaining raw human characters. But Maigo, adorned with huge spikes along her vertebral column, is soon trampling cities like Boston. And she possesses glowing, furnace-like membranes on her abdomen, which when punctured cast off a highly explosive orange-colored, gelatinous substance that ignites, blasting everything for miles around. Nemesis is also a vengeful goddess, because she remains intent on destroying her father—the one who murdered both her as a little girl, and her mother. The human portion of her mentality is able to 'connect' via extrasensory perception with Hudson, a mind-melding that becomes especially important later in the series. At the conclusion, her wrath unabated, Nemesis resides in the ocean—waiting, while the more-than-human General Gordon's fate is dealt with.

At risk of committing nuisance 'spoilers' for his novels, I can at least outline some key events later transpiring in Jeremy Robinson's 'Nemesis' series, representing the foreboding genetic threat. So, next, in 2013's *Project Maigo*, three additional kinds of gigantic monsters appear, intent on destroying civilization ... and Nemesis. They're 'hatched' from monster eggs found in an Alaskan cave where the original Nemesis Prime corpse was discovered. By this time though super-human General Gordon is teamed with the rapidly growing, monstrous newcomers—bonded to him through similarly shared DNA—while the 'good guys' now wield a brain-controlling counter measure chip device. Interestingly, Robinson offers a definition for "kaiju": "…an official term for any creature that is … not natural, with the understanding that it be reserved for things capable of mass destruction." (p.54) So by this time, in destroying the kaiju challengers, 350-foot tall Nemesis is taking on qualities protective toward the human race. And it is in this novel that Hudson temporarily, yet effectively 'becomes' a kaiju himself, mind-controlling one of Nemesis's adversaries! (p.243)

The monster rally mayhem involving "genetic monstrosities" continues in each of two subsequent novels, *Project 731* (2014), and *Project Hyperion* (2015). In the former, Robinson fingers certain scientists, "…the people who committed some of the worst crimes against humanity, and they weren't World

War II scientists. They were modern-day American scientists working for the government." (p.30) And the agency which so 'omnisciently' experiments and weaponizes this biological material is named Genetic Offense Directive, or, aptly, "GOD" for short.

As the American daikaiju battles inevitably wind down by the fourth volume—*Project Hyperion*— Robinson considers what genetic monsters might inherently be, and also how people might so easily be tempted to the dark side in a science still in its infancy. First, "Becoming a monster might be the only way the human race survives." Furthermore "... it might be time to reassess what defines a monster, seeing past what's on the outside and evaluating what is on the inside." (pp.170-171) While the latter phrase might sound philosophically and sociologically hopeful, Robinson is also keen to the horrors of *eugenics*, a possible social consequence of tampering with the gene, as many brilliant men have questioned through the years (and lesser men have committed). Yes—fear that irrecoverable descent along the 'slippery slope'! For by now, Robinson's saga has entered truly cosmic proportions, as we find that through millennia the human race has been genetically programmed (i.e. via "manipulated evolution") to commit genocide, by a star-faring alien race. (pp.212-213) Here lies the key downside to genetic engineering through which scientists and politicians might vitiate themselves after immersing in this tainted frankensteinian (biotech) soup.[4] This point isn't 'pounded on' by Robinson though, who instead is aiming for fun reader kaiju battling entertainment throughout, as opposed to truly dark thematic messaging. But it's still there, in black and white.

Jeremy Robinson regards 'Nemesis' as "America's first real iconic kaiju." He doesn't count Kong because "...he's just a big ape and a snack for Godzilla..." And he recognizes "...*Cloverfield,* but can anyone even remember what that monster looks like?" Certainly you wouldn't regard, say, a SYFY channel bioengineered 'Sharktopus,' or those bunny behemoths of *Night of the Lepus* as contenders for that lofty title ... would you? Regardless, *Project Nemesis* ranks as the "highest selling original (non-licensed) kaiju novel of all time." His novel series steers us cosmically beyond territory explored in dino-monster movies invoking genetic disaster, such as *Godzilla vs. Biollante* (1989), or even *Jurassic World* (2015) with its "Indominus rex" and those weaponized Raptors. Interestingly, Toho's "Biollante" predates *Jurassic Park*: authors Sean Rhoads and Brooke McCorkle note that in the former, by that time—1989, biotechnology had "... surpassed nuclear energy as the Godzilla's franchise's preeminent concern. ... Godzilla, a symbol embodying the threat of nuclear energy, pales in comparison to the might of genetic engineering represented by Biollante." (*Japan's Green Monsters*, pp.152. 154) Which fatefully leads us to our final 'Horseman.'

Environmental:

"... the science is damning and the clock is ticking."
(Justin Worland & Charlotte Alter on climate change in *Time*, 4/1/19, p.34)

Well, even the Smog Monster Hedorah rapidly mutates, but we'll soon come to that. Who doesn't love Gaia and Mother Nature—Earth is the only habitable planet for our current biological form. Given my three decades-plus years in the environmental chemistry sector, *and* coupled with a lifelong profound interest and fascination for both real and fictional dinosaurs, I parlayed a chapter of my 2016 book, *Dinosaurs Ever Evolving* into a study of how "prehistoric life spawned an environmental movement."[5] Therefore I was intrigued to find a recent scholarly book, *Japan's Green Monsters: Environmental Commentary in Kaiju Cinema* (2018), by 'Japanologists" Sean Rhoads and Brooke McCorkle. But first, a little more relevant background, off the beaten path.

Public concerns of deadly radiation within the body certainly escalated following a 1958 paper written by chemist Linus Pauling, warning of dangers posed by radioactive carbon-14, one of the byproducts of nuclear detonation. But few may know that biochemist and esteemed sci-fi writer Isaac Asimov had written on the same topic in 1954, a paper titled "The Radioactivity of the Human Body." However, it was Pauling's paper that "…played its part in the eventual agreement on the part of the three chief nuclear powers to suspend atmospheric testing," the first nuclear test ban.[6] At the time, however, the hazards of chemical smog forming in large urban centers had already been acknowledged. Human-caused pollution was on the rise and, it must have seemed, branching out as the chemical industrial complex—with resulting chemical products available for consumption and disposal, expanded. Yup— "Chemistry for better living."

During that same year when Asimov's paper was published, Universal Pictures released *Creature From the Black Lagoon* (1954), incorporating an inaugural scene suggestive of Man's unyielding pollution of the environment. For not only does 'Kay' (played by beautiful Julia Adams) flick her cigarette butt into the Devonian Gill Man's natural habitat (i.e. *his* 'Black Lagoon'), but also the male scientists haphazardly poison the waters with a toxic chemical— rotenone, in order to capture the 'prehistoric' living fossil.

To my knowledge Spencer Weart was the first to note how the dawning *environmental movement* (in America) stemmed from growing contemporary concerns over radiation poisoning via nuclear bomb ash fallout. For he stated in 1988, "There were many reasons for the rise of the environmental movement, but the basic themes—dismay with technological authorities and systems that seemed about to doom the entire world—had first been thrust upon the public by hydrogen bombs."[7] Weart further credits language used by Rachel Carson in her influential 1962 book, *Silent Spring*, bridging nuclear and chemical pollution/environmental themes and imagery. For Carson wrote of an imaginary USA town, presumably resultant of pesticide crop dusting, "In the gutters under the eaves and between the shingles of the roofs, a white granular powder still showed a few patches; some weeks before it had fallen like snow upon the roofs and the lawns, the fields and streams. No witchcraft, no enemy action had silenced the rebirth of new life in this stricken world. The people had done it themselves."[8] Clearly, nuclear fallout imagery—steeped in looming fear of the hydrogen bomb, our 'Sword of Damocles'—was used to emphasize a more chemically-driven phase of the environmental movement.

Eventually, by 1970, President Richard Nixon instituted the U.S. Environmental Protection Agency. Establishing the USEPA turned out to be a shrewd political move, even though Nixon didn't see it that way originally, yet one tempered by economic concerns as well. (The Japanese Environment Agency was instituted in 1971.) For in January 1973 Nixon stressed the necessity of maintaining, "... balance between economic growth and environmental protection."[9] While economics textbooks of the time didn't emphasize environmentalism, more recent texts, such as by Sean Flynn, do—through a principle flagged as the "Tragedy of the Commons," in which common resources, or commonly held property rights are overused, or 'overgrazed.' Ultimately, despite personal greed and incentives for personal gain, depleting said resource and despoiling its surroundings is not only unwise, but considered bad economic policy, whereas preservation of the resource is seen as beneficial. Thus overuse should be restrained because in the long run preserving valuable resources is good for the economy.

Which leads us to authors Rhoads' and McCorkle's new book. So, steering readers through the nuclear-radioactive-fallout-as-environmental-pollution topic (which won't be emphasized under this heading because the 'nuclear' theme was already addressed here previously), as presented in *Godzilla* (1954 and 1956), the authors move onto the core ecological/environmental arena—isolating topics that would seem most recognizable or relevant to western audiences. Rhoads presented at G-Fest in 2011 (a slide lecture which I attended), and he is an author of a past *G-Fan* article—both synopses of his Masters thesis concerning *Godzilla vs. the Smog Monster*. McCorkle is an assistant professor who has written for *Horror Studies*. And so, focusing on dino-daikaiju films addressed in their book ... they interestingly delve into a facet of 1964's *Mothra vs. Godzilla* under a subheading titled "Nature vs. Capitalism," a prominent kaiju-related topic that I hadn't thought really about before in depth.

Beyond background nuclear devastation resulting on Mothra's Infant Island due to prior nuclear bomb tests, thematically, this movie (which I originally watched, titled as *Godzilla vs. the Thing*), concerns greed of disagreeable capitalists intent on making a mint after appropriating Mothra's gigantic egg (i.e. the 'resource'). Rhoads and McCorkle make the interesting point that circumstances surrounding Infant Island are not unlike those concerning Kong's 'Skull Island,' in the sense that both of these island's indigenous giant monsters are exploited by insensitive money-grubbing businessmen. The giant egg serves as a stand-in for nature and the environment. Furthermore, "Mothra represents an alternative to big business and its wanton destruction of the environment … industrial capitalism, like science and militarism, repeatedly fails to consider the disastrous repercussions of certain actions." (p.70) And so it must seem heartening to see Godzilla (in his opening scene still retaining tangible radiation exposure threats to humans), the infamous hulking byproduct of American nuclear bomb tests, defeated by a "green" symbol of environmentalism. Finally in this nightmare, the 'good guy' team—(i.e. Mothra's caterpillar offspring) win. Significantly, Mothra is viewed as female—perhaps allied to Mother Earth. Enviro-concerns projected in the 1964 donnybrook were (to use another 'green' term …) somewhat 'recycled' in 1992's *Godzilla vs. Mothra*—with climatological-'Gaian' factors taking prominence, personified as a trio of giant monsters.[10] (The debated science of 'Gaia' as a 'living self-stabilizing Earth,' had become a popular notion then—something of a 'fad' in the western world.)

In their respective debuts (1954 and 1965), giant 'prehistoric' monsters Godzilla and Gamera both are put through paces amounting to a 'Nature cannot be *controlled* by man' theme. Like Dr. Serizawa's doomed fish in his aquarium, or the electrified wires guarding Tokyo melted by radioactive flame, that famous close-up shot of Godzilla roaring at the *caged* songbirds underscores this idea. A decade later, *Daikaiju Gamera* was released with a bevy of environmental messages. The monster Gamera itself (in the Japanese version, *not* the Americanized one, where it appears 'prehistoric' arising from glacial ice in aftermath of nuclear detonation), by then a familiar scenario to American audiences, appears to have grown and mutated from a pet turtle that was inadvertently released into the wild—a seemingly excusable accident, yet valid commentary on the rampant *invasive species* problem resulting in out-of-control ecological havoc—often dooming native flora and fauna. Gamera's actions in this film also portray the monster as an "insatiable consumer of natural resources," implying that "Japanese socio-economics during the high-speed economic growth period of the mid-1960s were equally monstrous." (p.99) Even a humorous scene where Gamera lands at Haneda, where it goes on a rampage, emphasizes noise and atmospheric pollution associated with the smoggy Haneda Airport. (Indeed, the USEPA ran a noise pollution program during the 1970s.)

Also highly interesting is a detailed chapter outlining "1970's Japan: A polluter's paradise," in which underlying wartime and postwar economic factors led to accelerated environmental devastation in Japan, as reflected in contemporary kaiju eiga films. As the authors document, "Throughout the postwar period, political and industrial leaders prioritized economic growth over ecological preservation, and the Japanese populace largely accepted the industrial pollution that accompanied rapid re-industrialization ... Severe environmental destruction on an astonishing scale accompanied ... economic growth ... This toxic environmental degradation of Japan's land, water and air became the focus of kaiju eiga productions in the early 1970s." (pp.105-106) Indeed, such movies emphasized environmental issues to greater extent and more poignantly than did contemporary western giant monster films. (Unless you want to include that 1970s flick, *Night of the Lepus*, which only made me laugh and giggle.)

Two additional, significant 'green' filmic dino-monster cases were 1967's *Son of Godzilla* and, of course 1971's *Godzilla vs. the Smog Monster*. In "Son," atomic/nuclear concerns are still entwined with the specter of pending ecological havoc, however, as a radioactive isotope opens "Pandora's Box" toward controlling global weather patterns, albeit for the alleged good of mankind. Given the tendency toward human overpopulation (another contemporary environmental issue recognized then by alarmist Paul Erlich by 1968), and diminished food resources to feed everyone, scientists intend to "sacrifice" ecological systems in less populated regions where climate is non-ideal for human habitat "for the sake of feeding an ever-expanding human population." Their dangerous experiment is carried out on Solgell Island, and of course it goes awry, rocketing temperatures on the island up to 160 degrees F, permitting gigantic bugs to hatch and attack Godzilla and his 'son.'

The fallacy exposed here is that while deserts, tundra and jungles are regarded as "useless" and agriculturally non-productive, and therefore worthy of transforming such that they can be inhabited by man, subjected to our industry, "... these terrains ... are nonetheless active and vibrant ecological regions, boasting wide arrays of plant and animal life." (p.78) A 'dark side' of producing ample food through clever technology means though is that, foreseeably on a planetary scale, this would lead to continuing population growth and therefore further destruction of eco-spaces and the 'natural' environment. The speciously noble concept of using technology to increase food quantities for human consumption re-emerged in 1989's *Godzilla vs. Biollante*, although then with its unique bioengineering twist.

What's "best" for mankind? Leave Mother Nature alone; that's the stance that Godzilla had taken by this time! But of course, we're only human and human nature seems opposed to Mother Nature.

Although environmentalism on various levels (beyond the underlying nuclear bane) is an omnipresent theme of mid-1960s Godzilla filmography, the dire pollution message reached a crescendo in the form of Hedorah—the Smog Monster, a film that Rhoads has carefully researched. How odd that (radioactive, H-Bomb derived) Godzilla is the catalyst for removing the threat of chemical (and social) Pollution Incarnate in Japan … this time! (One added thought here is how *both* Hedorah and Shin Godzilla are representative 'pollution monsters.') Although "Smog Monster"—viewed as a serious, sobering "protest film"—has been denigrated as one of Toho's worst in the Godzilla series, thematically it impresses, offering stark social commentary. Imagery on many levels fortifies how mankind is a wanton polluter of the air, land and sea, spewing waste chemicals and garbage into the pristine environment. The authors state, "… instances throughout the film that mimic the original (i.e. *Gojira*) indicate that industrial pollution, through Hedorah, should be considered the equivalent of the earlier anti-nuclear fears exhibited in *Godzilla*." (p.118) In the end, despite Hedorah's defeat, the potential for more environmental catastrophe remains unresolved. Likewise, by the later 1970s, Japan's environmental crisis had diminished, yet remained a steady societal concern. Through subsequent decades, on into the new millennium, evolving environmental "green" themes in both the Godzilla/Mothra and Gamera series reboots were not left unattended.

Perhaps due to Japan's geographical situation as an island nation, dependent on the maritime commerce and fisheries, subsequent kaiju eiga films (such as *Gamera vs. Zigra*, and *Godzilla vs. Gigan*) emphasized protection of the oceans and marine environment. This is distinctive, arguably, from western contemporary eco/enviro-themed sci-fi films, in which consequences of seas when polluted by chemical wastes (apart from nuclear rad-wastes) aren't as prominent a theme. If the oceans degrade, say, into what paleontologists refer to as a "Strangelove" condition – anoxic and void of life, circumstances which have nearly occurred in remote prehistory, then civilization is surely doomed. But mankind seems in denial. Already fresh potable water supplies are shrinking. Recent projections indicate by 2025, 3.5 billion people—half the world's population, will live in "water-scarce regions." (*Time*, 2/19/18, p.34) Overpopulation and adverse climate change will not alleviate such a situation. Globally, we dealt with averted the ozone layer crisis, bu so far seem unwilling to fully address human-caused climate change.

The End?

As presented in episode 2 of *Myths and Monsters* (2017), our science may be allegorical for ancient Greek visions of a 'forbidden' wilderness—landscape outside their reassuring city walls, places later viewed as "supernatural or parallel to our own"—a 'beyond' where there be 'dragons.'

Pandora's Box was a hodgepodge of horrors; *four* kinds of genies have escaped from the bottle, which can be symbolized today by paleo-monsters. In the case of the nuclear threat, yes, riffing Mr. Bogue, was it only a *pending* ... 'Apocalypse *Then*' .... but what about *'Now'*? Even as former Japanese battleship *Nagato*, used as an experimental nuclear bomb test structure at the Bikini Atoll, became a popular dive site—a stark reminder of what happened and how things could be, in recent years, the United States is poised to escalate its arsenal of nukes. Meanwhile, an increasingly acidified and synthetic world ocean, filling with plastic particulates and debris floating from continental areas threatens marine life. Seawater steadily absorbs latent heat from an increasingly carbon dioxide-laden atmosphere, sufficient someday perhaps to erupt megatons of a more potent greenhouse gas—methane from deposits of methane-hydrates, currently frozen on the seafloor, thus warming climate dramatically beyond comprehension. And, let's face it, do we really trust those guys in the white lab coats working on the scientific fringe, with genetic material and viruses? Where will their biotech researches lead?

Even if a 20-mile diameter asteroid doesn't pulverize Earth over the next 20,000 years or so, maybe the only way for some semblance of the human race to survive the coming dark age is—as Jeremy Robinson opined—to *become monsters ourselves*!

Notes to Chapter Twenty-four: (1) Horace Freeland Judson, *The Eighth Day of Creation: The Makers of the Revolution in Biology*, (New York: Simon and Schuster, 1979); (2) Refer to Chapter Ten, "Godzilla's Dinosaurian Origins," in my *Dinosaurs Ever Evolving: The Changing Face of Prehistoric Animals in Popular Culture*, (2016), pp.127-140; (3) Many have recognized the dinosaurian (theropod) body contour of the titular *Alien* (1979) creature, which is referred to sometimes as a 'bug' due to its revolting means of reproduction. However, like Nemesis, and despite deceiving body configuration, vaguely reminiscent of a fantasy dinosaur, both are not of the Earth. The *Alien* movie monster is derivative from 1950s science fictional outer space monster writings of A. E. Van Vogt. For more on Van Vogt in this context, see my articles in *Filmfax Plus* no. 104, Oct./Dec. 2004, pp.64-67, 132-133; and *Scary Monsters Magazine* no.50, April 2004, pp. 94-99; (4) For more on eugenics in the history of science, when contemplated as a 'solution' to society's ills, see Peter J. Bowler's *A History of the Future: Prophets of Progress from H. G. Wells* (2017), particularly pp. 199-203, under the heading "Breeding." (5) Allen A. Debus, *Dinosaurs Ever Evolving;* (6) Isaac Asimov, *In Joy Still Felt: The Autobiography of Isaac Asimov, Vol. 2 1954-1978,* 1980, pp.9-10; (7)

Spencer R. Weart, *Nuclear Fear: A History of Images*, 1988, p.325; (8) Rachel Carson, *Silent Spring*, 1962, p.14; (9) Meir Rinde, "Richard Nixon and the Rise of American Environmentalism," *Distillations Magazine,* Spring 2017, vol.3, no.1, pp.16-29; (10) In *Godzilla vs. Mothra*, (besides Godzilla's exertions), another giant bug—Battra "embodies nature's vengeance…" while Mothra "represents the beneficial and beautiful aspects of nature when it is respected." Rhoads and McCorkle, p.166.

Chapter Twenty-five — *Dinosaurs in Fantastic Fiction - Extras - III*

(Author's 2019 preface—I'm afraid this chapter represented one of my lesser executed fanzine ventures. Following the original article's printing in *Prehistoric Times* no.83, two letters from readers appeared in issue no. 84 (Winter 2008), informatively outlining additional fictional tales featuring prehistoric marine sea-monsters then unknown to me. I just hadn't done my homework thoroughly. The two stories mentioned by those readers were Wade Carmen and Alan Dean Foster. I've included their interesting letters in a footnote to this chapter.[1] Carmen mentioned a 1980 novel by Robin Brown, *Megalodon*, of which I wasn't aware, and Foster referred to his 1976 short story, "He." Also, although not outlined herein, long since publication of those "*PT*" issues, many readers have been thrilled by Max Hawthorne's 2013 novel, *Kronos Rising* (featuring a cryptozoological-Kronosaur), and Eric S. Brown's *Megalodon* (2015)—both entries with several sequels in print. With release of the 2018 motion picture, *The Meg*, and sequels to Steve Alten's *Meg* novel out there as well to paddle through, the topic of undersea, primeval horrors remains fair game territory for writers of fantastic fiction. Bring a harpoon!)

Due to my emphasis here on 'dino-monsters' of the aquatic realm, regarded as prehistoric and hence rightfully extinct (but which still *suddenly* rear their angry heads in the modernity of fantastic literature), I nearly titled this article "Crypto-

Creepy-Critters." Yet that working title sounded too much like an awful kind of cereal product so let's pretend like you never heard me 'say' that.

Let's blend a little fun fact with paleo-fiction, focusing on published stories which necessarily received little attention in my 2006 book, *Dinosaurs in Fantastic Fiction: A Thematic Survey* (McFarland Publishers). A motif here is that each 'critter' in my selected starring lineup of paleo-monsters bears relation to significant events in the history of paleontology. (Note—this was my third in the post-*DIFF* series.)

Time to dip your feet in and go for a swim. Hope you aren't afraid of the water, heh, heh... or, rather, what lurks below its undulating surface!

Pictures are worth a thousand words, or so some say, and so we need look no farther than early illustrations by artists Edouard Riou and Zdenek Burian to envision how astounding glimpses of marine paleo-monster combat would seem to our eyes. Riou had illustrated 1860s editions of French geologist Louis Figuier's *Earth Before the Deluge*, incorporating one striking image (Plate XV) showing a sea-battle between *Ichthyosaurus* and *Plesiosaurus*. Jules Verne, relying on Figuier's text for authenticity and science facts, imagined an even more terrific marine confrontation as witnessed by explorers of the subterranean prehistoric world, aboard a raft made of fossil wood, in his foundational 1864 novel, *Journey to the Center of the Earth*. Here, in this marvelous blend of imagetext, the two battling 100-foot long sea-monsters were again illustrated by Riou, although this time coupled with Verne's gripping verbal description of the science fictional encounter. A century later, Czech paleoartist Burian painted similar (non-sci-fi) scenes, seen for the first time by this author in my copies of Joseph Augusta's *Prehistoric Animals* (1956, Plate 20,), with another striking painting published in Augusta's *Prehistoric Sea Monsters* (1964, depicting a primeval battle between *Elasmosaurus* and *Tylosaurus* fought among foamy waves). Evidently Burian was influenced by Verne and Riou. Ever since, writers have striven to capture essence of these ancient sea monsters, and their mythical confrontations with man.

Mark Berry, author of *The Dinosaur Filmography* McFarland Publishers, 2002), has given us candid synopses of dino-monster films featuring 'Nessie,' the famed Loch Ness Monster (also see his articles in *Prehistoric Times* nos. 62, 63), and so it's appropriate to explore a pair of Nessie's appearances in paleo-literature too. But why all the fuss over an alleged creature so clearly (and rather cleverly) contrived?

While for centuries Scotland's legendary monster may have been relegated to ranks of one of those ordinary run-of-the-mill sea monsters cited by early mariners (as recounted by Charles Gould in his *Mythical Monsters*, 1886),

Nessie's true fame came to the fore following an amazing photograph produced by Col. R. K. Wilson, taken in 1934, in which the monster's plesiosaur-ish 'attributes' seemed evident. Of this photo it was aptly stated that, "...though it never actually proved the existence of the Loch Ness monster, it was sufficiently compelling to nourish the legend and shake the doubters." (*Chicago Tribune*, "Farewell to a Loch Ness Legend," 4/9/94) Since 1934, there have been several scientific missions to discover the monster in all its imagined glory, all to no avail. Then in April 1994, the famous Nessie photo's authenticity was falsified by one of its perpetrators—a deathbed, and death knell, confession by Christian Spurling. Still, as one individual exuberantly stated, "It's much more fun to believe there is a plesiosaur here than that there isn't." Well stated, for the possibility (or plausibility?) attracts millions of vacationers each year.

And so, in Jeffrey Konvitz's engaging *Monster: A Tale of Loch Ness* (1982) the famed monster is a complexly woven evil, vengeful, symbolically naturalistic, nationalistic and even canonical creature which (literally) rises from its murky lair when greedy petroleum engineers try to strike it rich in the Loch's geological deposits. Protagonist—American ex-pro football star, Scotty Bruce, proves he has a conscience after all and does what's right in the end, against all odds forcing his employer, Geminii Petroleum Corporation, to cease exploration of the Loch while freeing the otherwise doomed monster. It's a rather simple story, albeit with plot twists involving an interesting assortment of bad guys. But what of Konvitz's Nessie? Is it a huge relative of a relict, isolated population of plesiosaurs that somehow survived the Cretaceous-Paleogene ("K-Pg") boundary extinction?

Bruce's boss, an oil king named Whittenfeld, a 'Captain Ahab' type character who maniacally intends to drain Loch Ness of its riches, ironically refers to the Loch in words which would perhaps more appropriately characterize a living Nessie. "There it is! Loch Ness. The big prize. Sitting and waiting. Waiting to get angry... I've seen the fear it instills. But what intrigues me... Is what lies beneath. A colossal anomaly. A freak of nature. A renegade *oil field*." Is that guy really crazed, or what?

After one drilling rig is wrecked, Bruce surmises from sonar traces and other evidence (a *chewed* pipeline) there may be something alive down there which caused the devastation. Soon two crypto-zoologists who are absolutely obsessed with capturing and studying the Loch Ness monster arrive on the scene, ("...it's the greatest discovery in the history of science. Legend comes to life. And we are going to catch it. Catch it for all the world," declares one of them). But heroic Bruce's patience wears thin after realizing they've sold him out to ambitiously have their way.

One of the scientists claims from analysis of mucous tissues found on a section of chewed pipe salvaged from the wreckage that the monster is not a "categorizable group of species," yet "reptilian in derivation." They estimate the creature's length at 125 feet! Eventually Bruce witnesses the monster as it kills the parish priest on the waters of Loch Ness! But later Whittenfeld still doesn't believe him, because after all, "This is the twentieth century. Dinosaurs and sea monsters do not exist."

Fortuitously, a submersible equipped with underwater cameras that crumbles in the monster's jaws before inevitably sinking catches a hazy photo of the beast. Poring over this, the crypto-zoologists describe a scaly creature with vestigial flippers and a single huge claw at the end of the front (arm) flippers. But ultimately Bruce has the best, rare and final glimpse as he frees the roaring and frightening yet wounded 'innocent' creature after dynamiting a path through an underground cavern, allowing it to pass from the loch into the sea.

Oddly enough, I first learned about what is now Illinois' State Fossil (*Tullimonstrum gregarium* - or the 'Tully Monster' of Mazon Creek's Coal Age), circuitously, by way of an introduction to Loch Ness' Great Orm (an old Norse word meaning 'worm,' or 'dragon.') In the summer of 1970, I devoured a paperback edition of journalist and cryptozoologist Frederick W. Holiday's (1920-1979) *The Great Orm of Loch Ness*. Holiday asserted that Nessie wasn't a plesiosaur but a different kind of prehistorian—yes, a gigantic form of Tully Monster. Kudos to Holiday for this fascinating account, fortified as it was on the basis of 'actual' sightings, historical facts, biological logic, and other anecdotal information. It all seemed so verily and eerily plausible then, utterly 'goose-pimpling'! Not a reptile but a *marine invertebrate*—of course! No wonder Nessie was spotted so infrequently on the Loch, as it didn't need to surface for air!

But whereas a full grown Nessie grew anywhere from 40 to 100-feet long, 280-million-year-old Tully Monster fossils are typically less than one foot in length. And yet, ironically, the painting on the cover of Holiday's book (see image) presents a reptilian Nessie, looking sort of like an aquatically adapted *T. rex* - fingered plateosaur. Beyond the conventional plesiosaur paradigm and Tully Monster theory, crypto-zoologist Roy P. Mackal suggested that Nessie might be a surviving crypto-form of 'zeuglodont' derived from prehistoric whale stock. Few fiction writers of today, however, would suggest Nessie was a plesiosaur, let alone a mammal—although undercurrents of Holiday's Orm theory surface in Steve Alten's highly entertaining 2005 novel, *The Loch*.

As in Konvitz's earlier novel, oil and the idea of escape from Loch Ness figure prominently in Alten's *The Loch*. Marine biologist Zachary Wallace, a worthy descendant of Scotland's heroic William Wallace and of 19th century evolutionist Alfred Russell Wallace's bloodline, has a host of mysteries to solve at Loch Ness, and in so doing finds his own past inextricably linked with a terrible monster. In fact, after having fallen down and out on his luck, the *only* way to his personal (psychological and professional) salvation is to symbolically (and literally) slay the dragon. While a rival monster-hunting scientist likened to *King Kong's* Carl Denham caters to the paparazzi, illogically concluding that

Nessie is a feisty plesiosaur, instead Zachary sensibly maintains a low profile and an open mind, amassing evidence that strange 'drownings' in the Loch can be attributed to a mutant marine species which he'd encountered twice before with awful consequences.

Quite rationally, Zachary realizes that the species Nessie belongs to is a precursor form of Anguilla eel which got big because, "It's Nature's way. The ancestors are always larger. Then evolution takes over, it adapts, it makes adjustments, based on environment and competition, and the availability of prey." Furthermore, and here's where the oil seeps into the plot..."She must be a female. Female Anguilla grow really big... They leave Loch Ness when they're ready to spawn, returning to the Sargasso Sea. But Nessie's trapped, it's screwed up her biologic clock. She can't have spawn in freshwater, her DNA won't allow that, so instead, she just kept growing ... She's a mutant... and now she's become dangerous, her brain filled with lesions caused by an oil leak." Yes, oil has leaked into Nessie's lair from a broken North Sea petroleum pipeline, polluting the Loch's aquifer system.

When Zachary finally confronts the mutant Eel in an oil-saturated cavern far below Aldourie Castle, brandishing the Braveheart sword of the Templar Knights who have cloaked Nessie's existence for centuries, he notes, "The monster's head was colossal, its face a combination of giant eel and a vampire bat. Snub-nosed nostrils were upturned and pronounced, revealing a mouth filled with an assortment of elongated teeth that would put a *Tyrannosaurus rex* to shame." The slayed 'dragon' slides into a fiery inferno of burning oil (reminiscent of the climatic fin-backed 'Fire Monster' scene in Irwin Allen's *The Lost World* (1960) film production). All mysteries solved, Zachary's sanity is restored and he marries the voluptuous Scottish lass who has teased him since his youthful days. And he—not his deceased scientific rivals—gets to assign the former 'crypto-' species, now known to science, the tongue-twisting name "Anguilla giganticusnessensis."

Peter Benchley's 1975 novel *Jaws* set a tidal wave in motion... baiting viewers to a heart pounding film, and then by 1997 a pair of novels about an even more monstrous shark—the largest such genus known to science, *Carcharodon megalodon*—whose enormous teeth have been prized from southeastern U.S. coastal Miocene deposits. But lost in the cash cow feeding frenzy is how the *idea* of a *fossil* shark rests at the hallowed foundation of the geological sciences. Centuries before dinosaurs had been named and categorized, scholars had an inkling that mysterious animals resembling modern sharks known only from fossilized teeth, or 'tongue-stones' as they were commonly called, had lived an indeterminately long time ago. In fact it was Danish geologist Nicolas Steno (1638-1686) who quite rationally argued in his landmark1667 account titled "The head of a shark dissected" that isolated fossil shark teeth, thought by many to be inorganic products of nature, instead formed within the mouths of living sharks. Such research catalyzed more advanced ideology. For Steno's remarks on tongue-stones led to his most significant publication, *Forerunner* (1669) to his *Dissertation on a solid naturally enclosed within a solid*, which in turn paved the way toward modern understanding of depositional sedimentary processes and geological superposition.

If not for fossil shark teeth, which Steno realized were similar to former *living* sharks (as Swiss naturalist Conrad Gesner had also perceived a century earlier), the sciences of Geology and Paleontology might have languished for decades. But, despite their precociousness, neither Steno (nor Gesner) were equipped with the knowledge or savvy to write a real page-turner about demonic *prehistoric* sharks, like those penned in 1997 by writers Charles Wilson (*Extinct*) and Steve Alten (*Meg*).

While most readers in-the-know have probably heard of Alten's 'Jurassic Shark' *Meg*, few of you may know of Wilson's *Extinct*, published a few months earlier. Here, enormous sharks appear off the Mississippi coast where they find live humans appetizing. Meanwhile, Admiral Vandiver, who has a fascination for megalodons, speculates that the man-eaters appearing in coastal waters aren't ordinary 25-foot Great Whites! No, they're evidently much larger and their freshly shed teeth uncannily resemble the prehistoric variety. Vandiver theorizes that millions of years ago, the *Carcharodon megalodon* species descended into oceanic trenches, where with rare exceptions it has remained hidden from civilization. Now, however, with atomic testing by the French military in the South Pacific, the megalodons have reclaimed their former Mississippi River basin spawning ground. Shark fanatic Vandiver yearns to stare into the eyes of a living megalodon; he's awarded this awful opportunity, risking his life. Only after a rampaging 50-foot long specimen is finally subdued, does he spy a truly monstrous creature further out at sea—a *200-foot long* megalodon appearing like a 'cliff emerging' on the horizon when it surfaces to briefly survey mankind's folly.

Alten's first novel, *Meg*, isn't as artfully written as *The Loch*, although it is hugely absorbing, or one might say 'consuming' (like *Jaws*). We realize what a fine kettle of fish we're in for from the get-go through an introductory vignette set 70-million years ago; Megs hunt Rexes which stray off shore. Predator becomes unwitting prey in this unsettling turnabout scene. "For the first and last time in its life, the *Tyrannosaurus* felt the icy grip of fear." The 60-foot long prehistoric megalodon rips and chomps its way through a quivering, bloodied Rex which had the misfortune of pursuing tasty hadrosaurs into the waves. For since the K-Pg extinction event 66-million years ago, species of megalodons managed to survive in the isolated hydrothermal vents of the deepest ocean depths.

That is until Man invaded their dark lair in the Pacific Ocean's Mariana trench. Protagonist marine biologist/paleontologist Jonas Taylor's adventures with the Meg are inspired by the biblical tale of Jonah and the Whale. Taylor claims to have seen a megalodon before on a deep sea submersible dive in which lives were lost. (Indeed there are many plot parallels one may make between *The Loch* and *Meg*.) But of course nobody believes him. Now outcast Taylor is wary of the deeps. But soon he's enlisted into another expedition where a menacing, marauding Meg, "… the greatest predator, the greatest killing machine in the history of our planet" is lured to the surface. Soon the monster's weakened eyesight is blinded by searchlights which "… smashed through the darkness as if guided by the hand of God." But even the hand of God isn't sufficient, as the creature continues to dispatch whales, fishing boats, submersibles and even a nuclear sub with reckless abandon. (Devout Steno and Gesner would surely have been appalled by what they had bequeathed to science, unleashed through their insightful interpretations of fossil tongue-stones!)

Later Taylor, staring at the creature's open jaws, muses, "Coursing through its DNA was seventy-million-plus years of instinct. The predator could not think or choose; she could only react, each cell attuned to her environment, every response preconditioned. Nature itself had decided that the species would dominate the oceans, commanding it to perpetually hunt in order to survive." Then he whispers, "We should have left you alone … " But his lament comes too late for its many victims, both human and cetacean. Soon Taylor lodges his submersible craft into the belly of the beast, intent on destroying it. Lacking proper tools, ironically and symbolically, he must rely on his *own* fossilized tongue-stone—a finely serrated *Carcharodon* fossil tooth which he always carries for good luck—to cut his way out of the Meg's stomach and sever the heart arteries. That's not the end of the story, however, as this Meg was a female spawning offspring, one of which is hauntingly recognized, (once again too late by Jonas), sufficient for Alten's sequels. (Another short story involving a 'Meg' was published in 1992, Poul Anderson's well-paced "Unnatural Enemy.")

Is 'Meg' truly extinct? How strong is the case for extant Megs (which are thought to have expired about 2-million years ago)? No enlarged shark tongue-stones attributable to 'Meg' have been found in a pearly white, un-fossilized state. So until that fateful day when such a monster surfaces, 'Meg'-sized sharks will stray closest to living reality through restorations of artists such as William Stout (who illustrated Poul Anderson's story), and Charles R. Knight.

The Creature (or 'Gillman') from the Black Lagoon's compelling (dare I say 'universal'?) paleontological appeal stems from a sensational 1938 discovery of a genuine 'living fossil,' the modern Coelacanth (*Latimeria*) and, perhaps, two 1930s sci-fi/horror novels by Karel Capek (*War With the Newts*, 1936) and H. P. Lovecraft (*The Shadow Over Innsmouth*), plus an added intriguing plot element that the movie Gillman represented a Devonian fish - anthropoid ('missing link') cross-evolutionary relic surviving in the Amazon. (In his *Classic Movie Monsters* (1978), p.259, Donald F. Glut cites another crypto-factor.) However, a distinctly *dinosaurian* Gillman variant slashed its way through 'Carl Dreadstone's' *Creature From the Black Lagoon* (1977), a trashy yet gripping novel *inspired by* the classic film (based on a story by Maurice Zimm). The novel doesn't follow the movie script by any means, although the Amazonian setting may seem familiar.

After spying the gigantic 'Creature,' which is 30-feet tall and has a powerful tail, scientists onboard the *Rita* debate whether the animal is prehistorical (e.g. – interestingly a possible descendant from plesiosaur or diplodocine stock), or an atomic mutant. Does it represent our future or the past? Although they name the monster 'Mutantus giganticus,' its reptilian-amphibian characters seem predominant. Plus, it's highly intelligent and in a King Kong-like fashion holds strange reverence for the expedition's sole female, Kay. In the end, after the immensely strong creature has been subdued by a torpedo, Kay stares at its decapitated head lying on the ship. "Looking down at the eye, Kay screamed repeatedly, for the eye was accusing her, reproving her. The eye was the eye of a betrayed friend." Nature betrayed?[2]

Okay - so after reading all this, wouldn't you agree that Man should best remain in the shallows, avoiding the darkest depths and the oblivious uncharted waters of lost places where prehistoria dwell? Maybe so. But inevitably we'll continue to challenge raw, savage Nature – rabid red in tooth and claw, won't we... that is, until our bitter end! Mua hahahahahahaha!

Note to Chapter Twenty-five: (1) Wade Carment wrote, "Allen Debus forgot to mention a book in his feature (PT # 83). It was *Megalodon* (1980) by Robin Brown. This book is what inspired me to try for the Navy in 1990. Would've been accepted if it wasn't for flat feet. Anyhow, *Megalodon* has a meg attacking a US nuclear sub & a Soviet one. And … it's a juvie! The adults are over 2000 ft! Read this book four times." Then Alan Dean Foster wrote, "In response to Allen Debus' listing of fiction focused on megalodon, he might want to add my short story "He," which appeared in the June 1976 issue of the *Magazine of Fantasy and Science Fiction*, and centers on a South Pacific encounter with a surviving megalodon (reprinted in the collection *With Friends Like These* (Del Rey, 1977). Presumably. Steven Alten never saw the story, which predates his by many years. I'll be driving the area just south of the Marianas Trench in January and will keep an eye out for the 'meg.'… though since there are sperm whales in the area there's much more of an actual chance to encounter an Architeuthis. That would truly be a dream come true. …"

(*Prehistoric Times*, no.84, Winter 2008, p.7) (2) Vargo Statten's authorized 1954 novelization, *Creature From the Black Lagoon* is another important entry.

FIVE

1923 restoration by E.M. Fulda, showing "Homelife of the dinosaurs of the Cretaceous beds of Iren Dabusu. In the background iguanodonts are being attacked by deinodonts. In the foreground the Ostrich dinosaurs are running away in a cretaceous panic." Figure and quotation are from Roy Chapman Andrews' On the Trail of Ancient Man, Doubleday, Page & Co., Inc., 1926. See Chapter Twenty-eight for more.

Paleo-Pioneers

It would seem remiss to discuss pop-cultural paleontology without outlining several individuals who formatively shaped the essence of our current understanding of dinosaurs and paleoimagery. So, without further ado… The following chapters address certain men (for more on female pioneers, check out Chapters Thirty-five and Thirty-nine) who lived in the 20th century. Through discoveries, insightful observation, perseverance or invention, these men handed down to us modern concepts of geological time, vertebrate paleontology and dinosaurs in a golden 'renaissance' era, or—in an embryonic stage—in film, perhaps beginning where things rightfully should … with good old *Brontosaurus*!

Several of these chapters originally appeared in the following publications:
Chapter Twenty-six—*G-Fan* nos. 123-124, Spring 2019 (Part 1, pp.24-29) and Summer 2019 (Parts 2 in press).
Chapter Twenty-seven—previously unpublished.
Chapter Twenty-eight—previously unpublished.
Chapter Twenty-nine—*Earth Science News* (Earth Science Club of Northern Illinois) March 1984 to July-August 1984 (Parts 1 – 5), (pp. 17-19; 19-22; 25-27; 15-18; 11-13, respectively)
Chapter Thirty—*Fossil News: Journal of Avocational Paleontology* (circa 2006).
Chapter Thirty-one—*Prehistoric Times* nos. 69-71 (Parts 1 – 3), Dec./Jan. 2005 to April/May 2005, (pp.49-51; 49-51; 50-51, respectively).

Chapter Twenty-six — *When Brontosaurus was Monstrous: Three Pioneers' Enchantment with our First Filmic Paleo-Monster*

Before *Tyrannosaurus* and Godzilla came along to plunder our planet—threatening mankind's security in the universe, *Brontosaurus* reigned as king! Every artist needs his/her muse, and a century ago for Winsor McCay, Herbert M. Dawley and (arguably perhaps) Willis O'Brien that inaugural inspiration was apparently *Brontosaurus*, especially the 'giant' reconstructed *Brontosaurus* skeleton at New York's American Museum of Natural History, first displayed there in February 1905. These three talented individuals each independently created means for animating creatures—regarded as prehistoric and monstrous by captivated audiences—as shown 'live' on film.

But first—few *today* would regard gentle, lumbering *Brontosaurus* as a 'monster.' In its heyday, however, it clearly was, as recently reinforced in an objectively-minded panel survey (titled "Top Ten Sci-fi Prehistoric Monsters") I conducted at the 25th annual Godzilla convention "G-Fest" in July 2018. Here, *Brontosaurus* finished in a lofty 8th place (with a respectable 71 points out of 96), from a field of 47 candidates. Still having trouble imagining how & why *Brontosaurus* 'defeated' other worthy prehistoric monster entries—such as Gorgo, Anguirus, and the "Giant Behemoth" Paleosaurus?

How did classical monstrous 'Bronto's' impact on public consciousness and pop-cultural imagination prove pivotally inspirational to a trio of filmic paleo-monster pioneers? Times change, and so now let's explore circumstances of a century ago … when *Brontosaurus* was not mired in 8th place, but ruled as absolute King of the dino-monsters!

By the early 1890s, dinosaurs and other large extinct reptiles were often imagined as 'prehistoric monsters'—as popularized for example in the title of the Rev. H. N. Hutchinson's 'dinosaur' book *Extinct Monsters* (1893 ed.). Over decades the fascination mounted, for there was even a "Prehistoric Monster Bash" held in Butler, PA during the Summer of 2012! Yet how do we define a science fictional *prehistoric monster*? It's a term I've wrestled with over the past decade, defining it in my 2009 McFarland book—*Prehistoric Monsters*, and also in a 4-part series of articles published in *Scary Monsters Magazine* (January through June 2006). Essentially then, we recognize science fictional prehistoric monsters as those creatures that have been entombed in ice for eons—only to emerge as a result of nuclear bomb tests or global warming; 'living fossil' species that have gone on cryptozoologically living in 'lost world' hidden places of the globe, or deep *inside* Earth within its innermost, impenetrable caverns, where Time has mysteriously stood still for millennia; creatures that have been 'hibernating' under the sea—until disturbed, awakened somehow by mankind; creatures encountered by humans who slip backward (or sideways) in machines, say, to the Mesozoic where direct confrontations are possible; aberrant creatures with more direct genetic/DNA ties to long extinct species revived into our present via mad scientist bioengineering; or even creatures from other planets (such as Venus) where the evolutionary state remains primordial relative to Earth's.

In short, science fictional prehistoric monsters are what mankind *shouldn't* be living alongside because *they* should be dead, extinct! Such creatures are not necessarily *giant* (think of 1954's *Creature From the Black Lagoon's* (Universal, D: Jack Arnold) Gillman, for example), nor always 'dinosaurian,' and also not necessarily 'kaiju.' And yet sometimes real dinosaurs as formerly known to science—like influential *Brontosaurus*, living in misty, fog-swept swamps—also make tremendous science fictional monsters! As instilled through an 1898 painting by American Museum paleoartist Charles R, Knight, a century ago those huge lumbering brontosaur sauropods were also viewed as mainly aquatic, swamp-dwelling—like Gillman.

And they were so much larger than elephants—man's usual standard for comparison among the living. At first and above all, late 19th century fantastic artwork emphasized the awesome, gigantic sizes of the super long-necked 'sauropods,' the dinosaurian group to which *Brontosaurus*, or others such as *Diplodocus* and later known *Brachiosaurus* belonged. Sheer size immensity could then be allied to their supposed horrific nature, especially if such huge monsters would anachronistically confront and clash with mankind. Wouldn't such bulky brutes be unstoppable, invulnerable? Just look at their enormous museum skeletons and imagine how they would seem—*alive*, in the flesh!

And, decades before Stephen Speilberg's *Jurassic Park* (Universal/Amblin, D: Steven Spielberg, 1993) movie, that's exactly what Winsor McCay did.

Winsor McCay's Increasingly 'Monstrous' Brontosaurs:

Winsor McCay (1867-1934) is credited with having created the first dinosaur movie, initiating a trend continued through a century, eventually yielding every persuasion of filmic 'extinct monsters' terrorizing mankind, threatening civilization along the way. Fate and fossils, those of the American Museum's *Brontosaurus* especially, would conspire in McCay's favor—an indelible mark in history.

Although I'd been a major fan of dino-monster movies since the late 1950s, the first time I remember seeing McCay's *Gertie the Dinosaur* (D: Winsor McCay) animated cartoon short in its entirety (i.e. first released on February 2, 1914) was during Don Glut's "Dinosaurs in the Movies" seminar presented at Chicago's Field Museum of Natural History during the weekend of October 5th – 6th, 1985. "Gertie" began as a live vaudeville stage performance in 1912—with the cartoon serving as 'background' for McCay who interacted with his friendly dinosaur. The staged version was later filmed and released to theaters by 1914 including an introductory sequence showing the influential mounted museum skeleton with Knight's paleoart visible underneath the dinosaur's long neck. This movie may be seen on Youtube.

'Dinomania' popularity had only begun to sweep the nation as of 1896, according to dinosaur historian Paul D. Brinkman. By the 1910s, *Brontosaurus* had become America's most familiar dinosaur, thanks to its 1899 discovery at Wyoming's Nine Mile Crossing, extraction from Upper Jurassic strata, with later mounting of the reconstructed skeleton in New York (an arduous task completed in 1905). McCay's infatuation with this particular *Brontosaurus,* the "germ-cell" of his paleo-creations commenced in 1905, the "key year" when "inspired by the inauguration of a dinosaur skeleton (i.e. the *Brontosaurus* – author's insertion) in New York, (he) drew his first dinosaur strip," the start of a lifelong fascination." (Merkl, p.61)

Subsequently, manifestations of the American Museum sauropod (a generic 'brontosaur') began appearing in his print comic strips (e.g. "Dream of the Rarebit Fiend"), first as skeletons (March 4, 1905) and then in "Little Sammy Sneeze" (May 27, 1906), followed later by a fifth dinosaur incarnation—a live 'Gertie-esque' version in "Midsummer Day Dreams" (August 1911). All could be considered practice cameos for the eventual Gertie cartoon. Ulrich Merkl, author of the delightfully lavish *Dinomania*, speculates (p.78) that if the American Museum's *Tyrannosaurus rex* skeleton had been displayed *before* its brontosaur, perhaps McCay's dinosaur muse for print cartoons and his later movies would have instead been the former—maybe a live, city-attacking theropod version, "Gertie-saurus rex"? Henceforth, McCay also produced a bevy of print cartoons and editorial images over the years incorporating Gertie look-alike creatures, underscoring his enchantment with America's then foremost dinosaur icon, *Brontosaurus*.

McCay's work proved so successful as to invite imitation from New York animator John R. Bray (1869-1932), who according to Donald Glut made his own 'Gertie the Dinosaur' cartoon. However, "... this ersatz Gertie lacked the original's fluidity of movement and amiable personality." (Glut, *Scrapbook*, pp.168, 170) Merkl refers to Bray's 1915 (~ 7 minutes long) copycat production (also available on Youtube) as a "ripoff." (Merkl, p.281 n.133) McCay sued Bray and prevailed, but the latter was soon at it again with a Gertie-like *Diplodocus* cartoon character.

McCay's career as a pioneering dinosaur artist culminated in "Gertie." However, as the geopolitical spectrum moved on from the horrors of World War I toward the dawning days of Nazi power and its war machine he imagined and created other monsters. McCay's outlook became darker, more sinister, as illustrated by Merkl's discovery of his "lost art" 'Dino' cartoon of 1934 (line drawings under his pseudonym "Bob McCay" most likely produced between March 1933 and July 1934—his death). Dino the brontosaur is clearly Kong-inspired, borrowing concepts from his two 1921 films, "the Pet" and the evidently non-released "Gertie On Tour." McCay's sauropod 'idea' became more menacing and seemingly more monstrous in print and in film.

McCay's two movies of 1921 are rarely known, but Merkl regards *The Pet* (11:48 in length) as the first in the "giant monster attacks metropolitan city genre." (p.263) The monster in question isn't a dino-monster, but instead a dog-like creature: its accelerated growth anticipates 'Ymir' from 1957's *20 Million Miles to Earth* (Columbia Pictures, D: Nathan Juran). *The Pet*, a poorly distributed film, is available on Youtube. Secondly, his *Gertie On Tour* film was approximately 12-minutes long but possibly never released. The original nitrate film stock has mostly disintegrated, leaving only one minute of footage intact, in which Gertie roams about 'toying' with a train. And thanks to Merkl's diligence, in preliminary concept sketches reproduced in his book prepared for *Gertie On Tour*, now representing apparently lost footage, we see ideas for additional scenes showing Gertie spring-board pouncing upon the Brooklyn Bridge as if it were a trampoline. And another in which Gertie bites and uproots the Washington Monument as if it were a tall tree. As Merkl suggests, the sequence of Gertie's 1921 *On Tour* scenes may be considered precursors for McCay's later day "Dino" strip created during 1933 – 1934, which never made it to print.

Although resembling Gertie, from surviving sketches and Merkl's own inferred efforts, relying on Roger Langridge to reconstruct missing cartoon panels, McCay's 'Dino,' "lost art" dino-monster sauropod character—an evidently gentle creature yet utterly misplaced in modernity, seemed inadvertently bent on rampage in metropolitan areas. As Merkl opined, the Dino strip would have been "A ... masterpiece very nearly (making) its debut (i.e. in the fall of 1934): Winsor McCay's Dino, the odyssey of a dinosaur who awakens from a long, deep sleep and starts getting into trouble with human civilization. But it was not to be …" (Merkl, p.15)

As initiated, the Dino concept was as follows. Dino emerges from a rocky tomb at Granite Mountain Stone Quarry and lumbers toward New York City, then proceeds into New Jersey via the Holland Tunnel, clashing with landmarks along the way. McCay died early in concept creation and so Merkl is left in a quandary: "We can only speculate about the route, but also about possible sub-plots…. Might Dino be attacked by the military? Might a paleontologist take the stage to defend the dinosaur? Might another dinosaur emerge from the quarry, perhaps a *Tyrannosaurus*? Alas, we will never know …" (Merkl, p.40) To me, the Dino idea does anticipate many of the popular dino-monster themes of the mid-20th century and beyond. Based on today's pervading pop-culture, reliance on a sauropod with a gentle rather than a hostile persona seems misplaced. However, it does signify how influential was McCay's brontosaur skeleton 'muse' at the American Museum, first seen three decades earlier.

Dino was left unpublished during McCay's lifetime. Why? Merkl lists several possibilities in his 2015 book (p.57), the most plausible of which would seem (1.) that the perambulating dinosaur would have reached all the key landmarks within a limited number of newspaper installments and then have exhausted the kernel of the cartoon's theme, and (2.) of course—McCay's untimely death. Over the years "Dino"—a prototypical dino-monster in its own right that might have further fueled dinomania years before arrival of Rhedosaurus, Godzilla and *Jurassic Park's* cunning *Velociraptors*—faded into obscurity. Yet, intrigue surrounding the essence of the *idea* of 'bronto-monster' hugeness, permeated into prehistoric symbolism, remained at the forefront of human imagination through the war years.

Herbert M. Dawley's Romance of the American Museum's Bronto-Monster parlayed into Film History:

Most in the younger generation, 'millennials' and such, haven't (ever) watched a silent movie and don't seem to care for black and white films. Even if such films were the first of their kind involving those ever-popular dinosaurs.

But for those who do care about the subject, for many years Herbert M. Dawley (1880 – 1970) was at best only known as an obscure character in the early history of film-making, and an unworthy antagonist to our usual, accustomed 'hero'—Willis O'Brien (yes, that genius behind Kong's special effects). Yet, analogously, just as *minutely* detectable evidence of a global iridium boundary layer overturned conventional wisdom in the matter of Late

Cretaceous dinosaur gradualistic (uniformitarian) extinctions—since favoring a 'catastrophic' asteroid death projectile scenario, likewise, a mere 'smidgen' of photographic evidence fortuitously discovered by sculptor and paleoartist Stephen A. Czerkas utterly reversed opinions on that strained 'Dawley versus O'Brien' relationship. But first who was this guy Dawley, whose tarnished name rarely came up in the old books and articles we've read.

Dawley was an innovative, highly talented and inventive renaissance man who for a short while turned his interests to creating 'live' realistic footage of dinosaurs. And yes, like McCay, his interest in dinosaurs was also inspired by a fated visit to the American Museum of Natural History, although during the summer of 1917 when he spied that great *Brontosaurus* skeleton. Displayed underneath the long neck of the great skeleton were two seminal works by famed paleoartist Charles R. Knight: a miniature sculpture restored 'in the flesh,' and his 1898 painting of swamp-loving brontosaurs—the scientific paradigm then, while his painting also suggested that brontosaurs sometimes ambled about sluggishly on land too. Knight's artwork had been exhibited there possibly since 1905 (or at least by 1914 when McCay's *Gertie* was produced).

According to Czerkas (1951-2015), Knight's live-looking model haunted him throughout the day and night as "... Dawley became increasingly obsessed with the thought of being able to visualize live dinosaurs actually moving about." (Czerkas, *Dawley: An Artist's Life,* p.50) Further fueling his obsession, during the 1910s when Dawley worked on engineering and auto-design details for the Pierce-Arrow automobile company, he had met Thomas A. Edison "...specifically to learn about his motion picture business and how he could use it for educational promotions..." thus impregnating the germ of an idea. (Czerkas, *Dawley: An Artist's Life,* p.42)

So by 1917, Dawley had already made short films for Pierce-Arrow, familiarizing himself with the 'how-to' of motion picture making. But in order to make dinosaurs appear live on camera, choreographed, he needed small models, like Knight's *Brontosaurus*, except sufficiently flexible such that they could be incrementally moved and photographed with motions orchestrated in sequence frame by frame. When the reel was projected onto a screen the dinosaurs would then seem alive in their prehistoric settings.

According to Czerkas, Dawley suffered a "sleepless night ... literally tossing and turning in his bed, while his creative imagination ran wild." (Czerkas, *Dawley: An Artist's Life,* p.50) Inspired by Knight's artistic impressions—the *Brontosaurus* model, Dawley undertook the task of building his own miniature clay model of dinosaurs for film-making purposes. His mind was steeled for this venture as of September 3, 1917, when he revisited the American Museum, meeting with Curator, Dr. Clyde Fisher, who in turn introduced him to Frederick A. Lucas—perhaps *the* ideal scientist there at that time for learning how to recreate realistic-looking dinosaurs! For around the turn of the century, Lucas had published, both, a popular book *Animals of the Past* (1901), which I've referred to elsewhere as the first bona-fide 'American dinosaur book,' and also a pivotal paper concerning proper 'how-to' restoration of prehistoric animals. Mark Wolf and Sylvia Czerkas also opine that McCay's *Gertie the Dinosaur* (1914), "Although drawn animation, the fact that Gertie was a Brontosaurus very possibly inspired Dawley and/or Obie to think in terms of prehistoric creatures depicted on film by animation." (Stephen Czerkas, *Dawley: An Artist's Life*, p.335)

Throughout September, Dawley's encouragement escalated as he obtained photographs and drawings of various dinosaurs illustrated in the flesh, or in skeletal form, technical advice on their paleobiology, and expert guidance on how to sculpt them realistically in miniature. He also visited the Museum's laboratories and was allowed to make bone measurements firsthand—including of the *Brontosaurus*. Then after a month of preparation, on October 4, 1917, Dawley made a schematic diagram of a brontosaur, like Knight's museum model, but fortified with an internal *mechanical* armature. This conceptual phase concluded two months later in his first 3-foot long camera-ready dinosaur model! Czerkas, an expert stop-motion animator in his own right, explained how the articulated mechanism of the brontosaur was designed to be employed, as it was capable of being "... continuously repositioned and yet ... (able) to stay in whatever pose that it was put into. It was like making a skeleton for the animal. But where the joints connected they were held together by nuts and bolts. By adjusting the nuts and bolts to be not too tight and not too loose, the joints could be moved and yet hold their position. It was basically a simple process, but deceptively complex to work sufficiently well." (Czerkas, *Dawley: An Artist's Life,* p.51)

Next, Dawley cautiously began a more filmic experimental phase, taking a hundred or so photos of a brontosaur 'moving'—for the first time since the Jurassic Period—binding them into a small 'flip card' booklet. Flipping its pages with his thumb, he saw that the stop-motion photographic process would work convincingly enough. He showed the flip-book to a cousin, J. Searle Dawley, (who had directed Edison's 1910 film adaptation—*Frankenstein),* guiding further progress. (As I've mentioned in several other writings, odd associations between Mary Shelley's 1818 novel strangely linked with early considerations of dinosaurs do quite fascinate me!)

It seems likely that Dawley's same prototypical test *Brontosaurus* (which was later donated to the Natural History Museum of Los Angeles County—by Willis O'Brien) was also featured in his first dino-monster film—a major production for its time titled *The Ghost of Slumber Mountain* (World Cinema Distributing Corp., Prod.: Herbert M. Dawley, 1919). Although not granted much more than an extended cameo, albeit in a very short film, *Brontosaurus*—described therein as a 100-foot long "Thunder Lizard" from the mists of 40 million years ago—earns pride of place, appearing both in *Ghost's* opening credits, and is the first prehistoric creature spied by the film's story-teller (played by Dawley himself).

And the bronto-monster also appears in 42" X 72" color poster advertising for the movie, along with images of *Tyrannosaurus* and *Triceratops*. *Ghost* also featured a reenactment of a titanic struggle between *Tyrannosaurus* and *Triceratops*, clearly inspired by Knight's (circa 1906) American Museum painting and his 1904 painting of *Triceratops*. Quite ironically, considering how Dawley and O'Brien acrimoniously clashed over the movie's production, two of the film's characters row themselves down the ... (ahem!) "River of *Peace*." Yes—release of this movie brewed considerable contention between Dawley and his 'apprentice' animator Willis O'Brien (1886-1962), which led to legal and slanderous sparring (to be outlined shortly) over a more famous production, an adaptation of Arthur Conan Doyle's 1912 novel, *The Lost World* (First National Pictures, D: Harry O. Hoyt, 1925).

Following this, and O'Brien's eventual firing, Dawley went on to create additional stop-motion dinosaur 'puppets' for another stop-motion animated movie, long considered 'lost' by film historians until very recently, titled *Along the Moonbeam Trail* (B.Y.S. Films, Prod. Herbert M. Dawley, 1921) although, based on surviving footage, apparently lacking a bronto-monster. Like *Ghost of Slumber Mountain*, *Moonbeam Trail* also exudes dreamlike surrealistic qualities, fortified with science fictional elements—travel through outer space in an old biplane to another planet where prehistoric " frightful monsters from the dim days of creation," (*Stegosaurus,* a duckbill "Trachodon"—now referred to as *Anatotitan*, and *Tyrannosaurus*) are encountered. Of course, Earth trembles when the latter two savagely fight. Also a pterodactyl is encountered as they sail past Mars and the Pleiades, along said "moonbeam trail." This could be the earliest film in which both human actors and stop-motion animated puppets—enlarged through camera projection, appear *together* in the same frame. Notably, Dawley made all the puppet models and did all the smooth animation himself.

Today, we wouldn't suspect Dawley's significant contributions to early development of special effects techniques and sci-fi dinosaurs of filmland if it wasn't for Stephen Czerkas's keen observation of that photograph he came across in an internet auction that he was able to purchase for a king's ransom. This photo, showing Dawley removing a dinosaur model used in *The Ghost of Slumber Mountain* from a plaster mold, indicated that he had used a much different method than did Willis O'Brien in creating models for stop-motion animation purposes. O'Brien—using a "build-up" technique in shaping his dinosaurs—typically slathered clay over an armature, and then covered this with "rubber dental dam ... later add(ing) surface details." (Czerkas, *Dawley: An Artist's Life*, p.52*)* Dawley's procedure evidently involved further steps, that is, in making a plaster mold from his original (static) build-up clay model, then casting a replica using a pliant forming material, with the metal armature laid within the mold.

Here is Czerkas' analysis of Dawley's original method as published in his authoritative 2016 book, *Major Herbert M. Dawley: An Artist's Life*:

> "Dawley first made the jointed armature. Next he sculpted the figure in clay directly on the armature. Dawley then made a plaster mold in sections … The original clay model was cleaned off of the armature and discarded. To fill in and help maintain the bulk, the armature was padded with cotton and sponge rubber, built up to conform to the basic shape of the figure. Dawley then either poured a liquid rubber into the mold or pressed it into the mold while in a soft, pliable state. When the rubber cooled it became solid but still soft and pliable. This rubber-like material formed the outer 'skin,' reproducing the sculpted details and shape of the figure." (p.69)

One might judge Dawley's process of making these "articulated effigies"—which he patented in 1920—as more 'refined' than O'Brien's earlier efforts (although the latter's particular method for similar on-camera effect was developed independently and slightly before Dawley's).

Whereas O'Brien continued in the motion picture field, becoming associated with several film classics, Dawley's stop-motion technical innovations more or less dead-ended, as he moved on to other kinds of theatrical projects.

This aforementioned auctioned photo further indicated to Czerkas that Dawley's contributions to *Ghost* and possibly *Moonbeam* were far more original, enhanced and intensive than others had surmised. In short, had Dawley been dealt a rotten hand by history, while O'Brien's illustrious career had been celebrated? Through careful research Czerkas exposed what may have really happened between the two men.

Willis O'Brien's Bronto-Monsters—Dinosaur and the Missing Link, The Lost World and King Kong:

Shockingly, I learned recently that neither of my two 30-something daughters have seen the classic 1933 RKO *King Kong* movie. (Had I shirked by paternal duties?) Nor had they ever heard me utter Willis O'Brien's hallowed name.

Willis O'Brien also had inspirational ties to American Museum exhibits where he became deeply impressed by *Brontosaurus*, sufficiently so to cast it as a starring paleo-monster in three stop-motion animated movies. Film historian Mark Wolf suggests that McCay's *Gertie the Dinosaur* was the "catalyst" which inspired O'Brien to begin experimentation with brontosaur model-making in clay, cloth and wood. (Wolf in Czerkas, *Dawley: An Artist's Life*, p.336) O'Brien incorporated stop-motion animation puppets in his movie projects for Edison's Conquest Pictures independently of Dawley, as early as c. 1914—the year of *Gertie's* filmic release. O'Brien's first such production was *The Dinosaur and the Missing Link: A Prehistoric Tragedy* (1915, although according to the Library of Congress, 1917).

This was a short film (available on Youtube), released by Conquest Pictures. It is one of several shorter comedies he did involving prehistoric themes while affiliated with no less than Thomas A. Edison. To his credit O'Brien also did his homework, consulting with American Museum's renowned paleontologist Barnum Brown before building his puppets, as he did later in 1917 while working with Dawley on *Ghost of Slumber Mountain*, and then once more during production of *King Kong*. Mark Wolf considers it "… fascinating to consider that there's a chance that O'Bie and Dawley were at the museum at the same time and could have passed each other unknowingly." (Wolf in Czerkas, *Dawley: An Artist's Life,* p.336)

While the motion of a number of animated cavemen and cave women puppets is quite smooth in *Missing Link*, movieland's first appearance of a stop-motion animated brontosaur puppet (crudely constructed and poorly mobilized by O'Brien) takes place about four and a half minutes into the film. Here the two titular creatures square off, with the Missing Link succumbing to the Dinosaur in a mighty struggle. Mark Wolf suspects that this *Brontosaurus* puppet was "built for his initial test because it isn't very active; the legs are obviously not working very well." The movie "… was finished about one year after initial tests." Edison purchased this reel for $525. (Wolf in Czerkas, *Dawley: An Artist's Life,* p.336) The *Brontosaurus* model was reused in O'Brien's later *R.F.D. 10,000 B.C.* (1917). Audiences can see that Dawley's first puppet armatures clearly must have been far more sophisticated, involving ball joints and perforated metal—quite an upgrade over O'Brien's earliest efforts.

The significance of *Brontosaurus* in two other movies with which O'Brien is highly affiliated is also notable. First, understand there was no such creature written into Doyle's 1912 *The Lost World* novel. However, in the 1925 silent, Prof. Challenger and company ship a living brontosaur specimen (perhaps underscoring its personal significance to O'Brien) back to London after it tumbled from the 'lost' Amazon plateau, consequent to its battle with an *Allosaurus.* Then in a climactic scene *Brontosaurus* becomes the very first movie dino-monster to ravage a modern city. In a promotions booklet issued to accompany the 1925 film, impact of the bronto-monster is dramatized:

> "If you should meet a ferocious prehistoric monster bigger than eleven elephants--? …a terrifying giant of a beast, supposed to have been dead ten million years ago …The huge beast, a brontosaurus … one hundred feet in length and bigger than eleven elephants, ploughs down the busy

thoroughfare ... Impossible you say. His elephantine feet turn omnibuses and taxicabs into kindling wood. His long tail, terrific in strength, sweeps over monuments ... London bobbies with pistols and rifles fire at him. Their bullets are about as effective as a bean shooter. ... The monster continues his mad race for liberty from civilization and comes ... to Tower Bridge. The bridge collapses..."

Due to its great mass, the brontosaur smashes through the bridge into the Thames River; fortunately it can swim too! Such an 'impossible' scene in 1925 would have seemed just as exciting to moviegoers then, as did Godzilla's Tokyo stomp in 1954's *Gojira*.

Also, in *Lost World's* amazing dino-stampede scene several sauropods including brachiosaurs—then the largest known dinosaurs—are seen fleeing from a fiery volcanic eruption, And then of course, there's that terrifying swamp-dwelling brontosaur in 1933's *King Kong* that chases an unfortunate sailor up a tree, intent on devouring him—a scene that haunted me for years as a child! Readers are more familiar with these entries (i.e. than with the work of McCay and Dawley, or O'Brien's *Dinosaur and the Missing Link*, as previously outlined), so there's no need for plot-wise elaboration here. Of course in more sophisticated movies like *King Kong* and *Lost World* a single animator wouldn't do. O'Brien couldn't manage sole responsibility for stop-motion effects successes of movies so intricately complex involving so many puppets 'in play.' So, besides O'Brien, a *team* of talented technicians (including famed sculptor Marcel Delgado) was needed to create the many incredible dino-monsters starring in such memorable scenes.

Okay—some readers may be wondering why I am not exactly 'extolling' (let alone paying lip service toward) O'Brien's contributions to early stop-motion animation, at least as much as might be ordinarily expected. This is because of Czerkas's meticulously researched revision to the story, culminating in his landmark book, amending and correcting prior authoritative and conventionalized accounts as published by Don Shay and others between 1973 and 1982. Shay began researching O'Brien's career in film in 1962, publishing articles in *Focus on Film* (no. 16, Autumn 1973) and *Cinefex 7* (Jan. 1982). Meanwhile, Orville Goldner and George E. Turner published their *The Making of King Kong* (1975); during the 1970s Mark Wolf interviewed Willis's widow, Darlyne O'Brien. I was unaware of their works. Instead, I avidly read Jeff Rovin's *From the Land Beyond Beyond: The Films of Willis O'Brien and Ray Harryhausen* (1977), in which the, by then, traditional tale of Dawley—cast in a 'villainous' role, obstructing O'Brien's rightfulness, was retold. (Shay now acknowledges the story revision in Czerkas's recent publications.)

In a nutshell, here's the crux of the matter as based on Czerkas's publications which I strongly encourage readers to consult for all the sundry laid out details extending beyond scope of this chapter—the reconstructed history. O'Brien took more credit for his involvement in *Ghost of Slumber Mountain* than was warranted. Thereafter, things quickly unraveled. Dawley had already been generally dissatisfied with the quality of O'Brien's animation work on *Ghost*, even discarding some footage. But at the time of the movie's November 17, 1918 premier in New York City—by then O'Brien had opportunity to bill *Ghost* as "his" production—Dawley was astonished, mortified to see that he wasn't properly credited: O'Brien was indicated as sole Producer. Or, as Czerkas concluded, "Film historians have believed it was Dawley who had tried to take credit for O'Brien's work in *The Ghost of Slumber Mountain*. But the reality is that it was the other way around. O'Brien was trying to take credit from Dawley." (Czerkas, *Dawley: An Artist's Life,* p.337)

Why and how did this awkward turn of events come to pass? Well, cutting to the chase, fate intervened once more! On October 4, 1918, a month prior to *Ghost's* premier, Dawley, at the time a Major (i.e. commander of the 4th Battalion of the New Jersey Militia) was called into service following a catastrophic explosion killing about 100 people at a munitions ordinance manufacturing plant in Morgan, New Jersey. With Dawley obligated to duty, O'Brien was then assigned to wrap the project—thus using this opportunity to take lead credits. (When the cat's away, the mice will play!) Following *Ghost's* uncomfortable premier, things went downhill from there between the two film pioneers.

A repercussion was a several-years delay of 1925's *The Lost World*. Dawley, who wasn't aware of O'Brien's (unethical, if not illicit) efforts to work on *Lost World's* stop-motion animation—a project he aspired to do himself—until June 3,1922, after some footage was presented to a publicized meeting of the Society of American Magicians. So understandably Dawley was protective, and (arguably) vindictive concerning his well-earned 1920 patent rights, threatening a $100,000 law suit. Czerkas opines: "…film historians…given O'Brien's later prominence and Dawley's virtual obscurity … assume that Dawley was the aggressor who tried to make an outlandish claim at O'Brien's expense." (Czerkas, *Cinefex* 138, p.26) But it was Czerkas who recently uncovered reminiscences in Dawley's diary—entries for November 12, 1962 and June 26, 1969, documenting the reason for *Lost World's* offset production. Their dispute was finally settled out of court for a "substantial fee." (Czerkas, *Cinefex* 138, p.26; Czerkas, *Dawley: An Artist's Life*, p.86)

I leave this section with a tangle of stray thoughts—an upshot. If Dawley had been (perhaps rightfully) hired to do *Lost World* (with O'Brien out of the way, fired in disgrace—again see Czerkas's publications for more on this aspect), what kind of production would *that* have turned out to be, and would there ever have been a *King Kong*? Or if there was no influential Kong, would Godzilla have ever attacked Tokyo? Film historians and fans alike can speculate.

Brontosaurus – Symbolic Science-Fictional Monster's Progeny:

Brontosaurus and its sauropod kin were certainly 'hot property' then. And the bitter falling out between Dawley and O'Brien was only the second filmic sauropod-fueled feud of the time. For yet another legal war over copyright infringement of a patent claiming rights for construction design of the first life-size, mechanical *Brontosaurus* went to trial in January 1934. Except O'Brien and Dawley weren't involved: this time it was the Messmore and Damon (M&D) company versus a theatrical showman named Earl Carroll. M&D had built their first 49-foot long mechanical *Brontosaurus* (nicknamed "Dinah") in 1919, while records indicate Carroll's (nicknamed both "Sarah" and "Dinny") had come about later in 1931. George Messmore's patent was granted in 1933. My grandparents took my (then 7-year old) father to both the Chicago World's Fair in 1933/34, and later the New York World's Fair in 1939, where they saw Dinah and a host of other moving mechanical 'prehistoria' on display, such as M&D's "World A Million Years Ago" Chicago exhibits. The lawsuit was dismissed by the court, and Carroll's entry faded into obscurity.

Surely if anyone had thought a century ago to conduct a survey or poll as to which prehistoric monster was top dog then—*Brontosaurus* would have emerged victorious. However, to the general public, perhaps no less equal to the American Museum's awesome skeletal exhibits on imaginations were paleoart paintings and miniature models of prehistoric animals—restored 'in-the-flesh,' especially those by Charles R. Knight. For a time, *Brontosaurus* reigned supreme among those who conjured the Mesozoic world—lively in imagination, relying on such displays and restorations. For example, it's well known that Marcel Delgado's model puppets for *Lost World* and *King Kong* were based on Knight's paleoart.

Let's consider some other old films featuring contemporary, early footage of brontosaurs, otherwise lost to time. *Brontosaurus* was also featured in two contemporary (silent) film documentaries, Max Fleischer's *Evolution* (Red Seal Pictues/Inkwell Studios, D: Max Fleischer, 1923) and Service Film Corporations's *Fifty Million Years Ago* (1925)—the latter released only a few months following First National's *The Lost World*.[1]

In *Evolution* (about 1 hour, 23 min. long), partially to address the theme "How did we come to be upon this Earth?," (the short answer being evolution), 'lead-off' imagery of American Museum *Brontosauruses* are shown: a small-scale reconstructed skeletal armature 'dissolves' into Knight's sculpture of this dinosaur, and we're also shown his by then famous 1898 brontosaur painting. Then a portion of *Ghost's* animated sequence of Dawley's experimental *Brontosaurus* sculpt was spliced in (along with other scenes from *Ghost* showing his battling *Triceratops* and *Tyrannosaurus*) and more dinosaur statues and small sculptures. There's also an extended shot framing the girth of Josef Pallenberg's life-size sauropod statue, intended as *Diplodocus* but billed on an inter-title card as *Brontosaurus* in Fleischer's documentary. Pallenberg's sculpture built for the Tiergarten at Hagenbeck's Zoological Gardens in Stellinghen, Germany during the 1910s, was a first-of-its-kind statuesque restoration of a full-sized, life-like sauropod constructed anywhere. Further underscoring its institutional prominence, *Evolution's* science advisor was Edward J. Foyles—yes, of the American Museum.

Fifty Million Years Ago is a film recently discovered by Dr. Robert J. Kiss, and preserved in 2015 by the Academy of Motion Picture Arts and Sciences, now available for viewing online with a short explanatory article by J.B. Kaufman. Its first run was in July 1925, at the Piccadilly Theater. Like *Evolution*, this short (~ 7-minutes long—some footage is missing) film projects the ever-popular and by then traditionalized 'life-through-geological-time' theme, tracing from Earth's primordial origins to recent Ice Age extinctions. While the overall effect may not seem so 'monstrous' as in aforementioned science fictional entries, featured scenes strike as "museum dioramas come to life." (Kaufmann, National Film Preservation Foundation, "About the Preservation," online) It is unknown who did the stop-motion animation for each of the depicted prehistoric animals, but sure enough—*Brontosaurus* is the very first 'live' dinosaur shown, both on land and seashore settings. Other prehistoric animals are not quite so well executed, four of which were modeled after Charles Knight's artistry, and another based on an American Museum *Allosaurus* skeletal display. According to Kaufman, this short film would have seemed intriguing at the time, since the contentious Scopes trial over evolution had begun, and welcome because audiences may have seen *The Lost World* with its (far superior) stop-motion effects not long before.

So by now savvy readers are probably wondering, 'why is this guy (i.e. me) referring to 'Brontosaurus' when everybody in-the-know *knows* its real name was *Apatosaurus*,' right? Well, in a century's worth of time we find that real dinosaurs and thus their dino-monster doppelgangers are protean—as research ensues they're ever evolving. A century ago, for instance, sauropods were subject to scientific debate concerning their walking mobility. Land-lubbers versus aquatic swimmers ... tail-draggers versus whip-tails ...marsh-dwellers versus forest-dwelling ... sprawling posture versus columnar leg stance—given their massive body tonnage, were their legs sufficiently strong to enable them to walk on land? This story is recounted expertly in Chapter Five of Adrian J. Desmond's landmark book, *The Hot-Blooded Dinosaurs* (1975). Note that O'Brien's brontosaurs in *Lost World* and *King Kong* are both adeptly adapted to aquatic life and walking on land. Incidentally, in 1905, Knight's small accurately-scaled model was used for the first time in calculating the mass of a living *Brontosaurus*. (William King Gregory, "The Weight of a *Brontosaurus*," *Science* vol.22, (1905), p. 572) Even names and fossil skulls (e.g. *Time*, "Skull and bones at Yale," 11/9/81, p.86) switch and change. Since then, by 2016, *Brontosaurus's* good name was resurrected from the scrap heap of history: *Brontosaurus* is now seen as a valid genus apart from *Apatosaurus*—also still valid (since 1877).

Today, even though huge, more recently discovered and restored record-setting, massive sauropods like *Argentinosaurus*, *Dreadnoughtus* or *Patagotitan* render the once proud and mighty *Brontosaurus* puny by comparison, instead, theropods like *Tyrannosaurus* and *Spinosaurus*, or even Godzilla, have assumed top dog status in pop-cultural dino-monster iconic-ness—eclipsing sauropods. Meanwhile, the general sauropod/brontosaur 'long-neck' form refused to die in sci-fi monster films of the mid- 20th century, either as species known to science, or revivified in the guise of "paradinosauroids" (to be discussed shortly).

Filmic 'real' brontosaurs *have* survived since Kong's and Gertie's heyday, although usually playing less memorable, not as prominent, or more subordinate but generally less monstrous/scary roles. Just a few highlight examples, however, would include classic stop-motion animated brontosaurs appearing in *The Lost Continent* (Sigmund Neufeld Productions/Lippert Pictures, D: Samuel Newfield, 1951), and 1955's *Journey to the Beginning of Time* (Czech. 1955, American filmic release – 1966, D: Karel Zeman). *The Twilight Zone* episode "The Odyssey of Flight 33," first broadcast 2/24/1961, used the brontosaur puppet sculpted by Marcel Delgado, from 1960's *Dinosaurus!* (Universal, D: Irvin S. Yeaworth, Jr.), but with new animation. And I would be remiss by not mentioning *Brontosaurus* sculpted by Arthur Hayward, animated for Ray Harryhausen's *One Million Years B.C..* (20th Century-Fox, D: Don Chaffey, 1966)

A *Brontosaurus* museum (papier-mache) skeleton later reused in *The Beast From 20,000 Fathoms* is the central prop in 1938's *Bringing Up Baby* (RKO, D: Howard Hawks), a farcical comedy starring Cary Grant and Katherine Hepburn. And for those of you who might recall, 1985's *Baby ...Secret of the Lost Legend* (Disney/Buena Vista, D: B. W. L. Norton) concerning the mokele mbembe—envisioned as brontosaurs living in a remote region of Africa, was a sentimental journey into a remnant cryptozoological setting. The first theatrical release I took my older daughter to see was 1988's cartoon, *The Land Before Time* (Universal, D: Don Bluth), in which a (*Flintstones* 'Dino'-like) sauropod "Littlefoot" (a young, friendly brontosaur) becomes inspirational leader and hero, fending off a monstrous tyrannosaur.

And of course, the first living—inspiring awe not dread—dinosaur spied by fictional paleontologist Alan Grant in 1993's *Jurassic Park* is that majestic cgi-*Brachiosaurus*. Then a quarter century later, in *Jurassic World: Fallen Kingdom* (Amblin Entertainment/Legendary/World Pictures, D: J. A. Bayona, 2018), *Brachiosaurus* fleeing another one of those inevitable lost world-ish volcanic eruptions evoked compassion rather than dread as it burned to a crisp on a shipping loading dock. Clearly, a century after introduction of McCay's,

Dawley's and O'Brien's brontosaurs when super-sauropods reigned over the Mesozoic in popular imaginations, paleo-popculture had moved on toward the far greater thrills of blood-thirsty *T. rex* and vicious *Velociraptors*.

Size still matters, of course, but other 'super powers' seize our psyches too. So what about those mutated brontosaurs? Per Jose Luis Sanz's 2002 definition: "Paradinosauroids are quadruped animals similar to sauropods but with much shorter necks and with powerful heads, akin to those of the great carnivorous dinosaurs." (Sanz, *Starring T. Rex!*, p.114) The most 'sauropodean' of this particular class expressed in film would, to me, be the radioactive Paleosaurus starring in 1959's *The Giant Behemoth* (U.K. Allied Artists, D: Eugene Lourie). Another morphological mutant might be the Rhedosaurus in 1953's *The Beast From 20,000 Fathoms* (Warner Brothers, D: Eugene Lourie), both appearing as stop-motion animated puppets. William Schoell's novel *Saurian* (1988) introduced a malicious paradinosauroid to printed paleo-fiction. Resembling the Rhedosaurus or the Paleosaurus, Schoell's "Gargantosaurus" is an intelligent alien entity that melded itself within the body of a former dinosaur, thus controlling its mind, persisting through the Ages—while morphing into a paradinosauroid.

Although a familiar foe in published paleo-*fiction* during the first half of the 20th century in sci-fi literature and horror tales, stories involving more recognizable (e.g. 'non-mutated') forms of *Brontosaurus* (and its closest sauropod kin) as *the central* dino-monster antagonist are relatively far and few between. (An interesting 'find' though—only ancillary to our tale is H. G. Wells's Time Traveler's brief remark on a fossil skeleton of *Brontosaurus* ironically rotting away in a future version of the British Museum 802,701 years hence in 1895's *The Time Machine*.)

Another early entry was the marauding, monstrous Mayan 'god' "Cay," the last dinosaur alive, a living brontosaur in Frank Mackenzie Saville's 1901 novel, *Beyond the Great South Wall*. "Cay" eventually expires in an Antarctic volcanic eruption. Then an explorer is literally frightened to death at sight of a living *Diplodocus* he is tracking through a Bolivian jungle in Willis Knapp Jones's short story, "The Beast of the Yungas," (*Weird Tales* Sept. 1927). Later, Katherine Metcalf Roof's "A Million Years After," (*Weird Tales,* Nov. 1930) hit the pulp stands, one of the stories first in which an incubated museum dinosaur egg hatches a monstrous creature in civilized modernity; her bronto-monster goes on a city-wide rampage prior to succumbing to gunfire. Then in 1953, yet still in those pre-godzillean times, John Russell Fearn's novel tailored for younger adults, *The Genial Dinosaur*, concerned a carnivorous form of *Diplodocus*, also the last of its kind, emerging from Earth's interior: this sauropod eventually proves instrumental in averting an alien invasion.

'Brontosaurus's' (i.e. as 'Josasaurus') last symbolic bastion in printed fiction perhaps was in Sharon N. Farber's ("S. N. Dyer's") 1988 short story,

"The Last Thunder Horse West of the Mississippi," printed in *Isaac Asimov's Science Fiction Magazine*. Here, in this anachronistic 'throwback' tale set in the late 19th century, greedy feuding paleontologists search for and inadvertently kill the creature in the old American west. What a stark contrast between 'Josasaurus's' plight—viewed sympathetically as an innocent creature, versus the old-skyscraper-smashing bronto-monsters of yesteryear.

Epilogue:

So we've seen our brontosaurs first cast with gentle demeanor a century ago, then rather swiftly transformed into terrible Mesozoic monsters through a subsequent generation, while later evolving back into a new-millennial breed of more fragile, sympathetic creatures, yet inevitably doomed to extinction thanks to Man's erratic ways.

Movies and obscure stories-in-print come and go with the seasons, but for a time sauropods 'found' a way to survive through the Depression years and beyond World War II. V. T. Hamlin's cartoon strip *Alley Oop*, about a tribe of cave people, appeared in newspapers on Dec. 5, 1932; his faithful brontosaurian 'ride' "Dinny" made its debut several months later on August 12, 1933. According to Glut, "Oop meets Dinny by plopping onto the dinosaur's tail while escaping from an angry cave-bear." (Glut, *Carbon Dates*, p.163) And by the early 1930s, the Sinclair Refining Company adopted the image of a brontosaur for their logo. Thereafter, through the 1960s, Sinclair found many inventive ways to promote their petroleum products using dinosaur imagery. Being a young dinosaur enthusiast in the late 1950s and early 60's, I always encouraged my parents and grandparents to fill their gasoline tanks at Sinclair filling stations. There were often treasured dino-souvenirs to be had there, and child-sized outdoor brontosaur statues lured customers in off the road. Surely such daily, in-your-face personal newspaper or roadside encounters fortified which dinosaur truly was 'king' then. But times inevitably change, and by the mid-1950s brontosaur fandom would be eclipsed by visages of competing, more terrifying prehistoric monsters in our museums or magically out of Movie-land.

While both McCay's and Dawley's infatuations—if not absolute indebtedness, to American Museum staff and the inspiring *Brontosaurus* skeleton on display there is well documented, O'Brien's selection of brontosaurs for his movies may have been more a reflection of period zeitgeist. Brontosaurs simply became 'in' then, no doubt due to their mind-staggering sizes—and enormity of their even larger long-neck cousins (e.g. *Brachiosaurus*), relative to other types of (non-sauropod) dino-monsters.

Note that McCay, Dawley and O'Brien all were born in the late 19th century, so permit a tepid psychological inference. Each pioneer would have been exposed to a dawning American dinomania, possibly including sensational news transmitted in the December 11, 1898 edition of *New York Journal* telling of the "most colossal" dinosaur ever to ever walk on Earth (i.e. *Brontosaurus*), discovered and reported during our three pioneers' younger, perhaps more impressionable stages of adulthood—preconditioning for later American Museum experiences of the 1910s when they began their filmic creations.

For on the front page of that 1898 *New York Journal,* showing, at bottom, a brontosaur skeletal reconstruction, we also see an artist's arresting sci-fi-esque portrayal, with the caption, "How the *Brontosaurus giganteus* would look if it were alive and should try to peep into the eleventh story of the New York Life Building." Horrific if not downright creepy! Though not the first of its kind—surely a startling image signaling onset of the first phase of America's dinomania while foreshadowing exertions of Hollywood's and Japanese Toho Productions' dino-monsters in the later 20th century!

Key References:
Mark F. Berry, *The Dinosaur Filmography* (McFarland & Company, Inc., Publishers, 2002).
Paul D. Brinkman, *The Second Jurassic Dinosaur Rush: Museums & Paleontology in America at the Turn of the Twentieth Century* (The University of Chicago Press, 2010).
Stephen Czerkas, "O'Brien vs Dawley: The First Great Rivalry in Visual Effects," (*Cinefex* 138, July 2014, pp.13-27).
Stephen A. Czerkas, *Major Herbert M. Dawley: An Artist's Life – Dinosaurs, Movies, Show-Biz, & Pierce-Arrow Automobiles* (The Dinosaur Museum, Blanding UT, 2016).
Allen A. Debus, *Dinosaurs in Fantastic Fiction: A Thematic Survey* (McFarland & Company, Inc., Publishers, 2006).
Allen A. Debus, *Prehistoric Monsters: The Real and Imagined Creatures of the Past That We Love to Fear* (McFarland & Company, Inc., Publishers, 2009).
Allen A. Debus, *Dinosaur Memories II: Pop-Cultural Reflections on Dino-Daikaiju & PaleoImagery*, Chapters One, Six and Eighteen, (Amazon Createspace, 2017).
Adrian J. Desmond, *The Hot-Blooded Dinosaurs: A Revolution in Palaeontology*, Chapter Five, (The Dial Press/James Wade, New York, 1975).
Donald F. Glut, *The Dinosaur Scrapbook* (The Citadel Press, Secaucus NJ, 1980).
Donald F. Glut, *Carbon Dates: A Day by Day Almanac of Paleo Anniversaries and Dino Events* (McFarland & Company, Inc., Publishers, 1999).

Ulrich Merkl, *Dinomania: The Lost Art of Winsor McCay, the Secret Origins of King Kong, and the Urge to Destroy New York* (Fantagraphics Books, Seattle, Dec., 2015).
W. J. T. Mitchell, *The Last Dinosaur Book:* (The University of Chicago Press, 1998).
Neil Pettigrew, *The Stop-Motion Filmography: A Critical Guide to 297 Featues Using Puppet Animation* (McFarland & Company, Inc., Publishers, 1999).
Ronald Rainger, *An Agenda for Antiquity: Henry Fairfield Osborn and Vertebrate Paleontology at the American Museum of Natural History, 1890—1935* (The University of Alabama Press, Tuscaloosa, 1991).
Jeff Rovin, *From the Land Beyond Beyond: The Films of Willis O'Brien and Ray Harryhausen*, Chapter Two (Berkley Publishing Corp., 1977).
Jose Luis Sanz, *Starring T. Rex! Dinosaur Mythology and Popular Culture,* Chapter 22 (The Indiana University Press, 2002).

Note to Chapter Twenty-six: (1) Another obscure documentary featured a sample of Dawley's sculptural artistry. For many years an obscure photo printed in *Famous Monsters of Filmland* (no. 19, Sept. 1962) showing a pterodactyl was thought to have been animated for a 1931 documentary *The Mystery of Life*. However, since *Moonbeam's* rediscovery, now it is known to have instead originally appeared in Dawley's 1921 movie. Writing in the *Prehistoric Times* Letters column, Glut wrote: "Stephen Czerkas was indeed correct when he suggested that the pterodactyl from *Along the Moonbeam Trail* later appeared in the 1931 Universal documentary *Mystery of Life* (now apparently lost). In fact, scenes from both *Moonbeam* and *Ghost of Slumber Mountain* appeared in *Mystery of Life*. Also appearing as stock footage in *Mystery* were shots of the full-size cement dinosaur statues … on display at the Hagenbeck Zoo in Hamburg, Germany. … Presumably, Dawley just licensed the *Ghost* and *Moonbeam* footage to Universal. …" (*Prehistoric Times* no.123, p.6, Fall 2017) *Mystery of Life* examined the fraught debate over organic evolution.

Chapter Twenty-seven — *That 'Other' Holmes*

No—not Arthur Conan Doyle's fictional 'Sherlock,' but a genuine yet not as famous 'investigator'—geologist Arthur: *Arthur Holmes* (1890 - 1965), that is. But we'll come to him ... in *time*.

When I was a child, first expressing interests in paleontology, my mother (raised Catholic) related how the biblical story of Genesis with its Seven Days of Creation could serve (allegorically not literally) for what geologists had learned about Earth's origin and sequential evolution of life—corresponding to what transpired on each of those halcyon seven days. The length of each 'day' was not just 24 hours long, but to be thought of as 'extended' in geological terms, permitting a Creator sufficient time to get everything done. I was unconvinced. Later in adulthood I collected several old 19th century books written by individuals who today would be regarded as 'creationist geologists.' One of these volumes by geologist Arnold Guyot (1802-1884), *Creation or the Biblical Cosmogony in the Light of Modern Science* (1884) seems to answer where (or when) that old 'theory' my mother offered may have originated. Intriguingly, this unique book was also graced with several engraved reproductions of Benjamin Waterhouse Hawkins's 1870s paintings restoring scenes from the Silurian, Carboniferous, Cretaceous and Tertiary periods.

While the early heroes of Geology hammered out its basic tenets during the 19th century, the notion of the First Chapter of Genesis's 'seven days' as geological 'epochs' or ages might have occurred to any number of biblically-minded souls. But in 1884 Guyot documented how and when he conceived an inaugural version of this postulate during the winter of 1840—only months before the Dinosauria were named by Richard Owen. So, in 1840, during that 'pre-Darwinian' time, Guyot was clearly reconciling science with theology. Guyot eventually published an abbreviated version of his views in 1852 "as a series of abstracts from a public course of lectures (he was) delivering in New York." (*Cosmogony* p.ix).

In a nutshell, what were Guyot's beliefs? Well, he was anti-Darwinian to the core, perhaps explaining his eventual fellowship with like-minded Hawkins, and he strove to compress/restrict each major geological (and evolutionary) event then known to have occurred in Earth history within a biblical context. Regarding Earth's age, Guyot could only suggest for two of those allegorical 'days,' that "… the geological history of the creation of animals and man demonstrates that they are *long, indefinite periods of time*." (my italics, p.51) Guyot's book is a fascinating account, yet all so off kilter—archaic, especially for the year it was published—1884. For by then biblical explanations for Earth 'prehistory' and its duration were becoming superseded by enlightened consideration of fossils, strata and geological investigations. Meanwhile other scientists, one predominantly—Lord Kelvin, also known as William Thomson (1824 - 1907), were forging ahead with exact physical measurements for Earth's age, even though a proper means of dating the Earth wouldn't be devised for several decades.

Through millennia, numerous ideas concerning Earth's absolute age have been proffered, fewer still regarding how remote in time are each of its designated Eras, Periods and Epochs. But only one (generic) strategy has been developed for *accurately* measuring Earth's "abyss of time" and its ages—thus my segue to Arthur Holmes. But before a prototype of our modern Geological Time Table could be produced, methods and overarching ideology of preeminent physicist Lord Kelvin first needed to be falsified and overthrown.

Kelvin was a brilliant mathematician and inventor with expertise in the physics of heat flow through material bodies. His interests in measuring the age of the Earth surfaced in 1855. Between then and 1897 he repeatedly tackled the problem, publishing updated calculations based on scientific experimentation, revised assumptions and new geological evidence. Relative to modern age-dating results his estimates were quite low, not because they lacked mathematical sophistication but instead for a reason unknown to him at the time (until much later in life), because they were fundamentally, theoretically flawed . Due to his rank as a first rate physicist Kelvin's "physics of time"[1] remained overwhelmingly influential throughout the second half of the 19th century. But, as in the case of modern baseball sabermetrics, "…absolute figures (can) lie."

Kelvin's 1855 result—supported again in 1871, founded on a premise of the Sun's radiance (i.e. subject to physical laws of matter as then understood in stars) warming Earth, suggested 100-million years as an accurate absolute age—although with an upward margin of error extending to 500-million years. However, implicit in his calculations remained a possibility that life on our planet could have (only) existed for 20- to 25-million years corresponding to when our Sun sufficiently warmed. Interestingly, in 1871 Kelvin allied himself to a "star germ theory" of panspermia, wherein he accepted a possibility that life (e.g.

seeds or even insects) may have originated on other planets, which then primordially alighted on Earth "after which evolution ran its course."[2]

Another line of reasoning for age-dating, published as a "secular cooling of the Earth" approach in 1862, considered rates of outward terrestrial heat flow through rocks. Kelvin's core idea was that heat generated from gravitationally-induced bombardment of asteroids during Earth's formation as a molten planetary body had been radiating into space ever since. Kelvin's calculated time for crustal consolidation of rocks was estimated as spanning in the range of 98 to 200-million years! Afterward, rocks cooled sufficiently for life to proliferate by about 20-million years ago. Kelvin revised his estimates repeatedly, considering other models, although his results typically remained within the same general order of magnitude. In 1899, for example, he reduced the age of igneous rocks consolidation (based on revised measurements of their melting properties) to within 20 to 40-million years ago. This tighter range was published four years after the discovery of x-rays. Radioactivity offered a new strategy for dating geological strata—relying on ratios of radiogenic isotopic decay half-lives for elements such as uranium and lead as assayed in rock minerals, techniques which are now familiar to every freshman chemistry (or geology) student—for age-dating Earth (or the Moon)!

Ernest Rutherford (1871-1937) became a chief investigator of radioactivity by the turn of the 20th century, applying this discovery to planetary history. He reasoned that radioactive substances within the body of the Earth continually added to its heat gradient. Because Kelvin had not accounted for this important energy source, his estimates clearly were minimal ages. Meanwhile, Rutherford realized:

> "Thus it does not appear improbable that the temperature gradient observed in the earth may be due to the heat liberated by the radioactive matter distributed throughout ... the earth may have been at a temperature capable of supporting animal and vegetable life for a much longer time than estimated by Lord Kelvin from thermal data. ...thus increases the possible limit of the duration of life on this planet, and allows the time claimed by the geologist and biologist for the process of evolution."[3]

Years later Rutherford recollected his unease when advising a scientific gathering that Kelvin's measurements were indeed wrong. With the imposing figure of Kelvin seated in the audience at a 1904 Royal Institution meeting, as Rutherford recounted, the:

> "... room ... was half dark ... To my relief, Kelvin fell fast asleep, but as I came to the important point, I saw the old bird sit up, open an eye and cock a baleful glance at me! Then a sudden inspiration came, and I said Lord Kelvin had limited the age of the earth, *provided no new source (of energy)*

was discovered. That prophetic utterance refers to what we are now considering tonight, radium! Behold! The old boy beamed upon me."[4]

Enter geologist Arthur Holmes who a few years later in 1911, at only 21 years of age[5], utilized results of radiometric studies to prepare a first crude set of laboratory measurements representing several Periods of the Geological Time Table; Carboniferous—340 million years ago ("mya"); Devonian—370 mya; Ordovician/Silurian—430 mya; and Precambrian rocks—1,025 to 1,640 mya. Holmes prophesied that "… it is to be hoped that by the careful study of igneous complexes, data will be collected from which it will be possible to graduate the geological column with an ever-increasingly accurate time-scale."[6] By 1927, Holmes was instrumental in further refining the Geological Time Table, calibrated in years founded upon 23 uranium-lead and uranium-helium results. Further adjustments were published in an updated 1937 edition of Holmes's book *The Age of the Earth.* These newer values strike closely to current ages assigned to each of the geological periods of the Phanerozoic Era.[7] For instance his published 1937 result for time elapsed since the Cretaceous Period was remarkably posted as 68-million years: today's value is 66-million years.

While superficially it may seem that contemporary geologists and biologists were "vindicated" by results accruing after 1911, thus falsifying shorter lifetimes insisted upon by physicists and also some geologists 'conditioned' by Kelvin's work, the physicists were correct in their assertions that endless, cyclical "uniformitarian" (as early geologists such as Charles Lyell had declared) indefinite, steady state time could not rule the day. As Burchfield noted, by 1922, "… uniformity in geology was no longer a tenable assumption: the present is a period of extraordinarily rapid geological change; radioactivity provides the only non-variable clocks for measuring geological time and these point(ed) (then) to a terrestrial age of about 1.3-billion years; and finally, this estimate is compatible with both biology and astronomy."[8]

Relying on radiometric data, in 1931 geologist Charles Schuchert stated that Earth originated from the solar nebula 1.5 billion years ago.[9] Yet this is only one third of Earth's absolute age as currently known. One of my first year college geology textbooks was the massive 2nd edition of Holmes's *Principles of Physical Geology* (1965), outlining techniques and underlying science for radiometric age-dating of Earth, therein incorporating the 1959 version of The Geological Time-Scale (pp.156-57). Here "origin of the continental crust" was noted as occurring 4.5-billion years ago, a finding Kelvin would have found remarkable. (Refinements to the Time-Scale were published in 1989 and then again in 1995.)

Geologists and paleontologists had long held aspirations that their discipline could become elevated to an "exact science," like physics. And so the prospect of measuring Earth's age using known scientific principles seemed a worthy endeavor, tackled repeatedly from the late 18th century on into the 20th. But the turn of the 20th century advance in determining the age of the Earth has been characterized as revolutionary. Techniques steeped in applied mathematics, physics and traditional laboratory inductive experimentation wielded by physicists were 'overthrown,' in turn, by field geologists who instead often relied upon alternative logical means and methods of investigational pursuit—since wielded triumphantly and accepted by the scientific community. Anyway—that's how this scientific phase is sometimes characterized, as a Physics 'versus' Geology tussle.

By the 1910s several geologists who for many years had accepted Kelvin's disproven mathematical approaches, assumptions and results now expressed skepticism over an inconceivable expanding vastness of geological time as indicated by introduction of yet another non-geological (i.e. radiometric) method. Nevertheless, Holmes stated in 1927 that "It is one of the triumphs of geology to have shown beyond all possibility of doubt that there has been ample time for all the slow transformations of life and scene that the earth had witnessed."[10] But such a case for Physics *versus* Geology may be exaggerated because age-dating Earth became rather a joint enterprise. For Holmes—a pioneer/explorer of the abyss of time, working "almost alone on the problem,"[11] while forging ahead with geological aspects of age-dating, relied heavily on radiometric data provided by physicists.

Certainly the law of conservation of energy and the second law of thermodynamics (i.e. entropy), particularly—both employed early on by physicists in their 'low-ball' estimations of Earth's age—are universal absolutes. But heat generated from naturally occurring radioactive elements was the key missing piece of the puzzle. Once acknowledged this unanticipated factor permitted geologists (working in tandem with physicists by the early 1900s) to revise the Geological Time-Scale with ever more accurately measured ages. Using the new chronometer, Holmes's efforts supplanted Kelvin's overwhelming influence, facilitating paleontologists' abilities to envision Earth's successive geological stages of life.

And so I've often marveled at those *series* of paleoart 'portals,' paintings and murals peering into successive stages of prehistory such as portrayed by Charles R. Knight, Rudolph Zallinger, Josef Kuwasseg, Edouarrd Riou, Zdenek Burian, and yes—Benjamin Waterhouse Hawkins as published in Guyot's long-neglected in time 1884 volume. How differently must these brilliant artists have each separately pondered the meaning of geological time and Earth's span in the universe as they performed magic with their brushes and palettes. When I gaze upon their artistry—transcending time itself, I am reminded that we are each a product of our moment in history ... so surely their thoughts and beliefs must have differed considerably from my own.

Perhaps *the* most original, impressive and recent 'take' on the life through geological time theme was completed during the late 1990s by artist Barbara Page for the Paleontological Research Institution in Ithaca, New York. Her series of 544 separate painted panels, roughly one painting for each million year interval since the Cambrian "explosion" (spanning over a total 500 feet in length—placed end to end), portrays traces of life evident since Paleozoic Era's dawn. These "contiguous" panels, now on permanent display, are chronologically arranged, each showcasing a faux-fossilized 'record from deep time'—preserved moments of our past. More overtly than usual for such works representing planetary history, Page's masterpiece 'nods' to Charles Darwin's and Charles Lyell's realizations of the paleontological record's incompleteness (e.g. "imperfection") and geological time's immensity.

Of the inspiration for her work Page wrote:

> "Why not resurrect a lost landscape in the form of a journal where a day in the life of Mother Earth lasts a million years? A record of the past as it slides into the present, a map where Father Time dictates the spatial dimensions." (p.xviii) The panels are a "... translation of time into space," an arbitrary compression of a "million years into 121 square inches." (p.xxiii) "According to the date associated with each panel, these animals are still up and roaming. According to the matrix they're long gone. The paintings show what is happening in conjunction with the extreme old age of future—action, omen, disaster, and preservation—all in one." (p.xviii)

The remarkable 2001 book in which these paintings were reproduced, *Rock of Ages—Sands of Time*, with added commentary by paleontologist Warren Allmon and photographer Rosalind Wolff Purcell then becomes a "... natural prehistorical flipbook of creatures in the throes of life" (Purcell, p.ix).[12] Sudden mass extinctions (such as the K-Pg event of 66 million years ago) are suggested in panels by 'debris' deposited in ages corresponding to those cataclysms that doomed many species.

And yet, a touch of the 'magnificence' of certain species during their heyday—those which have become exalted in modern pop-culture—seems missing in this magnificent art form, somehow lost in the ceaseless flow of time. The fossils shown in panels appear as fragmented as is our imperfect (broken) record of the history of life through time. Where are reconstructions of the mightiest dinosaurs, savage *T. rex* or its adversary, a 'Knightian' *Triceratops*, mighty marine *Mosasaurus*, or restorations of those spritely Chinese feathered dinosaurs that have become all the rage lately? Instead we see so many broken, shelly remains of organisms many of which would be unrecognizable except to the most devoted of fossil hunters, buried in 'sediment.' And there, 'preserved' at 114-million years ago we spy fossilized evidence of an "unidentified cockroach." We only see glimpses of species that came before ... and left, hinting how they lived, represented as if '*in-situ*'—merging strata from around the world, throughout individual, sequential paintings.

It would be impossible to know or interpret what such fractured traces mean without fortification of over two centuries of paleontological science, and the elucidation of geological time—thanks to refinements in radioactive age-dating.

Skimming through latter-most pages in *Sands of Time*, suddenly the Pleistocene is upon us and then ... in a *flash* it's over—the end! Contrary to what Guyot, Hawkins and Kelvin may have believed, Earth's 'rock of ages' absolute age is immense, while man's time on Earth is but a sliver—remarkably short.

Notes to Chapter Twenty-seven: (1) By far the best source available on Kelvin's measurements of geological time in contemporary context with competing ideologies may be found in Joe D. Burchfield's classic *Lord Kelvin and the Age of the Earth* (1975). The phrase in quotes is borrowed from Burchfield, (p.21); (2) See Claude C. Albritton Jr.'s *The Abyss of Time: Changing Conceptions of the Earth's Antiquity after the Sixteenth Century* (1986), pp.184-185; (3) Rutherford, *Radio-activity*, Cambridge Univ. Press, 1904, p.657; (4) Rutherford quoted in Albritton, p.204; (5) To his credit, Holmes was also precociously-minded when it came to the subject of "continental drift" as proposed by Alfred Wegener a century ago. Historians of science note that Holmes was a "convert" to the idea, and even "... proposed a new mechanism for producing ... convection currents in the earth's mantle ...(that) would tend to produce mountain-building and continental drift." (See I. Bernard Cohen's *Revolution in Science*, 1985, pp.455, 620.) A chapter in Holmes's 1965 geology textbook also addressed the continental drift idea at a time when it was becoming more commonly known as the "sea-floor spreading" or "plate tectonics" theory; (6) Holmes ("The association of lead with uranium in rock minerals, and its application to the measurement of geologic time," *Royal Soc., Proc.*, ser. A, pp.248-56, 1911) quoted in Albritton, p.206; (7) See Albritton, pp.208-209 (8) Burchfield, *op. cit.*, p.199; (9) Schuchert, *Outlines of Historical Geology*, 1931, 2nd ed., p.15; (10) Burchfield *Ibid.*, Holmes quoted, p.201; (11) Burchfield, *Ibid.* p.198.(12) Page numbers in this paragraph all refer to *Rock of Ages—Sands of Time* (The University of Chicago Press, 2001).

Chapter Twenty-eight — *Ironical Dinosaurs—Dragon Bones of the Gobi: Roy Chapman Andrews, the Central Asiatic Expeditions, and the most celebrated Fossil 'Consolation Prize' ... Ever!*

Although during the 1920s, the American Museum's highly celebrated exploration of Mongolia's geological setting, paleontology and natural history during the 1920s became known perhaps as America's grandest and most exotic scientific venture (until then), ironically, those missions 'failed' to find proof of the *central* basis for organizing those expeditions—to discover fossils of mankind's origins, there—a theoretical notion proposed by Henry Fairfield Osborn. Instead many other kinds of fossils were uncovered and shipped to American institutions, thus forever linking mission organizer Roy Chapman Andrews' (1884-1920) with dinosaurs of the Gobi Desert.

And yet another, albeit lesser, irony is that Andrews—while an excellent field man, team leader and all-round hearty 'adventurer' *extraordinaire*, was not formally degreed in paleontology and geology, nor trained as fully in dinosaur science as were fellow museum participants Walter Granger and William D. Matthew (whose good names are not as widely known).

Decades ago, as a grade-schooler reading Andrews' book (intended for younger readers)—*All About Dinosaurs* (1954), my impression was to regard him as a chief dinosaur 'guy,' even though I really didn't know much about him then. And yet a third 'irony' is that publicized theoretical notions concerning their most famous discovery—that of '*Protoceratops* eggs' were ... well—as we shall see, 'inverted.' Nonetheless, Andrews' intrepid and daring expeditions (1922-1930) opened doors to what decades later has emerged, regionally, as the world's most enlightening series of Mesozoic fossil vertebrate discoveries to date, further fueling the dinosaur renaissance!

So who was this 'mythical' Andrews persona, how was Osborn's theory—his "brilliant prediction" falsified, and what (as a major consolation prize!) did his missions instead discover to widespread popular acclaim?

From the get-go, Andrews, raised in Wisconsin, was prone to an outdoors, sportsman's life. He loved being in the woods and thrilled at studying natural history and wildlife of his surroundings through hunting and taxidermy. In school he was challenged in several classes, but graduated from Beloit College, and eventually completed a Master's of Science thesis in 1913 at Columbia University on California Gray Whales, the latter degree following his employment at the American Museum of Natural History (AMNH) in New York. Interestingly when Andrews arrived at the AMNH in 1906 he vowed to do anything to earn a job there, even just mopping museum floors! But Andrews quickly impressed the staff, especially Osborn and Museum Director H. C. Bumpus. With his enthusiasm and yearning to learn he moved up the ranks, forever leaving his mop in the maintenance closet.

It was Bumpus who assigned Andrews his first major museum project—to assist in building a life-sized display model of a blue whale, which led to his reputation as a skilled zoologist in cetology positioned as Curator in the Department of Mammalogy. His researches, observing whales at sea and collecting whale skeletons, samples of whale tissues and other specimens, by 1910, further led him abroad to Japan, Korea and the Orient—circumstances which were to shape not only his future as an intrepid 'geographic' explorer of Outer Mongolia, but also that of paleontology—as well as elevating the world view status of American science.

But it was his mentor Osborn who green-lighted the expedition helmed by Andrews that would bring him fame and captivate the world! For a time in New York, the two were inseparable, Andrews regarding Osborn as "truly a foster father." Andrews appealed to Osborn's scientific sense with a means for testing the latter's hypothesis—as published in a 1900 *Science* article. Osborn had suggested that Central (northeastern) Asia was not only the mammalian fauna's "Garden of Eden," now distributed throughout the northern hemisphere in Europe and (via the Bering Strait) North America, but also represented a subsequent "Cradle of Mankind." Would fossils of the earliest "Missing Link" primate (e.g. between apes and modern man) be found there, or even origins of modern civilization? Andrews hoped to confirm Osborn's theory, which evolutionist Matthew also supported. And such exploration would also satisfy Andrews' restive nature.

The series of expeditions commencing in 1916 brought considerable hardship and danger that few today would be capable of imagining or experiencing in a single lifetime—which is why some have likened Andrews' persona to that of fictional film character, archaeologist-adventurer "Indiana Jones," (even though this 'origin' claim has been denied by movie mogul George Lucas). Dodging around political turmoil, upheaval and perpetual civil war raging throughout the route's exotic staging points in China and Mongolia, encounters with camp-infesting vipers and man/corpse-eating wild dogs, and desert environmental conditions—sub-zero freezing temperatures alternating with broiling hot days, or facing occasional hurricane gust, knee-buckling sand storms—would have crippled us with fear. Yet death-defying Andrews traveled there *multiple* times, craving more field adventures in search of fossils, despite encountering shifting political factions, rifle-toting bandits and thieves, skirting corruption on all levels and greedy officials who would tax the mission's purse and objectives.

Andrews always persevered, stressing "qualities of courage and endurance, the willingness to undergo hardships and to face death…" while afterward thirsting for more! He felt that 'adventures' were only suffered by those who were disorganized or unprepared. Andrews was always very prepared, but he still routinely experienced what most of us would regard as adventures. Yet he was far, *far* from the 'best' fossil collector in-house. As he recalled, although he would always be ready with pickaxe in hand… "When a valuable specimen had been discovered, (Granger) usually suggested that (Andrews) go on a wild-ass hunt or do *anything* that would take us as far as possible from the scene of his operations." (*On the Trail of Ancient Man*, p.134-135, 213) For Granger understood how delicate some of their finds were and therefore knew to necessarily employ less heavy-handed techniques to preserve fragile specimens.

By late 1921, Granger, who would make several key paleontological discoveries later in Mongolia, had identified "rewarding" deposits on reconnaissance near a village in China, Pleistocene vertebrates—elephants and rhinos, particularly. Then on April 24, 1922 at a site of sedimentary rocks, Granger presented Andrews with a pocketful of fossil bones and teeth, stating "Well, Roy, we've done it. The stuff is here. We picked up fifty pounds of bones in an hour." The expedition was 'on.'

Yes—as the team quickly discovered the *stuff* was there –fossils in the Gobi, that is. No—but not that fanciful "missing link" human sensationalized in Press headlines circulating widely after Andrews' expedition's aim was formally announced in October 1920. As one such news story reported "Scientists to seek Ape-man's bones: Natural History Museum will begin five-year quest for Missing Link."

Instead, one could argue that one such 'link' had already been found … although in *Africa*—Eugene DuBois's 1891 discovery of a 600,000 year old *Pithecanthropus* skull (popularly known as "Java Man")—even older than Europe's Neanderthal Man! Osborn, who unfortunately supported 'leanings' that would be frowned upon today—such as eugenics, hoped instead to find even older primate remains in Asia. But any suggestion of a missing evolutionary 'link' to be found somewhere offended anti-Darwinian Christian fundamentalist groups. Furthermore, scientific detractors opined that it would be impossible to find fossils in a desert, with outcrops covered in sand—hidden to human eyes. Regardless, in midst of this fanfare, Andrews was able to cast off in February 1921 on the "Third Asiatic Expedition" (acknowledging Andrews' prior travels to China and Mongolia in 1916-17 and 1919, respectively), although the mission's name was changed to "The Central Asiatic Expedition" in 1925.

Over the full run of their expeditions, Andrews, Granger and company made several highly significant fossil discoveries, significance of which perhaps may be categorized in tripartite fashion. First they found prehistoric mammals, especially those dating from the Tertiary Period. Secondly they found important Late Cretaceous dinosaur fossils. And finally—Osborn must have been somewhat satisfied after all with this outcome—they discovered traces of the earliest *placental* mammals then known, a definite 'link' to ourselves right out of primeval time although not conclusively situated in their/our 'birthplace'. (The first geologist-explorer of the region (during 1894-1896) to identify fossils in this region (i.e. rhino teeth) was the Russian V.A. Obruchev—also later author of a sci-fi novel, *Plutonia*,)

Perhaps the most impressive mammalian beast they uncovered were remains of what turned out to be the most gigantic land mammal genus that ever lived—an Oligocene rhino relative, 17-feet at shoulder height, which Osborn named *Baluchitherium*, (or whimsically as the "colossus of Mongolia")! Although previously described from scanty remains during the 1910s, and related to *Paraceratherium*, it is now known as *Indricotherium*. The four legs and feet of one specimen were found in a sandy deposit, as if the poor creature had succumbed to burial in quicksand—dying in upright-standing position. Granger envisioned a dramatic struggle as the creature came to drink from a pool of water only to become stuck. "Suddenly, it began to sink, settling back on its haunches, struggling to free its enormous body until the sediments suffocated it." (Gallencamp, p.202)

While upper skeletal segments of the fossil rhino had weathered away, on August 5, 1922 at a nearby locality Andrews' team also found another prize—a skull belonging to the *Baluchitherium* genus. These specimens arrived at the American Museum in December 1922, after which Charles R. Knight painted a stunning restoration. Osborn thought discovery and delivery of this 4-foot long skull alone represented "one of the greatest achievements in the history of paleontology." (Preston, p.123)

Besides Asian 'titanotheres'—creatures also famously known from South Dakota's "Badlands," subject to Osborn's truly *massive* 1929 monograph, Central Asiatic Expeditions also recovered remains of a Pliocene *Equus*, eggshell fragments belonging to a giant form of ostrich, a peculiar "clawed-hoofed animal" chalicothere, and a "shovel-tusker" mastodont elephant related to the *Platybelodon*. The impressively-sized fossil skull of an early Tertiary 'creodont' that Osborn christened *Andrewsarchus*, honoring Andrews, was also discovered. Osborn was also particularly impressed by Mongolian fossils belonging to the Eocene genus *Coryphodon*. The explorers wondered whether such genera—those magnificent titanotheres especially, evolutionary progenitors of mammals which later migrated into North America along the Bering Strait?

Rather unexpectedly, *their* exciting discoveries of dinosaur remains reverberated throughout the world! One of their first such 'finds' in 1922 at a locality now known as Byan-Dzak—or the "Flaming Cliffs" was that of the Late Cretaceous *Psittacosaurus*. Andrews described this locality in his best book, *On the Trail of Ancient Man:*

> "Everyone was enthusiastic over the beauty of the great flat-topped mesa on the border of the badlands basin. Its surface was covered with black lava but the sides were blood-red. ... The men said that against the sunset glow it surpassed anything ... that they had ever seen. Granger picked up a few bits of egg-shell which made us believe that we had discovered a late Tertiary deposit. These eventually proved to be bits of dinosaur egg-shell, and the great basin with its beautifully sculptured ramparts was the richest locality of the world from a palaeontological standpoint." (pp.179, 181)

Dinosaur bones were scattered everywhere there, prompting subsequent visits beginning in 1923. As geologist Charles P. Berkey commented, their initial 1922 discovery of dinosaur bone excitedly meant they "... were standing on Cretaceous strata of the upper part of the Age of Reptiles—the first Cretaceous strata, and the first dinosaur ever discovered in Asia north of the Himalaya Mountains." Andrews later opined that, collectively, these inaugural discoveries indicated "... that the theory upon which we had organized the expedition might be true; that Asia is the mother of the life of Europe and America." (Andrews, *On the Trail of Ancient Man*, pp.80-81) Out of these deposits also came the *Velociraptor*, described from scanty remains by Osborn in 1924.

Of course, *the* most celebrated fossils announced by Andrews in behalf of the Central Asiatic Expeditions were those (alleged) *Protoceratops* eggs, initially identified by George Olsen. Like many of us (then) youngsters, baby boomer dino-philes of the early 1960s, we learned of this discovery at the Flaming Cliffs "incubator" through Andrews's delightful 1953 book, *All About Dinosaurs* (with artwork by Thomas W. Voter), written for grade-schoolers and once available as a Scholastic Services paperback. (In my case, as well, the Field Museum had a terrific *Protoceratops* display.) From embryonic dinosaurs found in broken eggs—thus coining the term "palaeoembryology," on up through more mature 9-foot long specimens, Andrews' team was able to generate a "developmental series" for the genus.[1] This hornless variety of ceratopsian was judged older and ancestral to North America's *Triceratops*.

While numerous skeletons of this famed dinosaur genus had been found in 1922, those abundant "reptilian" eggs and egg clusters weren't until during the 1923 expedition—it didn't take long for scientists to tie these to an alleged egg-thief represented by bones of the bird-like *Oviraptor philopceratops* whose name—coined by Osborn, meant "egg seizer that loves ceratopsians." Scientists claimed it had died in a sandstorm while caught in the act of "raiding a *Protoceratops'* nest," a dramatic story which held for decades. The 'egg-thief's' skeleton was discovered lying atop a nest of eggs, although separated from these by 4 inches of sandstone. Osborn also identified and *named* bones of the *Velicraptor* (whose pop-cultural fame wouldn't be fully realized until after 1971, and especially following publication of Michael Crichton's *Jurassic Park* 1990 novel). Remains of *Saurornithoides*, and a tyrannosaur were also found.

Back in New York one dinosaur egg was auctioned in 1924 to raise money for their next expedition, but this business stunt egregiously backfired, infuriating officials in Mongolia who projected an 'assigned' monetary value to each egg. Had Andrews' team inadvertently become the 'egg-robbers,' not unlike *Oviraptor* millions of years before? As Andews recalled, "Nothing else so disastrous ever happened to the expedition." (Gallenkamp, p.183) In the meantime though, Andrews's visage appeared on the cover of *Time* magazine (Oct. 29, 1923 issue), and he became renowned in a number of other periodicals. The world was simply enthralled with his exploits and journeys in mysterious lands, and of course with detection of dinosaur eggs! But Andrews put a distinctive, nationalistic stance or 'spin' on their accomplishments, "This is an all-American Expedition. We are carrying American ideals, American science, and the American flag into one of the least known countries of the world. We are bringing back the fruits for the entire American people." (Gallenkanp, p.184)

Under Osborn's and Andrews' leadership, American paleontology was maturing rapidly during the post-World War I period. And yet that elusive evolutionary/anthropological Asian "Missing Link" was still on the minds of museum staff, particularly Osborn's. Where could it be? Well, as of 1926, speaking for members of the Central Asiatic Expedition team, Andrews opined, "There can be no question, from our discoveries already made, that Central Asia was the chief theatre of evolution, not only of the land *Mammalia,* but of the giant land *Reptilia* of the world." (*On the Trail of Ancient Man*, p.202)

Well, one fossil skull—just over an inch long, found in 1923, initially labeled as an "unidentified reptile," embedded in Cretaceous sandstone from the Flaming Cliffs hinted at what this regional 'theater' *could* reveal with further study. Other such fossil skulls and teeth were soon picked out of the bedrock. They were remains of small rodent-like creatures—multituberculates (which became extinct by around 50 million years ago), marsupials, and *voila*!!—

placental insectivores. (To place matters in proper context however, based on contemporary geological data, Andrews thought the time of the great dinosaur K-Pg extinction 66-million years ago happened only about 10-million years ago, while the *Protoceratops*- rich Flaming Cliffs formation was estimated to be 2 to 3-million years older.) Carried in a cigar box, Andrews claimed the tiny insectivore fossils were "... the crowning single discovery of our research in Asia," the "most important discovery of the expedition," that is—in with respect to Osborn's theory of mammalian origins. (Quotes from Andrews p.331 and Gallenkamp, p.201) Furthermore, from these a theory of rodent-size mammals devouring tasty dinosaur eggs—possibly leading to dinosaur extinctions, was fortified.

But what of that anthropoid 'link' that seemed still woefully 'missing'?

Echoing Osborn, Andrews' preliminary conclusion would seem rather odd today. He thought the earliest "... ancestors of man may ... be found in the same country " as the *Baluchitherium,* "... because we are now convinced that our human ancestors branched off from the other anthropoid-ape stock in Oligocene time, the very time when the *Baluchitherium* was flourishing ... In other words ... probably contemporary with our remote ancestors about the time that they were beginning to lead an independent existence and to move about in an erect or semi-erect position ... a relatively large-brained, erect-walking ancestral type of man ... found in Asia." (*On the Trail of Ancient Man*, pp.206-207) That would mean about 3 to 7-million years ago, according to Andrews' sense of the geological time scale—much older than the (then still generally accepted, but faux-fossil) Piltdown Man. In his sketched geological time-scale, p.xii in *On the Trail of Ancient Man, Baluchitherium* is assigned to a 7-million-year-old 'stratum,' whereas on p.290 of this same book Andrews describes the fossil of this genus they disinterred in June 1925 was 'only' 3-million years old.

Besides paleontological relics, the team also discovered archaeological artifacts, for instance those belonging to a race known as the Dune-Dwellers. But they lived during Neolithic times, or within the past several thousand years. However, no traces of early man were found on these expeditions—nothing age-comparable to Peking Man, or *Homo erectus*. Despite Osborn's and Andrews's ardent assertions, eventually ideology shifted toward an alternative that the African continent instead represented a more likely "Cradle of Mankind." So those tiny Late Cretaceous mammalian skulls—hinting at an early stage for development of mammalian life (although not necessarily representing our absolute geological origin, and far removed in time from the earliest anthropoid), an outcome not anticipated by Osborn, must serve as the mission's 'consolation prize' for lack of that still missing Asian anthropoid 'link.'

After the Central Asiatic Expeditions' work ceased in Mongolia, further attempts went afield to find fossils there, this time a Russian-led endeavor from Moscow headed by Dr. I. Efremov. During 1946 to 1949 these expeditions were particularly successful, logging 120 *tons* of dinosaur bones. Then a series of joint Polish-Mongolian investigations hunted for fossil bones there, beginning in 1964 (through 1971—the year the famous fighting pair of dinosaurs, *Velociraptor* and *Protoceratops* fossilized in a 'frozen' death combat pose—were found in Mongolia). A park located in Chorzow, Poland, replete with many life-sized dinosaur genera endemic to the Cretaceous of Mongolia, restored as statues was built, opening to the public in July 1975. Known as the Valley of Dinosaurs, this park commemorated successes of the Polish-Mongolian expeditions.[2] (See Chapter Thirty-nine.)

Later, North American teams ventured back to the haunting deserts 'tamed' by Andrews seven decades before. First the Sino-Canadian expeditions headed into Inner Mongolia for a five-year mission ending in 1990—intent on surveying extent to which Cretaceous fauna of Asia and North America are evolutionarily related (possibly via intercontinental migration routes). Then after Cold War tensions eased by 1991, American scientists from the American Museum, helmed by paleontologist Michael Novacek, teamed with Mongolian scientists on a joint expedition, returning to the Flaming Cliffs region in search of dinosaur fossils during the mid-1990s. They prospected a new incredibly fossil-rich quarry known as Ukhaa Tolgod, finding dinosaur eggs resembling those allegedly from *Protoceratops*. One egg, found in July 1993, was broken and an embryo could be seen inside, but startlingly it was not a *Protoceratops* embryo: it was oviraptorid! This new discovery falsified the older view of an oviraptorid egg-thief; instead that genus was considered the egg-layer and an avian-like egg 'incubator' as well. Don't blame Osborn for this error though. For as paleontologist Mark Norell noted, years ago, Osborn had cautioned that designation of *Oviraptor* as an egg-thief could "...entirely mislead us as to its feeding habits and belie its character." (Quote from Gallenkamp, p.319) Prophetic words indeed!

Novacek outlined key accomplishments of their 1990s expeditions in his 1996 popular book, *Dinosaurs of the Flaming Cliffs*. Particularly noteworthy was their discovery of numerous oviraptorid skeletons, including another brooding specimen, which died (and was preserved) overlying its nested eggs. In 2001, this genus (although described in 1999) was named *Citipati*. A stubby-armed, bird-like dinosaur, *Mononkyus*, also came forth from outcrops, as well as a genuine Cretaceous bird fossil, *Gobipteryx*. At Ukhaa Tolgod they found many more fossils of rare placental mammals than previously accounted for by prior expeditions. Indeed, as Novacek wrote: "Ukhaa Tolgod exceeds all other known Cretaceous localities in the abundance, concentration, diversity and cumulative quality of the fossil terrestrial vertebrate remains."[3]

Since Andrews's pioneering expeditions to the Gobi, many other paleontologists and geologists have combed the region for fossils—hitting 'pay dirt' along the way over the course of several decades. But Andrews got there first, most successfully showing how it can be done, while finding fossils that fired imaginations across several generations. Now, especially since the fall of 1996, furor over Lower Cretaceous Chinese feathered dinosaur fauna became all the rage—even eclipsing those 1990s discoveries made in Mongolia. And then we have Andrews' books to read, engaging accounts[3] of what it was like to prospect for dinosaurs back in those hardy times—dodging sandstorms while steering a Dodge vehicle, while enduring circumstances that would make any 'Indiana Jones' movie character seem 'soft' by comparison. No—Osborn's theory was not borne out: ancient man seems to have drifted out of Africa, not Asia.

But who cares about that when Mongolia brought us *Velociraptor*—now so prevalent in popular culture, mysteriously long-armed *Deinocheirus*[4], dinosaur eggs by the dozen, and mighty *Baluchitherium*!

Key References:
Roy Chapman Andrews, *On the Trail of Ancient Man,* (New York: G.P. Putnam's Sons, 1926).
Roy Chapman Andrews, *Under A Lucky Star,* (New York: Viking Press, 1943).
Roy Chapman Andrews, *All About Dinosaurs,* (New York: Random House, 1953).
Michael Benton, ed., (et. al.), *The Age of Dinosaurs in Russia and Mongolia,* (Cambridge University Press, 2000).
Charles Gallenkamp, *Dragon Hunter: Roy Chapman Andrews and the Central Asiatic Expeditions,* (New York: Viking).
Donald F. Glut, *Dinosaurs: The Encyclopedia, Supplement 3,* (Jefferson, NC: McFarland & Company, Inc., Publishing, 2003, entry for *Citipati*, pp.292-295)
John R. Lavas, *Dragons From the Dunes: The Search for Dinosaurs in the Gobi Desert*, 1993.
Michael Novacek, *Dinosaurs of the Flaming Cliffs,* (New York: Doubleday, 1996).
Douglas J. Preston, *Dinosaurs in the Attic: An Excursion into the American Museum of Natural History*, (New York: Ballantine Books, 1986).

Notes to Chapter Twenty-eight: (1) See Andrews' *On the Trail of Ancient Man*, p.231. Andrews claimed to have spotted dinosaur "embryos" visible within two broken eggs, but proper identification of these specimens seems to have been in error. Also see Lavas, pp.47, 124.; (2) See Allen A. Debus and Diane E. Debus *Paleoimagery: The Evolution of Dinosaurs in Art,* (Jefferson, NC: McFarland & Company, Inc., Publishing, 2002, pp.200-202; (3) Novacek, p. 287, excerpted from Dashzeveg, D. and Novacek, M. J., (et. al.) "Extraordinary preservation in a new vertebrate assemblage from the Late Cretaceous of Mongolia," *Nature* vol. 374, pp.446-49. Novacek also incorporated a part-fictional 'pen-picture' or image-text describing a disastrous day in the life of Gobi Desert dino-inhabitants, focusing on an encounter between a nesting oviraptorid and *Velociraptor* in the eye of a major dust storm 80-million years ago. See Novacek, pp.289-294; (4) For current views on the no-longer so mysterious *Deinocheirus* see Gregory S. Paul's masterful *The Princeton Field Guide to Dinosaurs,* 2nd ed., (Princeton University Press, 2016), pp.129-130.

Chapter Twenty-nine — *Edwin H. Colbert: Wandering Lands, the Triassic & Cold-blooded Archosaurs*

As silly as it may sound I first became acquainted with revered dinosaur popularizer Edwin H. Colbert through a Chicagoland children's program—*Garfield Goose and Friends*, hosted by Frazier Thomas, which was broadcast weekly afternoons during the 1960s. Every so often during the early 1960s, Thomas would play approximate 3 minute segments of Karel Zeman's (1955 Czech) film *Journey to the Beginning of Time*, in which a group of boys sail down the river of time into remote prehistory. The beginning of the 1960 Americanized version added introductory remarks by Colbert, filmed at the American Museum of Natural History. It took over a month of *Garfield Goose* episodes for the serial program to run through its entirety, but what a thrill to see all those stop-motion prehistoric animals and wonder at the end of the movie what lay 'beyond' that beckoning, primordial ocean of time. A little later, a copy of Colbert's, to me—enthralling *Dinosaurs, Their Discovery and Their World* reached my bedroom bookshelf.

But now decades later, the spirit of the dinosaur and extinct relatives has never been more alive. Popular consideration of prehistoric animals, particularly archosaurs—including the two traditional orders of dinosaurs (saurischian and ornithischia), winged pterosaurs, "thecodonts" (once thought possible ancestors of some dinosaurs and maybe also of birds), and crocodylians, increased dramatically in America during the early part of the 20th century. The inspiring wealth of dinosaur discoveries occurring in the western U.S. and Canada between 1877 and 1922 certainly contributed to this phenomenon. Between 1877 and 1893 Edward D. Cope of the Philadelphia Academy of Sciences and Othniel C. Marsh of Yale unearthed numerous gigantic specimens, notably those from the Upper Jurassic Morrison formation in Colorado and Wyoming. Many of these are now on display at Yale and the American Museum of Natural History. During their careers, Marsh described 19 genera of dinosaurs while Cope described 9 genera.

Inspired by their remarkable efforts and with intentions of developing museum exhibits that would attract public support, Henry F. Osborn became influential in organizing expeditions headed by Walter Granger and Barnum Brown to explore the Bone Cabin Quarry vicinity in Wyoming where more Morrison dinosaurs were discovered between 1889 and 1905. Osborn and W. D. Matthew, both of the American Museum also encouraged a young talented artist named Charles R. Knight to prepare paintings and illustrations enhancing museum fossil exhibits, showing dramatically how these creatures may have existed in life.

During 1909 to 1922 North American dinosaurs were discovered on two major fronts. Earl Douglass of the Carnegie Museum in Pittsburgh found more Morrison fauna, this time in Utah at a famous locality that was proclaimed Dinosaur National Monument in 1915. Farther north, Charles Sternberg and his three sons, affiliated with the National Museum of Canada in Ottawa, W. A. Parks of the Royal Ontario Museum in Toronto, and Barnum Brown of the American Museum discovered numerous Cretaceous dinosaurs along the Red Deer River in Alberta, between 1909 and 1917.

These did not represent the earliest (or latest) dinosaur discoveries in America, nor did these field excursions represent full extent of these individuals' careers. Nonetheless these scientists and their co-workers certainly provided museums with a great wealth and selection of dinosaurs, captivating public interest and curiosity. Furthermore, at that time, when it came to finding spectacular dinosaur bones and those of other 'prehistoria' from bedrock of ages, the American west was the place to go!

Adding to the romantic aura of the period, Sir Arthur Conan Doyle (of Sherlock Holmes fame) wrote an intriguing adventure story titled *The Lost World* (1912) concerning discovery of prehistoric animals surviving on a remote plateau in South America, which was later produced as a major motion picture in 1925. The Sinclair Refining Company even adapted the symbol of a dinosaur for its highly successful oil and gasoline products advertising campaign, providing several accurately portrayed, full-scale models for the 1933-34 Century of Progress Exposition held in Chicago. These were exciting times for those interested in popular culture of dinosaurs, perhaps only to be superseded by the current dinosaur revival!

Edwin H. Colbert was born into this world of dinosaurs and men in 1905 at Maryville, Missouri. Like many interested in paleontology, by age twelve, Colbert already had become interested in natural history. He collected Carboniferous fossils, camped with his Boy Scout troop and later, during a summer trip to Denver in 1922, became enchanted with the mountains. Although the following summer after high school graduation, an inspiring trip to Chicago's

Field Museum set him considering the idea of museum work, instead Colbert pursued his favorite ambition of working for the U.S. Forestry Service in Colorado during summer vacations from college.

Having enjoyed exhibits of Upper Cenozoic fossil mammals on visits to the museum at the University of Nebraska in Lincoln, Colbert decided that the study of fossils might be more intellectually rewarding than that of trees. In 1926, Colbert transferred to the University of Nebraska, training under E. H. Barbour, thus learning ways and means of preparing museum displays. Colbert spent much of his spare time assembling skeletal reconstructions of *Diceratherium*, *Moropus*, and *Dinohyus*, which are still on display, and collecting fossils in the field.

While undertaking graduate studies at Columbia University in New York, Colbert was assigned from 1930 to 1935 as scientific assistant to the eminent Osborn. He even traveled to England where he studied a large collection of mammalian fossils from the Siwalik formation of India—the subject of his doctoral dissertation. In 1935 he obtained his PhD, two years after marrying Margaret Matthew—daughter of paleontologist W. D. Matthew. She had taken a position as scientific illustrator at the American Museum in 1931.

Edwin H. Colbert's early professional, post-graduate career as a paleontologist contrasted significantly from his interests in later years. Initially he was devoted to the study of fossil mammals, especially those of the Asian continent. While studying the Mongolian mammalian fossil fauna retrieved by Roy Chapman Andrews' teams, Colbert became closely acquainted with reigning authorities in the field such as Walter Granger—Curator of Fossil Mammals at the American Museum, George Gaylord Simpson, and Bill Thompson. Colbert conducted important field work at Agate Springs Quarry in Nebraska, and at White River Badlands in South Dakota with these knowledgeable individuals, experiences that greatly honed understanding of fossil mammalian fauna.

Colbert then replaced Barnum Brown, who had retired in 1942, as Curator of Fossil Reptiles at the American Museum, but it wasn't until 1945 when he decided that study of Triassic fauna, such as that exposed in the Chinle formation of Arizona's picturesque Painted Desert and Petrified Forest, or at Ghost Ranch in New Mexico would become focus of his research—his specialty. From a paleontological perspective, the Triassic of the Mesozoic Era is a highly intriguing time period. During this time, animals were adapting to new niches, and repopulating niches vacated as a consequence of the geological revolution closing the Paleozoic Era. Remains of the earliest mammals and dinosaurs date from this period. Colbert was certainly well-qualified for this ambition, as his familiarity with mammalian paleontology would be valuable in examining fossil remains of 'transitory species' such as the class of mammal-like reptiles known as therapsids.

The Ghost Ranch Quarry in southeast New Mexico was proclaimed a National Landmark in 1977 by the U.S. Department of the Interior in consequence to the great abundance of Triassic fossils discovered in 1947. A small coelurosaurian (saurischian) known as *Coelophysis* had been found there in 1881 and described by Cope in 1887. In later years, a remarkable fossil was found—a small *Coelophysis* discovered within the body cavity of a larger specimen. Colbert stated that these small carnivorous dinosaurs may have been cannibalistic. (Other paleontologists suggested that this dinosaur may have given birth to live offspring.)

Numerous skeletons studied via Colbert fortified probable whereabouts of small 'bird-like' dinosaurs that made some of the famous tracks in the Connecticut Valley—that were first extensively researched by geologist Edward B. Hitchcock of Amherst College in 1835. Their initial find in 1802 by a farm boy named Pliny Moody represented earliest recorded evidence of discovery of dinosaurs in North America. Since skeletal remains were so scarce at Connecticut Valley, Hitchcock believed on basis of fossilized three-toed footprints that the tracks were made by ancient birds. The first birds do not appear much later in the fossil record until the Late Jurassic Period, however. Discovery of tracks such as these, when correlated with numerous skeletons found at Ghost Ranch offered rationale for why coelurosaurs were gregarious, herd-like animals—not unlike animals noted by Hitchcock a century before.

Beginning in 1953 Colbert was chiefly responsible for reorganizing H. F. Osborn's central dinosaur exhibits at the American Museum. In the Hall of Early Dinosaurs the *Apatosaurus, Stegosaurus* and *Allosaurus* skeletons were set in the center of the Hallway. The basic theme of this exhibit was a probable reconstruction or re-enactment of what may have happened at Glen Rose, Texas 80-million years ago. Here in 1938. Roland T. Bird discovered a fascinating trackway made by a large sauropod. Cast in limestone adjacent to the sauropod footprints are those of a large three-toed predator that presumably could have stalked its prey. Skeletal remains of animals that made either set of tracks were never found.

In assisting mounting of the tracks situated toward rear of the *Apatosaurus* skeleton, Colbert contracted Bird's services. Several feet behind the skeleton was situated the famous *Allosaurus* skeleton in the pose of tearing 'flesh' from a dead sauropod. This was a fascinating exhibition made even more so by illustrations prepared by Knight and other museum staff. An equally impressive display rearrangement was made in the Hall of Cretaceous Dinosaurs. These museum exhibits, originally mounted in other configurations by Osborn half a century earlier, became world famous—iconic.

Then Colbert became fortuitously involved in the mid-20th century's major geological controversy! In 1912, German meteorologist Alfred Wegener hypothesized the concept of Continental Drift, founded upon convincing geological, glacial and paleontological evidence. He demonstrated that it was possible to correlate geological structures and paleontological faunal horizons on both sides of the Atlantic Ocean. (Notably, the Gondwanan strata of India, southern Africa, Antarctica and South America show remarkable resemblances.) Fossils of seed ferns (*Glossopteris* and *Gangamopteris*), mammal-like reptiles (*Cynognathus*) and the aquatic reptile *Mesosaurus* were represented in the southern Gondwana.

The hypothesis stressed that until about 200-million years ago, all major continents were joined into a giant super-continent named Pangea. After this time they split and rifted apart (becoming the Laurasia landmass in the north and Gondwana in the south), forming the Atlantic Ocean. Wegener's hypothesis challenged the then prevalent theory of land-bridging, thought to have been effected by an ever-cooling, shrinking globe. Land-bridging was a theory commonly invoked at that time to explain worldwide paleontological distributions and geological occurrences.

The major problem with Wegener's theoretical argument was that he did not suitably account for the tremendous geotectonic forces involved in terrestrial processes, and his idea lacked a suitable 'mechanism' necessary for shoving large continental masses aside. Although Wegener was supported by African geologist Alex du Toit, many renowned geologists could not accept continental drift, believing that continents and ocean basins remained stable, situated where they always had been since the primordial crust cooled. "Once a continent, always a continent. Once a basin, always a basin" was the mantra of the time, riffing geologist Bailey Willis's perspectives of 1932.

Wegener's hypothesis of Continental Drift stirred great controversy during the 1920s. But it wasn't until after World War II when oceanographers, geophysicists and seismologists began accumulating highly convincing evidence about the ocean floor, as well as Earth's crust. Only then were Wegener's ideas seriously reconsidered. This resulted in the most dramatic scientific debate of the century, culminating in the late 1960s when scientists added more paleontological evidence supporting drift. Geologist David Elliott then stated: "Geophysical measurements and related studies are convincing, but nothing can dispute the authenticity of a fossil bone with a valid record as to its occurrence." It should be realized that today everyone generally accepts "Plate Tectonics" (a 1960s term replacing Continental Drift) as fact, as a result of Colbert's paleontological contribution, to be outlined below.

Colbert's 'drift' involvement began in 1962 when he examined deposits of lands which formerly may have been interconnected before alleged rifting occurred. In Africa, Colbert visited the Great Karroo Basis. There, in Lower Triassic deposits, a *Lystrosaurus* skeleton was discovered. In 1964, he collected another *Lystrosaurus* skeleton from the Lower Triassic Panchet formation in India. Later that year he observed dinosaur remains from the Cretaceous of Australia. (Earlier, in 1959, Colbert had discovered a Triassic rhynchosaur and other dicynodont therapsids in South America.)

Lystrosaurus was a herbivorous mammal-like reptile known as a therapsid. Because of the two large tusks present in its upper jaw it belongs to the dicynodont group as well. *Lystrosaurus* and other related therapsids such as rhynchosaurs 'dominated' the Lower Triassic landscape.

By the Middle Triassic another rapidly developing reptilian taxon known loosely as 'thecodonts'—some of which had adapted an erect, bipedal posture became established. Thecodonts presumably profited at expense of the therapsids, which became extinct in the Late Triassic. However, at this later time, thecodonts were in turn being replaced by their own descendants—dinosaurs, which were even better adapted to the prevailing climate, terrain and atmospheric-environmental conditions.

Although Colbert's firsthand discovery of *Lystrosaurus* remains and other early Triassic creatures and dinosaurs in the remote areas of India, South Africa, South America and Australia indicated that there was paleontological connection between these lands, the most conclusive discovery occurred in Antarctica!

In late 1967, a geological team exploring Triassic rocks in the Central Transantarctic Mountains came across an isolated jawbone belonging to a labyrinthodone amphibian. When the bone was presented to Colbert in 1968, paleontologists became highly interested in implications of this discovery. An expedition, including Colbert, was sent to Antarctica to search for fossils. On December 4, 1969, Colbert was cleaning some bones that had turned up earlier in the day at Coalsack Bluff. Suddenly, just after muttering to himself, "Why can't we find a *Lystrosaurus* skull," he recognized in his hand the tusk and jaw bone of ... *Lystrosaurus*! The geological team, including Jim Jensen of Brigham Young University, David Elliott of Ohio State and James Schopf, had discovered unequivocal evidence that Antarctica was once connected to other continents. It would have been impossible for land-bridging to have allowed animal migration to a continent as remote as Antarctica (i.e. in its present position relative to other continents). Therefore Antarctica *must* have rifted relative to other continents in closer proximity. Announcement of the discovery created a wave of sensation![1]

Announcement of their fossil was immediately praised on the front page of *The New York Times* (Dec. 6, 1969), messaging the discovery as: "... not only the most important fossil ever found in Antarctica but one of the truly great fossil finds of all time." Their landmark paper, "Triassic Tetrapods from Antarctica: Evidence for Continental Drift" (Elliott, Colbert, *et. al.*, *Science* vol.169, 9/18/1970) offered scientific details. And another article, "Lystrosaurus Zone (Triassic Fauna From Antarctica," (*Science* vol. 175, 2/4/1972, p. 524, penned by Kitching, Collinson, Elliott, and Colbert)) concisely pinned everything down—case closed:

"The discovery at Coalsack Bluff in 1969 of fossil bones of *Lystrosaurus* was a significant addition to the evidence for continental drift. The occurrence of this form in now-isolated Antarctica is not explainable except in terms of the existence during the Triassic of a former supercontinent, Gondwanaland, which later split into the present southern continents and peninsular India. ...abundant occurrences in Antarctica of fossils representing the *Lystrosaurus* fauna confirms the close connection between Antarctica and Africa in early Triassic time. Moreover, the fact that the *Lystrosaurus* fauna was nearly complete in Antarctica suggests the Triassic connection between the two continents was probably a broad one, rather than an isthmian link over which the effects of faunal filtering would be more evident. In short, the *Lystrosaurus* fauna as it is now known in Antarctica and in southern Africa constitutes one of the strongest lines of evidence as yet adduced in support of Gondwanaland and, correlatively of continental drift."

But shortly, here we'll see how, while himself a pioneer facilitating modern understanding of plate tectonics, when it came to the subject of dinosaurian physiologies, Colbert remained for the most part ...'older school.'

Discovery of dinosaurs in polar latitudes may have initially suggested a worldwide tropical climate during the Mesozoic Era. Why else, some thought, would *Iguanodon* remains have been discovered well within the Arctic Circle (in 1960), off the coast of Spitzbergen? Although Colbert believed that tropical conditions prevailed during the Mesozoic, months of winter darkness at the Arctic Circle would certainly have doomed supposedly Sun-basking, reptilian dinosaurs. Furthermore, how or why would *Iguanodons* have migrated such great distances to reach the Arctic? Dale Russell's interpretation of why Hadrosaurs fossils were discovered in 1973 within the Canadian Arctic Circle relied on a totally different physiological principle. This is the principle of warm-blooded dinosaurs, toward which Colbert remained generally skeptical, if not opposed. (Ironically, Russell was once a student of Colbert's at the American Museum.)

In 1944, after having become intrigued with R. B. Cowles' suggestion that dinosaurs became extinct at the end of the Mesozoic because of a global cooling phase, Colbert was prompted to conduct the first analysis modeling physiological characteristics in archosaurs. In this study, temperature regulatory patterns of modern cold-blooded alligators were extrapolated to dinosaurian proportions. It was observed that a larger alligator must bask in the Sun for longer periods in order for its internal body temperature to rise one degree, than was required for smaller individuals. Conversely, a larger alligator may reside in the shade for a longer duration than a smaller alligator before its body temperature decreases by one degree. Colbert concluded that larger dinosaurs need not be warm-blooded because their larger mass to body surface area ratios would have effectively maintained their body temperature within a comfortable range.

But the actual length of time necessary for a dinosaur's body temperature to fluctuate by one degree was debated. Originally, the estimate was 86 hours for a 30-ton (60,000 pound) *Apatosaurus*, which seemed excessively long. Another estimate was placed instead at only 2 to 3 hours, which would probably allow the animal to freeze if ambient temperature became sufficiently chilled at night. By basking too long in the Sun to regain this energy, such an animal's internal temperature may quickly reach a comfortable zone, but its external surface integumentary temperature may have also become raised to a dangerous level, baking it to a degree sufficient to kill the animal. This actually happened with one of Colbert's experimental alligators.

Colbert suggested that dinosaurs were also capable of regulating their body temperatures through other means. Thus, portions of dinosaurian bodies—those with small mass relative to surface areas could have cooled themselves rapidly via their extremities (e.g. legs, tail and neck). For example, it was hypothesized that *Stegosaurus'* spinal plates, presenting 'extra' surface area, which in life contained large blood vessels, were also utilized for temperature control.

Larger dinosaurs would have to eat more than smaller individuals. But, say, if *Apatosaurus* was warm-blooded as was suggested by Russell and Ostrom—another student of Colbert's at the American Museum, they would have eaten a ton of food per day, that is—extrapolating from the mass of food required by modern elephants which only weigh 10,000 pounds, in order to maintain their high body temperatures. Warm-blooded animals are more spontaneously energetic than sluggish, cold-blooded reptiles. However, warm-blooded animals such as mammals and birds pay the price of their ability to maintain constant body temperatures (without need for energy absorption from the Sun) by consuming much greater quantities of food. This extra food supplies energy by which they regulate their internal temperatures. By contrast, a *cold*-blooded *Apatosaurus* would have needed only less than 1,000 pounds of food per day.

How were newly hatched Apatosaurs able to survive if they hadn't reached a critical size where their internal body temperatures resisted fluctuations? Perhaps it really was sufficiently warm both day and night during the Mesozoic, or perhaps those young dinosaurs pursued different life habits until they reached maturity, as do modern crocodiles.

An early argument in defense of warm-blooded dinosaurs was that because dinosaurs and mammals both appeared during the Triassic, and because mammals were warm-blooded, then dinosaurs which dominated over mammals during the Mesozoic must also have been at least as physiologically developed. This argument suffered from use of vague, inaccurate terms such as "dominate" and by assuming that for all environments a warm-blooded metabolism would always be superior. Apparently during the Late Triassic a *dinosaurian* metabolism was favored.

Colbert maintained that pterosaurs, such as winged *Pteranodon*, were probably "reptilian vultures." This came in contrast to a more popular view promoted in Adrian J. Desmond's 1975 book, *The Hot-Blooded Dinosaurs*, where pterosaurs were described as soaring, warm-blooded animals, clothed in fur, occupying niches not unlike that of modern seagulls. Desmond claimed that on basis of convincing fossil evidence, pterosaurs were generally quite intelligent. Paleontologists have generally associated proportionally smaller brain sizes with sluggish (reptilian) life habits, while larger brain sizes have been associated with warm-bloodedness. Wann Langston of the University of Texas, who described the gigantic (40-foot wingspan) pterosaur *Quetzalcoatlus* in 1971 also maintained that pterosaurs were warm-blooded.

Nonetheless, Colbert—in associating relatively large brain sizes of later-evolving Upper Cretaceous carnivorous dinosaurs such as dromaeosaurids with possible warm-bloodedness, permitted an exception representing, for him, a point of departure from both his roots, and the mainstream. Back in 1914, Barnum Brown had described *Dromaeosaurus* known from Canada's Red Deer River. He only had scrappy remains to work with—yet among the fossils was a skull braincase. So in 1969 both Colbert and Russell teamed on an American Museum monograph reconsidering this small dinosaur. After examining the fossil, noting the braincase's "undinosaurian shape," Colbert and Russell concluded "If the posterior moiety of the brain even approximately filled this space ... it must have been very broad indeed," (i.e. much larger than a 'walnut').[2] Years later in his coffee table book, *Dinosaurs: An Illustrated History* (1983), Colbert diplomatically went further, commenting: "...certain dromaeosaurs had brains about six times as large in relation to body size as the brain in modern crocodiles. So perhaps dinosaurs such as these may have been endothermic, even though other large dinosaurs were not." (p.151) Furthermore, they stated, "*Dromaeosaurus,* with its relatively large brain, large eyes, and grasping hands, must have been a swift, skillful, and formidable hunter." It also may have disemboweled seized ornithischian prey with its enlarged claw (like *Deinonychus*).[3]

Still, such a view contrasted with an alternative, that warm-bloodedness in dinosaurs developed much earlier—first in ancestral thecodonts of the Triassic. Meanwhile, paleontologists such as Robert Bakker and John Ostrom proposed that bipedal motion must be powered by a warm-blooded metabolism. Since thecodonts such as *Euparkeria* of the Middle Triassic had already developed an erect bipedal posture in which the feet were placed close to the body's midline (as opposed to a reptilian, sprawling gait), therefore descendant dinosaurs must have inherited their warm-bloodedness from thecodont-like ancestors during the Triassic.

The matter of warm-bloodedness versus cold-bloodedness in dinosaurs has often been fought on peculiar terms. Colbert claimed that larger dinosaurs' food requirements would have been unrealistically enormous if they were warm-blooded, while others stated that only a warm-blooded dinosaur would have been sufficiently energetic to stand long enough eating even the daily requirement of a cold-blooded dinosaurs.

Colbert remained professionally active during the 1980s and 90s, as more bones and controversy turned up. In 1981 he described an Early Jurassic denizen—*Scutellosaurus*, which he described as "...one of the earliest and most primitive of the ornithischian dinosaurs ...protected by numerous little armor plates (and) ... characterized by an extraordinarily long tail, which may have served in part as a counter-balance to a body burdened by the heavy weight of its armor...."[4]

Then suddenly there was furor over the nature of the genus *Coelophysis*—the dinosaur upon which Colbert had staked popular dinosaurian phases of his career. Cutting to the chase, by 1991, several paleontologists suggested that the generic name *Coelophysis* was a 'dubious name' and that instead fossil specimens found at the Ghost Ranch New Mexico Quarry were re-described under the genus *Rioarribasaurus colberti*. However, Colbert challenged this name re-designation and by 1996, following a special International Commission on Zoological Nomenclature vote the original name became re-established. Their conclusion was that: "Of the nominal genera involved, *Coelophysis* is the most important in phylogenetic discussions and the name is well used in the literature. Therefore, to conserve 'common usage' it is necessary to approve the proposals of Colbert ..."[5]

Through his numerous articles, popular books, and long-term promotion of dinosaur science, Colbert—who died in 2001, became an iconic paleo-figure in mid-20th century America. In case of my fascination toward him, this was strangely in part because of my early exposure via an afternoon cartoon show—*Garfield Goose*, and its welcomed Czech 'dinosaur' film serial broadcast.

Notes to Chapter Twenty-nine: (1) Edwin H. Colbert, *A Fossil-Hunter's Notebook: My Life with Dinosaurs and Other Friends* (New York: E.P. Dutton, 1980, pp.224-227;(2) Adrian J. Desmond, *The Hot-Blooded Dinosaurs: A Revolution in Palaeontology* New York: The Dial Press, 1976, p.97; (3) Donald F. Glut, *Dinosaurs: The Encyclopedia* (Jefferson, NC: McFarland & Company, Inc., Publishers, 1997, p.365; (4) Colbert, *Dinosaurs: An Illustrated History* (Hammond Inc., 1983, pp.71-72; (5)

Details are provided in Donald F. Glut's *Dinosaurs: The Encyclopedia* (Jefferson, NC: McFarland & Company, Inc., Publishers, 1997, pp.302-304, 767-774; and *Dinosaurs: The Encyclopedia, Supplement 1* (Jefferson, NC: McFarland & Company, Inc., Publishers, 1999, pp.178-182. Quote is taken from p.182. Colbert also mentions discovery of the Ghost Ranch quarry in other publications, such as his engaging account, *Men and Dinosaurs: The Search in Field and Laboratory* (New York: E.P. Dutton, 1968, pp.139-142).

Chapter Thirty — *'Time' for Albritton's Catastrophism*

Certainly we'd agree that a concept of time must be considered a most fundamental geological tool. Without proper understanding of time, misinterpretation of events recorded in the rock record can easily result. So here's a synopsis of a late 1980s contemporary view concerning geological time, catastrophism and the fossil record relying on the writings of geologist Claude C. Albritton, Jr. for illustration.

American geologist Albritton's (1913-1988) posthumously-published *Catastrophic Episodes in Earth History* (Chapman and Hall, 1989) was a diplomatically presented, concise outline broaching all topics of relevance to discussion of mass extinctions and possibly associated catastrophism known to science by the late 1980s. (The only other survey book equally as encompassing was James Lawrence Powell's *Night Comes to the Cretaceous*, published a decade later in 1998). Early chapters in Albritton's *Catastrophic Episodes* focused on how geological time and process uniformity through time was viewed by early pioneering geologists. Emphasized were beliefs of obligatory catastrophists versus tenets upheld by process uniformitarian-ists—such as gradualists requiring vast excesses of time that were disallowed by late 19th century physicists.

After radioactivity was discovered and applied to age-dating of crustal rocks, Lyellian uniformitarianism was supposedly vindicated since, after all, it had been shown that Earth was very old indeed—perhaps old enough for the very gradual origin and evolution of life and development of terrestrial features. If Charles Lyell had been right, or at least more correct than Lord Kelvin about the Earth's age (old to the tune of a few billion years, as opposed to quite young as measured in a few tens of thousands of years), then perhaps his (i.e. Charles Lyell's) ideas concerning uniformity, gradualism, and exclusion of controlling extraterrestrial factors bore merit as well.

Geologists had effectively won the right to postulate enormous intervals of time; excess time that the more highly-esteemed physicists had discounted on basis of experiment and mathematical models. Years later, during the decade of the 1950s, geologists emerged 'victorious' once again in their contention that Earth was considerably older versus the astronomers' counterclaim that the *universe* itself must be only two billion years old (founded on astronomical evidence), a value which was only one-half to two-thirds of contemporary estimates for Earth's age as then measured on basis of radioactivity. In consequence, David M. Raup stated that "... geology has a curious moral authority over astrophysics." (*The Nemesis Affair,* 1986, p.132)

Albritton's fascination with exploration of geological time provided a focal point in the study of 'neocatastrophism,' since concepts of uniformity and rate of change exerted by either terrestrial or extraterrestrial controls are vitally linked to our understanding of time. In fact, one cannot correctly grasp essence of the uniformitarian/catastrophism debate (of the 19th century or, especially, the 1980s) until geological time has been comprehended.

Next, Albritton discussed matters leading to heightened 1980s controversy devoting brief, though thorough attention to all the major ideas and lines of evidence. (One sees that conventionally-minded scientists educated on Lyellian principles during the early-to mid-20th century played significant roles in upholding gradualist, uniformitarian dogma whenever controversial ideas concerning catastrophism emerged.) Albritton exemplified differences between meteorite craters, explosion, and cryptoexplosion structures. Conventional and unconventional theories for Late Cretaceous extinctions, which have received more scientific attention during the past half century than any other extinction event, were then compared in light of the great abundance and variety of evidence then at hand.

Reactions to catastrophist theories presented in *Catastrophic Episodes in Earth History* are now of interest to historians of science, as will Albritton's suggestion that 1980s amendments to geological knowledge may later be considered a revolution in Earth and Planetary Sciences. Whether or not the new catastrophism, or new uniformitarianism—Albritton disliked such terms—will ever qualify as a 'revolution' in thought (as opposed to an epidemic instance of 'science gone mad'), certainly it is possible to agree with him and others involved in, or observant of, the debates that the 1980s decade was indeed one of the "most intellectually stimulating times in the history of geology" as well as paleontology. (S. Dutch, "Rare events discussed," *Geotimes* vol. 31, no.4, 1986, quoted on p.178 in *Catastrophic Episodes*).

Albritton's book received a positive review in *Earth Sciences History* (vol.8, no.2, pp.196-97), written by Curt Teichert. However, Teichert was critical of Albritton's reluctance to discuss occurrence and statistical nature of fossil species extinctions. "Almost without exception, extinctions papers written by non-paleontologists are not based on the study of fossils, but on the interpretation of curves, showing the abundance of taxonomic families in the geological record ... many non-paleontologists do not realize that evolution proceeds on the species level and that genera and higher taxonomic categories are derived concepts that embody varying amounts of subjectivity."

Furthermore, Teichert noted that Albritton overemphasized dinosaurs, perhaps the least significant though most symbolic 'cult' case of all extinctions in the fossil record. Perhaps too, in other documented cases, Albritton wasn't sufficiently critical of geophysical evidence at hand. For instance, one of five craters possibly attributed to the Frasnian-Famennian extinctions boundary (Upper Devonian) is between 335 and 368-million years ago. "Sample calculations show that the ... bolide may have hit after the passage from the Frasnian to the Famennian. To connect ... the bolide with the terminal Frasnian extinction is, therefore, a purely speculative exercise. And this is not the worst case." (Teichert, *op. cit.*)

Teichert obviously sensed that Albritton had placed too much emphasis on various and intriguing catastrophist-astrophysical models developed during the 1980s (many of them *ad hoc*), while ignoring nature and meaning of the fossils themselves. Extinctions at generic and higher taxonomic levels had been dramatized because of such events' severity. "I am convinced that after thorough reviews by competent specialists, dozens, perhaps hundreds of families and hundreds, perhaps thousands of genera would disappear from the extinction lists. These basic flaws in the current extinction literature are accurately reflected in Albritton's book." Teichert noted how then-recent scientific discourse concerning the mass extinction controversy had been largely restricted to English-language publications, although there have been some notable non-Anglo, mid-20th century contributions (notably by German paleontologist Otto Schindewolf, subsequently translated into English).

It is somehow only fitting that a scientist of Albritton's stature, given his great interest in the nature and discovery of deep geological time should have written a book on the exhilarating controversy over mass extinctions phenomena. First, the nature of geological time has played a pivotal role in our understanding of terrestrial processes and Earth history, a circumstance which had been delightfully presented in Albritton's *The Abyss of Time* (1980). Secondly, as mentioned in the Preface to *Catastrophic Episodes* (p.xvii), he confessed,

> "... I have been asked simply to report and not to 'take sides' in arguments, pro and con. So at the outset I should confess to two biases. I was one of those early advocates of a meteoritic origin for certain cryptoexplosion structures (in 1936), as opposed to the then-current view that they were formed by explosive volcanic activity. I have found no reason to change that opinion. Secondly, I have a strong dislike for 'isms' ..."

In late 1989, an issue of *Earth Sciences: Journal of the Earth Sciences Society* (vol.8, no. 2) was dedicated to Albritton for his "life-long fascination with the concept of time." The issue included many articles presented at the 28th International Congress Symposium on the subject of the "Idea of Time," a symposium which Albritton had helped plan until his death.

A most prevalent issue in Albritton's *The Abyss of Time* was 'evolution' in man's conceptual understanding of time. Since the late 18th century, ideologies shifted from a non-directional, cyclical time concept (e.g. that of James Hutton) to a (biblical/theological) directional, linear perspective. Later Charles Lyell advocated a historically-founded cyclical nature of deep time. But difficulties were encountered in applying Lyellian uniformity to the past, even if one accepted his premise that 'extraordinary agencies' hadn't acted throughout the (now currently referred) Phanerozoic Era (although evidence for recurrent ice ages, the Coal Age, and strata restricted to the Lower Paleozoic such as greywacke and Ordovician black shales certainly hurt his cause). His approach to dealing with obvious problems created by an imperfect geological record and a highly incomplete fossil record didn't resolve all the difficulties, either. That is, rocks of earliest recorded time, the Precambrian—so different from those of other ages, and those of the lowermost Cambrian in which a proliferation of living shelled organisms 'suddenly invaded' sedimentary deposits—indicated decades ago that Lyellian uniformity could not be applied to Earth's beginning and formative stages, however long ago those times actually were.

Furthermore, composition of Earth's ocean-atmosphere system surely must have originated and changed since primordial times. According to Beryl Hamilton, (*Earth Sciences History, op. cit.,* p.141):

> "The Precambrian seemed to provide a major exception to the Lyellian doctrine of uniformitarianism, as study made it obvious that conditions in the early part of this period were unique and were never repeated. On the other hand, as the knowledge of the Precambrian developed, there was a growing realization that the later part of the Precambrian fitted the Lyellian paradigm very well."

Following partial vindication of Lyellian thought after falsification of Kelvin's experimentally-derived estimates for Earth's age and the rise of radiometric age dating, however, conventional thinkers turned to Lyell and Charles Darwin for their understanding of the rates of geological processes and their perceptions of how events changed in time.

Interestingly, although Lyell's principles of geology became dogmatized, few fully understood or bothered to resurrect his rigidly-structured view of cyclical time. More recently, geologists have reverted to the study of cyclical time. Both terrestrially-founded and 'extraordinary' events (those which have not been witnessed by man, historically) do recur, but whether they do so in an episodic, strictly periodic, totally random, or quasi-periodic pattern remains openly disputed.

> "Contemporary natural processes are the source of information for the codification of vestiges. But this does not imply any assumption about 'uniformity,' 'evolutionary,' or 'circularity' of nature ... The assumptions of uniformity, or continuity, or the choice of some kind of philosophy ... was a necessary preliminary step for overcoming the impossibility of inductive generalization from the present state of natural processes to the past ... It is our opinion that the idea of 'cycles' in geologic time must be considered as a methodological device applied to the study of natural processes in a geological perspective. Probably, recurrent or rhythmic patterns have been interpreted as having a cyclic nature." (pp. 121-122, *Earth Sciences History, op. cit.,* C. Pascholae and S. Figueriroa)

A short time after publication of Albritton's *Catastrophic Episodes,* geologists documented evidence for the Chicxulub Crater in Mexico's Yucatan, suggesting this was the site of the K-Pg impact boundary event 66-million years ago. While that evidence has been publicly debated, importantly, the identified crater is of sufficient size, shape and age. The 'smoking gun'—the site of the 'silver bullet—had been found! Thereafter, interest in 1980s celestial models conceivably giving rise to recurrent or cyclical events on Earth (i.e. Nemesis star, Planet X, and other galactic models periodically affecting the Oort Cloud of comets) faded. Although the crater's existence had been known probably since 1978, Albritton didn't live to hear of the Chixculub discovery's 'confirmation' as plausible site of the K-Pg event (at least there is no mention of it in his *Catastrophic Episodes*), but he would have appreciated its significance.

Albritton, who received his Ph.D. from Harvard in 1936, wasn't a paleontologist. He was renowned for his work in Quaternary geology, especially archaeological geology, for which he was presented the Geology Society of America's award in Archaeological Geology. Ironically, this happened on the day of his death, November 1, 1988. However, culturally and scientifically relevant books such as his *The Abyss of Time* and *Catastrophic Episodes in Earth History* can be read with enduring interest.

Chapter Thirty-one — *A Look 'Bakk-' Robert Bakker: Revolutionary Paleontologist*

"The concept of dinosaurs as avian- or mammalian-style endotherms has intuitive appeal to the layperson, and is infinitely preferable to the image of dinosaurs as the cultural icons of stupidity, sluggishness, extinction, and above all, failure.... An eager public has been extremely receptive to popularizers willing to offer a readily comprehensible account of dinosaurs as highly active, racing, dancing endotherms.... However, the view of dinosaurs that the public has embraced so eagerly is partisan and flawed....(This view) has derived from a philosophically... suspect concept (that endothermy is superior to ectothermy...), and from doctrinaire (views) that insist that all dinosaurs were endothermic that have failed to keep abreast with relevant developments in bio-physics." (J. R. Spotila et. al., 1991, *Modern Geology*, 16: 223-224 as quoted in *The Evolution and Extinction of the Dinosaurs,* by David E. Fastovsky and David B. Weishampel, Cambridge: Cambridge University Press, 1996, p.356)

Although the late 1960s and early 1970s will long be remembered as a period of socio-political upheaval, for me –at the time a young adult learning about the many wonders of geology and paleontology, it also represented a time of scientific revolution. Historians of science object to overstated, liberal use of the term 'revolution.' However, then, controversial ideas stirring over new interpretations of how dinosaurs appeared and lived seemed 'revolutionary' indeed. It wasn't until years later when I learned about an instigator, if you will, who promoted the fabulous new interpretations of dinosaurs, paleontologist Robert Bakker. By the early 1990s, Bakker had reached a zenith of his paleo-popularity—yet far more than just 15 minutes of fame.

Bakker formerly an adjunct curator in paleontology at the University of Colorado Museum has stated that his preoccupation with dinosaurs occurred at the age of 11. Bakker was admiring a scene from Rudolph Zallinger's *Age of Reptiles* mural prepared for the Peabody Museum at Yale in the September 7, 1953 issue of *Life* Magazine. "From that moment onward, I was a paleontologist.... My family and the doctor predicted it was just a hormonal stage that would pass. My parents still hope it will pass."

Fortunately, many of us have been able to share increased level of understanding about these fascinating animals, thanks to the investigations Bakker stimulated among his generation of paleontologists. So, it is all well and good that his affliction proved to be so consuming. (In case you're wondering, his all time favorite dinosaur is *Ceratosaurus*.)

One can trace the rise of the current general popularity in dinosaurs to the time when Bakker first began posing intriguing paleontological questions. In fact, this trend has been reflected in pictorial restorations of dinosaurs, a valuable means for conveying scientific opinions about how dinosaurs appeared and lived. To be sure, Bakker was on the leading edge of the modern movement in dinosaur restorations. On this point, Bakker remarked, "If (artists) want to paint dinosaurs, they have to learn anatomy and go the route of da Vinci by being both a scientist and an artist."

Bakker is artistically gifted with great powers of imagination, augmenting a rare skill for reconstructing fossil bones. For example, besides illustrating many of his own publications, Bakker also completed dinosaur restorations for the National Museums of Canada at Ottawa and Ontario. Several of these were reproduced in Donald F. Glut's 1982 book *The New Dinosaur Dictionary*.

The tempestuous phase of Bakker's career flared in 1968. At the time, dinosaurs were conventionally regarded as cold-blooded reptiles. Yet, there were deficiencies in the body of knowledge concerning dinosaurian behavior and social structure as well. Bakker's researches markedly changed all that. In the Spring of 1968, Bakker published an article titled "The Superiority of Dinosaurs," in an issue of *Discovery* (vol. 3, no. 2). This issue featured two of his now classic dinosaur restorations. A pair of elegant giraffe-like *Barosaurus* graced one page of his article, and two galloping *Chasmosaurus* with upright—not sprawling—front limb posture were featured on the journal's cover. Bakker also provided other line drawings contrasting 'modernized' sprinting ceratopsids with more conventional restorations in which they were depicted as sprawling lizards. Their stance resembled a human awkwardly "cheating at pushups."

Bakker noted that all living vertebrates with upright limb posture and 'long-leggedness' happen to be warm-blooded. Therefore, all bipedal vertebrates also are endothermic, which therefore may be prerequisite to obligate bipedalism. However, Thomas Henry Huxley first suggested that dinosaurs may have been warm-blooded, circa 1866-1867. According to Adrian J. Desmond, Huxley "... probably got the idea from a young palaeontological Ishmael and former piano-tuner's apprentice Harry Seeley, who had proposed that pterodactyls were 'hot-blooded'."

Then he formulated a simple question. Why is it that the dinosaurs apparently dominated Earth for well over a 100-million years, when their near relatives, the supposedly physiologically advanced mammal-like reptiles, from which the mammalian class arose, appeared first in the fossil record? "... if the later synapsids were such splendidly advanced animals with the improved physiology of mammals, and if dinosaurs were slow and sluggish, why were the mammal-like synapsids exterminated in competition with the first dinosaurs? And why didn't the mammals achieve a more significant diversification during the dinosaurs' reign?" This was a very bold issue indeed for a then mere graduate student to be addressing.

The mammal-like synapsids (and therapsids) may not have been as fully endothermic as modern mammals, and even so, they may *not* have lived in direct 'competition' for resources with (bipedal) thecodonts and early dinosaurs. In 1999 Paul Sereno noted there is a 15-million year gap between the advent, or 'initial radiation' of dinosaurs prior to their 'rise to global dominance.' Thus, whereas Bakker viewed the rise of dinosaurs as a 'competitive displacement,' Sereno characterized the pattern as an example of 'opportunistic replacement.' Bakker presumed warm-bloodedness is somehow metabolically 'superior' to cold-bloodedness, an assumption that doesn't necessarily hold true for all conceivable niches in every paleo-environment. Most living vertebrates happen to be cold-blooded. Four decades following his 'Superiority of Dinosaurs' article, Bakker remains unrelenting. "The biggest proof that dinosaurs were hot-blooded isn't the furry raptors from China... That's really strong proof of course......... but a bigger proof is that we mammals didn't really begin to expand until the dinosaurs finally disappeared. That's HUGE...... (dinosaurs) are superior. Rocks prove it. They beat us to the very end." (Quoted from *Prehistoric Times*, no. 46, Feb./March, 2001, pp. 49-52.)

Bakker's solution was that early on in their evolutionary development dinosaurs developed erect posture, a factor favoring dinosaurs over early mammalian lines possessing sprawling gait. This challenged conventional wisdom and distressed many of his colleagues because high running speeds and upright stance attributed to dinosaurs by Bakker implied an advanced warm-blooded condition. Bakker even suggested that some dinosaurs like giant sauropods may have maintained structural order within the herd, allowing them to better protect and rear young. To a degree, Bakker was supported by his Yale University advisor, Professor John Ostrom.

Bakker had assisted in a 1964 fossil collecting expedition to Montana where Ostrom had discovered in early Cretaceous sediments, remains of the carnivorous *Deinonychus*, which Ostrom viewed as a fierce, agile predator. From the appearance of its skeleton that included a weight counter-balancing tail probably held rigidly in life and long scythe-like talons on the second toes of the hind feet, Ostrom suggested that *Deinonychus* represented a possible case for warm-bloodedness (endothermy) in dinosaurs.

In retrospect, one may question whether Bakker's controversial ideas would have been regarded as seriously if Ostrom, then a more 'established' member of the scientific community, hadn't himself published such a remarkably 'new wave' view of how *Deinonychus* lived. Ostrom fueled support for the warm-blooded theory, founded in his 1969 published description of the *Deinonychus*, and in a paper delivered at Chicago's Field Museum of Natural History. Ostrom claimed dinosaurs weren't reliable indicators of climate—as cold-blooded reptiles would be indicative of warm climate because: "There is considerable evidence.... if not compelling, that many different kinds of ancient reptiles were characterized by mammalian or avian levels of metabolism."

Ostrom's monograph included Bakker's lively restoration and skeletal reconstruction of the *Deinonychus*, a contribution which Ostrom graciously complimented the artist for. "Bob was a very important member of the *Deinonychus* team because he has an unusual combination of talents—he is a very capable artist, as the cover of this issue demonstrates and he has good knowledge of vertebrate anatomy. Although *Deinonychus* may not have looked exactly as Bob has restored it, I think his sketch is probably quite close." However, in 1988, paleoartist Gregory Paul critiqued Bakker's classic study claiming, its head is too short in length and too tall proportionally and its belly too shallow.

In 1971, Bakker referred to dinosaur physiology as being 'homeothermic' in nature, although three years later he clarified that "I specifically stated that dinosaurs achieved homeothermy through bird-like endothermy and high yearly energy budgets." Due to heat retention, a large cold-blooded (i.e. 'ectothermic') animal—conceivably like sauropods—could be homeothermic, functioning with the high metabolism of an endotherm. Humans are 'endothermic-homeotherms.' And there are mammals such as bats, bears and hummingbirds that are endothermic although they can adjust their internal core body temperatures such as during hibernation when their core temperatures is lowered. Therefore, a bear is an example of an 'endothermic-poikilotherm,' which means warm-blooded but with the added ability to fluctuate core body temperature. Particularly large dinosaurs capable of retaining body heat due to their low surface area/volume body ratios, may have been cold-blooded 'ectothermic-homeotherms,' otherwise known as 'gigantothermy.' These maintained constant core body temperatures, even though they were cold-blooded, due to their great size.

Some dinosaur species may have had metabolisms that were intermediate in nature, being 'mammalian' endothermic-homeotherms during their juvenile years, while later resorting to a gigantothermy condition when they grew to large size in adulthood.) However, small *cold-blooded* dinosaurs would not ordinarily be expected to maintain homeothermic/endothermic metabolic levels. Bakker, who associated sprawling reptilian gait with ectothermy, believed dinosaurs possessed fully erect 'mammalian' limb posture. Therefore, dinosaurs, and ancestral thecodonts were warm-blooded endotherms.

Bakker's influence escalated rapidly during 1975, when he published a landmark discussion of the new view of dinosaurs in the April 1975 issue of *Scientific American*, titled 'Dinosaur Renaissance.' In this issue, now a collector's item among hunters of dinosaur memorabilia, Bakker presented lines of evidence supporting the warm-blooded dinosaur theory. The article featured a discussion of the predator-prey ratios in dinosaur communities which Bakker believed closely resembled those in modern mammalian populations.

And Bakker's ideas were peppered throughout Adrian J. Desmond's 1975 eye-openerr on the history of scientific dinosaurs, *The Hot-Blooded Dinosaurs: A Revolution in Palaeontology*, highlighting several of his new and conservative dinosaur restorations. These publications carried impact, "...to the dismay of the more conservative paleontologists, the general public joined the spirited debate and tended to side with Bakker's warm-blooded forces. People liked their dinosaurs lively, not languid." Throughout the 1970s, Bakker stirred paleontological *controversy*—the phenomenon that has historically contributed to popularization of the geological sciences. For shortly after Bakker and Ostrom ignited the debate over dinosaurian physiology, a new controversy emerged, this time over the relationship of birds to dinosaurs.

Although conventional wisdom dictated that birds were descended from dinosaurian ancestors, Ostrom claimed using convincing fossil evidence that birds were directly descended from dinosaurs. So, in 1974, Bakker, who maintained certain dinosaurs were warm-blooded, went one step further. He seized this opportunity to demonstrate that birds inherited their endothermy from warm-blooded ancestors, the dinosaurs. Bakker and Peter Galton then proposed a reclassification for birds and dinosaurs. They combined both into a new class named 'Dinosauria,' comprised of three subclasses, the saurischia, the ornithischia (the two already recognized major groups of dinosaurs) and 'aves' (birds). The new controversial system made sense if it could be definitively established that birds were truly descended from dinosaurs.

George Gaylord Simpson for instance, a respected authority on vertebrate evolution, referred to the new scheme as utter nonsense. Even though a modern consensus seems to be that birds descended from dinosaurs, it isn't at all generally agreed that birds' endothermy had been inherited from warm-blooded dinosaurs: "Theories may come and go, which is the dynamic of science, but the system of zoological classification should not be bent and warped in response to this ... the system should be like a constitution—subject to amendment only after the most rigorous and ponderous deliberation."

Well, Ostrom never doubted his claim that birds were descended from dinosaurs. And the modern consensus supports this position, despite the fact that Bakker and Galton's reclassification scheme proved unappealing to the scientific community at large.

By the late 1970s, however, Bakker's relations with Ostrom had become somewhat strained. Although the two had ushered in an era of enlivened discussion and scientific interpretation about dinosaurs and modern restorations, now there was conflict. The conservative Ostrom had treated the subject of warm-blooded dinosaurs somewhat cautiously, "... tempering the bold hypothesis with concessions that the evidence is less than conclusive." Meanwhile, Bakker insisted that dinosaurs were warm-blooded. "Bakker needed to establish dinosaur endothermy in order to reinforce his case explaining the ascendance of dinosaurs over mammals in the Mesozoic, suppressing their evolution for as long as they had to compete with dinosaurs." However, many paleontologists disagreed with Bakker, and even "... recoiled from his emotional rhetoric as the debate inflamed science through the 1970s."

Ostrom, once associated with the intensely driven, self-assured Bakker, yet now concerned for his profession, stepped away from the limelight. Ostrom stated in 1983: "I was never able to convince him that understatement was more powerful than overstatement.... My position really was, we may be right but we haven't proved it."

At the 1978 annual meeting of the American Association for the Advancement of Science in Washington, Bakker defended the warm-blooded theory in a somewhat uncompromising manner. Others present at the conference suggested that fairly constant body temperatures may be a trait shared by larger vertebrates regardless of ancestry. Instead of possessing mammalian physiology as Bakker suggested, the larger dinosaurs may have maintained nearly constant body temperatures because of their great size and life-style. Other issues were debated as well. However, Bakker remained unrelenting in his position that all dinosaurs (including sauropods) were warm-blooded.

Afterward, the consensus seemed to be that at least some types of dinosaurs like *Deinonychus* may have been endothermic. However, during another symposium this time held in February 1986 in Los Angeles, the atmosphere had changed significantly. "When I went to school it was indeed a heresy to think of dinosaurs as warm-blooded and light footed... And in 1979, the last time there was a major dinosaur symposium a lot of people still viewed these (ideas) as heresy. But at the L.A. meeting one could definitely see movement toward quick-footedness."

In the lead article, titled "The Return of the Dancing Dinosaurs," for Volume 1, of *Dinosaurs Past and Present*, documenting a symposium and traveling exhibition of dinosaur art which opened at the LA County Museum on February 15, 1986, Bakker claimed dinosaur's fossilized footprints as well as ligament and muscular marks left on their limb bones are signatures of endothermy. Bakker stated:

> "...dinosaurs cruised at warm-blooded speeds.... Most giant dinosaurs were stronger, faster and more maneuverable than the rhinos and elephants of today. The Dinosauria ruled by virtue of their fundamentally superior adaptive plan." Also, "...I am under no delusions .... that there is now a universal consensus about dinosaurs being entirely endothermic and endowed with as high a level of exercise metabolism as are rhinos. But at least the idea of hot-running dinosaurs is once again a very important and defensible line of thought...."

Many of the 1986 attendees found Bakker's evidence persuasive and his ideas convincing.

Bakker perceived the evolutionary pursuit of the warm-blooded condition during the Triassic and Upper Permian as a sometimes bloody and brutal struggle for predatory dominance, fought repetitively over many eons long before the first dinosaurs ever walked the earth. Then, "Once high metabolic adaptations had been introduced onto the ecological stage, no group of large land vertebrates could hope to achieve dominance without such physiological equipment." According to Bakker, their world was of the veritable 'nature red in tooth and claw' type envisioned by some 19th century naturalists and popularly regarded, until the cold-blooded reptilian orthodoxy set in during the 1920s.

By the 1990s, Bakker had built a strong case for vigorous, warm-blooded dinosaurs, viewed as creatures propelled through Mesozoic terrain on energy-efficient digestive 'after-burners' (e.g. gizzards and associated biochemical makeup) using several lines of evidence. Considerable substantiating evidence was summoned to defend the warm-blooded hypothesis, including studies of predator-prey biomass ratios in dinosaur communities favorably compared to other mammalian dominated ecosystems of both past and present, the presence of features such as growth rings and haversian canals in dinosaur bone initially thought to resemble that in other warm-blooded animals, and isotopic ratios of oxygen measured in fossil bones. Feathery insulation observed in some fossil dinosaur specimens (e.g. *Sinosauropteryx*) denoted a warm-blooded condition.

Furthermore, rapid evolutionary rates of dinosaur species, and even rapid growth rates of individuals within species have been interpreted as more characteristic of a warm-blooded rather than a cold-blooded condition, although evidently in certain dinosaurian cases, physiologically intermediate in nature.

Two excellent references neatly and authoritatively summarizing both the historical views and current status and extensive literature of the warm vs. cold-blooded dinosaur debate are Keith M. Parsons', *Drawing Out Leviathan: Dinosaurs and the Science Wars*, Bloomington: Indiana University Press, 2001, Chapter 2 'The Heresies of Dr. Bakker,' pp. 22-47, and David E. Fastovsky's and David B. Weishampel's *The Evolution and Extinction of Dinosaurs*, Cambridge: Cambridge University Press, 1996, - Chapter 14 'Dinosaur Endothermy: Some Like it Hot,' pp. 325-358. Many of Bakker's lines of evidence were challenged on firm grounds. Analysis of 'lines of arrested growth,' (LAGS) in dinosaur bone indicates that growth rates in dinosaurs such as *Troodon*, *Syntarsus* and *Massospondylus* represented cessation of growth caused by external seasonal temperatures. LAGs indicate a possible annual temporary reduction in growth rate, and are commonly observed in bones of extant ectotherms. Therefore, contrary to growth studies conducted on *Maiasaura* (which indicated a more rapid, endothermic growth rate, lacking LAGs), some dinosaur physiologies may have been 'intermediate between that of mammals and birds, and ectothermic creatures (such as crocodilians). As Parsons noted in referring to the most recent analyses of dinosaur bone histology conducted by Armand J. de Ricqles, dinosaurs fall somewhere within the extremes of endothermy and ectothermy. "Yes, dinosaur thermoregulatory physiology seems to have departed from the 'good reptile' model." But also, *"No, the evidence does not support the claim that dinosaurs were full endotherms in the mammalian or avian sense."* (Parsons, op. cit., p.41)

Of Bakker's interpretations on predator-prey biomass ratios, Fastovsky and Weishampel stated: "...is a brilliant study in creative thinking; however, so much uncertainty is introduced through his method that his results can hardly be conclusive." (p. 349) So the evidence remained equivocal. For instance, consideration of nasal respiratory turbinates, a mucous membrane effective in retaining body moisture (modern mammals do have this feature but dinosaurs evidently didn't), indicated dinosaurs *weren't* warm-blooded. However, an iron "concretion" interpreted as a fossilized four-chambered heart noted in a *Thescelosaurus* specimen *is* indicative of warm-bloodedness. (For more on this feature in *Thescelosaurus*, see Donald F. Glut, *Dinosaurs: The Encyclopedia - Supplement 2*, Jefferson, NC: McFarland & Company, Inc., 2002, pp. 549-552.) Parsons concluded, "So Bakker may have won a sort of backhanded victory. Though his endothermy hypothesis did not prevail, the stereotype of the dinosaurs as swamp-bound sluggard has not survived." (p. 47) Because both crocodilians and birds, the living creatures (i.e. "archosaurs") dinosaurs are thought to be most closely related to, have four-chambered hearts, dinosaurs and the common ancestral archosaurs may also have had four-chambered hearts.

Although usually downplayed by vertebrate paleontologists, because there is too much room for speculation, Bakker also advocated claims made about dinosaurian life styles based on presumed soft body anatomy or features not prone to fossilization. Thus, analyses of evidence such as that relating to physiology, sexual display tactics, and the highly supposed similarity of dinosaurs' breathing mechanisms to the avian lung sac system, has enhanced our visions of dinosaurs as lively, sporting creatures. Based on their lower encephalization quotients (brain to body mass ratio), Bakker conceded that sauropods, ankylosaurs and stegosaurs must have been rather dull-witted.

In the case of *Scipionyx* and *Sinosauropteryx*, soft tissues (esp. the lungs, liver and colon) visible under ultraviolet light preserved within the body cavity were positioned within the abdomen similarly to where they lie in modern crocodiles. This interpretation of the evidence not only challenged the warm-blooded dinosaur theory as well as the theropod dinosaur ancestry of modern birds, but also even the idea that birds are archosaurs. (See Glut, *Dinosaurs: The Encyclopedia, Sup. 2*, (2002), pp. 101-102.) However, in his *Supplement 2*, Glut referred to those who still disregard the evidence linking birds to theropod dinosaurs as the 'vocal minority.'

At the 1978 meeting, Bakker had suggested that the warm-blooded condition of dinosaurs would have contributed to their extinction. During Late Cretaceous cold spells, 'naked' dinosaurs adapted to tropical conditions would have suffered in their exposure to freezing temperatures. Many of them would have been too large to hibernate in caves or burrows as do many of small mammals and so would not have survived! "If dinosaurs were so hot, how come they're all dead? This question answers itself. Dinosaurs went extinct because they were hot...." However, dissenters voiced their opinion that at least some large, naked (i.e. unprotected by feathers or insulating 'fur') dinosaurs, or even smaller feathered varieties should have survived especially if they were warm-blooded, unless the thermal physiology of dinosaurs was indeed quite different than that of living endotherms, (e.g. placental mammals).

One major Late Cretaceous 'cold spell' happened at the Cretaceous-Tertiary (Paleogene) boundary, 66-million years ago. Bakker's view on the cause of Cretaceous-Tertiary boundary extinctions is generally conservative, with causes rooted in terrestrial rather than extraterrestrial processes. According to Bakker, in 1977, the reason for mass extinctions can be related to tectonic cycles that directly impact the variety and extent of niches available to species. Diverse environments are most likely to exist during periods of mountain building, orogeny and sea transgression. Such factors would contribute to rapid rates of speciation. And if speciation rates decline, as might happen during times of tectonic cessation or marine regression, then not enough new species will be generated to offset background extinction rates. Then a period of marine regression, such as during the Late Cretaceous, could be viewed later as one of mass extinction in the fossil record.

Instead of conjuring extraterrestrial causes, Bakker advanced a 'down-to-earth' theory encompassing both extinction of dinosaurs on land as well as their cousins, the great sea-faring reptiles. "The best answer for the extinction of the great sea animals is that their favorite haunts disappeared when the warm shallow seas drained off the continents. And the best answer for the extinction of the open water, deep sea creatures is that surface water becomes colder and more thoroughly mixed with deep water when the shallow seas drain off." Furthermore, as Bakker suggested, rejuvenating an older idea once proposed by paleontologist Henry Fairfield Osborn back in 1925, when sea levels subsided, dinosaurs may have been free to migrate across cooler continental regions formerly isolated by water. Spread of microbials among the intermingling animals competing for new niches in new terrains may have then caused proliferation of disease, further driving extinctions.

Which animals would have been most susceptible? Warm-blooded animals energetically capable of sustaining long migratory journeys! Furthermore, *"... tissue kept warm by high metabolism would be ideal from a parasite's point of view."*

Bakker claims only to champion "... heresies if they fit the facts better than orthodoxy." However, he has been accused of being too abrupt in his denouncement of extraterrestrial scenarios. Bakker criticized those advocating the asteroid impact scenario for mass extinctions, as quoted in David M. Raup's *The Nemesis Affair*:

> "The arrogance of those people is simply unbelievable. They know next to nothing about how real animals evolve, live and become extinct. But despite their ignorance, the geochemists feel that all you have to do is crank up some fancy machine and you've revolutionized science. The real reasons for the dinosaur extinctions have to do with temperature and sea level changes, the spread of diseases by migration and other complex events. But the catastrophe-people don't seem to think such things matter. In effect, they're saying this: 'We high-tech people have all the answers, and you paleontologists are just primitive rock hounds.'"

Bakker asserted that animals with high energy demands such as dinosaurs would have been more vulnerable to extinction than more lethargic, cold-blooded reptiles. Therefore, during stressful periods such as during famines or drought or any other climatic perturbation, warm-blooded types would be most sensitive and critically effected. Yet in 1986, Bakker conceded the possibility that an impact could have devastated the few remaining dinosaurs. Many paleontologists agree dinosaur diversity had already been in steady decline for millions of years prior to the end of the Cretaceous, representing the typical end stage of the waxing and waning of faunal dynasties. So maybe the impact wiped out the stragglers. Such phenomena occur repeatedly throughout the fossil record. "But the overwhelming share of the credit (or blame) for the grand rhythm of extinction and reflowering of species on land must surely go to the earth's own pulse and its natural biogeographical consequences."

Bakker also considered evolutionary rates exhibited by various early Cenozoic and Jurassic vertebrate fauna. Armed with this data, Bakker then challenged George Gaylord Simpson's contention that all species have evolved throughout geological time at rates which have remained in step with the pace of environmental change. "... most species don't evolve as fast as they should. And if this reading is the correct one, then all the reassuring notions about Nature fine tuned and elegant are bogus."

Bakker claimed that evolution is actually a very rare event proceeding rapidly once set in motion. However, species are apt to remain static showing little sign of gradual change for very long periods until the rapid (punctuated) speciation event. His evolutionary views generally coincide with those of another 'revolutionary' paleontologist, the late Stephen J. Gould, who along with Niles Eldredge introduced in 1972 the once controversial though now generally accepted framework known as 'punctuated equilibria.'

Bakker claimed "Fossil vertebrates are giving us disturbing messages: adaptive improvement is controlled by two rare events—budding off of new species, and replacement of parent by daughter species. The combined result is an evolutionary rate so slow that no species may ever reach perfection.... And where should evolutionary science go in the next decade?... back to the comprehensive study of real species through geologic time." In fact, other scientists have agreed that the ecosphere is in a continual state of disequilibrium caused by mass extinction events and major evolutionary innovation.

Robert Bakker will more than likely be remembered not only as a controversial though brilliant paleontologist and restorer of fossil vertebrates but also as the individual who instigated debate over interpretation of dinosaurs. "Those Bakkerian extrapolations! How many times have we dismissed his ideas and then had to reconsider them in light on new evidence?," stated Dale Russell in a March 1987 *Discover* interview while pondering Jack Horner's evidence for rapid growth rates in juveniles belonging to a genus of 'duck-billed' dinosaurs known as *Maiasaura*, a finding initially regarded as congruent with the warm-blooded dinosaur theory. Through Bakker's efforts we experienced perhaps the most enlivened period for dinosaur study since the Marsh-Cope era over a century ago. But whereas Othniel Marsh and Edward Cope sparred over collection of bones and priority of description, Bakker and his colleagues sparred over ideological matters.

Bakker's *The Dinosaur Heresies* (released in November 1986), featuring Bakker's superb illustrations throughout, prompted paleontologist Steven Stanley to state "...this is unquestionably the finest book ever written about dinosaurs." It has become a 'must-read' for all dinosaur aficionados. Bakker affirmed heretical ideas, introduced to laymen in the mid-1970s in popular fare such as Desmond's, *The Hot-Blooded Dinosaurs*, with new evidence. For instance, Bakker exposed the fallacious nature of swamp-dwelling giant brontosaurs, swimming duck-billed dinosaurs, and slow-footed, bow-legged horned dinosaurs.

While several concepts were enticing, others seemed almost too heterodox. Other ideas weren't even fresh, just ones that Bakker wanted to revitalize. Yes, controversy *was* stimulated by '*Heresies*,' but that's all in tune with Bakker's nature. "I'd be disappointed if this book didn't make some people angry... I have enormous respect for dinosaur paleontologists.... But on average, for the last 50 years, the field hasn't tested dinosaur orthodoxy severely enough." Regardless, many readers were left with a definite impression that all ideas presented in *The Dinosaur Heresies* are well founded on carefully evaluated evidence and sound reasoning, a tribute to Bakker's expertise in the field. Bakker diffused old, time-honored 'myths' explaining, for example, why dinosaurs couldn't have been swift runners, or why they couldn't have eaten enough to sustain themselves.

Bakker included his restoration of a pair of fighting, feathered *Deinonychus*. You may recall that Bakker's highly acclaimed 1969 *Deinonychus* restoration was feather-less. But science moves on, and Bakker now apparently believes that "... we would be justified in classifying them as either bird-like dinosaurs or dinosaur-like birds based on anatomical similarities." Bakker's new illustration calls to mind Charles R. Knight's famous 1897 restoration—two lively 'dryptosaurs,' one pouncing on the other, done under Cope's direction. Cope's idea was later viewed as improbable due to the dinosaurs' supposedly cold-blooded physiology and the fact that dinosaurs were shown to have possessed upright, rather than kangaroo-like stance.

Formerly it was believed that gentle 'armored tanks' of the Cretaceous, the nodosaurs, relied on their armor defensively, for passive protection. But Bakker suggested they were also geared for offense. Since these tanks would have been relatively easy to flip over when attacked by large carnivores, their short, strong legs could have pivoted relentlessly to ward off attackers, the nodosaurs awaiting opportunity to drive powerful elbows and knees forward. Bakker compares the devastating effect of its long murderous shoulder spikes driving into carnosaur flesh with the sharp horn jab from a charging *Triceratops*.

Old views also resurfaced in '*Heresies*,' such as a long-necked *Diplodocus* equipped with a tapir-like trunk adapted for grasping tree foliage. Dinosaur lips may not have been fleshy enough to easily evolve a long proboscis. So, maybe the nasal organs of Diplodocines were used instead as part of a resonating device used in "bellowing out Jurassic songs." But why simply pay 'lip-service'? Bakker encouraged young, aspiring vertebrate paleontologists to take up the study of dinosaur lips, an area of study with a promising future. Then Bakker used witty sarcasm, a tone prevalent throughout the book, to dispel notions that *Diplodocus* was an aquatic clam-eater. After presenting anatomical evidence that *Diplodocus'* head must have been held aloft about 20 to 30 feet vertically while eating, Bakker said, *"Not many clams are found at such heights."*

Bakker also postulated that sauropods gave birth to live offspring, while conversely, cold-blooded dinosaur supporters believed that (reptilian) sauropods hatched from eggs. By early 1987, however, an amateur fossil collector's discovery of eggshell fragments found south of Grand Junction, Colorado carried potential for falsifying Bakker's live-birth theory.

Another dated concept emerged in Bakker's description of the stegosaurs, which must correctly be viewed as animated suspension bridges. The large vertebral spines at the hips may have faciliated stegosaurs in rising to a tripodal stance for upright feeding, a concept first entertained by O. C. Marsh over a century ago, and later by a biophysicist named D'Arcy Thompson in 1924. "Just as the thick cables of the suspension bridge based on the central tower hoist the weight of the span, so did the thick ligaments of the stegosaurs' vertebral column based on the hips hoist the weight of its body."

Under attack, stegosaurian deltoid muscles pushed sideways to maneuver tail and dangerously armed hindquarters quickly into a threatening position. Their plates could be angled for deadly defensive/offensive purposes using powerful skin muscles. "Stegosaurus must have been a grand performer under attack—a five ton ballet dancer with an armor-plated tutu of flipping boney triangles and a swinging war club."

Conventional wisdom suggests that evolution of plant life, particularly development of flowering plants in the lower Cretaceous, had profound impact on dinosaurs. According to one unpopular, difficult to falsify theory, poisonous plants may have exterminated dinosaurs in the Late Cretaceous via terminal cases of constipation. Yet, as Bakker noted, herbivorous dinosaurs evolved at faster rates than plants they ate, and so at least to a certain extent, dinosaurs conversely must have exerted great selective pressures on Mesozoic vegetation. Bakker even suggested that evolution of angiosperms (flowering plants) may be attributed to shifting dietary habits of herbivorous dinosaurs, (i.e. from a 'high browsing' pattern in the Late Jurassic to 'low ground' feeding) in the Cretaceous. The fossil evidence is expertly presented in a delightful *Heresies* chapter titled, "When Dinosaurs Invented Flowers."

Sometimes the means of scientific discourse (leaden, technical terms interspersed through technical journals and trade publications) must seem as lumbering and stodgy as those cold-blooded dinosaurs themselves. Perhaps unsatisfied with conventional rites of scientific expression, throughout his career as a vertebrate paleontologist, in his popular writings Bakker offered alternative, tantalizing glimpses into his visions of the dinosaurian world. Seemingly outlandish scientific theories were fortified with artistic portrayals that to the uninitiated, appeared alien in aspect. We were exposed to lean and mean-looking dinosaurs that jumped, bounded, danced, and sprinted lively over ancient landscapes.

However, it isn't always readily possible to project scientific ideas and theories on canvas. Sometimes it is tempting or even inherently necessary to convey these messages in another medium, such as creative writing, which is what Bakker managed to do in his 1995 novel *Raptor Red*. Through his *Raptor Red* excursion, Bakker reintroduced the field of paleontological fantasy writing, something which ornithologist Gerhard Heilmann initiated 70 years ago in his book, *The Origin of Birds,* (1926). Heilmann provided short, in- depth "pen picture" descriptions of how early birds would have lived in the Mesozoic (see his chapter entitled, "Some Fossil Birds"). "I'm a method paleontologist," Bakker once said. "I want to be Jurassic. I want to smell what the megalosaur smells. I want to see what he sees." Bakker is an expert 'environmental vertebrate paleontologist,' taking particular care to understand paleoenvironments dinosaurs lived in. And his knowledge of the Cretaceous ecological system deeply enhances *Raptor Red.*

In spinning his tale of Early Cretaceous North American dinosaurian fauna, Bakker scored another strike for originality! This is not Disney, or "*The Land Before Time.*" Nor is it truly science fiction, as twice told masterfully by Michael Crichton in his "*Jurassic Park*" settings. Instead, readers may delight in an elaborate illustration of how dinosaurs lived their lives, conveying far more than would ever be possible in any painting or sculpture (which would merely focus on an *instant* in time). Bakker has woven many ideas concerning evolution and behavior of dinosaurs, albeit extrapolated from what is known of living species, into his story.

Much of what is said seems downright plausible, although at times the main 'characters,' a female *Utahraptor* identified as 'Raptor Red,' and several close kin seem a bit too human in nature. This apparent problem is a minor difficulty, however. Bakker, who is new to the field of fantasy writing, has only encountered one of the major paradoxes science fiction writers routinely encounter. This is *how* to convincingly write about an 'alien' race without revealing the fact that the writer is, after all, a human who is incapable of *thinking* like an alien (or in this case, a dinosaur)!

So, when Raptor Red expresses love-like emotions, the Reader must wonder, "can a dinosaur really *love* anything?" *Utahraptors* were probably among the most intelligent dinosaur species, yet who is to say they could love their offspring, or not? Well, to dinosaur aficionados, if it isn't strange reflecting on the fact that humans love dinosaurs, the idea of intelligent dinosaurs somehow "loving" their kin may not seem so bizarre. Actually, the characters were extrapolated from Bakker's ideas of how hawks would perceive their world had they been clothed in dinosaur flesh instead.

In providing a series of vignettes concerning how Raptor Red interacts with her environment through the course of a single year, Bakker leads us through the realm of Early Cretaceous North America. He effectively couples his original thoughts on dinosaur behavior to established ideas about genetic transmission and sensory adaptation in species. Realistic, believable dinosaur encounters are the end result.

Raptor Red relies on her wits and senses to hunt prey, and compete socially against other females in her quest for a mate. Prior to mating, the male must present himself in a way that proves instinctively appealing to Raptor Red, but also, as Bakker explains, so as to demonstrate lack of genetic defects, disease, or absence of parasites. The raptorial mating rituals may seem strikingly avian, yet today few would condemn the analogy's appropriateness. However, analogies are also made to African predators whenever the *Utahraptor* pack skillfully brings down prey in an orchestrated, 'feline' fashion.

Most interesting were Raptor Red's numerous encounters with non-dinosaurs, an assorted cast of turtles, crocodiles, fuzzy mammals, ammonoids, plesiosaurs, and pterosaurs, all known on the basis of fossil evidence to have coexisted with the star characters. Bakker's insights into just what a *Utahraptor* such as Raptor Red might have thought about other species in its environment are amusingly and enjoyably portrayed, and no stranger than similar speculations often made on televised wildlife shows.

Bakker illustrates just how harrowing and difficult it was for *Utahraptor* to survive in its hostile world. Floods, diseases and injuries sustained while hunting (or in curious play), plague the species. Being powerful, quick and blood-thirsty doesn't guarantee success in any ecosphere. Mere survival is hard won and risky. Yet there are occasions do exist for raptors to playfully explore and even savor their surroundings too.

If we detect a bit of Bakker himself woven into the story, it may be in the guise of a wise old pterodactyl character, who carries on a sort of symbiotic, peaceful existence with Raptor Red, soaring in the skies spying on the dinosaurs below.

Rather than killing off his tightly woven characters in an asteroid collision, volcanic eruption cliché', or battle to the death, Bakker provides a 'happy' ending. Both the 'dactyl' and Raptor Red have found new mates to continue their (genetic) struggle for existence.

Few writers have ever written about the natural history of 'whackity-whacks,' and yet Bakker has an entire, whimsically written chapter dedicated to the strange genus *Gastonia*, (i.e. a genus of nodosaurs which defensively "whackity-whacked" their bony armored tails). Raptor Red's curiosity over the queer little beast, finally deciding it isn't any good to eat, but fun to play with, is a delight.

In a scene staged on the shore of the epi-continental sea dividing western North America in the Early Cretaceous, Raptor Red encounters a *Kronosaurus*, which hurls itself through the surf in search of unsuspecting prey. Bakker has given us new insights as to how these whale-sized monsters may have hunted. The thought of a kronosaur lunging out at you on the beach seems truly frightening, and Bakker creates a dramatic scene, eventually pitting the wits of Raptor Red versus the bulk of mighty *Acrocanthosaurus*, which falls victim to the amphibious hunter.

As in the case of many of his other popular publications, *Raptor Red* came illustrated with many of Bakker's fine drawings. Bakker's thought provoking, artistic style lets us see his dinosaurs with feeling, enhancing the narration.

Bakker is truly a scientist of extraordinary vision and talent. And yet for all his notoriety, Bakker has described not even a handful of dinosaur genera that have withstood the test of time. For example, he described *Denversaurus, Drinker, Edmarka,* and *Nanotyrannus*. Although in Glut's *Dinosaurs: The Encyclopedia* (1997) reference, *Nanotyrannus* (described by Bakker, Williams and Currie in 1988) is described as a distinct genus, as a result of Tom Carr's recent analyses ("Craniofacial ontogeny in *Tyrannosaurus* (Dinosauria: Coelurosauria): *Journal of Vertebrate Paleontology* vol. 19, no. 3, 1999, pp. 497-520, and "FMNH PR308: (or analyzing an enigmatic tyrannosaurid specimen)," *Dinosaur World*, no. 6, 1999, pp. 16-18), *Nanotyrannus* was generally regarded as a junior synonym of *T. rex* (i.e. specimens are thought to be juvenile *T. rexes*).[1]

Bakker has been caricatured in film (i.e. as an ill-fated paleontologist devoured by *T. rex* in Steven Spielberg's *The Lost World* (1997, inspired by Michael Crichton's 1995 bestseller), and to my knowledge Bakker is the only paleontologist ever enlisted in a marketing campaign to sell dinosaur-related products (i.e. a Sega-Genesis computer game) in a mid-1990s televised commercial. After studying my own reflections on the subject of Bakker's full-on 'take' concerning warm versus cold-blooded dinosaurs, I see my attitudes closely mirroring the general attitudes; measured, yet welcomed skepticism in the early

1970s, more positive reception following publication of Desmond's classic and hesitant acceptance in the early 1980s.

Once known among his peers as an *'enfant terrible,'* Bakker'ss controversial persona seems to have mellowed. Now an old 'fossil,' he's also a paleontological treasure. Doubtless many of Bakker's bold ideas will continue to challenge and along the way, some will even be falsified. That is the way of science. But paleontology needs visionaries like Bakker, men with the audacity to dust off the cobwebs of convention, retest old ideas with new evidence and even stimulate new ways of thinking, if warranted.

In the midst of his self-acclaimed 'rebirth' of learning about dinosaurs, Bakker is a 'renaissance man.' Above all, he's shown us how to remain 'teachable.'

Note to Chapter Thirty-one: (1) In personal communication dated April 10, 2003, Steve Brusatte (then a student geologist) astutely summarized circumstances surrounding this rather controversial genus. "About *Nanotyrannus*... here's the brief story: Gilmore originally described the specimen in the 40's as *Gorgosaurus*. Russell, in 1970, sunk *Gorgosaurus* as a junior synonym of *Albertosaurus*, making the specimen *Albertosaurus*. Then Bakker, Currie and Williams came along in 1988, describing *Nanotyrannus* as its own genus. Much of this revolved around the interpretation of certain cranial sutures as closed, which indicated maturity to the authors. If the specimen was mature, that means its smaller size really was a unique feature that made it a separate genus. However, Thomas Carr addressed all of this in his seminal 1999 paper on tyrannosaurid craniofacial ontogeny. He dismissed *Nanotyrannus* as a juvenile *Tyrannosaurus*, noting that many of the sutures that Bakker, et. al., saw as closed really were either open or unable to be conclusively identified. He also listed numerous characters to link *Nanotyrannus* and *Tyrannosaurus*. To Carr, *Nanotyrannus* is not valid. However, in the summer of 2002, the Burpee Museum of Natural History found 'Jane,' their (mascot) tyrannosaurid specimen that is currently undergoing preparation in Rockford. It's from the Hell Creek of Montana and looks reasonably complete. Importantly, there seems to be good vertebral and appendicular material. Early reports (quotes in the popular press from Bakker and Pete Larson) indicated that the specimen was 'definitely an adult' because of fused sutures, but I (along with Carr and many others I've talked to) are very wary of these, because much of the material is still encased in rock. 'Jane' has the potential to overturn the entire story of *Nanotyrannus*. Only full preparation and study of the material will address the problem. According to the Burpee Museum newsletter, Bakker, Larson and Phil Currie will head up the 'Jane' research team, and they will submit a paper to Nature or Science within a few years." (Also see "Dinosaur's identity crisis may be solved," by Kathy Paur, *Chicago Tribune*, section 1, August 7, 2002, pp. 1, 18. Analysis of a young tyrannosaurid specimen collected in 1922, but not recognized for its significance until recently may also shed light on this issue. See 'Treasure unearthed at Field," by William Mullen, *Chicago Tribune*, section 4, March 17, 2002, pp. 1-2.) Tom Carr's interview in *Prehistoric Times* no. 126, (Summer 2018 pp.30-31) underscores these points—that both the Burpee Museum's "Jane" and *Nanotyrannus* are juvenile *T. rexes*.

SIX

The original Morlock: the picture in J. G. Wood's *Illustrated Natural History* [1872] that terrified Wells when he was eight. In Wood's book the picture is captioned: GORILLA.—*Troglodytes Gorilla*, a manifest association of underground (or cave) man and ape.

Image and quotation are from The Definitive Time Machine: A Critical Edition of H. G. Wells's Scientific Romance with Introduction and Notes by Harry M. Geduld, (Indiana University Press, 1987, p.28. See Chapter Thirty-four for more.

Odd Evolution

Any book discussing weird geo-monsters and paleo-kaiju would be remiss without further addressing some of our stranger, far out ideas concerning how such creatures could have conceivably evolved—at least I've always felt that way. This section is therefore dedicated to "odd evolutionary" processes and outcomes, as I've pondered them in past decades. Chapter Thirty-two originally appeared in fairly different form in *The Earth Science News* (vol. 41, Parts 1-2, March to April, 1990) although it was then titled, "Space Bugs and the Fossil Record." I deleted most passages on the paleontology of cockroaches when I updated it for *Fossil News* twenty years later. Chapters Thirty-three and Thirty-four are an homage to pioneer science fiction writers—chiefly Herbert George Wells. These writers' works have infiltrated our minds and so mightily shaped science fictional perspectives, while nodding to inescapable paleontological wonders since dawning of the genre.

And in a loose connection to the sci-fi writings of Wells, a February 26, 2019 program broadcast on the SCI network, "How the Universe Works: Hunt for Alien Life," discussed the panspermia idea. In this broadcast, it was suggested that terrestrial life might have originated on Mars—ultimately arriving at Earth aboard meteorites jettisoned from impacts on the Red Planet! Increasingly today, others (e.g. astronomers) consider that life in some 'primitive' form could have reached Earth via comet collisions.

Makes you think.

These chapters originally appeared in the following publications:
Chapter Thirty-two—*Fossil News: Journal of Avocational Paleontology,* Vol. 16, no.9 Nov/Dec 2010, pp.16-19; Vol. 17, no.1 Jan/Feb 2011, pp.7-11.
Chapter Thirty-three—*Mad Scientist* no.21, Spring 2010, pp.26-33.
Chapter Thirty-four—*Prehistoric Times* no.128, Winter 2019, pp.16-17, 19.

Chapter Thirty-two — *Far Out Evolution—OMG!*

No matter how blasphemous Robert Bakker's dinosaur 'heresies' of the 1970s and 1980s may once have seemed, the contemporary space happy ideas of two astro-scientists germinating the fields of biological evolution and paleontology would seem beyond compare. The names Sir Fred Hoyle (1915-2001) and Chandra Wickramasinghe (b. 1939) may be unfamiliar to most. However, some may vaguely recall, for example, their mid-1980s thesis that fossils of the Late Jurassic avian *Archaeopteryx* are forgeries, a premise since soundly refuted by experts faced with headaches of dissecting their challenge. But that's not all. Astronomer Hoyle and astrophysicist Wickramasinghe also 'explained' causes of the Cretaceous-Paleogene ('K-Pg') boundary extinctions, commented on Life's origins, offered uniquely alternate explanations for Life's history as recorded in the fossil record and outright repudiated Darwinian evolution in favor of immense "genetic storms" and "evolution by viral addition." Yes, we've been under attack by, well …alien life forms for billions of years! It sounds all very 'sci-fi,' and consequently I'm left feeling—so 'OMG!' But they aren't kidding. Do their peculiar ideas have any merit?

I've read three of their collaborations, *Evolution From Space* (1984), *Archaeopteryx, the Primordial Bird: A Case of Fossil Forgery* (1986), and *Our Place in the Cosmos: The Unfinished Revolution* (1993), sufficient to gain a most guarded appreciation for their outlandish ideas.

Hoyle's and Wickramasinghe's 'heretical' views on *Archaeopteryx* are entirely consistent (if not bent to conform) with their other-worldly interpretations of the fossil record and how terrestrial life may have been impacted by cosmic life debris. After elaborating on clues evident on the slab and adjoining counter slab of the famous 1861 fossil now housed at the British Museum (Natural History) ("BMNH"), they concluded that the old bird is a mere forgery, fabricated to fool Charles Darwin whose *The Origin of Species* (1st edition) had just recently been published (1859). But why would anyone stoop to such an elaborate hoax, and who? Well, you see it's analogous to what happened in the case of the infamous Piltdown Man fossil 'discovered' (or rather planted) within beds near the home of Sir Arthur Conan Doyle half a century later. In *Archaeopteryx's* case, anti-evolutionists of the time simply needed a 'Piltdown bird,' so to speak, in order to embarrass the Darwinians. One worth a fortune to lucky perpetrators who sold specimens!

Anecdotal information suggests that Darwin's colleague, Thomas Henry Huxley (who was not quite as totally wedded to 'Darwinian' evolution as one might expect), predicted that a reptilian-bird creature must have once existed as an evolutionary 'link.' Furthermore, according to both David Attenborough and Herbert Wendt, Huxley went so far as to project its probable appearance (even in a sketch, according to Wendt; *Before the Deluge*, 1968, pp. 228-235). Then, majestically, if not 'mysteriously,' voila! - just such a fossil of a reptilian bird becomes 'suddenly' discovered at a Solnhofen slate quarry. This timing is all highly suspicious, isn't it—claim Hoyle and Wickramasinghe? Then the story goes like this. The fossil was originally an (unfeathered) dinosaur, *Compsognathus*, to which Bavarian workmen added impressions of feathers. These were carefully impressed in hardening cement surrounding the actual preserved fossil bone. Because the fossil 'bird' fell into the hands of Sir Richard Owen (whom Huxley famously didn't see eye-to-eye with), Huxley was apprehensive of what had fortuitously fallen into their hands. Was it all too good to be true? Or as Hoyle and Wickramasinghe suggest, "In an age when fossil forgery was not a particularly uncommon phenomenon, the association of *Archaeopteryx* with Owen must have been sufficient to arouse suspicions of fraud." (Hoyle and Wickramasinghe, 1986, p.117)

True—even Charles Darwin didn't have too much to say about *Archaeopteryx* in later editions of his *Origin*. And isn't this rather strange considering one might expect the fossil to be championed by evolutionists? This general "lack of enthusiasm" for *the* magnificent fossil that should have been championed by evolutionists of the time, further 'proves' that leaders of the Darwinian movement sensed that both the BMNH and even the 1877 Berlin *Archaeopteryx* specimens were elaborate forgeries. So, given the uncertainty, best to remain tactfully cautious. However, my re-reading of relevant passages in Darwin's 6th (final) edition of his *Origin* (1872) tells that he felt anything but doubtful concerning the significance of *Archaeopteryx*.

One should really read and reason the odd logic within Hoyle's and Wickramasinghe's writings. They're eye-openers to be sure, although, to us, of a distorted sort. Near the conclusion of *Archaeopetryx, the Primordial Bird*, cosmic stuff re-enters the picture. Darwin insisted that nature does not make evolutionary leaps, but instead is painfully gradualistic, a tenet with which Huxley disagreed. By crediting other 19th century naturalists (e.g. such as Alfred Russell Wallace, Robert Chambers, Patrick Matthew, Edward Blyth, and of course Huxley) with most significant aspects of 'Darwinian' evolution, Hoyle and Wickramasinghe strip Darwin of any true, original discoveries whatsoever.

Yet gradualism isn't supported by the fossil record (or so they claim), especially so because little *Archaeopteryx* isn't even a true fossil 'link.' Instead, they insist, evolution has been driven by arrival of organismal debris from outer space filtering down upon Earth.

In their view, many non-evolving invertebrate forms (especially so-called 'living fossils') are rather immune to such celestial molecules, while much of mammalian evolution, as it unfolds, remains susceptible to a resultant genetic reprogramming. So then, genetic storms favor grafting of genetic segments into the genetic (DNA) makeup of terrestrial organisms, resulting in rapid 'speciation' events. (But, no, this is decidedly not tantamount to Stephen J. Gould's and Niles Eldredge's "punctuated equilibria.")

So the two cosmically-minded authors—Hoyle and Wickramasinghe—launched their mission during the late 1970s, insisting that life did not originate on Earth, but in outer space, and that bio-stuff intersecting Earth's orbit steadily trickles down through our atmosphere, and that this 'fertilization' process is the chief cause for alterations in the fossil record. They attempted to bolster their position in books like *Diseases From Space* (1981), as well as the aforementioned *Evolution From Space* and *Our Place in the Cosmos*. Here, they chide readers for our dim-witted, Earth-centric views on the origin of Life—a conundrum which Darwin never resolved concretely.

In their favor, astronomers *have* detected several dozens of relatively simple molecules in interstellar clouds, many of which are organic compounds. Several of these chemicals are believed to be present in comets. And it is commonly accepted that many of the biochemical precursors essential to life may have been transported to Earth aboard comets or via the vaporous trails left in the wake of a comet's trajectory as it rounded the Sun. In 1997, researchers determined that under the right conditions organic molecules could actually be synthesized during meteorite impacts. (*Science*, 4/18/97, vol. 276, pp.390-392) However, Hoyle and Wickramasinghe claimed years ago that not only are these simpler compounds continually arriving today on Earth, but that actual *living matter* adhering to interstellar grains is *also* raining down upon us from the heavens! Few evolutionary scientists have taken their views very seriously.

Let's consider their unusual idea in light of its relevancy to Darwinian evolution, as outlined in *Evolution From Space*. Imagine a steady rain of cosmic genetic material filtering from space into planetary atmospheres in the form of viruses and much larger bacterial cells. As many as $1.0 \text{ E } 18$ (e.g. or a billion-billion) cosmic bacteria and $1.0 \text{ E } 21$ (or a billion-trillion) viruses and viroids (which are minuscule, primitive husks, gene-containing sacks capable of specifying their reproduction in a host of organisms) arrive on Earth annually. People ingest as many as 1,000 of the former and 100,000 of the latter categories annually. The presence of microbes detected in the upper atmosphere is a result thought by some to be consistent with their theory. With so many space-faring germs floating toward Earth, is it little wonder that we haven't been able to lick the common cold! Has science gone mad?

Surprisingly, their theory, conceptually known as 'panspermia' is an old one. The 19th century physicist, Lord Kelvin, speculated on such possibilities in 1865. And in 1908, a Nobel Laureate chemist, Svante Arrhenius, professed that life on Earth may have originated as spores elsewhere in the galaxy. Propelled through interplanetary space by stellar radiation pressure, these organisms would have eventually anchored on primordial Earth, evolving into modern forms. Existence of extraterrestrial microbial invaders was also postulated by Francis Crick (yes, of DNA-discovery fame), and Leslie Orgel in the 1970s. They suggested that Earth and other initially sterile planets may have been deliberately 'seeded' by vastly superior intelligent beings from another world.

One obvious difficulty with such hypotheses is that the problem of Life's origin was never *ultimately* explained to anybody's satisfaction. This problem—one that Charles Darwin never resolved and which Hoyle and Wickramasinghe claim also *wasn't* answered by famous experimentation conducted by Stanley Miller and Harold Urey during 1952 to 1953, in which the alleged 'building blocks of life' were chemically synthesized from abiotic materials, is merely shifted to another time and place (or possibly even, *galaxy*). In contrast to Earth-bound Darwinian ideology, Hoyle and Wickramasinghe presumed that terrestrial life evolved from multiple beginnings, or from a number of separate and unrelated evolutionary descent lines of extraterrestrial origin. Whether or not all life originated somewhere from a *single organism* can never be ascertained.

Hoyle and Wickramasinghe attempted to demonstrate that evolution of the great diversity and complexity of life would be statistically improbable given the (relatively) short 4.5-billion year interval for it to develop here. They also discussed traits evident in a variety of microorganisms such as unexpected resistance to intense levels of stellar radiation and to freezing temperatures. The postulated adaptation of bacteria and spores to such conditions as would be encountered in interstellar space is fortified, they claimed, by these remarkably resilient characteristics.

In short, Hoyle and Wickramasinghe were hellbent to overthrow Darwinism: terrestrial life, the only kind we know of, didn't originate on Earth! And its broad pattern of evolution is controlled by genetic storms when Earth's motion through the galaxy encounters large molecular dust clouds at intervals of every 100-million years of so. These molecular clouds jostle comets from our Oort Cloud of comets at the periphery of the Solar System, casting them Sunward, where they spew germy dust and debris into Earth's orbital trajectory. This untidy dust, smeared with space viruses impacts terrestrial life, causing extinctions such as the Cretaceous-Paleogene event 66-million years ago, as well as precipitating major evolutionary bursts.

Hoyle and Wickramasinghe consider the Cretaceous-Paleogene boundary as the "... product of an immense genetic storm that wiped-out some creatures and drastically changed others, with the present mammalian orders emerging as grafts onto previously-existing stock. ... Reptilian ancestors to birds from before the Cretaceous-Paleogene boundary must certainly have existed, but what they were we do not know. What they were *not* we do know. They were not *Archaeopteryx* or *Ichthyornis*." (*Archaeopteryx, the Primordial Bird,* pp.31, 131)

(As of 2010, I've been unable to identify Hoyle's thoughts, if any, concerning the world famous Chinese feathered theropod dinosaurs, the first of which was discovered in 1996. Later, did they perhaps applaud "Archaeoraptor's" fraudulent case, when exposed?)

In *Our Place in the Cosmos* (p.144), Hoyle and Wickramasinghe claimed, "... principal episodes of evolution occurred quickly, almost too quickly to be captured at all in their details, and in immense bursts at intervals of the order of a geological period, about 100-million years." Later they said, referring to cometary debris dispersed throughout the inner solar system, "For a million years or so, the time for which such a large influx of comets persists, the halo of evaporated material becomes much denser, providing an intense supply of viruses to the Earth and a resulting genetic storm, producing both extinctions and rapid evolution among terrestrial plants and animals." (p.163) As we know however, geological periods are far shorter than 100-million years and both extinction and evolutionary events do not recur with 100-million year regularity ('periodicity').

Likewise, viral invasion triggered the Late Permian's 'Great Dying.' Furthermore, on a much shorter temporal scale—3 to 4 years, the oscillatory period of certain pathogens such as influenzas may be correlated with the orbital period of comets such as Encke's Comet, a celestial object Hoyle and Wickramasinghe assert spews germy debris within Earth's orbit around the Sun.

Most outrageously, in Sir Fred Hoyle's and Chandra Wickramasinghe's view, insects are an essentially alien life form. Their eggs, surviving atmospheric entry, may have descended to Earth from outer space aboard meteoritic debris hundreds of millions of years ago. (Once again, OMG!) Curiously, they stated, insect eyes are best adapted to a range of frequencies that are largely filtered by our atmosphere. (A conventional perspective instead might be that this unique adaptation represents a genetic clue to atmospheric conditions of the early Paleozoic.) Social insects may collectively represent a strange form of cosmic intelligence, or so they claimed. Unfortunately, though highly resistant and marvelously adapted to a diverse range of natural habitats, from a human perspective, insects don't seem very wise. 'Insidious' perhaps those nasty creepy-crawling, little biting bastards may be!, yet not 'wise.' But they *do* kinda look, well, icky & 'alien,' don't they?

Arthropods have certainly had a long and fruitful existence on Planet Earth. Consider for example, the Age of Trilobites, which opened the Phanerozoic Era some 540-million years ago. Here is how Hoyle and Wichramasinghe rationalized the so-called 'Cambrian Explosion.' "Just as the interiors of the meteorites that fall today can remain cool, so it was with members of the great swarm that fell 570-million years ago. A fraction contained not just bacteria and viruses, but also the frozen eggs of metazoan creatures. ... The metazoan explosion thus came from a swarm of such bodies falling onto the Earth, so that no long evolutionary period on the Earth was required and no explicit precursors were needed. ...the(ir) evolution was not on the Earth." The longevity of one 'living fossil'—our mighty Carboniferous Cockroach—is truly remarkable. But if insect eggs have been permeating through space since near the dawn of time, rather than marveling at the cockroaches' endurance, we should instead wonder why it took so long for the cockroach and his brethren to appear here. Why not during the Cambrian instead of the middle Paleozoic? And are there 'cockroach planets' out there elsewhere in the Universe? Eeeew!

However resistant or 'cosmic' insects may seem, their fossil record diversity pattern follows a general Phanerozoic trend. Their origination in the early/mid Devonian, subsequent radiation and relative sizes may even be reasoned by physiological adaptational changes evolved in consequence to fluctuating atmospheric oxygen content. Like all other taxa of organisms existing then, insect life even succumbed to (their only) mass extinction, during the Great Dying (i.e. Permian-Triassic transition), 251-million years ago. (*Science*, vol. 261, 7/16/93, pp.310-315)

Although certainly not supporting panspermia, record-setting 2.1-billion year old, 5-inch long metazoan fossils were recently reported from Gabon on the west African continent. Previously, paleontologists believed that only microorganisms could have existed on Earth that long ago. Amina Khan (*Chicago Tribune* correspondent) commented on July 1, 2010 that "The organisms, which don't resemble modern things, existed when Earth's atmosphere would have been uninhabitable for today's plants and animals." Did these entirely unanticipated creatures descend to our planet aboard meteorites? Probably not.

According to Hoyle's and Wickramasinghe's theory, after Earth cooled to a hospitable level, evolution of multicellular forms could proceed from simpler outer space-derived forms. However, to prevent 'stagnation' in the evolutionary process, a steady stream of genetic material continually trickled Earthward, only to be assimilated, utilized and manifested in descendant populations. Natural selection in the usual sense still weeded out the unfit. But space 'germs' could also quickly cause programmed demise of 'inferior' organisms. Variety among species would be stimulated by the 'space bugs,' which would spread rapidly among terrestrial populations through disease and infection. The rate of Darwinian mutation is too slow, they claimed, to produce sufficient variety for natural selection to act upon. "...Darwinians were eating their cake and keeping it. Evolution was to be slow enough to meet the biological requirements, but not slow enough to have been captured in the fossil record." (*Our Place in the Cosmos*, p.132) In their view, organisms with the greatest evolutionary potential would be those having the greatest tendency for communicable disease.

Proliferation of disease through epidemics would promote rapid evolution and origination of new species, perhaps occurring in only a few tens of thousands of generations. Over time, forms most highly adapted to terrestrial conditions would lose their cosmic traits and their potential for further evolution would decline ... that is until the next batch of cosmic genes became incorporated, thereby re-setting the evolutionary pendulum in motion once again.

It's easy to mock another's ideas, but fair scientific evaluation of a new hypothesis or theory should follow certain recognized 'rules of engagement.' Conversely, in order to warrant such scrutiny, a proffered theory must at a minimum explain key details of natural phenomena and provide an objective means for falsification. Yet like that business with the supposed *Archaeopteryx* forgery, it's challenging to assess evolution-from-space theories because it's all so '*ad hoc.*' (So, instead, I'll refer you to an enlightening book by Richard Dawkins, *The Ancestor's Tale*, 2004. Furthermore, in their popular 1985 book, *Comet*, astronomer Carl Sagan and Ann Druyan offered valid reasons for why Hoyle's and Wickramasinghe's theories remain unconvincing (pp. 310-316).)

Hoyle and Wichramasinghe were highly perturbed by the apparent 'suddenness' of speciation events, and by lack of fossil evidence linking (continuity of) lineages, tracing to a single origin rooted at Life's bushy tree. Diatoms, for example, "suddenly" appear in the fossil record 100-million years ago, suggesting that they (like insects) arrived from beyond Earth. "Many species particularly among invertebrates, have no ancestral branches in the fossil record at all." (*Our Place in the Cosmos*, p.135) Evolutionary 'tree' imagery and diagrams—supposedly showing genetic connections between organisms as a pattern of 'descent' through the ages, therefore, create erroneous impressions. However, contrary to their claims, a recent discovery posted online by *Scientific American* (May 3, 2010) headlined that according to biochemist Douglas Theobald, "The Proof is in the Proteins: Test Supports Universal Common Ancestor for All Life." (Note those words, '*universal* common ancestor.')

Most paleontologists would agree that fossil evidence cited by Hoyle and Wickramasinghe could be better explained by the occurrence of rapid evolutionary changes (i.e. "Punctuated Equilibria") advanced in 1972 by paleontologists Stephen J. Gould and Niles Eldredge. In their amendment to Darwinian evolution, it was demonstrated that evolutionary changes need not take place gradually and/or ever so incrementally. However, Hoyle and Wickramasinghe erroneously insisted in their stylistic presentation of panspermia that conventionally-minded paleobiologists demand that speciation events must necessarily always be slow and gradual (e.g. via classical Lyellian 'Darwinism') in order to be considered 'evolutionary' events. But since the fossil record doesn't support the 'slow & steady' gradualist theme (as Gould and Eldredge would have agreed), Hoyle and Wickramasinghe implicated *extraterrestrial* causal factors as evolution's main driving force. Very sudden 'grafting' of cosmic genetic material 'explains' the general absence of links in the fossil record. 'Invasion' of metazoans 'explains' sudden occurrences of other kinds of organisms, such as insects. Meanwhile, Gould remained silent on both panspermia as well as Hoyle's and Wickramasinghe's ideas in his mammoth tome, *The Structure of Evolutionary Theory* (2002).

Hoyle and Wickramasinghe also estimated the extreme improbability of creating spontaneous mutations leading to the development of 2,000 important biomolecules known as enzymes. However, they have made few allowances for the continued sequential and sometimes rapid modification of genetic systems, which is how evolutionary changes are thought to proceed through long ages of geological time. Indeed, one wonders how different the organic world would be today if evolution had to start over from a 'beginning' a second time on an initially sterile planet Earth. Certainly, as Hoyle and Wickramasinghe and others have suggested—Gould and George Gaylord Simpson included—there would be an infinitesimal probability of exactly replicating specific events shaping evolution of all modern organisms in the unique, historical pattern followed since the late Precambrian.

Because it seems so 'unlikely,' then there simply *must* be an extraterrestrial connection, right? Startlingly, Hoyle and Wickramasinghe claimed, rapidity of certain genetic manifestations could be further 'explained' if all the genes that have ever been terrestrially encoded were—ready to be activated or even 'programmed' by viruses arriving much later in geological time—already, that is, "...within the creatures which arrived at 570-million years ago." (*Our Place in the Cosmos*, p.178) Genetically, speaking, *everything* that has evolved here (and all that will evolve) was essentially ready and rearing to go, given proper stimulation, at the base of the Cambrian explosion! Now that's 'far out' thinking. And once again, OMG!

As an aside, in the early years of the NASA program yet prior to Hoyle's and Wickramasinghe's thesis on outer space germination of terrestrial life and influence on the resulting fossil record, geneticist Joshua Lederberg, head of NASA's "exobiology unit,"viewed his team's ancillary mission set to determine "the uniqueness of systems based on nucleic acids and proteins as bearers of life." His often-mocked unit dealt with possibilities that our space probes might contaminate other worlds and whether devices returning from outer space could contaminate Earth with exogenous microbes. (See Audra Wolfe's article in references.) It was paleobiologist George Gaylord Simpson, a self-professed sci-fi fan (and writer) himself, who wrote the most sobering and conventionalized view, claiming 'exobiology' is "...a curious development in view of the fact that this 'science' has yet to demonstrate that its subject matter exists!"

Contrary to panspermia theory predictions, no evidence for existence of alien microbes has yet been found in outer space, on the Moon, Venus, Mars—places which should by now be literally teeming with seeded life forms. Fear not Crichton's "Andromeda strain'? Remember the concern over the possibility that astronauts returning from the first Apollo lunar landing in July 1969 would have become infected by lunar microbes? Those mid-1970s Viking lander tests for microbial life on Mars were at best inconclusive, and more likely negative (but with more hard data presumably forthcoming). Perhaps, however, future astro-paleontologists will find fossils in ancient Martian sediments. Some astronomers remain optimistic that life may exist in the postulated nitrogen seas of Neptune's moon, Triton, in the crust of Saturn's moons, Europa and Titan, or even on Jupiter's moon, Io.

Life's mysteries will forever captivate the curious mind of Man. For instance, many philosophers have marveled at the progressive evolution of the brain through geological time, 'culminating,' as the story goes, in human intelligence. Man may be the first terrestrial, technological creature who questions, reasons and solves. And yet, postulating a definite organic ('orthogenetic') trend toward higher intelligence among terrestrial organisms, possibly leading to attainment (or driven by the influence) of a cosmic level of intelligence, as possibly exemplified by insects and/or computers today, (as Hoyle and Wickramasinghe claimed) seems highly conjectural given a present lack of knowledge concerning such matters. As several modern paleontologists have remarked, Intelligence itself may not even be a driving factor of evolution!

In their 1988 book, *The Anthropic Cosmological Principle*, (p. 22), John D. Barrow and Frank J. Tipler stated of Hoyle's curious leanings that his "... view would have been supported by the natural theologians of past centuries .... More recently it has been taken seriously by scientists who include the Harvard chemist Lawrence Henderson and the British astrophysicist Fred Hoyle, so impressed were they by the string of 'coincidences' that exist between ... constants of Nature without which life of any sort would be excluded." Certainly if Darwin's theory becomes falsified, modification or abandonment must follow. However, tenets of panspermia, founded largely in *ad hoc* reasoning, have so far not presented serious challenges to evolutionary thought.

Interpreting evolutionary bursts and extinction crises recorded in the fossil record in terms of evolving biogeochemistry and, especially, transformations of the ocean-atmosphere system, has become a new-millennial trend. Paleontologist Peter Ward, for instance, has written popular accounts, 'revisionist' in nature, offering intriguing explanations for why certain organisms survived and others evolved at certain junctures in Earth history. The 'driver' for most of these wide-sweeping changes is not extraterrestrial, but causally related to fluctuating levels of volatile chemicals—carbon dioxide, sulfide and particularly, oxygen—in oceans and atmospheres. Such geohistorical patterns and circumstances can be explained in terms of refined biogeochemical models (i.e. - such as "GEOCARBSULF"). Marine and terrestrial organisms therefore could either evolve or succumb to (anoxic or poisonous) conditions. In Ward's view, evolution of new species is essentially 'Darwinian'—not re-programmed genetically by molecules filtering from outer space. Evolutionary (and extinctions) events are, for example, likely triggered during episodes when Earth's atmospheric oxygen levels 'rapidly' descend, rise, or reach minima, (as 'predicted' retrospectively on basis of biogeochemical modeling). As supported by the fossil record, new organismal physiological adaptations are more prone to develop during pivotal or mass crisis episodes. (And, yes, a large bolide certainly struck the Earth about 66-million years ago too.)

It is already suspected if not proven that genetic reshuffling occurs continually, accelerating the rate of protein evolution by a factor of a million. The fossil record shows evidence of rapid speciation: details of biochemistry coming to light do support the case for rapid evolutionary turnover. As exciting may be the proposition, recent observations and information at hand have not proven a case for microbial space-invaders. Rather, in support of Ward's revisionist thinking, scientists recently determined that humans have rapidly adapted *genetically* to high altitude environments such as Tibet, within a 7,500 year period—a mere smidgen of geological time. ("Sequencing of 50 human exomes reveals adaptation to high altitude," Xin Yi, *et. al.*, *Science*, vol. 329, 7/2/10, pp.75-78) So far, in spite of Hoyle's and Wickramasinghe's assertions, there's apparently no need for 'evolution from outer space' ideology in our modern paradigm of how the fossil record played itself out.

A Final Word:

In order to underscore their theories of viral genetic storms from outer space, Hoyle and Wickramasinghe required a symbol of Darwinian evolution to 'assassinate' and during the 1980s that target became the BMNH *Archaeopteryx* specimen. However, Darwin, while most intrigued by the specimen, did not absolutely require a reptilian-bird 'link,' foundational to the validity of the evolution by natural selection hypothesis. Instead of mere forgery, as Ward suggests in *Out of Thin Air*, birds like *Archaeopteryx* gained a competitive edge. Their energetic flight adaptations developed from ancestral saurischian dinosaurs evolving during a stressed age of low atmospheric oxygen (~ 12%, as opposed to today's 21%, enjoyed at sea level). 'Evolve or perish' is the adage, and according to Ward, Late Triassic dinosaurians "had a lower extinction rate than any other terrestrial vertebrate group because of a competitively superior respiration system—the first air sac system." (p. 197)

Undoubtedly, *Archaeopteryx* is *not* a forgery but a genuine fossil avian, derived from theropod dinosaur ancestors. *Archaeopteryx* owes its means of flight (however rudimentary may have been the true capability in this genus) to prior ecological adversity, a condition of minimal atmospheric oxygen prevalent during the Late Triassic. Its ancestry hails from an age when vertebrates evolved metabolically and genetically in order to cope with globally adverse conditions. This hardy, surviving lineage certainly earned its stripes, or, rather, plumage!

Sir Frederick Hoyle versus paleontologists like Peter Ward: in this regard, who sounds more reasonable? Discuss!

Other References: (1) "Diseases from Space and the Giggle Factor," Biot Report # 456: August 25, 2007, posted on the Internet; (2) "Germs in Space: Joshua Lederberg, Exobiology, and the Public Imagination, 1958-1964," Audra J. Wolfe, *Isis*, vol. 93, no.2, June 2002, pp. 183-205; (3) "The Non-prevalence of Humanoids," George Gaylord Simpson, *Science*, vol. 143, Feb. 21, 1964, pp.769-775;(4) *Out of Thin Air: Dinosaurs, Birds, and Earth's Ancient Atmosphere*, (2006) Peter D. Ward;(5) *Under A Green Sky: Global Warming, the Mass Extinctions of the Past and What They Can Tell Us About Our Future*, (2007), Peter D. Ward.

Chapter Thirty-three — *Moreau: H. G. Wells's "Exercise in Youthful Blasphemy"*

Animals into Man; men into animals. In the horror/sci-fi realm, evolution, either forward or retro variety, is always fair game –as long as the transformation occurs lightning fast, for the thrill!

A staple theme in science fiction and horror genre is that of *accelerated biological evolution*, or even *de-evolution* of the human race—either upward or down the evolutionary 'ladder.' We dread the reality of our darkest, most primitive selves, as we do the unknowable future, particularly whatever will become of the human race. Perhaps the most spectacular, time-honored way of portraying how the past *might* have been, or of what humans *might* become is for sci-fi writers to display human biological evolution quickly 'on show,' in a story or film. Thus, in the course of a few minutes or several reading sessions we gain astounding insights, a fictional telling of, say, what will become of man. This trend in fantastic fiction inaugurated by Herbert George Wells, lingers today, albeit in derived form, through numerous books and films dealing with genetically-altered organisms forecasting our irrepressible future or representing our remote past.

Perhaps the most famous genre example readers may recall is that *Outer Limits* episode, "The Sixth Finger"(written by Ellis St. Joseph), starring David McCallum, first aired on Oct. 14, 1963. In this episode actor McCallum is subjected to a machine that accelerates mankind's evolution. McCallum's brain and mental powers are enhanced remarkably (so are his pointy ears!). *Outer Limits'* "The Sixth Finger" also recalls Edmond Hamilton's classic short story, "The Man Who Evolved," (*Wonder Stories*, 1931), in which a scientist subjected to intensified cosmic rays, transforms incrementally into an unrecognizable 'human" of over a quarter billion years hence. Hamilton's classic line, "... the last mutation of all! The road of man's evolution was a *circular* one, returning to its beginning!" is still thought-provoking. Here are excellent examples of mankind being projected 'instantaneously' for our viewing pleasure into subsequent evolutionary stages, not unlike what Wells conceived over a century ago with fantastic writings such as *The Time Machine* (1895), and "Man of the Year Million" (an essay published in 1893).

Evolutionary romances steeped in anthropological fiction became all the rage during the early years of the 20th century, as implications of Charles Darwin's theory of evolution for mankind were conflated, debated (e.g. the Scopes Trial of 1925) and publicized (however improperly) in prominent museum paleontological displays (e.g. at the American Museum of Natural History under Henry F. Osborn's directorship). Usually, the story's protagonist is simply 'there' on scene recording how organisms undergo rapid transformation, evolving just as they're 'supposed' to. Many of you are familiar, for example, with Edgar Rice Burroughs' 1918 trilogy *The Land That Time Forgot, The People That Time Forgot,* and *Out of Time's Abyss*, novels in which paleontological history, as then known to contemporary science, was rapidly replayed. A decade later, Olaf Stapledon projected many millions of years beyond Wells's 'Morlock' and 'Eloi' futuristic stages some 803,000 years hence, with mind-expanding novels, *Last and First Men* (1930)—mankind's future history outlining 16 species of men to come—and later, *Star Maker* (1937).

However, as opposed to simply witnessing how evolution is *supposed* to proceed, from our past leading into the far flung future, the theme of the mad scientist *altering* or tinkering with a ('ladder-like') prescribed course of evolutionary flow eventually became prominent in science fiction. Mankind's controlling free will became increasingly imposed over (evolutionary) determinism. Thus we have stories such as Kenneth Robeson's *Doc Savage: The Time Terror* (1943) involving an elixir synthesized for accelerating evolution. This 'mad evolutionary scientist' theme persists although taking on new form in recent years. Accordingly, today, we have novels such as Jonathan Green's *Unnatural History* (2007), relying on the gimmick of a "*de-evolution* formula," and movies in which extant DNA is altered producing retro-genetic monstrosities. (Think of Michael Crichton's *Jurassic Park*, for example, relying on frog DNA to fill in the dinosaurian genome.)

We owe this theme, ultimately, to H. G. Wells, as originally conveyed through his brilliant terrifying satire, *The Island of Dr. Moreau* (1896). (Yes—Dr. Frankenstein unleashed the awful *power* of restoring life to dead tissues, just as Crichton's *Jurassic Park* harnessed *genetic* forces, but Mary Shelley offered no evolutionary messages, *per se*, in her famous novel.) A faux de-evolution is also apparent in a peculiar movie—to be discussed shortly—*The Alligator People* (1959). In a deep sense, creatures spawned in *The Alligator People* are the opposite of the sort generated through vivisection on Dr. Moreau's island of torture in the Pacific. Yet, at least symbolically, *Alligator People* represents a minor yet instructive bastion in the modernistic retelling of *Dr. Moreau's* theme.

H. G. Wells, who studied under Thomas Henry Huxley (Darwin's 'bulldog disciple'), was an evolutionist of his time. At his acme, Wells tackled evolutionary themes, stressing brevity of man's existence in relation to the geological record's vastness. Both *The Time Machine* (1895), and *The War of the Worlds* (1898) bore significant evolutionary messages, for example. And *Dr. Moreau* is yet another in this vein. The premise is that mankind's vaunted civilization is not so removed from what we disparagingly refer to as the 'animal world.' To Wells, mankind is just another sort of (albeit haughty) animal, whose attainment of culture, religion and accumulated knowledge could, as a result of some global war, invention of superweapon, pandemic disease or natural catastrophe, quickly extinguish. And, as echoed time and again in Wells's writings, man is destined to suffer extinction not unlike the animals which have come and gone before. Although a smidgen of hope remains in our case, Wells instructed, because we, unlike other animals, "blessed and cursed with self-consciousness," have pronounced ability to *decide* our fate.

In *The Island of Dr. Moreau* (novel), the protagonist - narrator, named Charles Prendick, cast adrift at sea, is rescued and taken to the infamous Island. Prendick quickly notes the very strange appearance of some of Moreau's workers, who are also ordered about by an alcoholic, clearly 'human' assistant named Montgomery. For his own safety, Prendick is confined to his dwelling. He hears mournful cries issued from what is referred to by the disfigured islanders as the "house of pain." Prendick can barely stand listening to a tortured female puma's piteous wailing during his first night on the island.

Prendick has been trained in the sciences, although he doesn't quite put two and two together. Prendick erroneously thinks that natives are being transformed into animals in Dr. Moreau's laboratory. Vaguely recalling the fear-prickling name "Moreau," Prendick can't quite recall for what dastardly reasons he was ousted from England. Fearing that he will be next subjected to the awful surgical procedures, Prendick escapes. Exploring the jungle, Prendick encounters a predatory man-beast and later meets an anthropoidal "Sayer of the Law," who recites the Laws instituted by Moreau. As Prendick surmises, "A horrible fancy came into my head that Moreau, after animalizing these men, had infected their dwarfed brains with a kind of deification of himself." Prendick couldn't be farther from the truth; he's got things horribly reversed. Using vivisection, Moreau has been operating on animals, torturously changing them into 'men,' or pseudo-men. The Laws he's instilled in them are intended to civilize his creations. "Are we not men?" they chant, begging the reader to decide. (Well, are they?)

Thus, Moreau is, only in a sense as we examine matters today, speeding the evolutionary process, from animals into men. Yet, Wells seems to be taunting, are Moreau's creatures any more or less 'civilized' or 'feral' than *Homo sapiens*'? In treating them so cruelly, tyrannical Moreau certainly appears less than human and certainly not the godlike being he aspires to. Unknown to many who are only familiar with Hollywood versions of Wells's tale is that Moreau also made snakelike "footless things," which writhe along the ground throughout the jungle.

Finally, in chapter 14, "Doctor Moreau Explains" the disturbing nature of his experiments. Using skin and bone grafts, altering the larynx, "moulding" brain structures in his test subjects, giving blood transfusions and tampering with the "chemical rhythm of the creatures," Moreau has metamorphosed various animals into recognizable human-like form. Prendick is alarmed by the fact that these, "Monsters manufactured ... these animals talk!" Also, he questions why Moreau, who claims to be a religious man, "...had taken the human form as a model ... there seemed ... a strange wickedness for that choice." Moreau states that, beyond trivial suffering experienced by his transformed animals, all his life he's sought the laws of the world's Maker, "...to find the extreme limit of plasticity in a living shape." That limit is of course the human form. But, even after re-educating them, Moreau can't prevent the test subjects from reverting to their lower state, as "...somehow the things drift back again: the stubborn beast-flesh grows day by day back again ...I mean to conquer that, This puma ... " Moreau has not yet solved the impasse of 'stubborn beast-flesh' refusing to yield to his talents and intellect.

Hands, feet and nails rarely look human enough on his creations. Then there is the ready problem of reversion, especially if not reinforced in their brains, culturally, by the Law, a painful reminder that his 'men' are never supposed to act or even think like animals. Moreau constantly battles their deep seated bestial emotions, tendencies and cravings, however, controlled by the near indecipherable constitution of their brains. "First one animal trait, then another, creeps to the surface and stares out at me. But I will conquer yet! Each time I dip a living creature into the bath of burning pain, I say, 'This time I will burn out all the animal; this time I will make a rational creature of my own...As soon as my hand is taken from them the beast begins to creep back, begins to assert itself again." The puma-woman will be, Moreau declares, his pinnacle achievement, his *magnum opus*.

While Mary Shelley's Dr. Frankenstein doesn't take responsibility for his unholy human creation—his 'offspring' (which in the novel, interestingly, was fabricated at least partially from animal corpses), Dr. Moreau feels analogously remorseful toward *his* clan. He constantly creates but is only dejected by every outcome. Moreau says, "They only sicken me with a sense of failure." Yes, they're all mere 'experiments,' that is, until the day when his puma-female emerges triumphantly from the operating room as a human bereft of stubborn beast-flesh, with a brain cleansed of animalistic emotion (a fallacy, of course, as such 'lower' instincts are instilled in all humans). In a sense, perhaps, the puma-female is akin to Dr. Frankenstein's Monster's 'bride.' But we know how *that* experiment turned out too. In the novel, the puma-female never 'triumphantly' emerges in human form; in most 'advanced' form she has "...an awful face ...not human, not animal, but hellish."

Things rapidly unravel. Moreau must punish one of his creations, the Leopard-man, for breaking the Law. Soon, counter to one of the Laws imposed on his people, like a predatory animal rather than godlike entity, Moreau is *stalking* this offender through the jungle to its death. Next, the bloody, disfigured puma escapes from the laboratory and kills the pursuing Moreau. Eventually, without vivisectional treatments or punishing reenforcement of the Law, the beast-men steadily descend socially, behaviorally, physically and rather ironically into ... animals. The people kill Montgomery; months later, Prendick is finally rescued. But he's psychologically scarred by the experience, for, upon his return to London, Prendick can only see the 'animal' in his fellow civilized human beings—essence of Wells' ingenious satire. Prendick states, "My trouble took the strangest form. I could not persuade myself that the men and women I met were not also another Beast People, animals half wrought into the outward image of human souls, and that they would presently begin to revert, ... to show first this bestial mark and then that." Like Moreau, Prendick (a common gentleman of his time) errs in misunderstanding that there is not such great dichotomy between what is 'animal' and what constitutes 'man' after all.

As Frank McConnell (one of my former college English professors), states in *The Science Fiction of H. G. Wells*, (1980), "Even today, of Wells' five major scientific romances from 1895 through 1901, it is probably the least widely read or taught." Critics and readers alike found the story offensive, cutting too close to the quick. In *Dr. Moreau* there is wincing animal torture, mocking of religion, and the disturbing notion that civilized men were essentially talking animals well dressed in clothes. Furthermore, although it's a parable of how one demented soul (Moreau) aspires to become God, occurrences on the island have been likened to Nazi concentration camps. It's an objectionable formula that generally failed with a Victorian readership, that is, when compared to his lauded *Time Machine*. Wells referred to *Island of Dr. Moreau* as "an exercise in youthful blasphemy." Guilty as charged!

Essentially, *Island of Dr. Moreau* is Shelley's *Frankenstein* melded with Darwinian essence of Wells' *Time Machine*. Human creation of (sometimes hideous) anthropological beings, accented with evolutionary ambiance. As Brian Aldiss noted, "Moreau is ...Mary Shelley's protagonist in his maturity—Frankenstein Unbound." In order to ridicule the public notion that evolution culminated with Man's alleged, inevitable ascent, Wells relied on a *horizontal* sort of (mythological) 'transformism,' permitting—under Moreau's cruel scalpel—other animals to quickly become outwardly manlike (although, mirroring then prevalent Victorian racist attitudes, of lesser evolved, 'primitive' *mien*). Charles DePaolo refers to Wells' emphasis of this distorted horizontal 'speciation,' in contrast to the then more familiar (ladder-like, vertical) non-Darwinian 'scale of nature,' as "heterogony." *Dr. Moreau* is a parable about artificial evolution and, once the technology is invented, how it may be so easily misguided by irresponsible men. (Incidentally, evolution may be likened to a busy, branching tree, rather than rungs on a 'ladder.')

Shortly after publication of *Dr. Moreau*, Wells wrote an essay, "Human Evolution as an Artificial Process" (1896). Thinking ever forwardly (e.g. Moreau is pronounced 'Morrow' as in to-*morrow*), Wells cautioned at a time before the biochemical nature of what was then referred to as the "germ-plasm" (i.e. now 'DNA') was fully understood, how a true Darwinian god of the universal evolutionary process would not be benevolent, but essentially like Moreau, indifferent to human suffering, and relying on randomized natural 'experimentation' in the wild. As scholar Joseph Andriano comments, "If God exists, he would have to be a Moreau, indifferent to the pain of his creatures, many of which are failed experiments."

Over a century later, given awful powers of genetic engineering modern science is harnessing in the 21st century, one wonders how many malevolent 'Moreaus' are already out there?

To date, there have been three major motion picture adaptations of Wells' *Dr. Moreau* novel. Probably most of you are familiar with *The Island of Lost Souls* (Paramount Productions,1932). *Island of Lost Souls* (directed by Erle Kenton; screenplay by Philip Wylie and Waldemar Young) is a moody, atmospheric black and white production in early 1930s horror style, which, as most monster movie aficionados know, featured Bela Lugosi in a supporting role as Sayer of the Law. But, rather than Lugosi, Charles Laughton steals the show with his twisted, bent portrayal as Dr. Moreau incarnate, sadistic and evil. (Both Shelley's and Universal's Dr. Frankenstein was merely 'mad,' isolated and obsessed—yet arguably not inherently satanic. Laughton makes a far better, more convincing mad scientist than Colin Clive's Dr. Frankenstein.) The most significant cast addition, however, is Lota the Panther Woman, played by Kathleen Burke. Essentially, in this telling of the tale, Lota is Wells' Puma-woman, having emerged triumphantly human (and seductively female) from the House of Pain. Moreau introduces her to Edward Parker played by Richard Arlen (taking the place of the novel's "Prendick"), explaining her exotic features as "pure Polynesian."

This time, Parker is rescued at sea from the wreck of the *Lady Vain* onto Captain Davies' (played by veteran actor, Stanley Fields) vessel, the *Covena*. On board, Montgomery, who has a criminal past, refers to Moreau as "a brilliant man and a great scientist," although Davies claims Moreau is a "grave-robbing fool," likening him to Dr. Frankenstein. Also on board is one of the strange looking natives Parker will become soon become acquainted with, Montgomery's assistant, furry-eared Mling, whom he later ironically says is Moreau's "faithful dog," meant of course in a quite literal sense. Soon they land at the island, where Parker gets a first-hand look at Moreau's self-acclaimed, ahem, "experimental station for bio-anthropological research."

Once more, as in the novel, Parker presumes that, 'under the knife,' Moreau was converting men into animals. When he spies Montgomery and Moreau operating on one unfortunate victim crying out in mortal pain, Parker flees with Lota into the jungle, fearing "You and I may be next." But their freedom is cut short, as, entering the alleged natives' village, they're chased down by Montgomery and Moreau. Lugosi as Sayer of the Law leads chanting recitation of the Law, ("Not to run on all fours; Not to eat meat; Not to spill blood"), ending with mournful, reverberating references to Moreau himself, "His is the Hand that Makes; His is the Hand that heals; His is the House of Pain."

Eventually all devoted mad scientists of the movies spill the beans, that is, they must explain what they're doing and why, and here Moreau's explanation differs from that in Wells's chapter14 in the novel. Moreau has been experimenting for over a decade, managing to strip "hundreds of thousands of years of evolution" from the growth of orchids, other varieties of plants and man as well. Moreau (wrongly) theorizes "All animal life is tending to the human form.... Man ... is the present climax of a long process of organic evolution. All life is tending toward the human form." Of course, this is not true! In *Island of Lost Souls*, Moreau's ideology reflects the old Victorian ladder-like 'chain of being,' which Wells was trying to strike down and ridicule in his 1896 novel: to make people think. Then, Laughton's Moreau claims to have found a way to speed up evolution, that is, by altering the "germ plasm," thus allowing more 'primitive' animals to transform into 'higher' ones. The abject, proto-human failures are slavishly condemned to turn a millwheel to supply power to the compound, and we are shown a scene of that dismal, torturous affair.

Despite what is metaphorically implied, unlike Paramount's Moreau, Wells' Moreau was not explicitly intending to accelerate evolution. Also, Wells did not advocate (non-Darwinian) Lamarckism. For in chapter 15 of the novel, readers note "There was no evidence of the inheritance of their acquired human characteristics" passed along to progeny. In *Island of Lost Souls,* Moreau is clearly also a ruthless Frankensteinian figure, as when we hear him say, "Do you know what it means to feel like God?".

At first, Parker accepts Lota as a true human being, so much so that he becomes rather enamored with her voluptuousness, even though he is engaged to his fiancee', Ruth (played by the lovely Leila Hyams). Moreau deliberately wants

Parker to be attracted to Lota to test, well, what might arise. To him, they are merely sexual test subjects. Will they spawn and reproduce? Has Moreau succeeded at last? Will he someday be able to triumphantly return to London with Lota, his supreme creation, the woman He made?

The movie's experiment goes much farther than in the novel, but still ultimately fails. Andriano describes one passionate scene, a shared moment of intimacy between Parker and Lota, ruined by the resurgence of stubborn beast flesh. Lota and Parker kiss passionately by the pool. At one point "the rippling reflections of the couple shimmer into one another... This lingering shot seems to suggest that on some fundamental level the man and the woman are one, yet their union is fleeting." Instinctively responding (like an animal) to Lota's advances, 'used' Parker betrays Ruth, that is until "when Parker and Lota are at their most animal *and* their most human, that her talons begin to grow back." Now realizing Lota's (non-human) origin, Parker accuses Moreau, threatening to expose him. Creating men from pigs, hyenas and other creatures is rotten, but making a *woman* from an animal—now that's despicable.

Thwarted in this attempt, frustrated Moreau is almost prepared to give up, until he notices Lota shedding tears, the first one of his fabricated lot to do so. Professing that he's not beaten after all, Moreau vows this time to "burn out all the animal in her" in the House of Pain, once and for all. Then, predicts the voyeur, "time and monotony" will eventually bond Parker (whose rescue will be sabotaged) to Lota.

However, after fiancee' Ruth's unanticipated rescue party, commanded by the likeable Captain Donahue (Paul Hurst), eventually reaches the island, Moreau changes plans. Now he lasciviously intends to see whether a beast-man, henchman "Ouran" (Hans Steinke), can spawn with Ruth (i.e. rape). As Andriano asks, "...what kind of offspring would result?"

Rather than consenting to return to the House of Pain, Lota decides to flee with Parker, Montgomery and Ruth. Instinctively resorting to feline fighting tactics, Lota displays human character beyond wannabe-god Moreau's when she sacrifices herself to stop Ouran who is pursuing them with (in *defiance* of the Law, "Not to shed blood") orders to kill. Both Lota and Ouran die in the struggle. More blood will be spilled, however.

In the most ironic scene of all, Moreau becomes the monster, switching places with the beast-folk who display their most human qualities, first utilizing logic then exhibiting mob violence. Ouran has murdered Captain Donahue, carrying his corpse to the beast-folk village, where Sayer of the Law reasons there will be "Law no more." Killing is now acceptable, and Moreau's whiplashes no longer repel the awful, vengeful, "*Things—not* men, not beast—part men, part beast." Because Donahue is a man like Moreau, the beast-folk realize that Moreau can die as well. Time for vengeance: it is their turn to operate, to turn sharpened scissors, scalpel and other surgical implements on one of Hollywood's most infamous mad scientists.

Moreau would not be Hollywood's last, if not most insidious 'mad scientist.'

The Alligator People (Twentieth Century Fox, 1959) becomes an intriguing 'milestone' (I hesitate in referring to it as a 'golden spike') partially because of its period of release, its most interesting featured star—Lon Chaney (i.e. Jr.), but mainly due to its content. Long maligned as just another Hollywood 'cheapie,' few note that it was produced during the decade when DNA (Moreau's former "germ-plasm") was described (double helical structure proposed in 1953 by Watson, Crick and Wilkins), that is, when biochemical knowledge was growing by leaps and bounds. During the 1950s, science was rapidly changing with obvious impact upon the sci-fi/horror movie industry. *Alligator People* was one of those curious results.

Comparing *Island of Lost Souls* alongside *Alligator People* demonstrates how things can unravel, even when, reciprocally (as in the latter film), scientists strive for the public good. Moreau's handiwork was sadistic and evil-minded, whereas anthropoid byproducts resulting in *Alligator People* are unfortunate, accidental victims of science unwittingly gone awry. While Moreau was malevolent, *Alligator People's* Dr. Mark Sinclair (George Macready) intends to be benevolent; he's also merely human, prone to carrying things too far.

Also, that old shape-shifter himself, Lon "Wolfman" Chaney has a pivotal role, in a sense symbolically 'passing the baton.' The last time Chaney had been seen in a *serious* werewolf role, in Universal's *House of Dracula* (1945), he'd been cured of his lycanthropic condition by Dr. Edelman using offbeat biochemical science. And, in spite of and perhaps underscoring Chaney's presence, *Alligator People's* organic transformations are not supernatural or founded in haunted superstition, but instead wholly aided by molecular chemistry, (that is, late 20th century's threatening "chemistry for better living"). Increasingly thereafter in monster movies, science would trump the supernatural! However, as we know pure science can still become misused or even misapplied. For, in wielding scientific power, the inherent danger is that Man cannot possibly know everything. Mistakes will inevitably happen along the way—part of the research, or learning process.

While *Island of Lost Souls* was about animals being transformed into 'higher creatures'—men (as if men weren't animals anyway!), *Alligator People* is about men reverting into more primitive animals, in this case one of the most ancient vertebrate forms recorded in the fossil record—Archosaurs, from which arose the Dinosauria! *The Alligator People* was in a sense, 20th- Century Fox's answer to Universal's "Gillman" (especially as in 1956's *The Creature Walks*

Among Us). In an early scene, Dr. Mark Sinclair refers to the Louisiana swamp setting as primeval, a "vast swamp cradling life." Millions of years ago, most of the earth appeared like this. Not unlike how it was with Universal's Gillman, our murky origins may be traced to the alligators' symbolic swamp sediment, slime and ooze.

The screenplay was written by Orville H. Hampton; the sinister film was directed by Roy Del Ruth. The star was Beverly Garland who plays a distraught character suffering from amnesia, Joyce Webster. Under hypnosis (i.e. the film is essentially a long flashback sequence) she reveals to psychologists the tragedy of her deceased husband, Paul Webster (Richard Crane). Essentially, Joyce becomes an analog for Wells' "Prendick." And this time it is her beloved Paul who is afflicted with a terminal condition of "stubborn beast flesh."

Unknown to Joyce, Paul had been seriously injured in a plane crash, only to be miraculously healed by Dr. Sinclair at a secluded Louisiana clinic known as the Cypresses. Sinclair has extracted a steroid substance from the pituitary glands of alligators that he injected into Paul, curing his burns and other deformities resulting from the accident. During the 1930s, research into the steroidal compounds and the adrenal gland network escalated. By the 1950s, a Nobel Prize in Physiology and Medicine memorialized discovery of hormones of the adrenal cortex, their structure and biological effects. So, naturally by 1959, scientific reality proved fair game for science-gone-awry films such as *The Alligator People*. Anyway, thanks to these injections, Paul Webster has no scars; limbs intact and strong. But instead of trying out for a major league baseball team to break all the sacred homerun records decades before Barry Bonds, he weds Joyce, only to receive terrible news on their wedding night that an awful side effect has already manifested itself in several of Dr. Sinclair's accident patients.

After hearing the news of the side effect that will soon manifest itself, Paul shamefully disappears for many months, until Joyce tracks him to the Cypresses. There she sees what he's evolved into, a half-alligator creature! She meets Sinclair who is working on a cure for his several patients; x-ray and sunray lamp treatments seem to slow the reversion process, but so far no real cure has been devised. Sinclair hopes that more intense gamma radiation from cobalt-60 bomb isotopes will cure Paul and the other unfortunates, but he's still a long way from realizing that goal.

Meanwhile, Chaney plays a demented character named 'Mannon' who delights in firing his gun at alligators in the swamp when he gets liquored up in the evenings. This is because he's lost his left hand, replaced with a hook, to an alligator—perhaps prompting audiences to think of *Peter Pan's* 'Captain Hook.' So, understandably, he's also developed a psychotic aversion toward any gator type, including those awful yet unfortunate byproducts of Dr. Sinclair's mad science. Naturally, he dislikes what has become of Paul, futilely (and amusingly) yelling into the rainy swamp, "I'll kill you Alligator Man, just like I'd kill any four-legged gator." One wonders why Sinclair didn't also attempt to regenerate Mannon's hand using the serum.

At the climax, Sinclair is finally prepared to irradiate Paul with the supposed gamma -ray 'cure,' which must be administered for absolutely no longer than 30 seconds. While strapped on the traditional Hollywood operating slab, Paul forgives Dr. Sinclair, stating, "You're not God, Mark." Sinclair replies, "I feel as if I've played it, and am being punished." Of course, this is the moment that Mannon chooses to stalk Paul into the laboratory confines, where he interferes with the irradiation apparatus. Predictably, Paul receives a gamma overdose, and is rapidly transformed into a complete alligator-man, replete with cheesy-looking gator face, scaly glove-hands and upper suit torso. Mannon is accidentally electrocuted during a fight with Paul. Paul—now the alligator beast (played by Boyd Stockman), runs outside before the laboratory explodes. After seeing his reflection in the swamp water, Paul is beyond remorse. Venting his anger, he attacks an alligator swimming by, only to subsequently drown in quicksand as his wife Joyce watches helplessly.

Gimmicky gamma-rays had been relied on for catalyzing retro-transformations before, such as in Universal's *Monster On the Campus* (1958). A similar themed film was Global Productions/United Artists' *The Neanderthal Man* (1953). My point is that the late '40s and 1950s period is when sci-fi films explored every aspect of bio-molecular change in a variety of living organisms (Godzilla included!), even before the significance of a certain double-helical molecule (now out there in the everyday public lexicon) known as "DNA" was generally known. It's as if Dr. Moreau had been granted a new powerful set of laboratory tools (i.e. radiation, biochemicals, elixirs, etc.) with which to continue his monstrous handiwork, menacingly tinkering with the "germ-plasm." However, modern mad scientists still haven't learned how to safely control that resulting 'stubborn beast flesh.'

The Alligator People perhaps spawned a popular Marvel Comics character, Stan Lee's "The Lizard," introduced in *The Amazing Spider-Man* no.6 (Nov. 1963). Using reptilian DNA to regenerate limbs in amputees, Dr. Curtis Connors succumbs to the horrors of his experimental serum. The Lizard reappeared in numerous segments through the decades, (and you may read more of this amazing super-villain on the internet at Wikipedia).

If you had any doubt that discovery and description of DNA represents a major pop-cultural revolution, just consider the impact on sci-fi/horror genre, especially those films that we so love to see. I've compiled my own list of such films, but without divulging all titles here, just consider such movies produced, for example, between 20th Century-Fox's *The Fly* (1958) and 2009's *District 9*, all of which involve genetically altered men or other organisms, mutated into assorted monsters. That's just for starters. (How many can you name?) In a large sense, Dr. Moreau's, and hence Wells's stamp of authority is emblazoned upon them all.

Modern writers cite an apocryphal remark attributed to Wells, wherein the aging, then bitter story-teller allegedly answers a query concerning how he would compose his epitaph, "...just this—God damn you all; I told you so." Usually this remark is placed in context of the first and second world wars and development of mankind's first super-weapon, the atomic bomb (first used in warfare a year before Wells's death), awful outcomes, all of which in a sense he 'predicted.' Yes, for socio- and geo-political reasons, the world was spinning out of control rapidly then during the first half of the 20th century. But Wells's apocryphal remark gains new meaning today, although now in light of Dr. Moreau's darkest tendencies, given wherewithal of 21st century bioengineering.

Remember—he told us so!

Note - One might proclaim that Robert Louis Stevenson's story, *The Strange Case of Dr. Jekyll and Mr. Hyde* (1886), predating Wells' *The Island of Dr. Moreau*, is the original tale of a mad scientist influencing whole body organismal transformations aided by chemistry (that could be construed as 'retro-evolutionary'). I concede that is partially true, although to me, Stevenson's tale is more so a *psychological* thriller, emphasizing morality versus the baser element inherent within all men. Wells's Dr. Moreau was more intent on using *science* to simulate or experience what it was like to 'play God.' For within us all lies a propensity for evil, often suppressed, but which may conceivably come to the fore, as when aided by Dr. Jekyll's addictive and stimulating reagents. Although this is not to mean that Wells was not also interested in the dual psychological nature of mankind.

Select References: (1) Brian W. Aldiss with David Wingrove, *Trillion Year Spree: The History of Science Fiction* (Avon Books, 1986) (2) Joseph D. Andriano, *Immortal Monster: The Mythological Evolution of the Fantastic Beast in Modern Fiction and Film* (Greenwood Press, 1999); (3) Charles De Paolo, *Human Prehistory in Fiction* (McFarland & Company, 2003); (4) Frank McConnell, *The Science Fiction of H. G. Wells* (Oxford University Press, 1981); (5) Bill Warren, *Keep Watching the Skies!* Vol. 2, (McFarland & Company, 1982)

Chapter Thirty-four — *Wells versus Verne: Particulars in Paleo-Fiction*

Quick—can you name the three prehistoric vertebrate genera that are common to, that is, each mentioned in the publications of two eminent visionaries, Jules Verne's *Journey to the Center of the Earth,* and H. G. Wells's *The Time Machine*? (By the way, there's a fourth prehistoric animal—a dinosaur mentioned in *The Time Machine* (1895), making a more significant splash therein than the other three. Bonus points if you can name it right now!) It's curious that names of such extinct genera would appear in *both* stories given that Wells's masterpiece—as opposed to Verne's, conjuring mysterious denizens from Earth's distant *past*—concerned Man's remote *futurity*. While Verne was further intrigued with prospect of primeval/fossil man, Wells's classic tale of our foreboding future is much more than (simply) "anthropological fiction," as commonly interpreted. So, ironically, in contrast to Verne—due to the prominent overarching theme of his "JCE," one normally doesn't ordinarily associate Wells's persona, his fiction or his legacy with geological sciences, paleontology or prehistoric animals, which instead appears to be Verne's ruling 'turf 'as 'surveyor' of the deep past.

And yet, even though both stories involve modes of 'time travel,' how wrong is this perspective.

Besides Mary Shelley, Wells (1866 -1946) and Verne (1828 – 1905) are the two most celebrated science fiction writers of the 19th century (and early 20th) centuries. And although Verne gained reputation as the more –'geologically-minded' of the two—due to prominence of his 1864 JCE novel buttressed by fame and popularity of the 1959 20th Century-Fox movie with its special effects *Dimetrodons*, what a surprise to discover that Wells was far more the geological & paleontologically-cultured soul! Although directing his party of subterranean explorers in JCE 'backward' in time, Verne couldn't embrace 'Darwinian' evolutionary progress. Meanwhile, Wells demonstrated how evolution could proceed by sending his Time Traveler into the far-flung future, 802,701 years hence, (& beyond). And there, in Wells's bleak, forlorn vision, (modern) Man has become ironically as extinct as dinosaurs, symbolized by a fossil brontosaur skeleton rotting in the hulk of a variant futuristic British Museum, or the "Palace of Green Porcelain."

Although I intend to focus far more here on Wells than Verne, how may we process this counter-intuitive alignment of outcomes in two seemingly contrasting classic science fiction stories—one focusing on our Past, while the other, our harrowing Future?

During 1884 to 1887, Wells attended the Normal School of Science in South Kensington under a prize government scholarship. Here, he studied (i.e. in 1885) under the influential Thomas Henry Huxley—ardent evolutionist and science popularizer—(Charles) "Darwin's Bulldog." Wells later taught biology in 1891 and co-authored (with R. A. Gregory) a biology text book in 1893. During this formative period he was honing and expanding versions of (what would become) his seminal time travel novella, first appearing in 1888 as a crudely written short story, "The Chronic Argonauts." Armed with perspectives on organic evolution fueled by Huxley's philosophical cosmic pessimism, Wells's *non*-fictional essays turned a daring and dark corridor as we see in such samples as "On Extinction" (1893), "The Extinction of Man" (1894), and "The Man of the Year Million: A Scientific Forecast" (1893). Thanks to Huxley's influences, therein, messaging was stark, unsettling, in-your-face. For in their view, Nature was uncaring; the hostile universe and our Solar System were steadily winding down—dissipating energy true to physicist Lord Kelvin's considerations of the Second Law of Thermodynamics. And rather than always steadily 'advancing,' mankind and our precarious civilization are destined to degenerate and inevitably collapse. In May 1895, Wells sent Huxley a copy of his *Time Machine*, which the former perceived in homage to his mentor, as the "… central idea—of degeneration following security—… the outcome of a certain amount of biological study."

In utter contrast to Verne (who seems mostly indebted to 1860s French editions of Louis Figuier's *Earth Before the Deluge,* illustrated by Edouard Riou, for his familiarity with geology), most admirers of Wells's fantastic fiction aren't aware that the latter was very well steeped in geosciences. You see, Wells was an avid student of geology, comprehending the meaning of "deep time," although modestly professing no more than "elementary" knowledge. He failed one important exam (according to David Shackleton's 2017 article, "H.G. Wells, Geology and Ruins of Time"), but took first place in "second class honours in geology" toward his B.Sc. degree (awarded in 1890), later writing a manual—perhaps sort of a 'Cliffs Notes' booklet, titled *Honours Physiography* published in 1893. Wells's views on "imagining" geological deep time are peppered throughout his *Time Machine* novel—his richest ode to evolution & geology, even though only a few such passages reflect Earth's geological *past* (including one interesting segment that was omitted) that we'll come to. In particular, Shackleton convincingly demonstrates how Wells's Time Traveler's journey may be viewed as a fictional equivalent to the kind of "geological time travel described in (Charles) Lyell's *Principles of Geology,*" (or earlier through scientific writings of James Hutton or John Playfair).

Wells's Time Traveler's 'observed visuals' of subsequent terrains either spied or visited along his rather randomized journey through time are not simply similar to special effects scenes such as audiences witnessed in the classic 1960 *Time Machine* film. They're also highly analogous to the series of life-through-time 'portals' showing mere glimpses of geological periods and epochs traditionally portrayed in contemporary exhibit halls or paleontology books by paleoartists such as Riou—as dioramas or "scenes from deep time." A 'randomized' journey undertaken by the Time Traveler, allowing readers to briefly witness only a few scattered landscapes along the way, is also consistent with the old "uniformitarian" geologists' views on the rock record's fragmentary nature of incomplete preservation. In the future the Time Traveler notes changes in physiography, such as the Thames Valley having "shifted, perhaps, a mile from its present position," and he aptly describes decaying fossils once displayed in the Paleontological Gallery of that ruined "Victorian-style museum" remnant—the Palace of Green Porcelain, indicating that Wells's knowledge of geology is on rock solid foundation. As Shackleton claims, "Wells's Palace thereby stands as a monument to a longer geological history that has been marked by loss and extinction."

Whereas Verne's explorers experience the geological past both intimately and imaginatively without breaking the time barrier in Earth's sublime, supernal interior, Wells's Time Traveler possesses *technological* means of *also* journeying into prehistory—and so such a possibility beckons in *Time Machine* segments. For in the Epilogue, after the Time Traveler leaves the Present, forever, Wells's 'Outer Narrator' Hillyer muses, "Will he ever return? It may be that he was swept back into the past, and fell among the blood-drinking, hairy savages of the Age of Unpolished Stone, into the abyss of the Cretaceous Sea, or among the grotesque saurians, the huge reptilian brutes of the Jurassic times. He may even now ... be wandering on some plesiosaurus-haunted Oolitic coral reef, or beside the lonely saline lakes of the Triassic Age." And truthfully we don't know whether the Mesozoic became the Time Traveler's ultimate destination, rather than the future. (By the way—hint, notice that I've already provided one of the three prehistoric animal names common to both JCE and *Time Machine*. There are still two others, per my opening remark.)

Actually, Wells did pen a passage in which, after his haunting venture into Earth's future many millions of years from now, the Time Traveler, on a return journey yet "lost in time," overshoots his modern day period backward into prehistory. This deleted segment is dated April 4, 1894, and includes a brief visitation in the British Pliocene of 2-million years ago where he sees a large "hippopotamus" and other large unnamed beasts. This brief encounter happens plot-wise before he moves back toward the Present, prior to his final leaving 'forever.' But now consider—would Wells's perceptions of geological time and ultimately his *Time Machine* novel never have come to fruition were it not for Benjamin Waterhouse Hawkins's artistry?

An indication of Wells's fascination with prehistoric life may be gleaned from his 1934 *Autobiography*, describing a youthful visit (c. 1878-79) to see Benjamin Waterhouse Hawkins's 'Crystal Palace' paleo-sculptures displayed at Sydenham that left deep impressions. Wells wrote decades later how it dawned on him how Earth had an incredibly lengthy past: "I first became aware of that past at the gardens of the Crystal Palace at Sydenham; it came upon me as a complete surprise, embodied in vast plaster reconstructions of the megatherium and various dinosaurs and a toadlike labyrinthodon ... My mother explained that these were Antediluvian Animals ... there seemed something a little wrong in the suggestion that the ichthyosaurus had been drowned in a flood." Arrangement of these statues was intended as a 'life through geological time' exhibit. The Hawkins recreations are also mentioned in Well's 1905 novel, *Kipps,* within a social diatribe denouncing the "... ruling power of the land—Stupidity ...a monster ... like some great clumsy griffin ... like the Crystal Palace labyrinthodon." Furthermore, in his *Autobiography* Wells states how excited he'd been later in life when he imagined that an escarpment of a "denuded anticlinal" on the North Downs "... under my feet had once been slime at the bottom of a vanished Cretaceous sea."

Meanwhile, Wells delved further into evolutionary themes. Two of his other stories exhibiting paleontological or paleo-evolutionary ideas include his 1896 short story "In the Abyss" and a chapter titled "The Dreary Megatheria" within his 1928 neglected novel, *Mr. Blettsworthy on Rampole Island.* (Note – I will refrain from addressing Wells's classic *War of the Worlds*, or *The Island of Dr. Moreau* novels here, however, which are also evolutionarily-themed, yet not overtly with paleontological spin). In the 1896 tale, an explorer descends in a deep sea diving bell only to encounter, most disturbingly, a race of unknown marine, reptilian-like creatures—evolutionary "descendants like ourselves of the great Theriomorpha of the New Red Sandstone age." (i.e. 'Permo-Triassic') "In the Abyss" perhaps is Wells's story most thematically allied to his *Time Machine* novella. Then, published three decades later, his "The Dreary Megatheria" installment offers social commentary confounding the 'struggle for existence' idea, symbolized by a living crypto-herd of the Pleistocene beasts encountered by the protagonist (who finds the unfit creatures smelly, clumsy, tick-infested and downright disgusting), on titular Rampole Island. Here Wells (satirically) contemplates "futile evolution," leading organisms along "a graceless drift towards a dead end" (... not unlike Man's?).

Jules Verne's knowledge of geology and paleontology is more spectacularly apparent in his JCE novel, than in any of Wells's stories, even though the former's spin on such topics was far more Providentially-minded. So briefly, after descending into a volcanic shaft in Verne's subterranean realm three intrepid explorers led by Professor Lidenbrock, with nephew Axel (yes, their names differ from characterizations in movies), penetrate layers of time, encountering a succession of paleontological artifacts and paleo-specimen fossils, sometimes either 'living' in imagination, or in the flesh, along the way. (Most of us know the gist of the tale from the 1959 film starring James Mason and Pat Boone, but remember that's just a movie adaptation.) Especially after factoring in the JCE's 1867 edition when more paleontologically-themed chapters were added, there's a general movement through geological time evident in Verne's story, fortified with Riou's striking illustrations. Certainly as William Butcher notes, by 1865 prehistory had become a "major field of study," yet he further adds, "Verne apparently dislikes all theories of evolution." This may be so despite Axel's fantastic, pivotal "waking dream" where he envisions the totality of their time tour and Earth's geological history, a sequence that rings true to other contemporary imaginings of life-through-time, or such as popularized much later in Wells's own *The Outline of History*.

A 1930 3rd edition copy of Wells's *Outline of History* (1920) beckoning on my parents' bookshelf, not any of his fictional classics, served as my young introduction to his writings—here, his fascinating 'take' on geological concepts. Within *Outline's* first ~80 pages, Wells insightfully traced life through the Ages as subjected to the inexorably winnowing Darwinian process of natural selection, as read in the Record of the Rocks, thus constituting his preliminary 'outline of *pre*-history.' How peculiar that a man of such extraordinary vision, ability and evident command of paleontology would never write *the* quintessential sci-fi novel involving man's confrontation with living dinosaurs: that task was left to his great contemporary Arthur Conan Doyle.

Probably most readers are more familiar with Wells's *Time Machine* and Verne's JCE, not from their original writings but instead via Hollywood productions. Yet in spite of what you may (or may not) have been led to believe on basis of two classic, mid-20th century film adaptations, Wells was the ardent (cynical) evolutionist, while in his tale Verne's views remain (albeit ambiguously) more skeptical of Darwinism.

And the names of those three prehistoric vertebrates named in both JCE and *Time Machine*—my two all-time favorite sci-fi tales—as promised at my beginning? They are *Ichthyosaurus*, *Megatherium* and *Plesiosaurus*. Look them up in both stories, which are both well worth reading. And give yourself those bonus points if you caught my earlier reference to *Brontosaurus* in Wells's timeless classic.

Key References: *Jules Verne Journey to the Center of the Earth* (Translated and with a new Introduction and Notes by William Butcher), Oxford World Classics Paperback, 1998.; Allen A. Debus, *Dinosaurs in Fantastic Fiction: A Thematic Survey.* (McFarland & Company, Inc., Publishers), 2006.; Allen A. Debus, "Reframing the Science in Jules Verne's *Journey to the Center of the Earth, Science Fiction Studies*, #100, Vol.33, Part 3, Nov. 2006, pp.405-420.; Harry M. Geduld, *The Definitive Time Machine: A Critical Edition of H. G. Wells's Scientific Romance* (With Introduction and Notes by Harry M. Geduld), 1987.; Charles De Paolo, *Human Prehistory in Fiction,* Chapter 12, pp.127-141, (McFarland & Company, Inc., Publishers), 2003.; David Shackleton, "H. G. Wells, Geology, and Ruins of Time," 2017 (online)

SEVEN

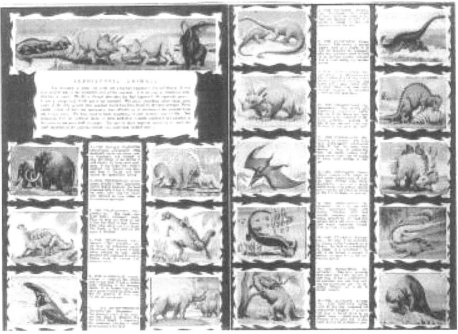

(Top) Jack Arata's imaginative 2002 drawing suggesting connections between Mary Shelley's Frankensteinian concept and Michael Crichton's Jurassic Park. (See Chapter Thirty-five.) (prepared under the author's direction for the author's 2003 article in Castle of Frankenstein). (Bottom) Dinosaur stickers from "Wonders of the Animal Kingdom." See Chapter Thirty-eight for more.

Dinosauriana

Yes, I'm borrowing a term coined by Robert Telleria and Joe DeMarco, as in their continually updated 2010s CD titled *Dinosauriana: The Essential Guide to Collectible, Figural Toy and Model Dinosaurs.* Although here I'm using the term as a 'catch-all' for the variety of dino-stuff that's out there, collectible & otherwise, including original forms or aspects of intellectual property. Surprisingly, there have been relatively few books or other compendia concerning the entire spectrum of dinosaur 'things' that individuals may desire to own or simply know about. Two of the earliest such compilations were Dana Cain's and Mike Fredericks' *Dinosaur Collectibles* (Antique Traders, 1999)—possibly now considered a collectible in and of itself; and John C. Fredriksen's *Dino-Guide: The Saurian Source Book* (Dollar Scholar Press, 1999). Over a decade later also came the lavish publication by Telleria titled *The Visual Guide to Scale Model Dinosaurs* (Blurb, 2012). Among those coveting 'dinosaurabilia,' writers have emphasized collecting of toys and models, but there are many facets to the madness. There have also been collecting guides covering all things 'Godzilla.' But the very best way to keep up on what's out there in this expansive genre is through informative periodicals such as *Prehistoric Times* and *G-Fan*.

While our house is littered with so many dinosaurian and paleontological items of every persuasion, I don't consider myself to be a 'collector.' However, in recent years I've written about so many varied aspects of the overall domain. Hence this 'paleo-potpourri' of topical items beginning with Chapter Thirty-five, offering connective thoughts between Mary Shelley's *Frankenstein* and Michael Crichton's *Jurassic Park*. (Over the years I've dabbled further into the Frankenstein mythos, for example, as in my article printed in *Mad Scientist* magazine—"Edgar Rice Burroughs' Mad Monster Armies," no.15, Spring 2007, concerning both ERB's 1929 novel, *The Monster Men* and 1940's *The Synthetic Men of Mars*.) Interests on the interplay between 'paleontological/evolutionary-themed' science fictional comic books and recordings of the late 1950s are outlined in Chapter Thirty-six.

And what would a section addressing "Dinosauriana" be without another excursion into facets of paleoart and paleoimagery—here addressed in Chapters Thirty-seven and Thirty-nine! Referring to the latter chapter, in her 1969 book, Zofia Kielan-Jaworowska aptly commented:

"The study of animals that lived on Earth millions of years ago is not merely a study of their anatomy, but first and foremost a study of the course of evolution on Earth and of the laws that govern it. All the philosophical systems ever developed by Man are essentially anthropocentric, with the tragic consequences for humanity which we know from history and which we are experiencing today. Studying the evolution of animals that inhabited Earth for millions of years and comparing this with the history of Mankind—so short on the geological scale—put Man's position in the living world in its proper perspective and help counteract anthropocentric ideas… Study of the reasons behind the extinction of species on our planet … may yet prove to be highly important to our own species in shaping its future development on Earth."

The latest, interesting paleoart controversy (though we will not delve deeper into this topic here) printed in pages of *Prehistoric Times* concerns whether non-ornithischian dinosaurs—like *Tyrannosaurus*—had fleshy cheeks and lips that functioned to conceal their sharp teeth. Paleoartist Gregory S. Paul believes they did (like modern Komodo Dragons), writing at length in an article to that effect in *PT* no. 127 (Fall 2018). Regular *PT* contributor Tracy Ford objects. However, Paul concurs that fish-eaters like spinosaurs would have had visible, snaggly-looking teeth, like modern crocs. The resultant of this disagreement would have great bearing on how faces of carnivorous dinosaurs will appear in future life restorations. Some fear that without their conventionalized protruding teeth & toothy grins, meat-eating dino-monsters would appear less 'menacing.' But really, then aren't we back to basics? Teeth in Charles R. Knight's earliest painting of *Tyrannosaurus* for the American Museum (circa 1906), shown approaching a *Triceratops* 'family,' are concealed within flesh, or 'lips.' Likewise, the most famous filmic tyrannosaur of all—that which Kong defeated in the 1933 RKO movie—bears a similar facial countenance to Knight's restoration. Knight and Kong; not bad company to be with.

Finally, there's a short piece on collectibles—*paper* collectibles—that are usually overlooked; several examples of dinosaur-themed stickers & stamps, prehistoric/geology-themed postcards, and of course those sought-after fanzines. And speaking of dinosaur stamps … the latest news concerns paleoartist Julius Csotonyi, who illustrated a new 'block of four – Forever' U.S. postage stamps commemorating *Tyrannosaurus* fossils to coincide with the opening of a 2019 Smithsonian exhibition—"The Nation's *T. rex*." You are forewarned though! His adult Rex appears on one stamp without those fleshy lips.

These chapters originally appeared in the following publications:

Chapter Thirty-five—Composited from *Castle of Frankenstein* no.34 Spring 2003, pp.32-37 and *Mad Scientist* no.31, Summer 2016, pp.19-26.

Chapter Thirty-six—Excerpted from *Mad Scientist* no.22 Fall 2010, pp.25-32.

Chapter Thirty-seven—*Prehistoric Times* no.129, Spring 2019, pp.22-24.

Chapter Thirty-eight—Composited in part from *Prehistoric Times* no. 58 (Feb/March 2003), pp.18-19, and no.74 (Oct/Nov 2005), p.57.

Chapter Thirty-nine—*Earth Science News* (Earth Science Club of Northern Illinois), Parts 1 to 2, Vol. 42, Oct. 1992, pp.11-16; Nov. 1992, pp.17-22.

Chapter Thirty-five — *"It's Alive"!!*

Welcome to 'Franken-Park': Fear Behind the Science:

To what extent can dinosaurs be regarded as 'monsters'? Many readers wouldn't guess that the roots of today's modernized, high-tech movieland 'dino-monsters' can indeed be traced to Mary Shelley's *Frankenstein, or the Modern Prometheus*—perhaps the first true sci-fi novel reflecting fears and futuristic hazards of 'bio-tech,' (written over a century before publication of Aldous Huxley's *Brave New World*, 1932).

Two monstrous pinnacles of fantasy literature, seemingly unassociated, are in fact from the same ideological mold—Mary Shelley's *Frankenstein: or Modern Prometheus*, (1818 & 1831 eds.) and Michael Crichton's *Jurassic Park* (1990). *Jurassic Park* can be viewed as a modernized version of *Frankenstein*, perhaps the tale Mary Shelley (1797-1851) would have written had she been born 150 years later. Fittingly, *Frankenstein* was written during an important period of discovery, when the first fossilized bones of extinct saurians, later to be classified as 'dinosaurs,' came to light. While Shelley dissuades use of contemporary, emerging science of electricity and chemistry to reveal God's benevolent design, Crichton exposes our fears and uncertainties over the misguided use, or abuse, of biotech. (For, if bloodthirsty and cunning dinosaurs could be cloned, why not the most contemptuous *humans*—modern 'Frankensteins' as well?).

What is a 'real monster'? Given steady bombardment of our senses by classic horror tale screen adaptations and monster fan literature, it is all too easy to accept 'traditional' views on the nature of 'real' monsters. In Shelley's story, Frankenstein's 'monster' was horrific to behold, a murderer. It was an unholy, rejected 'offspring' of its creator, instilled with unrealized potential. (In the 1931 film starring Boris Karloff as the 'monster,' however, Victor Frankenstein's creation proved psychologically unstable due to deterministic causes, basic human error—the inadvertent substitution of a criminal brain for that of a superior specimen of mankind.)

A modern renaissance in the science and art of dinosaurs budded during the heyday of 1960s monster fandom. Because they were once real animals—like lions & tigers & bears of today (Oh my!), dinosaurs are rarely given their due as 'true' monsters. Yet from human perspective, the largest and most ferocious of

Jurassic Park's carnivores (i.e. *T. rex* and vicious *Velociraptors*, a.k.a. 'raptors') would have seemed scarier if encountered in-the-flesh than would have Shelley's original conception of Frankenstein's monster. Frankenstein's monster even spoke *French*! If you could politely avert your eyes from his ghastly countenance, it would have been possible to sit down to dinner with Frankenstein's monster, engaged in artful conversation, although *Jurassic Park's* 'raptors' rudely disembowel their dinner 'guests.'

Jurassic Park's dinosaurs are encoded from DNA biomolecules, analogous to how Frankenstein's monster was resurrected from lifeless matter. As depicted on-screen and in his novel, Crichton's creatures certainly seem monstrous and terrifying. However, his 'raptors' and *T. rexes* may also be viewed as malicious, modernized, genetic reincarnations of Mary Shelley's *Frankenstein*. Billed as idealized restorations of our planet's former 'lords of creation,' much like Frankenstein's monster, stitched together from corpses, was intended to be a model human being, *Jurassic Park's* dinosaurs quickly run out of control. The moral is clear. Ultimately, with their vile (and soon disowned) creations on the loose creating unspeakable terror, men of science and greed are transmogrified into the real monsters! After all, havoc, mayhem and even chaos are anticipated consequences whenever scientists, stricken with hubris, corrupt natural order.

To be 'frank,' the first time I read Mary Shelley's *Frankenstein* years ago, I found it rather 'chewy.' After all, it *was* written nearly two centuries ago and popular writing style has eased since then. Crichton, of course, knows how to write a first-rate, modern 'page-turner.' So, in order to gain further understanding of the relation between these two great pieces of fantasy literature, I reread them both side-by-side, sort of in a 'Frankenstein vs. Jurassic Park' vein.

Metaphorically, "Frankenstein" (used hereafter to refer to Frankenstein's creation) is the 'King' of all monsters (yes, even outranking Godzilla) while from a pop-cultural perspective dinosaurs (particularly *T. rex* and evolutionarily related carnivorous theropods) reign as the lords of (prehistoric) creation. When viewed on the silver screen, such 'cult' creatures evoke feelings of horror and unholiness. An extraordinary lineage of beloved fantasy extending to Crichton's tale of bio-engineered dinosaurs has evolved from ideas first expressed in Mary Shelley's seminal novel. Since production of Thomas Edison's 1910 filmed production of Shelley's novel, a 15 minute screen adaptation starring Charles Ogle as the creature, Frankenstein or its 'derivatives' have been resurrected on film more often than any other 'monster.' Frankenstein is arguably the most influential monster of all time.

We've become so accustomed to Universal's classic universe of Frankensteinian horror, (even more so perhaps than to Hammer's influential Frankenstein-universe settings), that many of us may have forgotten the inherent fear-behind-the-science of Mary Shelley's novel. In weaving *Frankenstein* from a recalled nightmarish dreamscape, Shelley was influenced by the 'experiments' of Erasmus Darwin (1731-1802)—Charles Darwin's grandfather. Erasmus was a pioneer in the philosophical understanding of evolutionary principles. He also invented new types of pumps, carriages and other mechanical devices, proposed (if not actually tested) a means of rocket propulsion using hydrogen and oxygen, and reflected on a 'Big Bang' theory explaining the origin of stars from Chaos. But Shelley recalled in 1831 one particular experiment where:

> "(Erasmus) Darwin....preserved a piece of vermicelli in a glass case till by some extraordinary means it began to move with involuntary motion. Not thus, after all, would life be given. Perhaps a corpse would be re-animated; galvanism had given token of such things. Perhaps the component parts of a creature might be manufactured, brought together, and endued (sic - 'endowed)'with vital warmth."

Shelley's protagonist, the ingenious Victor Frankenstein, studies medieval 'alchemical' texts of Paracelsus and Albertus Magnus prior to gaining the 'modern' scientific knowledge necessary for stimulating life from cadavers. Interestingly, (according to Historian of Science Allen G. Debus), "Paracelsus wrote about generating the 'homunculus' (i.e. a miniaturized 'essence' of a human being) in a chemical flask, and Albertus Magnus was thought to have fashioned a head of brass that talked." Today, we can't speculate to what degree Victor, through Mary Shelley, was shaped by knowledge of their unusual, alleged preoccupations (and there were *other* influences as well).

Apart from his unfortunate predilection, Victor Frankenstein's ultimate failure stems from psychological character flaws. Victor is a loner, on the level of a Jeffrey Dahmer-type, a real weirdo anyone with a teenage daughter should be wary of! He is delusional, eventually branding himself a 'madman.' At first, he has prototypical visions of grandeur, desiring of his creation that, "*A new species would bless me as its creator and source.*" He rejoices in building his 8 foot tall creature. Given that the materials for his giant were furnished both from the dissecting room as well as the "slaughterhouse," one wonders whether he used any animal parts too, perhaps foreshadowing H. G. Wells' "The Island of Dr. Moreau" (1896). Later, when the 2 year project reaches fruition, Victor, stricken with 'breathless horror and disgust,' condemns the nightmarish apparition. "A mummy again endured with animation could not be so hideous as that wretch."

Taking its first convulsive breath, the creature joins the living, although without benefit of the electrical gadgetry and laboratory special effects seen in the 1931 film, and without Victor's dramatic exultation, "*It's alive!*"[1] Quite conversely, in the novel, by omitting details on how to reanimate the dead, Shelley opted to suspend disbelief in Victor's ability to do the unthinkable. Then, haunted by his ghastly creature, Victor requires medical asylum. After a prolonged convalescence, Victor (himself a 'modern Prometheus' like Benjamin Franklin who experimented with lightning) finally spies his murderous creation framed in preternatural lightning glare on a dark and stormy night. Mired in despair, Victor laments his strange progeny, a "...living monument of presumption and rash ignorance which I had let loose upon the world," much like John Hammond would 17 decades later in *Jurassic Park's* fictional setting. Ultimately, his maniacal mission to destroy the monster becomes all-consuming, ending tragically in Victor's death. (Rather analogously, *Jurassic Park's T. rex* emerges from its confines during an electrified lightning storm.).

Whereas in *Frankenstein*, Victor suffers for boldly going where no man should, in *Jurassic Park* many people are eaten and trampled by dinosaurian monsters set in motion by a chaotic turn of events. Victor pays mightily for his sin, playing God, while in *Jurassic Park* chaos, unpredictability and catastrophe become natural manifestations of the Universe's controlling force. *Jurassic Park's* somber message is that no ordered system is failsafe, no matter how many safeguards are employed, especially when creatures that should no longer be alive (i.e. bloodthirsty dinosaurs and other prehistoric animals) enter the picture, becoming integral components of that system. Chaos-personified will even pass judgment on those who have disturbed natural order. In the 1993 movie, this message comes across as, 'don't mess with Mother Nature, or you'll be devoured.'

Frankenstein and *Jurassic Park* share two other themes, those of consumption and replication.

Jurassic Park's dinosaurs go on a rampage eating everything in sight, including humans. After all, eating is a dinosaurian preoccupation. And from the outset of our fascination with such 'prehistoria,' humans *have* been 'consumed' within the 'belly of the beast'—by dinosaurs and other prehistoric horrors. In fact, in 1853, British Victorian scientists celebrated New Year's Eve wining and dining themselves into the wee hours of the morning within the life-sized mold of a dinosaur (*Iguanodon*) sculpted and cast for the world's first such exhibition of prehistoric animal statues sculpted by Benjamin Waterhouse Hawkins. To be consumed with (or by) dinosaurs is not far removed, metaphorically, from how fully Victor Frankenstein becomes self-absorbed in his maniacal quest to create a person from non-living body parts. In *Frankenstein*, self-absorption leads to 'enslavement' by the creature, and ultimately... death. In a sense, tormented Victor has been psychologically *consumed* by his monster, not unlike Crichton's victims have been savagely devoured by rampaging rexes and raptors.

The first true 'American monster,' every bit as imaginary as Frankenstein's monster, predates Shelley's invention. This, as claimed by Paul Semonin in his fascinating account, *American Monster* (2000) was early perception of a prehistoric beast—our American Mastodon, originally regarded as a fearsome carnivore. Intriguingly, on Christmas 1801, participants celebrated at a dinner staged *within* the first reconstructed skeleton of an American Mastodon, foreshadowing that British dinner-inside-the-*Iguanodon* half a century later! Thus, metaphorically, we sense a continuum, from the first real 'American monster' through Mary Shelley's *Frankenstein*, leading to *Jurassic Park's* raptors, and beyond.

Perhaps the gravest, ultimate crisis facing mankind today is overpopulation, yet is the least acknowledged environmental crisis. But when monsters begin spawning in an out-of-control rate, then we have reason to fear. So, in relative safety we face down our fears of overpopulation through favored sci-fi books and movies. For instance, how can we readily forget the horrific 'alien' replication featured in Ridley Scott's *Alien* (1979, and sequels). From self-replicating robots of the sci-fi pulps, frightening vampires of Stephen King's *Salem's Lot* which multiply at a geometric rate, vast clutches of baby-godzilla hatchlings invading New York City in Tri-Star's *Godzilla* (1998), to armies of human clones in a recent 'Star Wars' prequel.... we've seen em' all. Also, in his *The Last Dinosaur Book*, (1998), W. J. T. Mitchell discusses Czech writer Karel Capek's invasion of technologically advanced human-sized 'newts' (i.e. *War with the Newts*, 1937). Capek is also credited with coining the term, "robot," first used in 1921.

And *Jurassic Park* was no exception. Here, dinosaurs were bio-engineered to be all females, to *prevent* their replication. But (sigh!) because missing strands of dinosaur DNA were selected from modern frog DNA (incidentally, bird DNA would have obviated the replication disaster!), and simply because "life finds a way," male dinosaurs also hatched allowing fearsome dinosaurs, vicious 'raptors' and poisonous 'compys' especially, to reproduce and spread, like a 'viral epidemic,' at an alarming rate. In *Frankenstein*, 13 decades prior to the discovery of DNA, Mary Shelley's Victor creates a female companion for his creation. But in revulsion he destroys the 'bride' before the experiment reaches the point of no return. If he hadn't, conceivably, an unholy union of reanimated supermen and women would replicate the world over. In one passage of Shelley's novel, Victor Frankenstein reflects on the demon seed to be spread:

> ".....if they were to leave Europe and inhabit the deserts of the new world... one of the first results of those sympathies for which the demon thirsted would be children, and a race of devils would be propagated upon the earth who might make the very existence of the species of man a condition precarious and full of terror. Had I (the) right.... to inflict this curse upon everlasting generations?"

(Of course, because both the monster and his bride are wholly human—genetically, their progeny would be ordinary humans, not supermen or even monsters, although we can't fault Mary Shelley for not knowing this.)

Had Shelley written a 'sequel' to *Frankenstein*, wouldn't that have been the natural plot—a 'tale of futuristic' frankensteinian 'generations' challenging our species for supremacy? (Incidentally, sequels to *Jurassic Park* –Crichton's

The Lost World (1995) and Spielberg's 2001 film "*Jurassic Park III*" et. al., as well as all the Universal and Hammer Films movie spinoffs following the 1931 film *Frankenstein*, are in themselves forms of literary and filmic replication.) In monster stories, replacement of the human race by races of super-beings (genetically reconstructed dinosaurs or 'supermen') remains a conceivable outcome of 'replication.'

Even *Jurassic Park* (1990) was anticipated by (at least) four terrifying sci-fi tales. One notes the essential similarities of Donald Glut's novel *Spawn* (1976) to Crichton's 1990 novel. Except that in *Spawn* the pseudo-dinosaurs were brought alive to Earth to inhabit an amusement park; they were not regenerated from fossilized DNA. Robert Wells's 1969 novel *The Parasaurians* is among the earliest brimming with sci-fi, genetically-restored dinosaurs. Sci-fi legend Robert Silverberg's short story "Our Lady of the Sauropods," a 1980 entry recently reprinted in a 1996 anthology—*Dinosaurs!*, edited by Martin H. Greenberg, also relied on dino-DNA reconstructions, just like Crichton did a decade later. Harry Adam Knight's gripping 1984 novel *Carnosaur* and Allen Steele's novella "Trembling Earth" (*Isaac Asimov's Science Fiction Magazine*, Nov. 1990), both employed monstrous DNA-regenerated 'raptor' dinosaurs. (Knight's absorbing thriller caused me to miss my train stop one night on the way home from work, the only time that ever happened to me in 31 years of commuting.)

Following on the heels of *Jurassic Park's* book store and box office successes, other sci-fi tales appeared in the pulps, including Victor Appleton's *Tom Swift - The DNA Disaster* (1991), and Ian McDowell's "Bernie" (*Asimov's Science Fiction,* August, 1994), the latter a horrifying spoof on "Barney" –the alarming, TV-land purple dinosaur. Also, in Thomas Hopp's *Dinosaur Wars* (2001) a technologically advanced races of space-faring dinosaurs attack Earth from the Moon after being reconstituted from dino-DNA preserved in a lunar base. For those desiring peace of mind, Rob DeSalle and David Lindley clarify in their book *The Science of Jurassic Park and the Lost World* (1997), that it's pretty darn near impossible to regenerate a dinosaur the way Crichton speculated it could be done. (To my knowledge, "The Science of Frankenstein" has yet to be written.)

Many realize *Jurassic Park's* premise, that unpredictable, catastrophic chaos can arise from complicated ordered systems, (related to the natural extinction of dinosaurs in Crichton's sequel *The Lost World*) is flawed in its presentation. For Ian Malcolm, a mathematical 'chaotician,' cautions that dinosaurs shouldn't be cloned from ancient DNA, although by then it is too late. According to paleontologist Stephen J. Gould, Steven Spielberg's:

> "...Malcolm ....preaches the opposite of chaos theory...." and is given, "...the oldest diatribe ... and predictable staple of every Hollywood film since Frankenstein: man... must not disturb the proper and given course of nature; man must not tinker in God's realm."

Yet Gould was mistaken in tracing the origins of this time-honored theme to Hollywood. For a century earlier, in 1831, Mary Shelley penned these words following a terrifying nightmare in which she:

> "... saw the pale student of unhallowed arts kneeling beside the thing he had put together. I saw the hideous phantasm of a man stretched out, and then, on the working of some powerful engine, show signs of life and stir with an uneasy, half vital motion. Frightful must it be for supremely frightful would be the effect of any human endeavor to mock the stupendous mechanism of the Creator of the world."

The chaos which follows the unleashing of Frankenstein's monster on the world in Shelley's novel certainly rivals the chaotic impact of Crichton's dinosaurs.

Connections between Mary Shelley's *Frankenstein* and Crichton's *Jurassic Park* are inescapable. Dinosaurs and Frankenstein's monster(s) (i.e. including its 'bride') are forever linked thematically, historically and metaphorically. We acknowledge the dawning association between monsters of literature and the first known fossil reptile later claiming status as a dinosaur. For Gideon Mantell, co-discoverer (with his wife Mary) of the bones of *Iguanodon* claimed of its probable life appearance as he attempted an early 1830s reconstruction how he was appalled at the gigantic "..being which rose from his meditations..... like *Frankenstein*." *Iguanodon*, perhaps the 'Frankenstein' of dinosaurs, became the creature immortalized through the famous New Year's Eve celebration of 1853, with humans—those consumed engaged in consumption, reconstruction from inanimate (fossilized) remains and even replication (i.e. from cast/sculpted dinosaurs), dining within its restored abdomen. Thus, from early perceptions of America's Mastodon, through Shelley's *Frankenstein*, through an inaccurately restored *Iguanodon* of Victorian times, into the 'Jurassic Park' era of cgi-replicated dinosaurs, the spirit of 'real monsters' lives on......... *"It's alive"!!!!!!!!*[1]

Who Will be the LAST MAN? Mary Shelley's 'other' legacy:

Most would correctly couple the name Mary Wollstonecraft Shelley (1797-1851) with *Frankenstein*, written in 1816 when she was only nineteen. But if you asked readers what *else* did she write during her career, you'd probably be greeted with stone-cold silence. In truth, *Frankenstein, or the Modern Prometheus* (1818, 1831) proved to be the highlight of her career as an author. But besides lugubriousness, she had other writing interests delving into travel books, historical novels and editing a collection of her deceased husband Percy Bysshe Shelley's poetry. She even considered and proposed authoring books on geology and prehistoric archaeology! Indeed, in 1830 she had befriended geologist Roderick Murchison, and during the mid-1840s was a neighbor and friend of Dr. Gideon Mantell, who two decades prior had described the second extinct vertebrate assigned in 1842 to the Dinosauria, *Iguanodon*. Mantell even casually referred to this dinosaur as a "Frankensteinian" creature arising from his scholarly meditations.

Mary's writings, and adverse personal matters, diluted her time to further explore horror, science fiction or the supernatural. But she succeeded with one other important novel belonging to the horror, science fictional genre. While the story itself has largely been neglected, *thematically* the idea has become greatly favored by today's Hollywood producers and writers of dystopia, transcending its early 19th century gothic/romantic roots. I refer to the unsettling idea of the 'last man' on Earth, an end to civilization as we know it, anticipated by Mary Shelley in her 1826 novel, *The Last Man*.

Before we come to her 'last man' entry, however, let's briefly peruse Mary Shelley's additional shorter stories which were of science fictional or other-worldly, fantastic flavor. Yes – during 1819 to 1833, she did write five such tales. (She also wrote an article "On Ghosts" in 1824 for *London Magazine*, unseen by this author.) Two fantastic novels and five short supernatural tales published during those word processor-less times hardly qualifies her as a Victorian age 'Stephen King,' but in her themes of reanimation and also elimination of people from the globe, resultant of catastrophe, she was building a foundation for ensuing decades (and centuries) of popular readership.

Shelley's short fantastic fiction:

Perhaps Mary's best known (and last) genre short story is "The Mortal Immortal: A Tale," published in *The Keepsake for MDCCCXXXV* in 1833, for its time clearly science fictional in nature. Although the writing style no longer rings 'modern' to the ear, its first line draws you in: "July 16, 1833 – This is a memorable anniversary for me; on it I complete my three hundred and twenty-third year!" This is an exceptional tale about immortality, found in an elixir. Although dark alchemy lies at the heart of Shelley's tale, those who enjoyed Stephen King's *The Green Mile* might be intrigued by this early precursor. Young protagonist 'Winzy' (derived from the Scottish word, 'winze' meaning 'curse'), is alchemist/philosopher Henry Cornelius Agrippa's assistant and pupil. Readers familiar with her *Frankenstein*, may recognize the historical Agrippa's (c. 1486-1535) name from that novel; he is one of the old authors initially read by Victor Frankenstein as he misguidedly researches how to reanimate the dead.) Agrippa has found the formula for immortality, but when the reaction is completed, Winzy quaffs half of the potion, misunderstanding that it will cure him of his jealous, unrequited love for the pretty Bertha.

Hundreds of years later, Winzy, who has not aged in appearance throughout, marvels how much longer his natural life should extend, especially considering that he only consumed *half* of the liquid. In Frankensteinian fashion,

Winzy laments that he has not found a 'bride' for himself who also suffers from the same affliction, and ponders wandering to the cold planetary regions (i.e. as did Frankenstein's monster at the end of her most famous novel), where he might eventually succumb to the "powers of frost." Is Winzy then destined to become the Last Man?

The earliest short tale Shelley penned, "Valerius: The Reanimated Roman" (1819), was short because it represents the beginning to, perhaps, an unfinished novel. Mary had spent considerable time living in Italy, learned how to speak the language and ardently studied its history, collecting considerable information that would appear in her writings, such as "Valerius." (In today's parlance, Mary would rank as an accomplished scholar of the Classics (Latin and Greek) and qualify as a very capable student of the Romance Languages, perhaps attaining a PhD in Humanities.) Although not actually explaining how Valerius was reanimated from his 1st century tomb, the Roman republican expresses dissatisfaction with what imperial Rome has degenerated to in (then) modernity. During his rambles among the ancient ruins with a Lady Isabell Harley, she's finally forced to acknowledge that while "his semblance was life, yet he belonged to the dead."

It should be added that in his delightful 1996 tale, "The Unexpected Visit of a Reanimated Englishman," sci-fi writer Michael Bishop speculates that both Valerius (and Roger Dodsworth – see below) may be regarded as having *thawed* to life. Furthermore, Bishop suggests that in such writings Mary may have "pioneered" the alternative history story, that is, in juxtapositioning or "translating figures from the past into other time periods and (citing Virgil in his *Aeneid*) turning them 'naked of knowledge into this world'."

Shelley 'experimented' with the consequences and idea of reanimation once more in her 1826 tale, "Roger Dodsworth: The Reanimated Englishman," evidently left unpublished until 1863. The titular Dodsworth had died at 37 years of age in 1654, but was exhumed from under avalanche debris in July 1826, remarkably in suspended animation, "hermetically sealed" due to "action of the frost," and remarkably resuscitated. Mary conjectures, "Now we do not believe that any contradiction or impossibility is attached to the adventures of this youthful antique. Animation (I believe physiologists agree) can as easily be suspended for a hundred years to two years, as for as many seconds." An embodied anachronism, Dodsworth no longer fits in well with society. When he ultimately disappears mysteriously, this rather suspends disbelief as to whether the incredible tale was true, as it's now impossible to verify (or falsify).

Mary's next published tale was "The Transformation" (1830) published in *The Keepsake for MDCCCXXXI*. After a near drowning of a misshapen dwarf – referred to as "all-powerful monster"—who possesses magical powers, he and a wretched yet handsome man named Guido roving the Italian shore in desperation—who has been profligate in his youthful ways, agree to exchange bodies, but supposedly only for three days! Guido (now as the dwarf) receives treasure for temporary loan of his body, while the dwarf (now masquerading in

Guido's strong body) romances Guido's beautiful former fiancee', Juliet, (who earlier had spurned him for his utter irresponsibility).

To Juliet, that would-be 'Romeo' Guido now appears reformed and penitent, acceptable for marriage, which piques the real Guido (as dwarf) into a jealous rage. But, furthermore, the lying dwarf has reneged on the 3-day body loan deal, and so Guido (as the dwarf) kills the dwarf (residing in Guido's body) during a sword and dagger fight. Awakening in his own rightful body once more (e.g. effecting the titular 'transformation'), Guido now understands and deplores the folly of his former ways. This is another Frankensteinian tale, as the near drowning recalls that of Mary's husband Percy (in 1822), and the monstrous dwarf's (i.e. Guido's soul) visitation upon Juliet and Guido (as the dwarf) near the time of their wedding night recalls key passages from *Frankenstein*.

Mary's "The Dream" (1831), published in *The Keepsake for MDCCCXXXII*, an atmospheric fantasy set in historical France, concerns a lady in mourning, ghostly knights of the realm and a spiritual dream. Although supernaturally themed, it is often neglected by her biographers and historians of science fiction.

Grainville vs. Shelley:

Although the best known and arguably most influential example dating from this early period, *The Last Man* (1826) novel actually wasn't the first of its kind, and Mary probably borrowed from others' prior inspirations (possibly including poetic sources). However, there was (at least) one other novel originally published in French as *Le Dernier Homme* in 1805, written by Jean-Baptiste Francois Xavier Cousin de Grainville. Grainville's novel was translated into English as *The Last Man (or Omegarus and Syderia, a Romance in Futurity)* the following year. According to science fiction historian I. F. Clarke writing in 2002, Grainville's is the foundational apocalyptic fiction novel. It differs in cause for the end of man significantly from Shelley's later entry, in that a referential source of insurmountable problems faced by lovers Omegarus and Syderia lies in the biblical Book of Revelations (e.g. *not* as resultant of a global pandemic).

As stated by Clarke, "In effect, Grainville wrote his own sequel to Genesis...." (p.xxx) But like Shelley's, Grainville's novel (despite its biblical characterizations) is also regarded as highly 'secular' in nature, therefore casting it into the apocalyptical science fictional realm. Meanwhile, Paul Alkon considers how Grainville "secularizes the Apocalypse *without* discarding its theological framework." (p.175) Whereas in Shelley's novel the last man eventually expires and the world simply moves on ... after, but without man, in Grainville's more sweeping futuristic vision the dead return to life as zombies, a geological upheaval ensues, and even the Sun and light of the starry heavens are extinguished.

But although Grainville's novel may represent the proto-Precambrian beginnings of apocalyptical futuristic fiction, Shelley's has been more widely recognized as a bare bones benchmark ancestor of the genre. Brian Aldiss, for instance, catalogues Shelley's 1826 tale as "scientific romance" (p.49) while neglecting Grainville (p.448). Likewise, Morton D. Paley doesn't acknowledge Grainville, however denying that Shelley's *The Last Man* is "… itself a science-fiction novel (although) the genre has made the reading public acquainted with imagining an end of things … Lastness."

So, in a very different sense, has the experience of living with the Bomb. In a time more willing than Mary Shelley's to consider the subject of Lastness seriously, her novel has been republished and has received increasing critical attention. It has rightly become regarded as its author's most important work after *Frankenstein*." (p.xxiii – *Last Man* novel)

As she often copiously did, Shelley's intensive researches into her 1826 novel, (among other things), may have been influenced by contemporary events, a real plague of 1818 spreading northward from Calcutta, reaching the Caspian Sea area by 1823 (and sweeping into Britain by 1831), as well as books such as Daniel Defoe's – *A Journal of the Plague Year* (1722), and *Robinson Crusoe* (1719) with its abject isolationist theme. Shelley's plague isn't supernatural, but there are valid reasons to regard it plague as female-personified. (Paley, pp.xix-xx) Nor is her vision of late 21st century's technology much advanced from her own period. "Unlike Grainville's attempt to combine epic, biblical, and novelistic structures, Shelley's *Last Man* is firmly rooted in the realistic tradition of Defoe, although there are strong affinities with the gothic as in *Frankenstein*." (Alkon, p.190) Victor Frankenstein's Monster itself, while not a "last man," certainly was a wretched kind of lonely 'only man.'

Spencer R. Weart notes analogies between the two authors, Grainville and Mary Shelley:

> "… Grainville … referred to the Judgment Day of Christian tradition, but the bulk of his story was unlike anything published before; it showed the human race dwindling through natural processes to a lonely end. Probably the author was contributing something personal, for he was himself a desperately lonely man – late one winter night after finishing his book, he drowned himself. His unhappy image of a last man survived …. The most substantial of these was a *Last Man* novel that … Shelley published in 1826. Her long book was filled with melancholy passages about how war and plague extinguished humanity … According to her biographers, this author too had reasons for writing about desolation, for she had grown up motherless, and before writing this novel she had lost her husband and other(s) … through tragic early deaths. Perhaps such authors used tales of world devastation to stand for more personal problems?" (p.19)

Shelley's main character, Lionel Verney, who is an alter ego for herself, wanders at length, eventually reaching Rome (a monumental city and ancient civilization she revered from her travels and historical readings), before eventually expiring after realizing he has become the last man alive.

I've read *Frankenstein* 3 or 4 times, and appreciate Shelley's style of writing therein. But the first third of her *The Last Man* is downright tedious. This is probably why, while considered "impressive," her lengthy 1826 novel was poorly received. According to Sunstein, "One reviewer called the book an 'elaborate piece of gloomy folly.'" (p.271) Unfortunately, Shelley's editor "rushed the work into publication on Jan. 23, 1826, preventing Mary from making the revisions she knew it needed. ... reception of the book into which she had poured herself so demoralized (her) that she did not begin another novel for a year and a half." (pp.269-272) Sunstein also suggests that, "With *Frankenstein* she founded the genre we call science fiction ... and enlarged its possibilities in *The Last Man*, the first futurist catastrophe novel and one of the most ambitious novels ever undertaken by a woman." (p.4)

In Shelley's *The Last Man*, when an explosion detonated in a war between the Greeks and the Turks causes a toxic, plague-tainted cloud to spread over the planet, over the course of several years, Verney watches comrades and innocents alike succumb to affliction, then die. He writes the 'Last Man' document, for us readers to enjoy vicariously, before himself succumbing to the plague. Indeed, Mary may not have fully understood how cholera would be organically transmitted (via flies and feces—causes of such affliction); for the most part, her fictional plague isn't (generally) contagious. But eventually it accomplishes its contemptible work –destruction of the human race. Although it isn't a purpose of recent apocalyptical fiction to offer accurate futuristic forecasts, it's usually intended to suggest disquieting problems with the present day. It isn't clear to me, though, whether Mary had such a coherent vision in her *Last Man*. Perhaps her book served mainly as a powerful cathartic, a tomb-like exercise for sorting out sorrowful episodes in her own life using alter ego figures in the story representing people she had loved in life and lost.

Last Man Artistic & Literary examples:

Throughout the annals of sci-fi genre there have been numerous "last men," too many to recount here, but we must endeavor to explore Mary Shelley's (and to some extent Grainville's) strangely unrecognized, unheralded legacy, manifesting afterward in various media. One of the earliest was revered artist John Martin's painting titled "An Ideal Design of the Last Man," the first of 3 pictures in his series on the sublime subject, portrayals done between 1826 and 1849. But the idea of 'Lastness' and the experience of being Earth's 'last man' increasingly creeps into popular, apocalyptical fiction thereafter. A (c. 1840) play by George Dibdin Pitt was advertised as "partly founded on Mrs. Shelley's thrilling novel." (Sunstein, p.366)

An oft-cited last man tale is astronomer Camille Flammarion's *La Fin du monde* (1894), in which Earth's Last Man shivers in a greenhouse for warmth, because in the distant future the ever-diminishing rays of a continually cooling

Sun no longer heat our planet sufficiently. Extreme frigidity also eventuates in G. Peyton Wertenbaker's June 1926 *Amazing Tales* story, "The Coming of the Ice." Here, millions of years hence, a man, made immortal via fantastic surgery performed in 1930, writes a chronicle no one will read (other than those avid *Amazing Tales* subscribers, of course), of how he came to be Earth's Last Man. After living through countless millennia, the Sun finally dims signaling return of an Ice Age. And from the protagonist's final record: "The end of the world. And I – I, the last man ... the last man ... I am cold – cold ..." (p.66)

And if there is ice, then there must also be celestial fire, according to the poets. Accordingly, in Robert Duncan Milne's "Into the Sun," published in 1884, we read of a solar flare-up, resulting from a comet colliding with the Sun, raising the Sun's radiance to lethal levels. Here, our soon to be extinguished last man, attempting to escape the boiling temperatures below, pilots a hot air balloon, at last uncontrollably descending toward a molten, "igneous" Earth! (Milne's tale would perhaps prompt recall that eerie 1961 *Twilight Zone* episode, "The Midnight Sun.")

Two creatively spun paleo-novels cast another kind of (non-apocalyptical) last man, but who is also a 'first man,' living in the Mesozoic Era, having traveled there (1.) via a one-way time machine trip, and (2.) via inadvertent time slippage or "dechronization." These, respectively, are Douglas Lebeck's *Memories of a Dinosaur Hunter* (2002), and paleontologist George Gaylord Simpson's *The Dechronization of Sam Magruder* (1996). Both men are irrevocably exiled in the past, but turn out to be hardy Robinson Crusoe-like survivors of their ordeals. At least they have many real Late Cretaceous dinosaurs to observe during their golden years. And then a more controlled, scientific time travel excursion to the asteroid impact event of 66-million years ago, in Will Hubbell's *Cretaceous Sea* (2002), results both in a last man and last woman surviving in that prehistoric time frame—after the rest of their investigative team expires. They witness the 'actual' horrific impact, but eventually they're rescued, returning to modernity.

However, we encounter our most moving literary 'last man' example near the end of H. G. Wells's *The Time Machine* (1895), in which the Time Traveler advances to the far flung future. Long after human species has evolutionarily split and then expired and the Sun has entered a weakened red-giant stellar phase; Earth is in decrepitude. Therefore, by observing this oppressive scene the Time Traveler has become an anachronistic 'last man,' that is before he returns to his present, in Earth's by then forgotten past. (Yes – this fascinating literary scene was omitted from the 1960 film starring Rod Taylor and Yvette Mimieux.) Then, F. Bridge's *Amazing Tales* story, "Via the Time Accelerator" (1931), concerns a time traveler who becomes Earth's last man alive, over a million years from now, by moving forward through time. And the title of another sci-fi story in a sense pays homage to Grainville's odd juxtapositioning of first and last men (i.e. Adam and Omegarus key characters in his 1805 novel). For the title of Olaf Stapledon's 1931 novel, *Last and First Men*, a tale about our remote evolutionary futurity, certainly conjures Grainville, figuratively. But this short outline is merely the tip of that old proverbial iceberg! (For example, Arthur C. Clarke's 1953 novel, *Childhood's End,* creatively introduces a last *Homo sapiens*.[2])

Four (or maybe only three) Varieties of 'Last Man' TV shows & Films:

Eventually, after invention of the hydrogen bomb, when 1950s public fears of global thermonuclear war escalated, 'last man'—themed pocalyptical tales became increasingly popular, permeating to the cinema. There's quite a filmic assortment, scores of movies, but perhaps one may discern a general four-fold categorization. First, there were those last men resulting from a nuclear world war, such as in *On the Beach* based on Nevil Shute's 1957 novel, and we have that quintessential *Twilight Zone* episode, "Time Enough at Last" starring Burgess Meredith whose sad character only wants to read books. In *Gojira*, (1954), a metaphorical living manifestation of the Bomb threatened to destroy civilization down to the last man, thus rendering man extinct.

And, in a second category, there also were normal (immune) 'last men' facing metaphorical terrifying creatures who had mutated from fellow ordinary men under ravages of a plague, such as in *The Last Man On Earth* (starring Vincent Price), or *The Omega Man* (starring Charlton Heston) – both based on Richard Matheson's 1954 novel, *I Am Legend* – or 1968's gory *Night of the Living Dead*, also inspired by Matheson. In *The Day the World Ended*, mutated humanoids, who perish in uncontaminated rain water, menace a dwindling number of normal unaffected nuclear war survivors. These ideas have carried down to the present day, as eventually one of the humans cast in *The Walking Dead,* or SYFY's *Z Nation* series seems, conceptually, destined to become Earth's last normal human being, (while the *Mad Max* series rages on screen).

A possible third category, potentially, regards humanity on the brink, in the face of overwhelming natural disaster, such as a massive planetoid collision with Earth, or sudden drastic climate change. These movies were inaugurated perhaps by science fictional printed stories, the likes of *When Worlds Collide* (1933 novel by Philip Wylie and Edwin Balmer). This classic set a trend toward numerous, analogous outer space-triggered disaster films (including 1951's *When Worlds Collide*). Such dramatic fare became even more popular after it was discovered that the dinosaurs expired in the Late Cretaceous following a 6-mile diameter asteroid impact. How many humans would have survived a natural catastrophe of this magnitude, and the ensuing "impact winter," if any? Usually, however, at the cinema, even following the convulsive unleashing of nearly inconceivable, irreversible, Era-ending natural forces, humanity devises ingenious means of survival. So afterward there is no identifiable 'last man.' While this category remains a plausible one for 'last man' intrigue, offhand I cannot name a natural disaster-themed flick inevitably resulting in one *definitive* Last Man.

Perhaps a final *filmic* category presents the solitary mutated human – with mournful story related more or less from his (or her) perspective, in which his altered persona presages the inevitable forthcoming 'last (altered) human' event, symbolically resultant of nuclear war, as in films like *The Amazing Colossal Man*, or *The Incredible Shrinking Man* (based on Matheson's 1956 novel, *The Shrinking Man*. But there are so many fantastic films out there and it's difficult to 'shrink' them all down categorically under a mere three or four headings. Movies like 1968's nuclear war-themed *Planet of the Apes,* or *The Stand* (based on the Stephen King novel) involving a world sweeping plague, certainly flirted with last man apocalyptical concepts but in the end, even before ensuing climactic events, numerous humans are left living. And although, sometimes, reduced to savagery, in such stories it's clear that humans won't be *completely* exterminated. The producers of modern apocalyptical movies just can't seem to exterminate humans as efficiently as Mary Shelley, nearly two centuries ago.

Mary Shelley – the persona:

Throughout her young life, Mary Shelley was rather an outrider when it came to social mores. She lived a controversial lifestyle, fated as an outcast. During her youth – late teen years, especially while she wrote *Frankenstein,* I regard her somewhat as a hippie type mindset decades before such gentle folk came out in droves, as a wave or mid-1960s politically motivated demographic 'force.'(Yet she was also a largely self-educated, well-read intellectual: I'm sure that in polite discourse she could put most recent college graduates to shame with her knowledge, powers of imagination and deeply rooted level of thinking.) She was born in a year (1797) of freakish atmospheric circumstances, most likely due to volcanic eruptions in the Ecuadorean Andes, the likes of wouldn't be witnessed again until 1816 – the year she crafted her *Frankenstein* novel (resulting from the explosion of Indonesia's Mt. Tambora). Like conditions of her twin volcanic 'birthings,' for better or worse, Mary's life thereafter would forever be noteworthy, if not notorious.

She considered herself an atheist, and her societal reputation was forever tainted for her teenaged affair with the older (married) Percy, and shacking up with him in Geneva (before they eventually married). For this and other indignities she became a scurrilous source of gossip, considered a voluptuous center of the free love-espousing "League of Incest," a party pf comrades who roamed, 'summering' for several years from Switzerland, Germany, Italy and France. Mary became branded for life until her London death in 1851 from a brain tumor. Mary Shelley's fantastic writings accrued over a creative, formative 15-year period.

But following 1833's well written "The Mortal Immortal," then in her mid-30s, she seems to have retired from crafting sci-fi supernatural tales. *The Last Man* proved to be her final science romance novel. And yet she remained prolific. (See Sunstein, pp.409-415 for a complete listing of her works.) There was still writing to do, or rather – fresh ink left in her inkwell. But there would be no more tales in the sci-fi or horror vein. One wonders why. For one, she believed that she was more adept at writing nonfiction as opposed to fiction. And although she craved intellectual conversation, too often, people – who were perhaps of less classy distinction—were often rude to her in social settings, which she resented. By the 1840s, she was so despondent over the apparent "social ban" that even "she wished never to publish again, 'that never my name might be mentioned in a world that oppresses me.'" (Sunstein, p.361) Huh!? Yeah, wow! While she had high aspirations in beginning her *Last Man* novel, its disappointing reception probably chilled her alacrity for continuing in the newly created craft of science fiction writing, a genre which in a large sense, and ironically, she had created with her undying *Frankenstein* novel.

Select References:
Paul K. Alkon, *Origins of Futuristic Fiction*, (The University of Georgia Press, 1987).
Brian W. Aldiss, (with David Wingrove), *The Trillion Year Spree* (1988).
Lisa Debus, "The *Real* Castle of Frankenstein," *Scary Monsters* no.82, April 2012, pp.85-86.
Radu Florescu, *In Search of Frankenstein*, (New York Graphic Soc., 1975).
Donald F. Glut, *The Frankenstein Legend*, (Scarecrow Press, 1973).
Jean-Baptiste Francois Xavier Cousin de Grainville, *The Last Man* (Wesleyan University Press ed., 2002), translated by I.F. Clarke and M. Clarke; Introduction and Critical Materials by I.F. Clarke.
Samuel Rosenberg, "The Horrible Truth about Frankenstein," *Life*, vol.64, no.11, March 15, 1968, pp.74B-84.
Mary Shelley, *The Last Man* (Oxford University Press ed., 1998), with an Introduction by Morton D. Paley.
Mary Shelley, *The Immortal Mortal: The Complete Supernatural Short Fiction*, compiled by and with a narrative Introduction by Michael Bishop, (Tachyon Publications, 1996).
Emily W. Sunstein, *Mary Shelley: Romance and Reality* (Little, Brown and Co., 1989).
Spencer R. Weart, *Nuclear Fear: A History of Images*, (Harvard Univ. Press, 1988).
G. Peyton Wertenbaker, "The Coming of the Ice," (1926) in Michael Ashley, ed., *The History of the Science Fiction Magazine* (1926-1935), vol. 1 (Henry Regnery Co., 1976).

Notes to Chapter Thirty-five: (1) In 1928, several years prior to release of the Universal film, *Frankenstein*, author Max Begouen wrote a novel concerning the 'resurrection' of a Siberian mammoth, using electricity.

"...The galvanic currents were applied to the nerve centers, as biogenic rays bathed everything in an intense purple light....Suddenly, the animal's legs quivered. The bent trunk seemed to lengthen. The tiny eye, once dull and lifeless, appeared to swell and come alive, gazing vaguely about. The body temperature rose, quickly.... then the chest appeared to rise. 'Oh, look,' said Nadia, pointing at the faint cloud of steam coming out of the trunk's twin openings in rhythmic puffs. Nadia seized Mougin's hands. 'It's alive,' she cried. At that very moment, the mammoth's eyes blinked and a powerful blast of air roared from its rising trunk...." Quotation found in *The Fate of the Mammoth*, by Claudine Cohen, The University of Chicago Press, 2002, pp. 13-14. (2) "Sci-fi is apocalyptic," stated actress Gillian Anderson of TV's *The X-Files* in a 2016 interview. In Arthur C. Clarke's novel, the final generation of humans evolves into another enhanced species linking to the universal "Overmind," while human character Jan Rodricks, who becomes subjected to space-time relativity following an 80 year round trip journey to another planet becomes by proxy upon his return, Earth's last man. A significant & more recent example of the 'apocalyptical' phenomenon is kaiju-writing.

Chapter Thirty-six — *A Record of Space-Time*

Do you recall when we *listened* to the famous monsters and classic science fiction tales, rather than principally *viewing* them? If you do, you're reaching 'old codger' status by now. Here, I'm talking about vintage long-playing (LP) vinyl records featuring voices and sounds of the fantastic!

We owned four such LP vinyl records: "Famous Monsters Speak" (1963)—scripts written by Cherney Berg, highlighting the voices of Gabriel Dell; "Scary Tales, featuring John Zacherley" (1962); "Monster Mash, featuring John Zacherle" (1964); and "Space Stories and Sounds," (1959) dramatically narrated by Bill Stern (1907-1971), an American actor and radio sportscaster. In the midst of our monster model bedroom floor 'museum,' (accentuated by plastic model skeletons of *Tyrannosaurus*, *Brontosaurus* and plated *Stegosaurus*—yes, all inhabitants of Kong's 'Skull Island'), what better way to enhance the creepiness or the bizarre than in time to the 'actual' voices of Count Dracula, or in syncopation to Zacherley's lyrics of the macabre, or 'other-wordly' sounds of Morlocks machinery! Beatles bea-damned! Instead, using our record player, we would 'twist again' when it was time to squeeze the mummy for its succulent juices!

In retrospect, how we obtained these recordings is no real surprise. My father was a renowned collector of antique records and phonographs. Perhaps he sought to instill his kids with a love for such things. (In fact, I wrote about the oldest monster record in his collection. See *FilmFax plus*, 2006, no. 109, "Out of Hyde-ing," pp.86-87, 130.) I also covered the old 'Monster Mash' style recordings in an article printed in *Scary Monsters Presents Monster Memories 2010, Yearbook*, "Monstrous Recordings of the Past.")

But the earliest (and arguably best) recording that we received from our dad, listened-to avidly, was the science fictionally-themed *Space Stories and Sounds*, printed by Lion Full Fidelity (catalog L-70086). Side 1 of this LP contained narrations of H. G. Wells' (a.) "The First Men in the Moon", and (b.) "The War of the Worlds." Side 2 presented my two favorites on the LP (a.) Wells' "The Time Machine," followed by (b.) Jules Verne's "A Journey to the Center of the Earth." Each recorded segment is approximately 8 ½ minutes long. Scripts were adapted from *Classics Illustrated Magazine* by Roberta Strauss. Music was by Bill Simon; the album was produced by Arthur Pine. More particularly, organ music was played by Simon, percussion by Herb Harris and "sound effects" were attributed to Walter Gustafson. On the back cover of the LP, we read that "a coupon is included with this album which offers the purchaser a special money saving offer for a subscription to *Classics Illustrated Magazine*."

To youngsters as we were, the sounds on this album were absolutely mesmerizing—not unlike the variety of alien modulations and intonations in *Forbidden Planet*—and often startling or even terrifying at that! Stern's dramatic narration simply astonished, forever mind-melding my nervous system to hard sci-fi! As the back album cover advertised, "... the narrative is underscored by strange, other-worldly sound effects that will help the child's own imagination to conjure up the sights described for him." Being a former 'child,' I'm here to tell you that such acoustical interplay really worked. Back then, unlike with movies, which we were allowed paid admission to see only once (or until the whims of television station broadcasters would permit rescreenings), this magical record could be played repeatedly, forever imprinting Stern's voice and the imaginations of these marvelous science fiction tellers. Coupled with Stern's words, strange musical effects whisked us along from one exciting scene to the next; from one fantastic story into another mind-bending tale.

What we did not appreciate then was the immediate basis for the recorded tales—the four aforementioned *Classic Illustrated* magazines, "featuring stories by the world's greatest authors," which we didn't own. Front cover illustrations of these comics were displayed on the cover of the LP record. Fifty years later, I decided to find copies of those elusive *Classics Illustrated* magazines. Now at age 56 (i.e. in the fall of 2010), I can appreciate them far more than I would have then. While here, refraining from comparing the illustrated stories to either the blockbuster contemporary films, respectively, launched from Hollywood, or the masterful novels originally written by Wells and Verne, let's gain appreciation for each of these four illustrated magazines and the 1959 recordings they inspired. (Although for purposes of this book I've only excerpted detailed discussion for two of these 'books' from the original article, corresponding to "Side Two" of *Space Stories and Sounds*.)

Classics Illustrated # 124, "The War of the Worlds," was first printed in January 1955. Afterward came issues # 133 ("The Time Machine" 1st ed., July 1956), # 138 ("A Journey to the Center of the Earth" 1st ed., May 1957), and # 144 ("The First Men in the Moon" 1st. ed., May 1958). At the conclusion of each fantastic tale, the editors inserted an encouraging, edifying statement, "Now that you have read the classics *Illustrated* edition, don't miss the added enjoyment of reading the original, obtainable at your school or public library." At age 5 ½, for me, 'chapter book' reading such as this would have been challenging.

Bill Stern (1907-1971) was a sports enthusiast, excelling in football, basketball and polo. He became "intrigued by show business," trying out a career in Hollywood and later working as an usher at New York's Roxy Theatre. He became a stage assistant and assistant stage manager, before 1934, when he found his calling as a radio announcer and sports commentator. Stern played a small part in the film, *Pride of the Yankees*. He received several awards for excelling in radio journalism. We can spy his portrait on the front cover of the LP album, and there's even an entry about his life and career posted at Wikipedia.

The two recorded versions of tales featured on side 2 of *Space Stories and Sounds* happen to be my two absolute favorite science fiction stories, "The Time Machine" (based on Wells' 1895 tale), and "A Journey to the Center of the Earth," (based on Jules Verne's novel first published in 1864, which Verne later expanded into a newer 1867 edition). But of course Bill Stern's presentation of these stories was more immediately founded on the *Classics Illustrated* versions. As I've noted elsewhere, in their 'originally' published form, both stories do involve time travel (i.e. in one sense or another). And once again, *Space Stories and Sounds* offers expert interplay of words with sounds, accentuating *Classics Illustrated* versions.

One might presume the *Classics Illustrated* version of "The Time Machine" was based on the outstanding 1960 film starring Rod Taylor: it wasn't. The Time Traveler depicted on the cover of the issue even resembles Taylor, but the machine looks more like a strange kind of gyroscope than the ornate 'easy chair' design featured in the film. Time travel is perhaps the most 'time-honored' and popular sci-fi theme. And, here, *Classics Illustrated* doesn't disappoint! The tale follows Wells' original rather faithfully.

If you're reading this, you already know the basics, delightfully depicted by *Classics Illustrated*. The Time Traveler enters his study where his gentleman friends are startled to see him appear, dressed raggedly and disheveled. Why? He announces that he's traveled through time, living 8 days in the far future. Most of the rest of the story then is a 'flashback,' or should I have said 'flash-*forward*,' concerning what he did during those incredible 8 future days. Advancing the lever on his machine, the time traveler breezed, ghostlike, through futuristic wars, cities and then ....braked hard, joltingly in the world of 800,000 years AD. Panic stricken in foreboding, alien, rain-drenched terrain, and haunted by visage of a monstrous, sphinx-like statue adorning a tomb-like 'Pedestal,' the time traveler articulates the central idea of Wells' tale: "Do men still live on the Earth? ...Are they still human?"

A partial answer comes shortly, as the happy and beautiful, gentle 'Eloi' welcome him to their communal abode. Eventually the Time Traveler learns how to communicate with the Eloi, but is soon alarmed to find that the Eloi, or so he initially believes, have stolen his time machine, dragging it out of reach through a door leading inside the Pedestal, to underlying the Sphinx sculpture. Wretchedly, the time traveler puzzles over why the Eloi, who seem so "weak and childish" could have committed such an act. So he establishes what can be gleaned concerning the Eloi and their surroundings. Spying shafts in the countryside, from which echo the sounds of heavy pounding machinery, the Time Traveler soon encounters a *second* race of (to him) hideous futuristic men, known as

Morlocks, who live underground. Most horrifically, Morlocks emerge from their shafts to the surface world at night, and although unseen in the *Classics Illustrated* book, shepherd their subservient 'cattle'—the Eloi, to their doom.

After a near fatal skirmish with Morlocks, the time traveler meets pretty, blond 'Weena,' who warns him of the Morlocks. Without overemphasis, doubtless to avoid risking the objection of parents who purchased these illustrated classics for their children, in one frame, Weena states that Morlocks *eat* Eloi. Nightmarish! The Time Traveler deduces that the technologically-founded Morlocks stole his time machine, not the Eloi.

Weena takes him to visit relics out of her dim past ... an ancient museum (in the distance rather resembling the emerald city of Oz), which will be of the time traveler's distant future! The Time Traveler discovers that Morlocks fear the light emitted by fire, and so, taking stock of matches and a jar of inflammable camphor, he prepares to give the Morlocks 'what-for' in retrieving his time machine. A forest fire erupts, forcing both the Morlocks and the Time Traveler to flee in terror. Ultimately, the Morlocks lure the time traveler within the Pedestal, using his time machine as bait. The traveler desperately escapes into his nurturing past, carrying a pocketful of flowers lovingly given to him by Weena 800,000 years hence.

Although, to a savvy youngish readership, *Classics Illustrated's* "Time Machine" certainly implies that occurrences of the future resulted from evolutionary divergence of mankind, the word 'evolution'; doesn't appear in the text. However, in one frame the time traveler claims that "... descendants of man have become two species, the Eloi who are gentle and helpless and these other creatures who dwell in darkness beneath the earth." But that's as 'Darwinian' as it gets. Presumably the flowers brought back by the time traveler to his 'present' are of a species unknown in the late 19th century, because they haven't evolved yet, although this isn't so stated. The illustrated version also omits Wells' description of the time traveler's harrowing, inadvertent passage to Earth's near lifeless end billions of years in the future. *Classics Illustrated* did not present the Time Traveler's return to the future, (and that thought-provoking business of which books does he take along that return journey?—not written by Wells, was scripted inspiringly four years later for the 1960s film). Wells's 'perfect story,' was certainly watered down for *Classics Illustrated*. Evolutionary topics perhaps were still considered too revolutionary for the 'time.'

Now in the *recorded* version, *Space Stories and Sounds*, Bill Stern begins with the words, "Time is only a kind of space." Therefore, it's quite reasonable to perceive how it's possible to move either forward or backward in time just as we may move forward and backward in space. With this introduction, we're whisked along to a scene where (Wells as) the Time Traveler/narrator clambers aboard his newly invented time machine, pressing the levers forward, moving off into time. There's a sudden clap of thunder as he jarringly halts in 802,701 years AD! The haunting sound of an oboe accompanies Stern's hair-raising description of the "...colossal figure of a white marble sphinx on a Pedestal." Frightened, the time traveler nearly returns 'immediately' to his Present time, but is tempted to remain long enough to answer that vexing question whether men still live on Earth in this future.

After confronting the Eloi, who seem sufficiently 'human,' Stern dramatically (if not 'sternly'—sorry!) describes strange footprints alongside markings where the time machine was dragged within the Pedestal. And they aren't Eloi footprints! Later, when the forest fire erupts, we thrillingly *listen to* sounds of trees bursting into flame, and sense palpable panic as Stern dramatizes the nightmare of being nearly captured by the awful Morlocks, and then getting lost in a blazing forest. Ultimately, when the Morlocks spring their trap, first showing the time traveler his machine inside the opened doors of the Pedestal, the time traveler dejectedly decides that he doesn't want to "...live there anymore because ... man had become either useless, or terrible." We hear the Pedestal doors slamming shut, but he gets away hurtling through corridors of time back ... back to his laboratory, "finally, in the Present." Weena—so pivotal to Wells's original story and the 1960 film—isn't a character in the 1959 recording.

Space Stories and Sounds continues with one more tale of the utter fantastic, beckoning us even further within the Earth's dark, yawning passageways, beyond and below where Morlocks dwell, although not in our future.

For on the cover of *Classics Illustrated'* "A Journey to the Center of the Earth," we see two monstrous fighting aquatic reptiles, with two people in peril aboard a tiny raft on the frothy high seas. This may be the only instance where *Classics Illustrated* featured prehistoric animals (i.e. *Plesiosaurus* and *Ichthyosaurus*) on a cover. This time the 'splash page' is simple, yet effective highlighting (fictional) 16th century scientist Arne Saknussemm's fateful words, "Go Down the Crater of Sneffel, that the shadow of Scartaris softly touches before the beginning of July, Brave Traveler, and you will come to the center of the Earth. I did it." The illustrated book follows Verne's 2nd circa 1867 (French) edition, and not the doctored-up, circa 1870 English edition, to which a fictional encounter with the 'Ape Gigans' and the 'Shark-crocodile' was added by an unknown author.

Soon Axel (the narrator in Verne's original novel) and his Uncle Otto Lidenbrock are seeking means of following in Saknussemm's footsteps, (while along the way testing the theory of the Earth's central fire). Off they go, with their 'guide' Hans, to the volcanic fields of Iceland, where they find another rock inscribed by Arne Saknussemm. Scartaris' shadow indicates which opening will lead to their subterranean goal. After several weary days of underground hiking, Lidenbrock calculates that they've descended 48 miles into the Earth (while moving laterally 255 miles from their entry point on Sneffels). Curiously, the temperature remains only moderate, permitting life, thus 'disproving' that the Earth's center is very hot. Yet Axel despairs that, at the rate they're proceeding, it will take them over five years to reach the very center. Lidenbrock, however, lives on faith, claiming that "What Saknussemm has done, we can do!"

Next, Axel gets lost but is rescued by Hans and his Uncle who in the meantime have already discovered a great ocean within the bowels of the Earth, illuminated from the cavernous ceiling above by electrical storms. (In the 1st English edition, this is named the "Central Sea," whereas in original French editions, it is named the "Lidenbrock Sea.") There are also giant mushrooms and, faithful to the novel, bones of "prehistoric animals" scattered along the shore. Axel worries whether such "terrible monsters ... (are) still wandering about?" But, unlike the 1959 film starring James Mason, giant finned *Dimetrodons* neither appear in Verne's novel, nor in the *Classics Illustrated* edition. Hans builds a raft and soon they've caught a prehistoric fish on a hook, presaging the gigantic battle that soon ensues between the two greatest aquatic sea reptiles known in Verne's day, *Ichthyosaurus* vs. *Plesiosaurus*. The latter is mortally wounded, but then the adventurers are menaced by an intense electrical storm threatening to blow their gun powder!

They awaken on a calm shore, now at great distance below the Mediterranean Sea. More bones of extinct animals are strewn about, and among them Lidenbrock discovers a human skull (although evolutionary implications aren't explored in the *Illustrated Classics* edition), foreshadowing what happens next. For, in perhaps the most dramatic scene of all, Axel and Lidenbrock spy woolly mammoths being herded by a gigantic 12-foot tall prehistoric man. While these are suggested as phantasmagoric in the 1867 novel edition (this particular scene, therein, depicted by 19th century paleoartist Edouard Riou), in the 1957 *Classics Illustrated* edition they're intended to be more literalistic, fully realized as in-the-flesh. The rest of the story is devoted to setting a gun powder charge that should jettison them further within the Earth. But the explosion instead causes them to ascend to the surface aboard their raft, through a volcanic shaft on Stromboli.

Besides Stern's voice, the 1959 *Space Stories and Sounds* recording prominently features Bill Simon's grinding organ fugue accenting the 'atmosphere' of the Earth's foreboding and mysterious interior. The only noticeable 'flaw' (even to me then at age 5 ½)in this recording happens when Stern mispronounces the names of the two prehistoric reptiles, as "Plesia-cer-as" and "Itch-eeya-cer-as." Rhythmic bass drum beats roll as "A Journey" commences. In May 1863, Stern announces, taking the role of Axel as narrator both he and his Uncle found a message in an old Icelandic book. After journeying to Iceland, they venture through tunnels, to the accompaniment of high end piano keys and drum-building anticipating. While sailing across the ocean, "stretching farther than the eye could see," they attempt to determine its depth, only to gasp in wonder at tooth marks bitten into the heavy metal pick let down on a rope. There's a crescendo in sound reflecting, first, the great marine saurian combat, followed by the storm raging for three days. As Stern solemnly declares with awe and wonder, "they're living in a world of noise and explosions." The sound intensity builds, pounding chaotically, cataclysmically when at the story's climax, they rise from the volcanic shaft, erupting onto

Stromboli (3,000 miles from their starting point), without having reached the center of the Earth.

In this short, recorded segment there was either insufficient time to dramatize the adventurers' encounter with Mammoths and Paleolithic Man, or *Classics Illustrated* instead relied on Verne's 1864 novel edition, which did not include that particular scene.

Certainly, in the century to century and a half since Wells and Verne spun their classic tales, science has progressed exponentially. To a boorish critic, their stories would now seem old-fashioned, grossly inaccurate scientifically, or even trite. Yet, Wells and Verne inspired with their works, causing younger minds to dwell where Man had not gone. This is why, even after half a century, the (four) dramatic tales recorded in *Space Stories and Sounds* still entertain, refreshingly.

Of course, recording with lack of visual component, even coupled with illustrated (soundless) comic book, isn't equivalent to a movie. Yet, Stern's four tracks recorded on *Space Stories and Sounds* remain mesmerizing and astonishingly good, even today without the visual element.

Chapter Thirty-seven — *A Paleoart 'Revolution'?*

If there are acknowledged revolutions in science, may we also detect parallel 'revolutions' in paleoart? Here I shall refrain from referring to formal 'stages' or 'periods' in art history familiar to cultural historians and iconographers. Supposing the term 'revolution' isn't appropriate for describing developmental changes in paleoart, then possibly for our purposes, another term—'paradigm shift' may instead suffice. I shall use both terms synonymously.

First, however, we must suggest definitions for terms, especially "paleoart" and "revolution in science." Although I've offered more formal definitions for paleoart in three of my books[1], here I'll add further perspective. There's much 'gray' here. Paleoart is just a certain, perhaps narrower niche, style or form of an over-arching 'umbrella' paleoimagery. Instead of bright divisional lines, consider a 'continuum' merging paleoart—derived from the more science-oriented forms, and more copious, branching, popular examples of paleoimagery. Sometimes efforts to portray paleoart in some media do fall short of an 'ivory tower' scientifically-minded goal, but that doesn't necessarily mean those cases (e.g. *Jurassic Park* movie; *Walking With Dinosaurs* series) should be ignored (or disparaged). If an intended, sometimes edifying or entertaining objective was met then all is good, even if the paleoart end of the spectrum wasn't fully achieved. There are different consumer 'needs' for certain types of visual or even literary paleoimagery.[2]

Paleoart in our dinosaur renaissance period is practiced and philosophically approached differently than how paleoart was practiced back in the late 18th, or 19th or even early 20th centuries. Although from what I've been able to unearth through my readings the term "paleo-artist" in its modern parlance wasn't coined until 1870, post-dating Benjamin Waterhouse Hawkins's (1807-1894) most memorable creations. No artist shifted the popular course of paleoimagery during the 19th century more than Hawkins, tapping the hearts and minds of millions across generations.

If paradigm shifts or revolutions in paleoart are identifiable, then wouldn't we expect that, historically, they would lag behind respective, scientific revolutions in Earth science? In reexamining paleontological circumstances of the mid- to late-19th century—when the predominant paleoartist was Benjamin Waterhouse Hawkins, one would have such a notion. Thanks—quite fittingly to

Valerie Bramwell's—Hawkins's great, great, great grand-daughter, and Robert M. Peck's masterful *All in the Bones: A Biography of Benjamin Waterhouse Hawkins* (2008), we may further assess whether this was the case, then. (Theirs' is the most comprehensive, illuminating study on Hawkins published to date.)

For decades Hawkins's landmark dinosaur models sculpted for the Crystal Palace exhibition of 1854 were scorned as exemplary of how artistic visions in restoring prehistoric animals from fragmentary remains could go awry. And yet recorded in these models coupled with Hawkins's other efforts to reconstruct and convey impressions of the prehistoric world, I would contend is an early (Victorian) paleoart manifestation of a revolution in science, perhaps a first wave later culminating with our current "dinosaur renaissance" phase. These quaint and antiquated models while significant in their own right even today, do overshadow Hawkins's genius as a paleoartist, his work grounded in comparative anatomical studies—a crucial knowledge base for professional paleoartists. A synopsis of Hawkins's achievements would be enlightening.

Hawkins, perhaps the quintessential artist of his time delving into prehistoric themes, often seems to have been in the right place and at a 'right' time. In 1836, he sketched the famous, "exotic" giraffe in Paris at the Jardin des Plantes brought to Paris by Etienne Geoffroy Saint-Hilaire. The Giraffe stimulated conflict between naturalists who debated how its long neck appendage manifested. And in 1847 Hawkins did a color wash sketch of the Quagga—extinct since 1883. He also assisted Thomas Henry Huxley with illustrations comparing skeletal anatomy of primates, incorporating the then recently discovered Gorilla, one of which (ironically, because Hawkins remained steadfastly opposed to Darwinian evolution) would, a century later, inspire Rudolph Zallinger toward his oft-cited illustration of human evolution via descent from ape ancestors. Hawkins also worked with and studied fossil bones, including those of *Elasmosaurus* and *Hadrosaurus*.

During intervening years he produced numerous illustrations for important zoological publications, sculpted several extant animals for display and exhibition, and for his efforts was elected as a member of the Royal Society of Arts (in 1848) and later as a Fellow of the Geological Survey (in 1854). He also received an honorary Doctor of Science degree from Princeton University in 1874. Due to his preeminence as a sculptor and restorer of prehistoric vertebrates, by the 1850s he became a 'go-to' influential figure conveying knowledge—both visually and via lectures, concerning paleontological subjects, especially Mesozoic creatures and saurians. Hawkins's presumed role as 'scientist' then may seem off-putting to some, yet—like other paleoartists since—his natural ability to confidently unravel anatomical mysteries while relying on innate visual acuity transcended purely scientific perception.

The story of Hawkins's herculean creation of the Crystal Palace's prehistoric menagerie, thereafter followed by his paleoart efforts from the late 1860s through the late 1870s in America, has been retold in various articles and books. There were other artists then producing prehistoric animal reconstructions, restorations and "scenes from deep time," conveying Earth's succession of geological ages, but Hawkins's output was prominent, and most recognizable—even today. Initially there was due consternation that the life-sized, quadrupedal 'mammalian' Crystal Palace dinosaur statues especially (i.e. *Hylaeosaurus, Megalosaurus* and *Iguanodon*) Hawkins built from scratch while working closely with Richard Owen—*the* scientist who named the clade "Dinosauria" in 1841/42, were founded on gross speculation, a "lack of 'hard' science" (Bramwell, pp.23, 51note74). Unperturbed, Hawkins—by then promoted to Director of the Fossil Department of the Crystal Palace, boldly accomplished this major, first-of-its-kind feat, anyway. In the afterglow of Hawkins's eventually acclaimed success, commissions for other paleoart followed, including designs for elaborate sculptural displays.

Two decades later, Hawkins diverted his paleontological expertise toward America, at a time when vistas of the Mesozoic world and its denizens were in flux and new fossil discoveries were on the horizon. Now aged into his 60s, Hawkins's undertaking to build a "Palaeozoic Museum" in New York's Central Park was frustratingly ruined. However, 150 years ago in Philadelphia he did reconstruct bones of *Hadrosaurus,* casting several replicas (posed in bipedal stance), per Joseph Leidy's guidance. But then another grand paleontological exhibition, this time for the Smithsonian Institution, failed to reach fruition. Making *Hadrosaurus* recreations was to become a later life, recurrent theme in Hawkins's paleoart following his interactions with Leidy. Certainly, Hawkins's endeavors upon popular culture first in Britain with his sculptural creations and later in America with his *Hadrosaurus* skeletal reconstruction of 1868 exhibited in Philadelphia (Peck, p.79) were unprecedented.

However, Hawkins's most *lasting* impact (e.g. in terms of survival) on American paleoart was a series of 17 oil paintings (of which two are deemed lost), completed during the 1870s for Princeton University's former Elizabeth Marsh Museum of Geology and Archaeology, mounted proudly high up on walls in Nassau Hall, where I had the pleasure of seeing them in September 1972. Given my familiarity with Charles R. Knight's treasured Field Museum murals surveying span of the Phanerozoic Era, I certainly was predisposed to Hawkins's much earlier yet similar 'take' on the theme.

Two decades later I encountered these paintings again, once on Jan. 13, 1990—his two dignified dinosaur portrayals "Cretaceous Life of New Jersey," and "Jurassic Life of Europe" witnessed at the Field Museum as part of the "Dinosaurs Past and Present' traveling paleoart exhibition, and later in October 1993 when I visited archives of the Princeton University Art Library, courtesy of Maureen McCormick. This series of paintings, coupled in context with the theme of Hawkins's sculptural "antediluvians" displayed at Sydenham and plans for New York's Palaeozoic Museum, project ideas concerning 'life through geological time,' collections of portals and landscapes in which dinosaurs always play a central role in casting. This, then becoming highly traditionalized theme—

of which Hawkins was at the forefront, had also been elegantly conveyed through paleo-artistic renderings by Josef Kuwasseg and Edouard Riou.

The two recognized revolutions in science most closely allied to paleontology are (1.) the Darwinian revolution of the 19th century, and (2.) the Continental Drift/Plate Tectonics revolution of the 20th century.[3] Yes—'fixity' versus transformation in consideration of (both) species, and land masses/ocean basins! Certainly, before the Darwinian revolution, Cuvier's paleontological discoveries constituted a significant revolution in Earth Science as well, fully dramatized by his theory's fundamental reliance on successive, catastrophic geological revolutions. Conceptually, catastrophism then had its day during the 1980s uproar over the nature of the Cretaceous-Paleogene extinction boundary 66-million years ago. This latter historical scientific episode proved an important facet of our ongoing Dinosaur Renaissance ('rebirth' in our understanding of the dinosaur age), with scientifically-grounded ideology prevailing over gradualist theories.

One striking feature of scientific revolutions, as outlined in I. Bernard Cohen's *Revolution in Science* (1985), is how they may gestate over an *extended* duration. Scientific revolutions usually don't transpire 'overnight': decades or more may elapse (pp.37, 275). Relying on his four-fold definition for how scientific revolutions transpire, Cohen notes that "Just as there was no revolution in astronomy until more than half a century after publication of Copernicus's book in 1543, so there was no revolution in geological science until about a half-century after publication of (Alfred) Wegener's original paper and his book." (i.e. note – his publications on continental drift dated 1915 and 1924, respectively; p.465) Likewise, an "intellectual" phase of the Darwinian revolution in biology was documented in 1837, two decades prior to first announcement of the theory of natural selection via differential survival, in 1858.

Another significant aspect of scientific revolutions (e.g. such as the "Copernican revolution" or that concerning Wegener's "continental drift" idea), is how they "transform" over decades such that the 'final' textbook version may appear markedly different from original ideology at conception (Cohen, pp.38-39). For example, Cohen posits that the 20th century wave of evolutionary thought (encompassing genetics, an 'evolutionary synthesis' and punctuated equilibria, etc.) may represent either a "second Darwinian revolution or a second stage …" (p.297) If so, conceivably, what may be discernible as an overarching Darwinian revolution is over a century old!

So, consider that when it comes to dinosaur science our evidential and overall knowledge, while exponentially ballooning today, remains *in toto* a relatively *young* discipline. Yet there do appear to be marked 'phases' in its development, including stagnant periods of diminished scientific advancement concerning dinosaurs, corresponding to dormancy in paleoart—when 'state-of-the-art' themes, and 'preferred' styles of anatomical restorations seize our souls, and become dominant, stabilize and/or are recycled longer term[4] (e.g. from the 1920s through the mid-1960s, Charles Knight's awesome Field Museum murals and Zallinger's magnificent 1947 "Age of Reptiles" notwithstanding). During such phases, it may *appear* that we've finally gotten to an apogee of truth, with no further ideological changes anticipated. Warning—as 17th century Dutch painter, Samuel Van Hoogstraten stated, "a good painting was like a 'mirror of nature' that could reflect back to the viewer a more perfect form."[5] But what if said "mirror's" reflection is illusory, a common (historical) circumstance in interpreting fossils.

Academics unfairly ridiculed Hawkins's Crystal Palace dinosaurs due to their outmoded appearances—known to be incorrect even within his lifetime. However, such mockery has been characterized as "…thinly veiled criticism of Richard Owen." (Peck, p.73) Right or wrong scientifically, Hawkins's impressions of dinosaurs launched an early flowering or merging of visual art *with* paleontology. But dinosaur science, as portrayed via restoration, can seem 'protean,'… so often in flux on heels of new discovery. True knowledge creeps forward in fits & starts. All things considered, I am tempted to label Hawkins's mid-19th century visual contributions to paleontology a (first) 'revolution in paleoart.' Zoe Lescaze and Walton Ford, authors of the lavish and scholarly 2017 book, *Paleoart: Visions of the Prehistoric Past*, would concur (p.69). Then does our current dinosaur renaissance wave reflect a second stage of this revolution, encompassed within framework of the 19th century episode?

As proclaimed in the 2012 BBC documentary, *Planet Dinosaur*, we're living in a "golden age of dinosaur discoveries." In 1975, evidence for the hot-blooded' dinosaur model was heralded as a "revolution in palaeontology," the subtitle of Adrian J. Desmond's landmark UK book. Meanwhile, in April 1975 *Scientific American* referred to the recent paradigm shift as a "dinosaur renaissance." I am not a historian of science, but fortunately I knew at least one person who was—my dad. So when pressed (by me) in early 2001, he weighed-in on appropriateness of the term "dinosaur renaissance." My dad, Allen G. Debus, concluded that for our *current* phase of scientifically-fueled dinomania and paleontological intrigue the term "renaissance" is appropriate. (See Chapter 36 titled "Renaissance Dinosaurs" in my *Paleoimagery* book for details.)

As Albert Einstein stated, "The greatest scientists are artists as well." However, in revolutionary cases (e.g. that in Hawkins's heyday, and in modernity), key paleontological advances, evolutionary reconsiderations and geological discoveries usually precede associated, perceivable paleoart revolutions. However, science and art always go hand in hand here because paleoart is scientifically founded, with artistic visions dependent upon, even if lagging behind, scientific interpretations.

Notes to Chapter Thirty-seven: (1) See several chapters in my *Paleoimagery* (2002), Chapter Two in *Dinosaur Sculpting, 2nd ed.* (2013), and Chapter Nine in *Dinosaur Memories II* (2017); (2) Verbal (literary) descriptions of dinosaurs, or pseudo-dinosauria may also be considered as paleoimagery, or if scientifically imbued—possibly even as paleoart; (3) For example, in March 1969, Bjorn Kurten published a *Scientific American* article merging these ideas titled "Continental Drift and Evolution."; (4) Despite their popularity, with minimal innovation, or simply 'regurgitated' copy-cat, even though scientific debate may reign over certain matters, such as proper limb posture in sauropods—debated between 1905 to 1910; (5) Elisabeth Berry Drago, "The Masters of Nature," *Distillations Magazine* v.4, 2018, p.25.

Chapter Thirty-eight — *Collecting Paleo-paper*

Dino-Stuff:

While I've never considered myself a true-blue "dinosaur collector" type, certainly, by the late 1990s I had accumulated a LOT of paleo-stuff—some items being far more important than others. As noted in over 120 issues of *Prehistoric Times*, dinosaur toys, replicas and prehistoric animal collectibles of nearly every persuasion are highly sought after by a dedicated, select group of collectors. Our house is stuffed quite fully with a surfeit of such items, but there's a limit; we can't bring in anymore—there's no available shelf space! And have no desire to fully convert our house into an "Acker-mansion" of dinosaur things. *Although* I have seen two such residences, and, well, they really are cool!

Occasionally I would dabble in articles addressing collectible types that few would have thought of collecting, or even writing about. (My first such piece was on dinosaur stamps, of which a little more will be said later, written nearly 40 years ago – not included here because it is *so* dated.) So while this penultimate section is rather far from the hearts of you mainstream dino-collectors, at least it serves as a 'window' upon the rarely saluted fringe of dinosaurabilia.

The Lure of Dinosaur Fanzines:

So you think you'd like to produce and edit your very own dinosaur fanzine, huh? Think of *all* those perks and advantages! Soon the postman will be delivering scores of 'free' dinosaur books and dinosaur toy products for you-the-dinosaur-mag-editor to review, not to mention all that 'extra' cash you'll have rolling in from an avalanche of subscribers who can barely hold their collective breaths until that fabled, 'next' issue pounds into their mailboxes. And, lord willing, you will have enough space to cram all that expensive, exorbitantly paid-for, color advertising into each lavish issue! Soon too, you'll be proudly wearing your 'Press badge,' brushing shoulders and marching lockstep with famous paleontologists and celebrity paleoartists who will invite you-the-dinosaur-mag-editor to witness their latest projects firsthand in the field, studio or lab, right? Paleo-mags—can it get any better?

To clarify, I'm not alluding to those enduring science-oriented or science history, refereed journals (e.g. *Nature, Science, Journal of Vertebrate*

Paleontology, or *Earth Sciences History*), or the utmost slick, popular science and natural history magazines (such as *Scientific American, National Geographic, Natural History,* and *Discover*). Nope. Here we're referring to those largely self-edited magazines and newsletters primarily distributed in hard copy as opposed to electronic form, that appeared on the scene during the 1990s, many of which were inspired by successes of the 1993 movie, *Jurassic Park.* Many of these paleo-mag publications were created for a readership consisting of vertebrate paleontology enthusiasts, especially non-scientist types appreciating pop-cultural aspects of 'dinosaurology.'

Hearing the names of these old ('hard copy') publications may stimulate some alpha-waves in the minds of dinophiles who enjoyed receiving them, that is, before they became 'extinct': *Archosaurian Archive,* George Olshevsky's *Archosaurian Articulations, Dinosaurus Magazine, Earth, Paleo Horizons: Dinamation International Society, Model Dinosaur, Dinotimes, DinoPress, Dinosaur Collectors' Club Newsletter, Dinonews,* UK Dinosaur Society - *Quarterly, Dinosaur Discoveries* and *Dinosaur World.* That's probably not even a complete listing of all the dino-zines and paleo-mags, several of which sprang into existence on the heels of Stephen Spielberg's *Jurassic Park.* Some zines were slick and glossy drug store magazines, several were xeroxed bulletins, others were printed on cheaper newsprint. But at the time, their contents certainly seemed 'golden' to readers who paid for subscriptions, no matter how blotchy were the printed images, despite occurrences of typos, no matter how offbeat were the featured articles.

So, you're still fantasizing about starting up your own little paleo-magazine are you? The odds are against long-term success, ya' know. With exception of *Prehistoric Times,* a refereed journal—*The Mosasaur,* and *Fossil News,* my library accumulation of paleo-mags, each issue now enveloped in plastic protector sleeves and stacked neatly on book shelves, bears testimony that few survive the test of time. My own *Dinosaur World's* short 'reign' on Earth fizzled after 9 issues, even though a 10th, undistributed issue was completed and later scrapped due to insufficient funding.

A short-lived publication, *Archosaurian Archive* ("AA"), which lasted through 5 interesting issues, became a quasi-template for *Dinosaur World's* content and style. AA evidently bit the dust months before yours truly suggested to fellow co-conspirator Gary Williams (in August 1996) that they might consider doing their own dino-mag, which then carried the working title, "Dinosaur Country." A long-distance, mid-September telephone verbal agreement affirmed their commitment to collaborate on DC, and a month later the two met in Hanover Park, Illinois to sort out logistical matters before making any formal announcements. But there was no time to dally, for there was an available niche for DW to seize, or so it seemed to its self-appointed editors. But enough about DW; what about the other publications?

There was a perceived niche of the most worthy kind! First, there was the U.S. Dinosaur Society. The Dinosaur Society donated a healthy sum of dollars for printing pages of dinosaur-related articles in the *Journal of Vertebrate Paleontology*, and helped subsidize dinosaur research digs. Furthermore, delegated representatives of the U.S. Dinosaur Society reviewed and approved new dinosaur-related products, (e.g. books, toys, games, etc.) for scientific accuracy. Those receiving favorable review were granted the Society's 'seal of approval.' Authors (Allen A. Debus, with Bob Morales and Diane Debus) recall feeling ecstatic when our first book, *Dinosaur Sculpting: A Complete Beginners' Guide* (1995), received Society approval, meriting inclusion within the 1996 edition of its annual *'Everything Dinosaur Catalog.'* The *'Everything Dinosaur Catalog,'* mailed to all members, advertised new products certified by the Society. Their classy "*Dinonews*," a first rate, informative newsletter mailed to Society members, as well as their child-oriented paper, *Dino Times*, were highly regarded.

But, sadly, the U.S. Dinosaur Society went under in 1998. Ironically, the last issue of *Dinonews* we received, a 20-pager dated Winter 1997, sported the headline, "Going Once...Going Twice..Gone Forever," really a news feature about the auctioning of the 'Sue' *T. rex* fossil, but foretelling what was on the horizon for the Society's paleo-mags.

Contacted in August 2002, writer Don Lessem, recalled:

"Peter Dodson, Dave Weishampel and I founded the Society in 1990. I donated my Dino Times kids' paper to be a monthly and I then wrote for free the adult Dinosaur Society quarterly for grownups. Weishampel and I resigned when our terms were up in 1992, and Steve Gittelman took over. With my connection to Michael Crichton and Steven Spielberg and advising on the movie (gratis) money from Peter May's building a Jurassic Park exhibit and Universal's help in publishing one million special editions of the kids paper, the kids' paper circulation run to up to about 8,000 with Thom Holmes by then the paid-editor. By 1996, Gittelman had spent all the Society's income with only a third going to support dinosaur research. The publishers of the kids' paper whom I made the mistake of hiring, took $250,000 or more and didn't give it back, even when mandated to do so in court, and the Society folded its publications—though I offered then-president Don Wolberg $10,000 of my own money to complete the subscription years of the kids whose Dino Times was cancelled. The Society still exists, though it hasn't filed required government forms or reorganized legally."

Too bad. Complaints about *The Dinosaur Report* still reverberated years after its demise.

Another favored paleo-mag of the early publishing days was *Prehistoric Times* ("PT") which by leaps and bounds, attracted even more subscribers than those who received *Dinonews*. By the fall of 1996, PT's editor Mike Fredericks had produced 22 issues, the last 9 of which had sported color front cover dinosaur restorations. PT hadn't yet been enhanced into the slick and glossy paleo-mag it later became. But, arguably, with a present serial run of over 125 (mostly) glossy and color-filled, each 'getting better and better' than the last, *Prehistoric Times* has become the foremost dinosaur magazine out there today. Anybody who's enchanted by dinosaurology these days has heard of PT, and so we won't digress further about Mike's top notch paleo-mag.

The crisp monthly *Fossil News: Journal of Amateur Paleontology* (aka 'Fossil News' or "FN"), (later, in April 2000, FN's subtitle was changed to *Journal of Avocational Paleontology*), had been in circulation since January 1995. FN published a formidable array of articles on every conceivable aspect of paleontology, with an emphasis on fossil collecting. FN filled a void created with the mid 1980s demise of another similar publication, *Earth Science*, formerly published in Downers Grove, Illinois.

FN's founder, Joe Small recollected his early pre-FN publishing years:

"FN...'evolved' from a simple one page newsletter that I began sending to a few friends when I devoted a summer to fossil hunting in the region surrounding Havre, Montana. The adventure was documented in those early issues of 'Bone Bug Journal' a choice of title I subsequently found to be insufficient. As increasing numbers of people asked to be included in the monthly distribution list, I determined to make a business of it, and within a few months had embraced the excuse to abandon an unsatisfactory (and unsatisfying) 'regular' job in favor of the excitement and glamour of publishing. With rare exception, I mailed a new issue each month. The first issue with the name Bone Bug Journal was March 1993. The first FN was January 1, 1995. (My) last issue was March 1998..... Lynne Clos resumed publication shortly thereafter (i.e. in October 1998.)"

FN's second editor, the attractive Lynne Clos, who wouldn't allow FN to die prematurely, became one of the few paleo-mag editors with professional scientific standing in the paleontological field. A degreed engineer, with a Master's degree in Museum Science, years of fossil collecting experience and credited with an article in a refereed publication—*Journal of Vertebrate Paleontology*, (vol. 15, no. 2, pp. 254-267, 1995), Clos, who eventually changed the name of her zine to *The Journal of Avocational Paleontology*, stated in 1995, "Paleontology takes a lot more dedication when you are not being paid for it." No kidding, and not to mention the routine headaches of being an editor, dealing with unruly subscribers and or complaints from readers. FN continued its tour through time (12 issues per year) with features for children, fossil collectors, idiosyncratic essays for young and old at heart, and news updates for those who desired the latest summaries from the paleontological literature, until 2012 when it fizzled—only to become resurrected by another editor, Wendell Ricketts, in early 2016. During the 1990s Lynne also co-edited a monthly newsletter, *Trilobite Tales*, available to members of the Western Interior Paleontological Society.

There were other notable publications too, including another carefully edited journal, *The Mosasaur,* published intermittently by the Delaware Valley Paleontological Society, *Dinonews* - published by the Dinosaur Club of the Museum of Western Australia, Perth, and the UK Dinosaur Society's *Quarterly.* Each contains articles on dinosaurology, as well as other forms of prehistoric life, or on the history of the science of paleontology. It would be improper to neglect mention of other contemporary paleo-mags which were also 'out there,' available to dino-enthusiasts although not as popularly regarded because they were less publicized. And so let us quickly mention *Bones*, published by the Wyoming Dinosaur Center of Thermopolis, WY. While early issues of *Bones* (e.g. vol. 1, no. 3 - May 1996) were 10 pages in length, by the time of issue no. 2, vol. 5, came out in March 2000, *Bones* had shrank to a 4 page newsletter.

Dino-fans could also salivate over *Gakken Mook - Dino-Frontline*, later *DinoPress*, published in Japan and in Japanese, as well as George Olshevsky's *Archosaurian Articulations: Dedicated to the Dissemination of Dinosaurian Discourse* (first appearing in July 1988, a 10th issue of which had been printed by September 1990).

The pricy-looking *Dinosaurus Magazine* (DM) became a real downer, for most dinophiles, after its impressive advertising announcement on the back cover of the November 1994 issue of *Earth*. Perhaps *Dinosaurus Magazine* had many other dinophiles thinking of that 1960 film of same title, *Dinosaurus!* Gee—a new paleo-mag tailored for just one movie? Alright, joking aside, this was the second paleo-mag to go glossy and develop a slick-looking format and the first 'major' to flame out leaving many subscribers without promised issues. Published by editor Kim 'Bones' Nilson, the ranks of DM subscribers quickly swelled to about 2,000 almost overnight. But 'Bones' couldn't live up to expectations as issues were produced with increasing irregularity. His child-oriented paleo-mag folded in late 1997 after only a handful of glossy issues had reached most, or some subscribers, usually months behind schedule. Many subscribers felt swindled by DM's lack of devotion to customer service. Although Nilson & company intended to print 10 issues per year, for a one year subscription of $18.95, only 7 issues actually reached our mailbox over a three year duration.

Gakken-Mook - Dino-Frontline (GM), a slick and thick heavily illustrated color magazine appearing in 1992, was born out of the 1990 Great Dinosaur Exposition held in Japan, with its late editor, Masaaki Inoue, buoyed by a circulation of some 40,000 strong. By 1996, however, unlucky *Dino-Frontline* terminated with its 13th issue. Inoue later reestablished a new serial named *DinoPress*, published by Aurora Oval, in August 2000. Both GM and its successor *DinoPress* for a time were regarded by many as the most 'scientific' of all the pop-cultural dino-mags. Except I suspect these opinions were expressed by individuals who couldn't read the Japanese text and who preferred to luxuriate in the pictures. However, we did receive an English translation of George Olshevsky's and Tracy Ford's very thorough article on stegosaur evolution (published in a 1993 GM issue) illustrating just how comprehensive were the articles printed in GM. Mr. Inoue e-mailed us in June of 2001, hinting that DinoPress oversea sales weren't as good as anticipated, even though the magazines contained an inserted English transcript.

Sadly, Mr. Inoue passed away a year later, at the age of 52. The spirit of his impressive magazine seems to have also died with him. *DinoPress* also was one of the few 1990s paleo-magazines that paid writers for published material, (payment that I was elated to have received for one published article contribution therein).

Beginning in 1993, *Dinosaurs! Discover the Giants of the Prehistoric World* (Atlas Editions) ran, incredibly, through about 70 issues! Catering to a juvenile audience, like *Dinosaurus Magazine*, subscribers were offered special glasses in the premier issue for viewing '3-D' centerfold restorations of dinosaurs. Each colorful issue proved to be a mine of information for children, fortified with Dr. David Norman's 'Ask the Expert' column addressing questions of interest to young readers printed on the back cover. Features from the colorful series were later compiled into a thick children's book, *The Humongous Book of Dinosaurs*, (NewYork: Stewart, Tabori, & Chang, 1997), covering an A to Z listing of 'all known dinosaurs,' from *Abelisaurus* to *Zizhongosaurus*.

Dinosaur Collectors' Club Newsletter (DCC) was a 4 page newsletter produced by British paleontologist Mike Howgate, chief organizer of the Dinosaur Collectors' Club. Formed in 1993, the last issue we received (no. 32) is dated July 2000. Articles emphasized dinosaur toys, or restorations which had inspired toys and other forms of dino-collectibles, although the production value could never match its chief competitor across the Atlantic covering similar territory, the far superior PT. DCC happened to be the only fanzine featuring an autobiographical article on my own 'doings' with dinosaurology.

Especially when compared to its sister British 'paleo-mags,' DCC, and *Model Dinosaur,* a 4 page newsletter printed by Ray Rimell featuring, exclusively, reviews of new dinosaur model kits—the UK Dinosaur Society *Quarterly* publication was an elaborate affair. The *Quarterly* typically contained articles by paleontologists concerning Mesozoic vertebrates, reviews of new 'dinosaur' books, and other 'prehistoria' of general interest to paleo-connoisseurs. The UK Dinosaur Society was organized in 1993 on the heels of the U.S. Dinosaur Society successes, and formally organized in August 1994. Reportedly by 2002, *Quarterly's* availability was restricted to on-line subscription and circulation only.

I learned of the Australian *Dinonews* inaugural issue through a classified ad placed in a 1990 issue of the *Society of Vertebrate Paleontology Newsbulletin.* This glossy periodical, edited by paleontologist Dr. John Long, was published through the Dinosaur Club of the Museum of Western Australia, Perth. Initially, Long was the sole author, illustrator and editor, but soon much of the 'dirty work' was delegated to others, including paleontologist and artist Brian Choo.

The high-spirited *Dinonews* was still going strong as of early 1999, and may still be circulated. (I had received 12 issues as of July 1998.)

The Mosasaur was not so much a 'fanzine' but a high quality refereed journal published by the Delaware Valley Paleontological Society, although the last four of its seven altogether issues/(i.e. 'volumes') were published between 1986 and 2004. Its contents offered articles by both professional and amateur paleontologists covering a variety of topical areas. Articles addressed anything from description of new fossil invertebrate species, vertebrate fossils (including Mesozoic reptiles) and the history of paleontology. Although irregularly published, it was always a thrill to (finally) receive those meaty *Mosasaur* issues—(especially vol. 6, May 1999, with an article by yours truly).

In 1994, one other, 'one-shot' publication, *Fossil Tracks Newsletter*, became available to fans of Don Glut's dinosaur-related musical creations. The focus of this publication was on how the recordings of his musical "Iridium Band" were creatively inspired, played and produced. While tracking down an original copy might be difficult today, fortunately, the entire contents of Don's newsletter were reprinted in his 2001 book, *Jurassic Classics: A Collection of Saurian Essays and Mesozoic Musings*, (pp. 132-144).

So, into that challenging field entered ....*Dinosaur World*! But here my story ends...........

Curiously, there have been no new successful *print* paleomag entries since the late-1990s. But now we have the internet.

Sticker-saurs stick around:

As the years roll on by, how easy it is to forget the significance of curious prized possessions from our formative years. Later, when we rediscover such treasures, they're regarded as collectible, such as those stickers featuring prehistoric animals.

It was the very early 1960s—we'd recently moved to Park Forest, Illinois from Cambridge, Mass., when my Dad introduced my brother and I to the fourth[1] sticker dinosaur album we'd feasted our eyes on. This happened to be a booklet "Wonders of the Animal Kingdom" (published a couple years earlier in 1959), which my Dad bought from the local supermarket one day. Each week the supermarket would distribute a new packet of stamps which we'd carefully paste into the album. Since the first two or three sets of stickers (as *opposed* to *postage* stamps) involved prehistoric animals, compiling the booklet took on a certain sense of urgency.

There were 400 stickers in all, of which only 16 stamps—all having descriptions printed on their reverse sides—dealt with prehistoric genera. Diehard dinophiles may recognize artistic inspirations behind many of the prehistoric animal sticker paintings. "Wonders of the Animal Kingdom" appears to have been a Spanish-Mexican publication originally. Artwork was provided by artists Benita Bartolome, Filomene Marquez and Vicenzo Palma.

This welcome weekly activity of pasting in the new set of stickers represented my second such experience. In 1959, (at age 5), I had assisted my grandparents with "Sinclair's Dinosaur Stamp Album," with new stickers issued when we filled-up with gasoline at a Sinclair station. Then in 1965, as more of you may recall, the National Audobon Society issued a series of Nature Program guide booklets—the first of which featured "Prehistoric Life," authored by Bryan Patterson. In addition, a small box of labeled fossils was issued with each booklet. The stickers, which my brother and I avidly (and rather sloppily) pasted into our booklets, featured paleoart on display at the Field Museum of Natural History.

Decades passed by ever more swiftly, and then in 1986 I found a new set of superb stickers, this time issued by Panini, titled "Dinosaurs, Prehistoric Animals Stickers Album." It took some effort, but I managed to collect each of the 240 stickers as a keepsake for my then very young daughters. This activity often provided reminiscences of my Dad and I teaming up decades before with "Wonders of the Animal Kingdom." (And often, it recalled my baseball card collecting days too.)

Will there perhaps be a new set of prehistoric animal stickers available some years hence to pique the curiosity of my grandchildren who may someday enter this world? Well, hopefully, so why don't *Prehistoric Times* readers start dreaming up a new series for the next generation of sticker-saurs?

Postcards from the Edge (of Time):

I fell in love with dinosaur postcards at an impressionable age. It was sometime in the late 1950s when a friend of my father's who knew of my growing interest in dinosaurs mailed me a "Giant" set of American Museum postcards featuring the art of Charles R. Knight and George Geselschap. The color paintings on each postcard showing such marvels as Knight's 1897 restoration of swamp-dwelling apatosaurs, Geselschap's *Stegosaurus* 'family' and others were beautiful in themselves. But it was the way in which the four (15 ½ cm. by 22 ½ cm.) postcards arrived at our 62 Fayerweather Street address in Cambridge, Mass. that made them all the more mysterious. For my father's friend intriguingly wrote a single word on the flip/message side of each card, adjacent to the three, green 1 cent George Washington postage stamps. In sequence the backs of these four postcards read ...." Guess..... Who....Sent ....These?"

By the mid-1960s I had become a juvenile 'collector' of what few dinosaur postcards were available then. My father actually made the purchases at museums we visited, but soon we had secured sets of paleo-postcards from the

Field Museum of Natural History, featuring the magnificent paleoart of Charles R. Knight, and Frederick Blaschke, and of course the famous '*Gorgosaurus* with *Lambeosaurus*,' "Predator & Prey" skeletal display as it formerly was displayed in the Stanley Field Hall. During a late 1950s trip to England, we also purchased British Museum postcards, familiarizing us with Neave Parker's and Maurice Wilson's paleo-artistry. These, coupled with old, vintage black and white postcards purchased at the Field Museum, the Denver Museum of Natural History, The Science Museum of the St. Paul Institute, and a handful of other institutions during the 1930s by my grandparents for my father, formed the happy nucleus of a paleo-postcard collection truly marvelous to behold! Each card was carefully mounted in a scrapbook, the pages of which were often turned in admiration.

This article is a natural segue from another written back in 1981, for the Earth Science News (October 1981, vol. 32), titled "Paleophilately - Stamps, Stones & Stories," one of the first articles ever printed about dinosaur postage stamps.[2] (Yes, folks, in case you don't readily recall, before the days of Skype, E-mail and .jpg file attachments, the U.S. Postal 'snail-mail' Service would really deliver 'hard-copy' dinosaur postcards, sometimes with affixed dinosaur postage stamps to boot!) From stamps to the items dinosaur enthusiasts would naturally place postage stamps on—namely postcards—this write-up turned out to be way long overdue!

But before delving into the history of dino-'deltiology,' a term we'll introduce shortly, here's another fateful association for you to ponder, one involving paleo-postcards. Over twenty years ago, on the first day of Allen's job with an environmental agency, he noticed a dinosaur postcard pinned on the wall of a co-worker's cubicle. This was a card not then accounted for in the 'Debus collection,' nor one he had seen before. It was a card showing a life-sized *T. rex* battling *Stegosaurus* on a rocky prominence, photographed in gory, reddish 'haze,' as displayed at Disneyland's "Primeval World." (For more on Disneyland's display, readers are referred to Don Glut's *The Dinosaur Scrapbook* (1980).) It was a beacon-like card, beckoning fabulous vistas to Allen's future, as well into that special co-worker's past, a fateful joining of world-lines. Yes, that card certainly proved enchanting for more reasons than one! Never underestimate the power of curious association. For then I made the acquaintance of one, Ms. Diane Schlitz, and as the story goes, one thing led to another, and well, she, i.e. the person whose cubicle in which that dinosaur postcard resided, later became my wife in late 1982! Diane's postcard now has special pride-of-place in the growing 'Debus dinosaur postcard collection.'[3]

Can you name the third most voluminous collecting hobby in the world today? Most dinosaur collectors would be surprised to learn the answer—postcard collecting, otherwise known as 'deltiology.' In the U.S., postcard collecting is ranked fourth ahead of stamp and coin collecting, yet behind baseball card collecting. Back in 1981, while researching the little known history of dinosaur postcards for an article (before the easy, armchair days of internet searching) I contacted a society of postcard collectors. At the time their society published a guide to all the 'recognized categories' of cards valued by collectors. And guess what? There was no category, or subcategory, listed pertaining to prehistoric animals or dinosaurs. There was an 'animal' category, but the deltiology society representative knew of nobody then (besides myself) who exclusively collected postcards having a prehistoric animal theme. So the prehistoric trail seemed to end there. Nevertheless, when the spirit moved, I continued to collect postcards from more exotic places around the world through the only meager means then possible, by writing letters to museum gift shops inquiring as to whether they had any souvenir postcards for sale featuring displays indigenous to their institution.

Switch to the summer of 2000! The Debus postcard collection had grown considerably, but while magazines dedicated to all manner of dinosaur 'stuff' such as *Prehistoric Times* had only printed one brief article on dinosaur postcard collecting (see issue no. 13), surprisingly, dinosaur postcard-collecting still seemed to be a neglected topic. By the year 2000, however, the internet had become a resource of information and opportunity, not only for collectors of dinosaur postcards, but also for information concerning the history of postcard collecting!

As can be gleaned from internet sources, postcards have evolved through several 'eras.' The first cards regarded as 'postcard-like,' were printed during the Civil War, becoming the copyright of H.L. Lipman. These 'Lipman Postal Cards' generally featured patriotic images. But the first 'true postcard,' was printed during the 'Pioneer Era' (circa 1870 - 1898), by the Hungarian government in 1869, per the suggestion of one, Dr. Emanuel Hermann. The early years of postcard history have an interesting Chicago connection. Did you know that the earliest known exposition postcard memorialized Chicago's Inter-State Industrial Exposition of 1873? And the very first postcards printed intentionally as 'souvenir' cards—their primary purpose today, celebrated the Columbian Exposition in Chicago in 1893. Because there were prehistoric animal displays at both events, by searching for these old collectibles, as did fellow dinophile Jack Arata, one may uncover rare, vintage prehistoric animal postcards from the edge (of time).

By the early 1900s, American publishers were selling cards that could be mailed privately for one cents postage, provided there was an inscription on the cards stating, "Private Mailing Card, Authorized by Act of Congress on May 19, 1898." At first, writing could only appear on the front, image side of these cards. By 1901, postcards had divided backs (as they do now) so that purchasers could write notes on the reverse sides, leaving the front/image side unblemished by pen markings. As noted by postcard historians, "These changes ushered in the 'Golden Age' of postcards, "as millions were sold and used," a time swiftly ending during the escalation of World War I.

American printing technology finally caught up to the Europeans by about 1916, when postcards featuring 'views,' instead of greetings rose in popularity. Because so many of these published cards were printed with white borders, the 'era' from 1916 to 1930 is referred to as the 'Early Modern (White Border) Era.' Thereafter, improving technology permitted printing on linen type stock paper, which allowed printing of more vividly colored photographs. By 1939, photochrome cards were in widespread use, thanks to Union Oil, which launched its own series of postcards that year, and the rest is 'history.'

Surprisingly, there is a museum devoted to postcards in Wauconda, Illinois I learned of in August 2000. Named the Lake County Museum, Curt Teich Postcard Archives, there is a wealth of information within about the history of postcards awaiting serious dino-deltiologists. That is if you know what to ask for and how to search. For, (guess what?) the Curt Teich Museum, named after Curt Otto Teich (1877-1974), has no 'index' or 'category' devoted exclusively to prehistoric animals. (Sigh!) Still there are millions of published postcard images there to search through, and the Museum is also the repository of several historically valued collections of albums filled with rare postcards. For the record, the Curt Teich Company of Chicago, operating from 1898 to 1978, was the world's largest producer of advertising postcards. Dinosaur postcards are most likely there in the collections and archives too, awaiting patient researchers, although you'll have to search hard for them. The phone number for the Curt Teich Postcard Archives is (847) 526-8638.

Now about those problematic categories of postcards. We suggest that we dino-deltiologists propose an 'animal' subcategory devoted to prehistoric animals, or paleontology. This would include subdivided categories of postcards we would ordinarily collect, such as 'Fossils & Skeletons,' 'Museum Displays' (sorted perhaps by Institution)' 'Dinosaur 'Theme Parks,' 'Dinosaur Restorations' (sorted by artist or institution), 'World's Fair Exhibits,' 'Non-dinosaur, Prehistoric Animals,' 'Privately produced or Promotional' (by artist or theme), and a catch-bag, 'Miscellaneous.' Then somebody must impress upon the ranks of deltiologists that these categories should be formally recognized, as there are fine folks out there, such as ourselves, who favor them as collectibles.

Happy hunting—good luck with your own collecting efforts!

Notes to Chapter Thirty-eight: (1) I said 'fourth' in this paragraph because by then three treasured Sinclair prehistoric animal sticker complete albums were already present in our household. The first two were compiled by my father (aided of course by my grandparents) back in the mid-1930s). The first was titled "Sinclair Dinosaur Stamp Album with Contemporary Reptiles" (1935), containing 24 stickers; a second titled "Sinclair Dinosaur Stamp Album no.2 with Other Ancient Reptiles" also containing spots for 24 stickers was issued in 1938. I should emphasize that during the 1930s, postage stamp collecting was a very popular hobby, which probably made the concept of collecting *postage*-like stickers appealing. The third Sinclair dinosaur stamp album was issued in 1959, and I managed to collect a complete set thanks to my grandparents and their gas-guzzling vehicles. (2) For a time I believed I'd published the first article on dinosaur stamps, my "Paleophilately – Stamps, Stones and Stories," which was printed in the October 1981 issue (Vol. 32) of *Earth Science News* (Bulletin of the Earth Science Club of Northern Illinois). This was also one of my earliest writings to see print in some form. All (or most) of the paleontologically-themed postage stamps available at that time (*sans* the stamp issued by China in 1958) were described and discussed. It wasn't until over a decade later while reading Stuart Baldwin's and Beverly Halstead's exceptionally nice 1991 booklet, *Dinosaur Stamps of the World* that I noticed their reference to another article predating mine—P. Manning's "Earth Science on Stamps" (printed privately, 1979). More recently, two *Prehistoric Times* articles by Jon Noad retold the story of dinosaurs on postage stamps, ("Diamond Dinoversary," PT no.125 Spring 2018, pp.28-29, and "Discosaur Stamps of the early Seventies," PT no.128, Winter 2019, pp.13-15. By the late 1990s, I had accumulated many such dinosaur postage stamps, since added to my collection. (3) So yes, I've told this fateful tale three times now—first in my original article on dino-postcards, this paragraph of which was repeated in a biographical segment of my *Dinosaur Memories II* (2017), and now here.

Chapter Thirty-nine — *Into the Valley of Dinosaurs*

Doesn't the title to this chapter sounds appropriate for a science fiction or fantasy tale? But the actual subject actually did involve a very real "Valley of Dinosaurs," one that scientists (not sci-fi writers) helped reconstruct. In the tradition of any good sf story, intrepid explorers encountered treacherous lands and strange prehistoric reptiles. But because those 'reptiles' were extinct, the lands posed far more menacing circumstances for the scientists. Scientists who came to the land were not in search of adventure, but of facts that could be gleaned scientifically and through discovery. Undaunted by the immediate prospect for discomfort and interruption caused by austere climatological conditions to which they were subjected (in the present world), the scientific team had come to investigate harsher landscapes that existed long ago.

The Valley of Dinosaurs is actually part of a Zoological Park located in Chorzow, Poland, dedicated to the efforts of the Polish-Mongolian joint paleontological investigations of Mongolia conducted during 1963 to 1971. Several remarkable discoveries were made then, and in order to commemorate successes of these expeditions, plans were made to create a dinosaur park 'inhabited' by full-scale models of many impressive genera collected in the field as fossils.

There are many life-sized dinosaur sculptures and statues displayed around the world for people to see. In an out of the ordinary fashion, however, it is satisfying that Poland's Valley of Dinosaurs is *commemorative*—saluting national scientific achievement. So before discussing the Park itself, let's first consider significance of the Polish-Mongolian expeditions.

The stark land of Mongolia symbolizes that 'romantic' yet utter, extreme desolation that intrepid fossil hunters are 'supposed' to encounter. Others had tread over its baking (and sometimes freezing) sand dunes before. During the 1920s, American Museum expeditions headed by Roy Chapman Andrews (1884-1960) and Walter Granger (1872-1941)—ironically not deliberately in search of dinosaurs but with an intention of testing Henry F. Osborn's theory concerning the possible Asian origin of certain mammalian genera—made headlines following sensationalized discoveries of dinosaur eggs and nests in the Gobi Desert. Then, during the 1930s, a Swedish-Chinese team found Mongolian dinosaurs in the Gobi. After World War II (during 1946 and 1948-49) Russian paleontologists investigated fossiliferous deposits there, shipping 120-tons of fossil-bearing rock to Moscow. A Russian-Chinese team returned in 1959-60, and thereafter of and on since 1971. (Some details on latter US-led expeditions to the Gobi are outlined in Chapter Twenty-eight.)

When Andrews set forth in search of the earliest human progenitor, railroads and road-worthy highways didn't exist along expedition routes. Rather, camels and Dodge-made 'motor trucks' had to be relied on to carry the expedition's baggage, fuel and other supplies, and personnel. The "missing link expedition" as it became referred-to by popular press, came prepared for harsh climatological conditions in a unforgiving land that was politically unstable. With careful, savvy planning, these hardships were overcome. Years later, Andrews vividly stated accounts of the trail-blazing mission inspired a young female paleontologist named Zofia Kielan-Jaworowska, in planning the Polish-led expeditions. (Transportation routes into Mongolia had considerably improved by the mid-1960s, though travel within the Gobi Desert was still rugged.)

Those prior American Museum expeditions to Mongolia of the 1920s employed services of male scientists exclusively. However, in a rather unprecedented fashion, during the Poles' visits, several female paleontologists, besides Kielan-Jaworowska, participated, including Teresa Maryanska—an expert on bone-headed and armored dinosaurs, and Halszka Osmolska—another dinosaur specialist. In order to generally survey intended areas of study, the first Polish reconnaissance mission took place in 1963: then Maryanska and Osmolska first ventured to Mongolia two years later, in 1965.

Other participating paleontologists included Mongolian paleontologists—yes there were several with proper credentials by that time—Barsbold and Dashzeveg, and Poles—Wojtek Skarzynski and D. Walknowski. Together they braved vipers, blinding sandstorms, and unshaded 100-degree temperatures. Stinging plaster was wind-blown into their eyes, water proved scarce almost everywhere, and then when water was accessible—the scientists were subjected to fierce fly and mosquito attacks.

By the 1960s, however, numerous changes had occurred within Mongolia. First and most important to the Polish team, a railroad network could be relied on for transporting fossils out of the Gobi. Powerful trucks could also be used for hauling supplies and other freight instead of camel teams. The political turmoil and revolutions Andrews and Granger experienced had ended. Still, added cultural and political changes had transformed Mongolia.

During his visit to Inner Mongolia during the 1920s, Andrews noted centuries-old, nomadic lifestyle of the local populace. However, by the time Russian scientist Efremov arrived, Mongolia was transforming rapidly. The capitol city, Ulan Bator, was becoming modernized, both industrially and culturally. For instance, although dinosaurs and fossils were curious objects to shepherds that Andrews encountered, a quarter century later Mongolian scientists had developed enthusiastic interest in their native prehistory.

By 1962, a line of demarcation through the Gobi Desert separated Soviet-oriented Outer Mongolia from China. Missile launching bases were situated in the Gobi's sands—one sad testimony to human 'progress.' But the harsh wilderness of Mongolia remained oblivious to the plight of humanity. As has been the case for thousands of millennia, in the remote areas of the Nemegtu Basin, Bayn Dzak, and Altai Ula, fossils eroded from weathered rock, exposing creatures of 'other days' to the Sun's scorching rays ... for the first time in millions of years.

There were several objectives for conducting paleontological investigations. As stated by Colbert in his *Men and Dinosaurs* (1968, p.252), Mongolian scientists were interested in surveying regional natural resources including fossil riches of their country. "For the proper development of their plans, some outside help would be helpful to the Mongolian paleontologists."

Among other kinds of extinct Tertiary mammalian fauna, Andrews's teams had discovered skeletons of impressively large Tertiary rhinos and 'titanotheres.' However, arguably the most significant achievement of the American Museum Central Asiatic Expeditions had been discovery of the earliest known Mesozoic *placental* mammals. Kielan-Jaworowska wanted to extend the search for these (although by then Osborn's theory of human origins had been largely falsified—see Chapter Twenty-eight). Additionally, discovery of dinosaur skeletons permitted the Polish-Mongolian teams to make significant contributions to scientific knowledge concerning these extinct vertebrates.

The prospect of being able to address tantalizing questions such as 'what happened to the great Mesozoic beasts, and why did the mammals replace them?' with hard sought data secured in the field was immensely satisfying. One particular extinction hypothesis could be tested: "Could it be that the disappearance of certain environments here ... because of physical factors, was responsible for the extinction of dinosaurs?" Professor Stanislow Dzulynski had noted that types of continental deposits typical of the Mesozoic Era no longer appear in Tertiary mountainous terrain, a circumstance that may not have been conducive to survival of large animals. "The disappearance of shallow lakes and wide open coastal spaces would have caused extinction of herbivorous dinosaurs, and that of the carnivorous dinosaurs as well ..." (*Hunting For Dinosaurs*, 1969, p.20)

It cannot be understated that the Cretaceous fauna of Mongolia included a very interesting assortment of dinosaur genera. Of the 300 or so recognized dinosaur genera (i.e. as of 1992), about one sixth have been found in Mongolian Cretaceous deposits. As of 1990, of the 47 known Mongolian genera (again as of circa 1992), 31 were described after 1966. Of those 31 dinosaur genera described since 1966, 23 have been described by Polish and Mongolian paleontologists— 14 by the Poles. It is clear that Polish workers and particularly the Polish-Mongolian expeditions have been influential in uncovering the Mesozoic vertebrate geo-history of Mongolia. Efforts of Maryanska and Osmolska, collectively, have resulted in 13 of the Polish dinosaur descriptions. (However, not all of the genera discovered by the Poles are represented in sculptures at the Valley of Dinosaurs—which we will come to shortly.)

Kielan-Jaworowska presented her account of their expeditions through 1965 in her book, *Hunting For Dinosaurs* (M.I.T. Press, 1969). (Chapters in a more recently published book by Don Lessem, *Kings of Creation,* 1992 also outline the Polish-Mongolian contributions to dinosaur science.) Without understanding their scientific significance, some 'prize' discoveries made by the Polish-Mongolian teams can be briefly summarized. The tyrannosaur genus *Tarbosaurus* (i.e. sometimes referred to the genus/species *Tyrannosaurus bataar*) had been collected and studied before by Russian scientists. However, it is truly remarkable how many skeletons and skulls belonging to this genus were found both juvenile and adult specimens, sometimes at even rather inconvenient moments when the crew needed to devote attention to other project details, during the first two collecting seasons. Skeletons of this large carnivore are now mounted in Warsaw, Ulan Bator and Moscow. One 80-million-year-old specimen is 25-feet long—the first dinosaur to be exhibited in Warsaw—was found by the Poles, excavated from hard sandstone deposits at Tsagan Khushu.

In 1971, embedded within Djadochta strata near the famous Flaming Cliffs at Bayn Dzak (now presumed to have been a semidesert during the Upper Cretaceous, instead of a lakeshore environment), an amazing fossil of two 'wrestling dinosaurs—*Protoceratops* and *Velociraptor* was excavated. "The two skeletons were in their death poses, with the hands of *Veliciraptor* grasping the skull of *Protoceratops* with its sickle claw of the foot positioned as if tearing into *Protoceratops*' body, It would appear that *Protoceratops*, in turn, had buried its beak within the body of *Velociraptor*. The two antagonists, now exhibited at Ulan Bator, seemingly had killed each other and died, locked together." (Edwin H. Colbert, *Dinosaurs: An Illustrated History*, 1983, p.133) Kielan-Jaworowska had originally suggested that the two had "died as they fought ... embedded in quicksand." Their bodies were quickly covered by wind born sand in the arid landscape.

Also in that same year, scientists excavated the 23-foot long, armored *Saichania,* a name meaning "beautiful," from deposits at Khulsan. A variety of armored dinosaurs, such as *Shamosaurus, Tarchia,* and *Pinacosaurus* inhabited the Upper Cretaceous Mongolian landscape. However, the ceratopsid lineage seems to have terminated in Asia with 'primitive'-looking genera, such as *Bagaceratops* and *Protoceratops.*

Remains of long-necked sauropods were also recovered. Kielan-Jaworowska described systematic excavation of one such creature—discovered by R. Gradzinski on June 4, 1965, at a site named Altan Ula IV. Although other skeletons were simultaneously being disinterred elsewhere by the team, 12 tons of sauropod bearing rock were removed in "record time of not quite four weeks." There were 35 crates, the largest weighing about one ton. Sauropod remains were of special interest, not only because two genera new to science were found (i.e. *Opisthocooelicaudia* and *Nemegtosaurus*), but also because these genera were

Upper Cretaceous in age. In North America, for example, sauropod dinosaurs like 'brontosaurs' were much more common during the Jurassic Period. Unfortunately however, the *Opisthocoelicaudia* specimen had no preserved neck or head, while *Nemegtosaurus* was identified solely from its skull. Caution is usually urged against compositing bones excavated from different localities.

Perhaps the strangest discovery of all was made in the Nemegt Basin of the Gobi Desert (Altan Ula III) by Kielan-Jaworowska on July 11, 1965, when she spied a pair of huge claws with articulated forelimbs eroding from an outcrop. The curved claws were about one-foot in length. The arms were about 8-feet long! Although at first it remained sheer guesswork to speculate how this mysterious monster—*Deinocheirus* appeared in life, finally additional remains surfaced half a century later, thus permitting a more accurate restoration.[1]

The Polish-Mongolian team also found more dinosaur eggs and several skeletons of herbivorous dinosaurs (including those belonging to fast-running 'ornithomimids,'—the first found in Asia). One of these, *Gallimimus*, was a toothless, 'ostrich dinosaur' which grew up to 20 feet in length. Several herbivorous 'bone-headed' pachycephalosaurs were also found.

Furthermore, their expeditions falsified claims of a Soviet scientist, Novozhilov. Because Mesozoic mammal skulls hadn't been found during their 1940s visit to Bayn Dzak, Novozhilov stated that mammal skulls found there by Andrews were of Paleocene rather than Cretaceous age, originating through secondary (re-)deposition. A total of 150 Cretaceous mammal skulls were discovered by the Polish-Mongolian team, including the first primitive mammalian fauna ever found in the Nemegt valley.

Kielan-Jaworowska concluded her popular 1969 account with this summary:

> "We are now faced with the most interesting part of our Gobi adventure: the scientific interpretation of our findings. But is it really the most interesting part? Was it not more exciting to travel through endless Mongolian steppes and extract skeletons of extinct animals in the burning sun of the Gobi than to make tedious anatomical studies, measure the bones, describe their shapes, look for nerve and vein openings in skulls, and pore over the abundant literature on these subjects? ... No scientist familiar with the intellectual adventure of studying animals from times long past will have any hesitation in affirming that to travel millions of years back into the past, which is what paleontological study amounts to, is much more fascinating than the most exotic geographical travel we are able to undertake today." (*Hunting For Dinosaurs*, p.176)

One research area introduced by the Polish-Mongolian expeditions concerns how Mongolian dinosaurs relate, in an evolutionary sense, to other Asian genera, as well as those of North America and Europe—and to what extent land connections between Asia and the rest of the world permitted faunal exchange during the Upper Cretaceous.

In 1980, a Polish commemorative postage stamp was issued featuring an in-situ *Tarbosaurus* skeleton alongside a map showing the Polish-Mongolian collecting sites. Quite fittingly, in 1991, at an August meeting of the British Association for the Advancement of Science held at Plymouth, England in commemoration of the "150th Anniversary" of the naming of Sir Richard Owen's proposed "Dinosauria" classification, Zofia Kielan-Jaworowska lectured on results of the Polish-Mongolian Paleontological expeditions. (Personal communication dated Jan. 7, 1992)

Having an interest in prehistoric animal sculptures, in early 1990 I inquired by letter about the dinosaur sculptures on display in Chorzow's "Valley of Dinosaurs." Interestingly, in a written reply from Polish paleontologist Karol Sabath, I learned that all was not well (then) with their dinosaurs.

Sabath, associated with the Zaklad Paleobiologii PAN (Institute of Paleobiology) in Warsaw, Poland graciously sent me a wonderful set of photographs, news clippings and postcards concerning creation and history of the Valley of Dinosaurs in Chrozow. A portion of the Silesian Zoological Gardens is devoted to exhibition of several full-sized dinosaur models commemorating the series of highly successful fossil hunting expeditions to Mongolia, conducted jointly between Mongolian and Polish scientists during 1963-1971.

Sabath explained how the models were made by talented artists and submitted photographs of the model-building operation, completed over a number of years. The statues represent genera native to the Mongolian area of the Cretaceous period. When the park opened on July 22, 1975 there were models of nine different dinosaur genera. Following conclusion of opening ceremonies, Sabath, then age 12—literally the first visitor to enter the park—ran down into the completed park followed by staff of the Institute of Paleobiology, Zoo personnel and local authorities. Excited by construction of the Valley of Dinosaurs, Mr. Sabath later entered the field of paleontology vocationally.

Dinosaurs were not part of Poland's geological heritage, as Sabath noted. However, this is not the reason for the vandalism and environmental pollution that has degraded the models. Environmental factors, primarily industrial atmospheric emissions are largely unavoidable. Vandalism should be controllable, but has so far resisted control. Most unfortunately, the models are "... subject to aggressive behavior of some visitors (especially youth (in) packs ... or drunk ..." Quite ironically we find the Valley of Dinosaurs is becoming threatened by a manmade form of 'extinction.'

Deterioration of a full-scale park reconstruction featuring an encounter between *Protoceratops* and *Velociraptor* based on the famous fossil discovery made in August 1971, was documented in one set of photographs taken between 1975 and 1985. First, the model is shown beautifully restored by artists quite familiar with the fossil remains. Later, however, we find that the head of the

Velociraptor has been battered and severed. Its arms have been broken too. This has been a gradual process stemming from years of abuse.

But *Velociraptor* has not been the only 'victim.' Teeth of the large carnivore *Tarbosaurus* have been shattered by hurled rocks. The head of a *Saurolophus* has been repeatedly smashed and a model of *Sauronithoides* has been hopelessly ruined. A *Gallimimus* statue, thought to have been securely anchored to a large rock has been destroyed. Conservation of the park has led to some repairs, but maintaining them in a state of good condition has proven difficult.

Such problems may be symptomatic of political strife in Poland. For instance, because university graduates were earning less than the "unqualified labor force," it is difficult to stimulate young minds into a scientific profession. This is because paleontology was then much less popularly regarded than politics, (according to Sabath).

Consequently Sabath, who often lectures to Museum visitors on paleontology claimed:

> "… interest for the dinosaurs (and science in general) among teenagers seems to be gradually wiped out by the dull school programs and growing anti-intellectual atmosphere (and the university graduates or even specialists are underpaid in relation to the unqualified labor force, and this undermines need to learn anything more than the absolute minimum). So the youth is now oriented rather to political affairs than science. Most questions about the dinosaurs come from 8 to 12-year old students, but they are of course rather simplistic … and it is not easy to keep them interested."

In responding to Mr. Sabath I suggested the models should be considered a national treasure due to their significance in memorializing an important facet of the history of vertebrate paleontology. The Victorian dinosaurs at the Crystal Palace at Sydenham, England remain displayed after nearly a century and a half (i.e. as of 1992), for example, and the Pallenberg models at the Hagenbeck Zoological Park in Hamburg, Germany have been exhibited for a century. One hopes that well into the next millennium our great grandchildren would have opportunities to see the Valley of Dinosaurs intact too.

Sabath generously granted permission to reproduce photos, illustrations and portions of his letter to interested parties. Therefore it is fitting to present in his own words a description of the Valley of Dinosaurs, and how it was built. (Text by Karol Sabath, excerpted from correspondence to Allen A. Debus—dated July 31, 1990, pp.2-6.)

> "The Valley was useless for a long time, since it was formed by collapsing of emptied coal mines and thus was rather unstable. Thus the earlier ideas to fill the deeper part with water and keep there some pinnipeds were abandoned as the water would leak out due to seismic

damages induced by mining in the area. So it was not until more than 20 years after establishing the Park and the Zoo (in the 1950s) when the idea of turning the valley into the dinosaur park appeared. The Park Director, Mr. Julian Koba and Zoo Director Z. Zielinski, inspired by the Hamburg Zoo dinosaurs and Warsaw dinosaur exhibit, signed an agreement with our institute in 1972 ...

"The findings of Polish-Mongolian Paleontological expeditions were first presented to the Polish public in the late 1960s. Besides TV films, books (two written by Prof. Zofia Kielan-Jaworowska—scientific leader of the expeditions, and two by Maciej Kucszynski, B. Eng, technical leader) and press accounts, a temporary exhibit "Dinosaurs from the Gobi Desert: opened in the Palace of Culture and Science in Warsaw in June 1968. It included ... replicas of skeletons, also some restorations (juvenile ornithomimid *Gallimimus*, ankylosaur *Pinacosaurus*, *Protoceratops*). The biggest was a natural size *Tarbosaurus* model made of Styrofoam. The smaller models, after the exhibit was closed down, were transferred to the Museum of the Earth in Warsaw, and since 1985 they have been integrated into the "Evolution on Land" exhibit in the Museum of Evolution in the Palace of Culture and Science, Warsaw, where also other Mongolian specimens are on display. Those models were sculpted in plaster of Paris or paper pulp by Mr. Wojciech Skarzynski. He has been (an) employee of the Institute of Paleobiology (then Inst. Of Paleontology), Polish Academy of Sciences, in Warsaw, as senior technical assistant. He was also (an) active member of the Mongolian expeditions, and author of numerous widely reproduced photographs from the field work in the Gobi Desert.

It was also Mr. Skarzynski who produced the small scale models for the Valley of Dinosaurs project in 1972-74. The models were made to 1:10 scale (the big theropods, sauropods, ornithopods) or 1:5, 1:4, or 1:2 for the smaller dinosaurs. Also, some details (*Tarbosaurus*' forelimbs, skin textures) were rendered separately actual-sized. All (i.e. prototype smaller-scale) models were built on steel wire frame covered with paper pulp standing on wooden bases. They were painted with oil paints. The sauropods and pachycephalosaurs were brownish; saurolophs and ornithomimids green with rusty-orange underside, tarbosaurs bluish-gray with orange-yellow underside, small sickle-clawed theropods dark indigo with purple spots, and *Protoceratopses* as well as the ankylosaur matching the red color of Djadokhta formation sandstone beds (they were living probably on red semi-desert) with a darker pattern. The original models (except one sauroloph) had to be destroyed several years after completing the exhibit since they proved to be severely infected by boring insects. The remaining one decorates the hall in the Zoo administration building.

The scenario of the Valley was prepared by Prof. Teresa Maryanska (Museum of the Earth, Warsaw) and Eng. Maciej Kuczynski. Other scientific consultants involved were Prof. Z. Kielan-Jaworowska, Halszka Osmolska and Magdalena Borsuk-Bialynicka (who described the genus, *Opisthocoelicaudia* in 1977).

The (e.g. full-sized) dinosaur statues for (display in) the Valley were made by local Silesian sculptors from Katowice Art Workshop ... Henryk Fudali, Waldemar Madej, and Jozef Sawicki. Katowice-based architect, Jan Kosarz, not only prepared the blueprint of the Valley, but also lent a hand to the rendering of statues. The five largest models (sauropods, tarbosaurs and adult *Saurolophus*) were built in-situ, and the remaining ones were transported from the sculptors' steliers in neighboring towns (Bytom, Piekary, Slaskie). First, construction frames made of steel tubings were erected. The largest constructions were made by 'Nostastal' Enterprise. Then a set of grid coordinates was taken from each scale model and scaled up to the final size in form of welded steel wire lattice to which layers of wire mesh were fixed ... Then concrete was applied and thus the rigid steel-concrete shell was formed. ('Budokop' mining construction enterprise from Nyslowice helped to finish the largest models at this stage.) The interior is hollow, and in the largest sauropod a trapdoor was installed to permit monitoring of the steel construction preservation. The surface was then sculpted in additional layers of concrete. Finally, imprints of scaly skin texture were added with latex matrix. The models were left unpainted so they are all gray. It is also to indicate that we do not know the actual coloring of the dinosaurs. There was an idea to cover them with a neutral-colored resin layer, but for technical and financial reasons it was abandoned."

On its opening day, July 22, 1975:

"The Valley contained the following models of dinosaurs: a big tyrannosaurid—*Tarbosaurus bataar* chasing two ornithomimids—*Gallimimus bullatus* ... a larger one (i.e. ornithomimid) that had stopped for a moment and rose his head to look behind and the other one running with head and neck stretched horizontally; the third *Gallimimus* digging out some food; the big diplodocid *Nemegtosaurus mongoliensis* (25-meters long), an adult duck-billed *Saurolophus angustirostris* standing by the pond in bipedal posture (7-meters in height), and a younger one in quadrupedal stance; an ankylosaur *Saichania chusanensis*; three individuals of primitive ceratopsians *Protoceratops andrewsi*—the large adult displaying small horn on the nose, medium one running with half open mouth thus showing dentition, and the smallest pulling a plant with his beak; a pair of butting males of dome-headed *Prenocephale prenes*; a small troodontid coelurosaur *Sauronithoides mongoliensis* capturing a small multituberculate mammal with his forelimbs; a *Tarbosaurus* fighting with a *Nemegtosaurus* and trying to push him downslope.

The model of the famous fighting dinosaurs—*Velociraptor* and *Protoceratops* was added later. Thus the final number of dinosaurs in the Valley is 18 ...

There were also plans to extend the scope of the exhibit (additional open air models of other Mongolian Upper Cretaceous fauna, e.g. oviraptors and crocodiles, lizards, turtles, dinosaur nests, exposed dinosaur skeleton in-situ, etc.) and to build a pavilion at the entrance to the Valley ... to present an exhibit about the paleontological expeditions, and the fossil specimens (scientifically supporting each) restoration, etc., but lack of funds had not permitted such a development of the project (final number of models was intended to be 33). The total cost of the project was ... several hundred thousand dollars.

Instead only the information boards were mounted at the entrance and duplicated at the sightseeing platform over a rim of the Valley. They present a history of the expeditions with a map showing the fossil-bearing localities and information about the Valley ... In the mid-1980s, I renewed the old ones and made another set of information boards on three aluminum plates ... which were mounted at the second exit from the Valley ... These boards present colorful phylogenetic trees of the amniotes and of the dinosaurs in detail, showing their (evolutionary) relation with birds, and also some spectacular archosaurian record holders ... During construction of the dinosaurs and after the inauguration, numerous articles, pictures and short notes were appearing in press, mostly in the local newspapers and weeklies, many of them with nation-wide circulation."

In his detailed letter Mr. Sabath—an accomplished paleoartist, also described other facets of how the Polish people regard paleontology at the popular level.

Dinosaurs are enjoyable creatures of the imagination, both in our dreams and nightmares. However, despite popularization afforded by the media (augmented by blatant commercialization), it must not be forgotten that dinosaurs remain an important research vehicle for scientific investigation. We note that dinosaurs and their extinct brethren provide important lessons in our understanding of fundamental biological and geological concepts: evolution of vertebrates during one very long interval of Earth history, the pattern of continental movement, and pace of ecological change. Dinosaurs of the Gobi Desert and other Asian localities (such as the Middle Jurassic fauna of Dashanpu. China) merit full attention of western scientists. Lately numbers of new Chinese dinosaur genera have been outstripping those recently disinterred from North American deposits. (Note—this essay was first printed in 1992, as the ranks of significant Chinese dinosaur fauna were escalating.)

In the future more significant dinosaur discoveries will be made. Cultural exchange will reveal more to the western world and its paleontology enthusiasts about the rich and fossiliferous dinosaur-bearing bone beds of Asia, as well as the women and men who made these startling discoveries (including the Poles). It is very evident that many Japanese also love dinosaurs and perhaps stemming from the ageless Chinese tradition of dragon symbolism and lore, dinosaur popularity in Asian grows. During the 1980s and 1990s Chinese fossil exhibits including dinosaur skeletons and casts were well received in several American cities, Tokyo, Cardiff, Sydney, Stockholm and London.

Interested parties of the Asian world may in turn become more fully acquainted with our longstanding western dinosaur tradition, stemming from the days of Mary Anning, Georges Cuvier and both Gideon and Mary Mantell. (I was recently surprised – i.e. in 1990—to hear from American sculptor David Thomas that he made full-size dinosaur models of North American dinosaur genera *Tyrannosaurus* and *Pentaceratops* for a natural history museum in Taiwan.)

It is easy to chide the Polish public for their nonchalance toward paleontology, a science that appeals mainly to those who philosophize about abstractions like deep time and extinct creatures, while offering few material rewards to the business man (unless he can somehow exploit the field) or politicians. But then it is also distressing that in our country celebrities in various sectors of the entertainment industry are lavishly rewarded, instead of skilled scientists who may be able to benefit the welfare of mankind in some definite way. In Poland, rebellious youth occasionally wreck dinosaur statues. In the USA, businessmen mass-produce dinosaur toys for profit. In both countries too few care about deep time and the majesty of the fossil record.

Dinosaur popularity will bloom and wane somewhat with the vicissitudes of worldwide affairs. I wonder how imaginations of late 21st century dinosaur enthusiasts will be fired by exhibited fossil remains of ancient saurians. I suspect that, although presentation of facts may differ by then, the thrill and romanticism one senses in learning about dinosaurs and other kinds of prehistoric life—at least at an early impressionable age, will be inherently similar. One would also hope that tomorrow's world of science, epitomized by Poland's endangered Valley of Dinosaurs, shall not be extinguished by political oppression or a disinterested populace.

Other Key References: Dong Zhiming, *Dinosaurs From China* (1987, English trans.); Don Lessem, *Kings of Creation* (1992, Chapters 5 and 9); Personal correspondence from Karol Sabath (M.Sc.) dated July 31, 1990 and Dec. 12, 1990).

Note to Chapter Thirty-nine: (1) See Gregory S. Paul's *The Princeton Field Guide to Dinosaurs, 2nd ed.* (2016) for a new fleshed-out restoration of *Deinocheirus*.

EIGHT

Fantastic Adventures, copyright 1945, Ziff-Davis Publishing Co., (Author's collection), painting by Frank R. Paul. For me, G-Fest experiences, sci-fi conventions and comic-cons are always ... 'fantastic adventures.'

G-Fest Panels

I've always enjoyed G-Fest, although every time in the aftermath I'm bereft with feeling that there was so much more to do there than my participation allowed. The first G-Fest our family attended (I couldn't go myself but now don't remember why), Diane arrived at the 'G-Con' hotel near O-Hare (in 1997) listening for a while to a couple actors—one was the famous Haruo Nakajima who had played Godzilla. She was also thrilled to meet John Lees there on that occasion, although her ulterior motive that day was to hawk copies of our 1995 *Dinosaur Sculpting* book.

On another G-Fest occasion, I gave an introductory speech outlining Don Glut's career just before he was presented with the Mangled Skyscraper Award. Since then, I've been more or less a regular attendee, and have participated on quite a few panels. Every time I attend, I'm inspired to write new material, and so this section presents two such 'self-assignments' emerging from a couple of my G-Fest Panel experiences, presented here as Chapters Forty and Forty-one.

As an aside, there are many who *don't* care to think of Godzilla as some sort of dinosaurian derivation, or abstraction. Over the years I've presented considerable documentation and evidence to show why that monster indubitably is a *dino*-monster, and I argue the case has been settled! (But read Henning Strauss's intriguing article in *G-Fan* no.123, Spring 2019, establishing alliances between Herman Melville's *Moby Dick* and Godzilla. Strauss concludes these two "… fictional characters with origins firmly rooted in folklore, both having achieved iconic status in popular culture, a modern mythos … mine the same rich vein of contemporary public consciousness." p.53)

These chapters originally appeared in the following publications:
Chapter Forty—*G-Fan* no. 122, Winter 2018, pp.32-36.
Chapter Forty-one—*Mad Scientist* no.31, Summer 2016, pp.35-42.

Chapter Forty — *All-time Greatest Prehistoric Monsters of Science Fiction: A Panel Experience*

No, this certainly wasn't my first time giving a presentation at G-Fest, or participating on a G-Fest panel, although after 20 years of attending G-Fests in Chicago it *was* the first time I'd led a panel.[1] During the mid-2010s, the crucial duty of organizing lists of panels proposed for each upcoming convention fell to Martin Arlt. My panel idea with its synopsis for "Top Ten Sci-fi Prehistoric Monsters," floated to Martin in the fall of 2017, made the cut for G-Gest XXV (2018). Now I needed to find some willing, intrepid accomplices who would face the music in front of an audience as we tackled the thorny problem of which science fictional prehistoric monsters were very best of all-time. It's relatively easy tossing a rough idea for a G-Fest panel into the hat, but when push comes to shove in actually getting the work done the situation can seem quite time-consuming and daunting. Somehow I had to make it happen. Now what do I do? I definitely needed HELP.

One of those obvious 'daunting issues' was in the title itself … '*top ten sci-fi prehistoric monsters?*' "What're you doing—trying to start a 'rumble' in here?" Yes—that scene from one of my favorite *The Big Bang Theory* episodes, where Stuart the comic book shop manager pleads with Penny & girlfriends not to openly debate the greatest comic book heroes while in his store, came to me as I was figuring out what to do. When it comes to monsters and their cinematic on-screen exertions, lists of 'favorites' have been compiled before, dating back to 1973 when fans voted Godzilla the number one monster—outranking Frankenstein's monster and Dracula for the honor in a poll conducted by a now extinct zine—*The Monster Times*. Truly 'Showa' Godzilla was 'king' then: even Elvis would have had a hard time overthrowing him! But what about today?

Decades later, on at least two occasions, J.D. Lees printed ballots in *G-Fan* requesting readers to vote, for their favorite kaiju films and associated monsters. But in our case the key tag was not 'kaiju,' nor simply 'giant monsters,' but *prehistoric* monsters—at least those which were prehistoric-aspect in nature or perhaps anachronistically prehistoric in our present, confronting man. Also, I wasn't interested in conducting a simple popularity poll—those had been done. Instead I envisioned a more 'objective' approach, while realizing that at best the panel's decision might only yield a (semi-subjective) *illusion* of objectivity. Shortcomings aside, at least it might be fun to try and see what happens.

Actually, I had compiled something like this already, over a decade before for issues of Dennis Druktenis' *Scary Monsters* magazine. In 2006 I floated an idea to Dennis about preparing an extended listing of the 'greatest prehistoric monsters of all-time.' He replied enthusiastically that such a manuscript could be printed in serial format (which would potentially keep readers thirsting for more). So over the course of several weeks I compiled my list, which was presented *alphabetically* through each issue, not in any particular hierarchy as to which entry was 'best.' The 4-part series ran from January through June 2006. Then, 3 years later I incorporated a far more concise adaptation of the *Scary Monsters* article series as an abbreviated Appendix in my 2009 book, *Prehistoric Monsters: The Real and Imagined Creatures of the Past that we Love to Fear* (McFarland, pp.269-278). Except I retitled the Appendix listing to "The Most *Famous* Prehistoric Monsters in Popular Culture." 'Famous,' Greatest,' and 'Favorites'—and now a proposed 'Top Ten'! How does one definitively decide such things? Beyond my own musings on the matter, perhaps a panel of experts might develop a more refined listing founded upon more concretely set criteria?

So I approached two g-fans who I know enjoy this sort of thing—Jason Croghan and Greg Noneman (the latter of dinosaurian Prehistory Traveler's Guide Podcasts fame). Greg was instrumental in setting up a Facebook chat forum and inviting others who facilitated strategy in the early going—Matt Dennion, Samuel Iler, and David Prus. Samuel was eventually added as a panelist, while Matt and David bowed out in later stages of organization. My fellow panelists now consisted of Jason, Greg and Samuel. Our mission? We had volunteered to—in a sense—gauge how ingeniously imaginers of the past and present had transformed the familiar into the utterly fantastic! Each of us brought a particular knowledge base and skill set to the table, and I am indeed indebted to my co-participants' ready facilitation.

Based on early Facebook group chat postings and idea sharing, Matt took the first stab at preparing a "rubric" for awarding points to any nominated creature. I modified this rubric device further to reflect parameters that would be germane to assessing whether any particular prehistoric monsters had enough 'it' or 'wow' factors to warrant being in anybody's top ten list. In short, these *major* factors became (1.) "Influence on Popular Culture" (2.) Influence on Genre" (3.) How effectively has creature been represented in story-telling." More on these categories shortly. Quite simply, in this system prehistoric monsters awarded the most points wind up at or near the very top.

The entries/nominees need not be giant creatures, nor must they be 'real' dinosaurs known to science, and they certainly could be famous *both* in film and/or literature or even other media. We decided for simplicity's sake—to make matters more manageable, that any prehistoric monster's appeal should be founded upon the aggregate or sum of all its appearances we were readily aware of. For example, the monster Rodan considered on basis of all its Toho movie roles ... or *Tyrannosaurus* in all its various pop-cultural, science fictional (or horror) manifestations, from before 1933's *King Kong* to beyond *Jurassic Park*. All manifestations in popular consciousness were fair game for carrying our nominees forward for further, objective consideration.

Next, each of us generated our own individual lists of potential top ten candidates, which, when compiled, would comprise a 'master list' of 47 'significant' monster names that we would vote upon using more objective criteria which we would create. We expected that certain names such as Godzilla would be repeated on our individual lists. However, no 'extra' points were awarded for such concurrencies. Ballots were distributed and voting (by myself, Greg, Jason, Samuel, Matt and David) took place in the early spring of 2018. Compiled results were at least partially surprising ... but I'll get to that.

An issue that we propounded was how do we define what is a 'Prehistoric' monster—after all, shouldn't they all be dead—totally extinct, founded only upon genuine fossils. I had dealt with that perplexity before in the *Scary Monsters* articles. Not only our G-Fest audience, but also panelists as well needed to comprehend our perspective. After all, a panel of experts on prehistoric monsters ought to know what a science fictional prehistoric monster is—right?[2] Were the monsters entombed in ice for eons, surviving on lost world places where time has mysteriously stood still, hibernating under the sea before being awakened by man, or had they arrived from an outer space prehistoric planet? Without further elaboration here, in a nutshell, 'prehistoric' meant any creature, regardless of size that may have been in anachronistic contact with humans either in our present (or if modern humans travelled backward in time to *their* present), that struck us somehow as 'prehistoric' in aspect or appearance—some are kaiju, some aren't. Not all are giant-sized, or even 'real' dinosaurs. Quite a loose definition, yet for our purposes it sufficed.

Of the three Major Criteria, given six people voting, a total maximum tally of up to 96 points—a 'perfect' score—could be awarded to any particular monster. Voters were allowed to award zero points under certain subcategories of criteria if we lacked familiarity with any particular monster. Ultimately, the range of points cast was from a high of 94 down to a low of 20 points (for a creature—and here I apologize for my lack of recent pop-cultural familiarity—named Cebralral). But, hey, just to be nominated and 'make' the list was an honor in itself. The age range of us voters was 'multi-generational'—with me as the token 'oldster,' permitting perspective on a range of prehistoric monsters from 19th/early 20th century 'classics,' (my forte'), up through those that others may have encountered on computer games (which I don't play). Turned out I was also the toughest voter.

Now back to the three major criteria in that aforementioned "rubric"... under "*Influence on Popular Culture*," the iconic-ness of the prehistoric monster was under consideration—explored through five gradational stages (maximal points if the monster is recognized globally, but only one point if it is vaguely known). For "*Influence on Genre*", maximal points were awarded if the monster is a landmark, foundational creature, but only one point if it has been relatively unnoticed by fandom. And under "*How Effectively has Creature been Represented in Story-telling,*" six fortifying sub-criteria were stated (i.e. "embellishing the plot," "offering a satisfying and convincing suspension of disbelief," "symbolizing or foreshadowing through metaphor the human plight or condition." "adding pivotally to plot thrill and excitement," "imbuing feeling of audience 'connectedness' to the monster," "enhancing story through its visually 'arresting' nature – powerful, 'gosh-wow' entrances and ensuing action in memorable scenes." One point was awarded for each of the six sub-criteria that was effectively demonstrated in the prehistoric monster's story-telling 'arc.' It was possible in some circumstances to award only half a point for certain sub-criteria.

Once more—it didn't matter whether a monster was kaiju or gigantic (or not), or dinosaurian to be included on the master list, earning votes. Although being 'scary' might not gain too many extra points, overall 'awesome-ness' certainly helped. However, the bottom line was that in order to be under our evaluation each monster should strike one as somehow 'prehistoric,' or have primeval ambiance.

The weeks flew by and G-Fest XXV was suddenly upon us. Thanks to his artistic eye, Jason—inspired by our mission, had perfected a short but flashy slide presentation outlining what we had done, culminating with our final results. (Greg prepared a follow-up slide show, just in case there was more time at the end to cover why certain monsters didn't make the final cut.) Our panel was scheduled for Friday, July 13th at 3 PM. Uh oh—did I just say *Friday the Thirteenth*? Time to panic? What if the audience didn't like our results, or what if the slide projector computer system doesn't work? What if we run out of things to say all too soon, or if we're stricken with stage fright? Admittedly, we did experience technical difficulties doing a pre-check dry run of the slide show on Webmaster about one week prior to G-Fest—even though we all got to see and critique the show. Surely it would have been possible to do an entire 'panel' on any one of our top ten finalists, so how would it be verily possible to compress our discussions of each of them into just a few minutes? Lord—give us strength!

We'd planned on first introducing a handful of 'surprise' voting results and prehistoric monsters that were 'near – misses,' those falling just short of the top ten. I was surprised, for example, that plated *Stegosaurus*—a highly recognizable prehistoric creature—once famous for its role, for instance, in 1933's *King Kong* didn't even make the master list. As some savvy g-fans know, *Stegosaurus'* spinal plates (osteoderms) inspired Godzilla's jagged 'fins.'

Then there was fin-backed *Dimetrodon*, another familiar denizen out of prehistory, which fell short of the top ten (i.e. with a decent tally of 52.5 points). Distinctive *Dimetrodon*, of course, was featured in popular sci-fi filmic fare of the 1940s and 1950s, (e.g. 1940's *One Million B.C.* and 1959's *Journey to the Center of the Earth*) as a gigantic beast 'costumed' by gluing a rubbery flexible fin onto the spinal column of real reptiles—a feature that always seems to garner audience approval. Then came more familiar dino-monsters—Godzilla's first prehistoric foe, Anguirus (aka 'Angilas' 57 points), which was sort of a chimera between a ceratopsian horned dinosaur and an ankylosaur, the 'British Godzilla'—Gorgo (66.5 points), and Ray Harryhausen's near dino-look alike—the stop-motion animated Venusian "Ymir" from 1957's *Twenty Million Miles to Earth* (which received 68 points, missing the top ten by a mere half point that would've forced a tie with Gamera ... and immortality). As I explained to the audience, during the early to mid-20th century the planet Venus was considered, under contemporary astronomical theory, to be a 'younger' planet than Earth, since it may have emerged from the shrinking Solar nebula later than did Earth and Mars. Therefore, wasn't it possible that Venus may still have only been in its 'Mesozoic' stage of life?

Now it was time for the grand 'reveal'! We took turns unveiling our top ten prehistoric monsters, from bottom to top, filling in with remarks and info along the way as needed. In 10th place was Daiei's mutant chelonian Gamera (no stranger to a G-Fest audience) with 68.5 points. Next came the generic 'Caveman' and *Brontosaurus* (tied with 71 points each). These were older, vintage prehistoric monsters that would have fascinated audiences and readers of prior generations. To a 19th century populace, the Darwinian evolutionary concept proved especially troubling; the (scientific) idea that humans had a savage, primitive stone-age past was met with scorn and derision. So for many years, cavemen in science fiction were depicted as monstrous in appearance and bearing—creatures that certainly seemed appalling, yet from which one could simply not turn away. Certainly sci-fi and screenplay writers didn't waste opportunities exploiting such threatening creatures.

And for many years during the early 20th century, when living dinosaurs were imagined, it was their gigantic sizes (i.e. for some of them) that stirred awe and admiration ... and often abject horror. *Brontosaurus* was an early popular skeletal reconstruction at New York's American Museum, and soon others followed—*Diplodocus* and eventually towering *Brachiosaurus*. People were foremost fascinated with the seemingly impossible huge sizes attained by these Mesozoic giants. Sci-fi writers, cartoonists and film writers wasted no time populating their stories and screen plays with such menacing, city-smashing, man-eating sauropods, perhaps the preeminent examples of the day being 1925's *The Lost World* (First National), and *King Kong* (1933, RKO). Incidentally, *Brontosaurus* starred in the very first dinosaur film, a 1912-1914 cartoon—Winsor McCay's "Gertie the Dinosaur." Not long after, the Sinclair Refining Company adopted the symbol of a brontosaur as their company logo.

We still marvel at the huge sizes of dinosaurs—best exemplified by the 'sauropoda' several of which were much larger than *Brontosaurus*. Two decades ago, "Seismosaurus" tipped the scales of real dinosaurian gigantism, but now we have *Patagotitan*, an enormous sauropod exhibited as a Field Museum reconstruction. But not all real prehistoric monsters were so terrifyingly large (albeit still much smaller than Godzilla). While sauropods were once thought of as hostile, scary monsters, today they're regarded as gentle giants, eclipsed by other kinds of real and fictional prehistoric monsters in overall popularity.

Velociraptor (next up the ladder with 74.5 points) is one of those science fictional monsters that seems to have sprung out of nowhere into pop-cultural prominence. *Velociraptor* was a poorly known dinosaur for nearly half a century, until a spectacular fossil was discovered in the 1970s wherein it was preserved for all time in a supposed death grip pose, fossil-fused with its potential prey, *Protoceratops*. Still, for many years *Velociraptor* was overshadowed by its earlier evolutionary cousin—*Deinonychus* (another small, lightly built dinosaur with sickle-shaped claws on each foot) whose discovery and scientific description during the mid-1960s presaged the current wave known as the "dinosaur renaissance." Since release of 1993's *Jurassic Park* movie, thanks to Michael Crichton's 1990 novel, these (DNA-bioengineered) creatures have been portrayed as blood-thirsty, clever and vicious genetically brewed monsters freed from Mesozoic amber into modernity by mad scientists. During the past quarter century, *Velociraptor* (both non-feathered and feathered varieties) have left a significant impact on popular culture.

The classic Gillman (76 points) was the feature creature in 1954's influential *Creature From the Black Lagoon*. A living fossil relic from the Devonian Period, Gillman may be regarded as *the* most primeval form of 'caveman' imaginable, especially as portrayed in scenes where in a climactic scene he carries its prize, "Kay" played actress Julia Adams to his 'man'-cave lair, a stony grotto concealed by lagoon waters. If Gillman were considerably larger like the next two prehistoric monsters we voted for, no doubt it would qualify as kaiju like the next two higher ranking prehistoric monsters. These familiar kaiju-saur suspects, Rodan (77 points) and the Rhedosaurus (78 points), the latter featured in Ray Harryhausen's 1953 film *The Beast From 20,000 Fathoms,* need little further introduction here. Rodan, the largest fictional species of pterodactyl, flies at supersonic speeds buoyed on a 500-foot wing span—ten times longer than that in the largest real pterosaurs known to science. Its fame stems largely from its association with Godzilla, although its 1956/57 filmic debut (*sans* Godzilla) is most acclaimed. Rhedosaurus is a quadrupedal 'beast' released from hibernation in Arctic ice during a nuclear bomb test. It carries a germ deadly to humans in its blood. It was stop-motion animated by Ray Harryhausen; as a precursor to Toho's *Gojira* it has attained iconic status among fans of the genre.

However, the highest ranking dino-monster based on real fossils was *Tyrannosaurus* founded upon all its pop-cultural and science fictional manifestations, in 3rd place with a hefty 80 points. *Tyrannosaurus,* perhaps the epitome of 'real' prehistoric monsters was described during the 1900s. Ever since its skeleton was reconstructed at New York's American Museum, combined with Charles R. Knight's legendary American Museum (circa 1906) painting and magnificent Field Museum (circa 1930) mural restoration of this godlike monster striding toward its predestined adversary—*Triceratops*—in Late Cretaceous Valhalla, *T. rex* as it's fondly referred to, has through decades since struck a chord in public consciousness. A bastion of science fiction storytelling and drama, from RKO's *King Kong* and Ray Bradbury's famous 1952 short story "A Sound of Thunder," through *Jurassic Park* (and movie sequels), *T. rex* certainly earned its lofty place on our polling. Long live 'Rex,' the King! But wait, there's more; two in fact.

So by now our audience could probably guess which prehistoric monsters occupied the top two spots, although most wouldn't have anticipated the surprise—that is, how they ranked relative to one another. Rather amazingly, the last two monsters on our 'top ten' list were Godzilla and Kong, *tied* with 94 votes out of 96. Think of that—six individuals of disparate ages voting independently over several weeks' time, and collectively we agreed that Godzilla and Kong (each factoring in all their, respective, many pop-cultural manifestations)—shared equal rank as 'greatest science fictional prehistoric monsters.'[3]

To me, this was a highly satisfying result given my love of the 1963 Americanized film, *King Kong vs. Godzilla*, while anticipating their titanic Warner Brothers rematch to be released in 2020. But would our audience merely accept a tie? Too wimpy a result—shouldn't we somehow declare a clear winner? Yes, and so we did, asking for a show by hands, and the results fell fully in favor of Godzilla. As one audience member aptly summed, "There's no such thing as 'K-Fest.'" Or a 'K-Con.'

No—only G-Fest, and this very positive panel experience wouldn't have been possible, ultimately, without Godzilla.[4]

Notes to Chapter Forty: (1) Actually, at G-Fest in 2012 I presented a *solo*—not really a 'panel' experience—slide presentation on the 1960s *Gorgo* comic book series issued by Charlton. (2) In the Jan. 2006 issue of *Scary Monsters* (no. 57, p. 68), I defined what a 'prehistoric monster' was, to me, in context of the article series. '"It's difficult to place boundaries on what's truly prehistoric in a modern age of monsters. For many of the beasts listed in this survey live anachronistically in our present, having survived cryptozoologically for millennia hidden from man's view. So I've simplified matters by listing any creatures which, in film, lore or fiction, are presumed to have direct 'prehistoric connections, such as dinosaurs living in places where 'time stands still,' or—with certain limitations—where monsters' appearances somehow seem, well, 'prehistoric-looking.' I've extended the definition to other planets besides Earth, although therein limited the entries to species which may be categorized under paleontologist Jose Luis Sanz's 'exo-dinosauroid' definition, or cases where such 'prehistoria' commingled with Earth's ancient fauna. A prime example of the former (exo-dinosauroid) would be Ghidrah, while an example of the latter might be Lovecraft's Chthulu race. Just some stray thoughts ... Several of the prehistoric animals listed aren't actually related to any lineage known to science, sometimes their biological makeup even defies scientific principles (e.g. Godzilla or Gorgo). Here I'm not including, however, creatures which represent modern species mutated through exposures to radiation or chemical wastes into larger, more grotesque forms that only appear 'prehistoric' ... well, that's unless the species originally effected by the radiation otherwise seems 'dinosaurian'—a term used quite loosely here." Before voting commenced, this article was shared with all six voters via our Facebook connection. So as you can surmise, when it came to judging which monsters were considered 'prehistoric,' there was quite a bit of hand-wringing as to which ones were clearly 'in' and those that were 'out.' To keep things simple in 2018 though, our bottom line, loose definition was, 'does the monster look kinda' prehistoric?' (3) I have discussed Kong's prehistoric provenance in a pairing of *G-Fan* articles (nos. 117 and 121, see my "Modes of Survival" Parts 1 – 2) and Chapter Nineteen in this volume, as well as Chapter Ten ("Godzilla's Dinosaurian Origins") of my 2016 book, *Dinosaurs Ever Evolving*. Of course, Godzilla's original 1954 costume was highly influenced by tyrannosaur imagery (particularly that depicted in Rudolph Zallinger's *Age of Reptiles* mural, completed in 1947 for Yale University's Peabody Museum), and Kong's greatest symbolic prehistoric battle (prior to his on-screen tussle with Godzilla) came *versus* a gigantic *Tyrannosaurus* in his 1933 RKO debut performance. (4) The panel session is now available on Youtube, as filmed by David Prus. Gosh darn-it—if only I was a more dynamic public speaker!

Chapter Forty-one — *Two Maligned Dino-Monster Films*

Panel discussions at G-Fest always provoke intrigue and mental perambulations about matters that otherwise might not occur. And so, while during the *Godzilla Raids Again* panel at G-Fest XXII in July 2015, I found myself wondering how one movie that I bonded with during those good old formative years, and another of more recent origin – so reviled by scores of g-Fans – could wind up more or less in the same 'bottom of the barrel' category, according to *G-Fan* magazine opinion polling. Maybe the sagacity of my objective critiquing powers is merely an illusion? So first, what's wrong about *Godzilla Raids Again* (or as I first became acquainted with it in its Americanized version – *Gigantis the Fire Monster*), and then, as an aside, what seems so right, today, among growing ranks of giant monster fans about TriStar's 1998 *Godzilla*?

In *G-Fan* # 49, Jan/Feb. 2001, *Godzilla. Raids Again* (GRA) ranked 28th out of 36 rated movies. (1998's *Godzilla* was last.) There was no separate rating for *Gigantis, the Fire Monster*. "Anguirus" then was our 6th favorite rated monster. Much later in *G-Fan* # 105, March 2014, GRA rated next to last at 28th out of 29 movies (just above 1969's "Godzilla's Revenge"). Yikes! Angurus ranked 6th favorite godzillean foe, out of 12. These rankings are rather astonishing to me because GRA – either the American or Japanese original, rates as my second favorite in the Godzilla series. (Note – the Japanese original *is* better, but the Americanized version still has its finer qualities.) As of July 14, 2015, at the International Movie Data Base (IMDB), GRA scored a 6.0 (out of 10) star rating – not bad, considering that the 1954 *Gojira* rated a solid 7.5. For comparison, *Godzilla's Revenge* (which we *know* is inherently bad – right?) got only 3.9 stars, and the Americanized *Godzilla, King of the Monsters* (1956) scored 6.7 out of 10. There was no separate rating for *Gigantis* at IMDB.

Meanwhile, as stated above, g-fans detested TriStar's version of Godzilla-in-name-only ("GINO," aka "Zilla") in 2001, and by 2014, the verdict was that its giant dino-monster doesn't even qualify as a variety of "Godzilla," as it wound up dead last out of 28 in the "Ranking the Non-Godzilla Movies" category. Even *The Giant Claw* prevailed over faux-Godzilla GINO. (But, hey – as a child I really liked that one too.) Curiously though, *Godzilla* (1998) received a far from abysmal 5.3 out of 10 stars at IMDB, but still not as good as

GRA. What seemed more perplexing were the far vaster numbers of individuals voting in the case of *Godzilla* (1998); 142,178 people voted. In *Gojira's* case, only 17,169 did, and only 2,699 people offered opinion ratings for GRA. I presume that's simply a 'pull of the recent' phenomenon.

Steve Ryfle (p.62) had this to say about the commercial success of GRA:

> "It was made more quickly and apparently somewhat more cheaply than the first, and as a result it lacked much of what made Godzilla so special, except for Tsuburaya's special effects, and even they failed to fully meet expectations. The film fared well commercially but had little impact in Japan."

Furthermore, changing the title in the Americanized version stunted "… commercial potential. As a result of all this, GRA has always been a somewhat overlooked entry…"

One might take issue with Ryfle's comment on GRA's special effects, however, which I thought were quite captivating overall.

As mentioned during the panel, what drew me to "Gigantis" early on during my impressionable years of the early 1960s were those 'mad-science-speak' scenes, and the furiously fought, primeval epic dino-monster battles, chiefly waged in Osaka. This film's release came at a time when Toho was still struggling with origin stories for its monsters, and in translation the Americanized version only conflated matters even more. Also, I've always been a 'dinosaur guy,' and so while I knew creatures Gigantis and Anguirus were not real dinosaurs, visuals of their fighting seemed as 'authentic' as any dinosaur battle rightfully should. (And in my initial viewing at a young age, what a wonderful thrill to discover that "Gigantis" was really a Godzilla!)

Let's first consider the strange phylogeny of the two monsters. Anguirus (sometimes misspelled "Angurus"), Anzilla, Angilas, Anguillasaurus: is it my imagination or does this particular 'porcupiney' creature have more synonymous names than any giant dino-monster? And what is this 'new' introduced, chimeric Angilas (or however it should properly be referred to), exactly – part ankylosaur, and part horned dinosaur, part stegosaur (e.g. multiple brains – one in pelvis)? In *Gigantis*, Dr. Tadokoro professes that Angurus has (at least) three brains (in its head, abdomen and chest). But in the 'Japanese' GRA version, from the 'new dinosaur book' by fictional Dr. Preterry Hawdon of Poland, we learn that Anguirus is identified as an "ankylosaur" that had *two* brains (in the breast and abdomen), not three. During the first half of the 20th century it was a popular (although incorrect) notion that *Stegosaurus* had two 'brains,' including that second one in its pelvis, accentuating ability to swing its weaponized spikey tail.

Furthermore, as I mentioned during the panel, the arrangement of horns on Angilas' head shield resembles that of the North American *Monoclonius* or 'Agathaumas', the latter made famous by Charles R. Knight in an often-copied, 1897 *Century Magazine* painting. (Of course, Agathaumas-Monoclonius was herbivorous and had a pointy 'beak,' whereas Anguirus has fangy, carnivorous-looking sharp teeth.)

And we learn from that crazy phylogeny session, that bipedal Gigantis is a member of the "Anguirus family." We are also treated to a curious film sequence spliced into the 1959 "Gigantis" movie, conveying Earth's evolutionary derivation of these two "fire monsters." But while in the Japanese original, it's clear that a second Godzilla has emerged, in the 1959 American version, this same animal is named "Gigantis" instead. Why the name change? Steve Ryfle concludes, "In a gross underestimation of Godzilla's marquee value," Godzilla's name was changed to "bamboozle the public into believing this was an 'all-new' monster. It was a stupid mistake, and it backfired." (p.67)

Overall, now into my 7th decade, I can still say that the plot was simple (and at its worst, quite excusable). Despite flaws in its construction, the Angilas suit looked really cool! Although I was mesmerized by buck-fanged Gigantis' suit, several fellow G-Fest panel attendees (rightfully) remarked that it was inferior to that appearing in the original 1954 film.

An emphasis on fire in the evolution and gestation of such monsters was carried to the extreme in 1959's "Gigantis." (It may have even been influential toward naming of a finned creature called a 'fire monster,' appearing at the climax of Irwin Allen's *The Lost World* (1960).) One peculiarity concerns whether Anguirus breathed fire, like its adversary. I have vague recollection of seeing brief footage of Anguirus doing this, issuing from its mouth as a misty spray, but after the G-Fest Panel, attendee David Prus questioned whether this was just a "myth." What I vaguely remembered is seeing an Osaka battle closeup shot of the Anguirus head breathing thin misty fire outward toward its adversary, in a short ~ 2 second sequence, shown probably twice.

Rechecking my DVDs of both films, I could not find any moment in which Angurus breathed fire. I have two circumstantial lines of evidence supporting my vague recollection that it did, however. First is a poster in my collection for the *Gigantis the Fire Monster* movie, showing crudely drawn Anguirus and Gigantis, *both* of which are spouting flames toward one another. And there are supporting statements in Don Glut's, *The Dinosaur Scrapbook* (1980), p.151. Here, Don states, "Anguirus is a quadrupedal, fire-breathing creature with a horned head and a spiny armored carapace ... Gigantis eventually slays his fire-spouting enemy ..." So – there's a bit of mystery. If I halfway recall seeing the fire-spouting, and so did Don, then what happened to the footage? In the Japanese original, we can see that Angilas'/Anguirus' roar was mostly substituted in for Godzilla's more distinctive, characteristic growl. Is it possible then that there were, likewise, 'reciprocal scenes' filmed investing a fire-breathing property onto Anguirus? A stretch, I know.[1]

The absolute best part of this film is the giant monster battle. Yes, there *are* a few (forgivable) 'problematica' with it, in the way brief segments were filmed and how actors managed their difficult suitmation acting, which I will not harp on here. Rather, to me, the surprisingly haunting opening Monster Island (Iwatco Island) battle sequence, realizing that Godzilla – (remember, I first saw *Gigantis* (televised) decades before I was able to watch GRA)—or something very much like it, had returned, ending with their dramatic dive into the sea, merely previews what is to come in the prolonged giant Monster battle staged in Osaka Then the 9-minute thrill-fest battle scene begins at ~ 36 minutes in Osaka with all that roaring, sounding distinct from Godzilla in the 1956 movie, and vigorous tussling to the death, which seemed somehow so 'realistic' to my then 8-year old mind. This was way much cooler than Universal's much hyped *Frankenstein vs. the Wolfman* fight of a decade earlier. And perhaps the best dino-monster fight I'd ever seen. The best battle scene rival I'd seen *until* then was at a downtown Chicago theater in 1960's *The Lost World*, but even to a 5 ½ year old at the time, Irwin Allen's 'dinosaurs' appeared anything but. Toho's giant monster fight scenes were staged in far more foreboding nighttime, black and white sequence, during an imposed city blackout. Flames licking toward the sky and a gloomy musical score accentuates a sense of impending doom—as was ultimately spelled for doomed Angurus. This may not have been Godzilla's best giant monster battle ever, but even now it strikes me as the most 'dinosaurian' of all his encounters.

I suppose a third vital reason why I enjoy this film so much today is entrenched in nostalgia, as it was the second Godzilla film I'd seen, and at a time when only two had been made. Strange thought, huh. I first watched *Gigantis* on Chicagoland monster movie programs, years before Svengoolie's dawn. Chicagoland area shows where I likely was first able to see shows like this (after we returned to the area in August 1961) included "Thrillerama," which aired on Saturday nights at 10:15 PM; but more frequently on either "The Big Show," featuring science fiction-themed movies on Mondays from 4:00 to 5:30, first airing on Sept. 16, 1963, or "The Early Show," hosting 'science fiction Wednesdays,' first airing as a program April 29, 1964. The Early Show ran from 4:00 to 5:30 PM. (Years later, in Feb. 1970, "Creature Features" Saturday evening show came along, but by then I was a high school sophomore.) Two films I most wanted to see on those early 1960s programs, which I yearned to see repeatedly after I had seen them that addictive first time were *Godzilla, King of the Monsters* and *Gigantis, the Fire Monster*. First time I saw *Gigantis*, I had already seen the Raymond Burr Godzilla. So *Gigantis* simply fooled me because of the lead monster's disguising name… but, as stated earlier, got off to a terrific bang with that almost immediate monster battle on the deserted island, starting early at 8 minutes into the film.

Now (e.g. in *G-Fan* magazine and at the G-Fest convention) I'd like to share two photos from my mother's family scrapbooks that revealingly document a small element from those old monster times. First, even though the early 1960s was a time when widespread enthusiasm, trending publicity and love of movie monsters, horror movies and science fiction television programs was at an all-time high, most curiously, one could hardly find any information on Godzilla or about related Japanese films such as *Rodan, the Flying Monster*, *The Mysterians*, or *Mothra*. This is in stark contrast to anyone's experience who has wandered through the G-Fest Dealer's Room, where you're surrounded by godzillean regalia. So, back then, having one's own photographic record of Godzilla to gaze upon as you waited for that interminable months-long interval between broadcastings of Godzilla or Gigantis was truly significant.

To my knowledge no one else took photos of these movies *as they played* on Chicagoland television, in our living rooms. But as you can see here my dad did take two such photos, from our living room vantage point, as my brother and I watched *Godzilla* and *Gigantis*, thus allowing us to maintain a visual record of these incredible creatures. But my dad rather facetiously scribbled in captions under each photo in the album, in this case stating, in *circa* Christmas 1964, during another long-awaited screening of our favorite monster, "In sharp contrast to the joys of the yuletide season was Allen's addiction to the monsters of movie-dom. Here is the final gasp of that synthesis of plastic and rubber called Godzilla." Then a few months later during the spring of 1965, he noted, "In the meantime, at home all was well while Gigantis stalked the screen, but temper tantrums came when (school) reports were due." Well, I suppose I *did* have my 'priorities' even then. Egad, how I hated writing stuff (which is difficult now to rationalize after nearly 4 decades of 'off & on' fanzine writing)!

What else strikes me about this film? Well, just some random thoughts before moving on to my GINO dilemma. Hidemi's view of the distant Osaka conflagration appears like a huge mushroom cloud. The scene where Gigantis' tail swipes the lighthouse, toppling it, is in homage to Ray Bradbury ("The Fog Horn") and Ray Harryhausen (*The Beast From 20,000 Fathoms*). (And yes, there's an interconnectivity between the two 'Rays', detailed in my *Dinosaurs in Fantastic Fiction* (2006).) Tension between fire and ice, with the last of the fire monsters quenched and buried in an icy avalanche, is also rather, well (if one may say so), 'poetic.'

Some interesting Differences between 1955 and 1959 version:

(a.) In the Japanese original version we learn from the get-go that it is clearly a "2nd Godzilla" plaguing Japan again, instead of some sci-fi nonsense about Gigantis being a member of the Anguirus family or connections to fire monsters.

(b.) Toho's original version clarifies that these monsters originally roamed Earth 70 to 150 million years ago. (In *Gigantis* it's vaguely stated that these monsters were "born millions of years ago.") Also in the Japanese original, there is also no corresponding allusion to "Fire Monsters" and

Yamane's footage film is not 'life-through-geological-time' oriented, and is silent without sound narration.

(c.) There is no H-bomb introductory sequence prefacing the Japanese version, as in *Gigantis*, although these concerns show up later when Yamane refers to "radiation-containing Godzilla," and when he declares that the world is now faced with a greater threat than posed by even nuclear weapons.

That famous *Gigantis* movie poster highlighting Anguirus' flame-breathing also suggests an Americanized version that never came off. This feature would have been billed as "The Volcano Monsters" (instead of *Gigantis, the Fire Monster*). For in the 'Gigantis' movie poster one spies metaphorical volcanoes, which one wouldn't anticipate in the immediate San Francisco area, erupting in the distance. As many g-fans know, Toho shipped the original suits used in GRA to the United States just in case producers here wished to film additional scenes for an Americanized version – such as "Volcano Monsters." (See Ryfle, pp.70-71 for more on this unmade film.) 'Lore' surrounding these suits is that they were lost, stolen or simply forgotten somewhere in a warehouse. Steve Ryfle (p.69) documented in a 1996 interview with Bob Burns how the suits were actually identified decades ago shortly after they arrived in the US, during filming of *Invasion of the Saucermen* (1957).

Burns recalled:

> "...I asked ... if we could look in (i.e. the marked crates) ... ... And it turned out to be the original Godzilla suit. ... The head appeared to be over some kind of wire armature. ... He had a little green in him, but was basically a dark gray color. The suit was in really bad shape, it looked like they were going to really have to patch it up, because it had burn holes all over it – you could really see where they had shot that thing up, and where they had run lines when they were shooting fireworks into it."

The Anguirus suit was present also, although they never removed it from its crate for examination. However, "The Angilas suit had more color, he had blue, a little red." The suits were intended to be used for shooting "inserts." Burns concluded, "I never knew what happened to those suits."

Okay – although not discussed during the G-Fest panel, let's move on to a more recent movie that many g-fans regard as inherently bad, even though it sneakily isn't popularly regarded quite that way. Yes, I refer to TriStar's infamous *Godzilla* (1998), scoring 5.3 stars out of 10 at IMDB (as of 7/14/15), but with a whopping 142,178 reviewers, followers, and/or supporters. While so many have seen the film, doubtless far fewer have ever read Stephen Molstad's 1998 novelization based on the screenplay written by Dean Devlin and Roland Emmerich. (See Chapter Five for more on "Zilla," or "GINO" in film and print.)

While I revere GRA (but mainly in its sillier Americanized "Gigantis" format – the version I originally became acquainted with), surprisingly to me, most g-Fans do not count it among their favorites. However, while I don't absolutely scorn the newer TriStar entry, many newer generation G-Fans, including perhaps many who don't (regularly) attend G-Fest, or sci-fi movie fans in general, actually happen to kinda like it. Both movies rate consistently low in *G-fan* polling, however. So does it all come down to nostalgia!?

Millennials can't even imagine what it was like sitting on a living room floor half a century ago, restively waiting for that long-awaited "Gigantis" broadcast to *finally* happen again. And when it aired on Thrillerama or some other local program, it seemed magical. Reciprocally, with so much overly used cgi 'out there' today, on DVDs readily available via Amazon, or Netflix, and with my 6-decade old memory fueled in 'real' Godzilla mythos, it's difficult to discern that TriStar's movie is somehow a cut above any of the others, even though many people younger than I do enjoy many of its aspects. The TriStar version (and novelization tie-in) must be part of *their* nostalgic makeup and imprinting, however, even if denigrated as 'GINO' by those who would claim to be more in the know

Meanwhile, might Legendary reintroduce Anguirus in a movie *Godzilla* sequel someday? I hope so, whether or not the 'new' monster breathes fire. Maybe we'll hear some character yell, "It's An-GIL-as, you moron."

Other References: (1) Mike Bogue, "Kaiju Korner – 'It's not Go-Jee-Ra, you Moron': The American Godzilla 15 Years Later," in *Scary Monsters* no. 89, Oct. 2013, pp.44-48; (2) Ted Okuda and Mark Yurkiv, "*Chicago TV Horror Movie Shows: From Shock Theater to Svengoolie*" (2007); (3) Steve Ryfle, *Japan's Favorite Mon-Star: The Unofficial Biography of 'The Big G."* Toronto: ECW Press, 1998.

Note to Chapter Forty-one: (1.) A comment at the Wikizilla website for the Anguirus entry states, "It appears that according to a 1955 Japanese manga adaptation of GRA, which was contemporary with the film, Anguirus may have been originally conceived as a fire-breather as he is depicted several times breathing some sort of flame." And there *are* scenes in *Rodan, the Flying Monster* (1957) in which the pterosaur(s) appear to breathe a misty steam or 'fire.' Have we over time somehow conflated scenes in these movies?

NINE

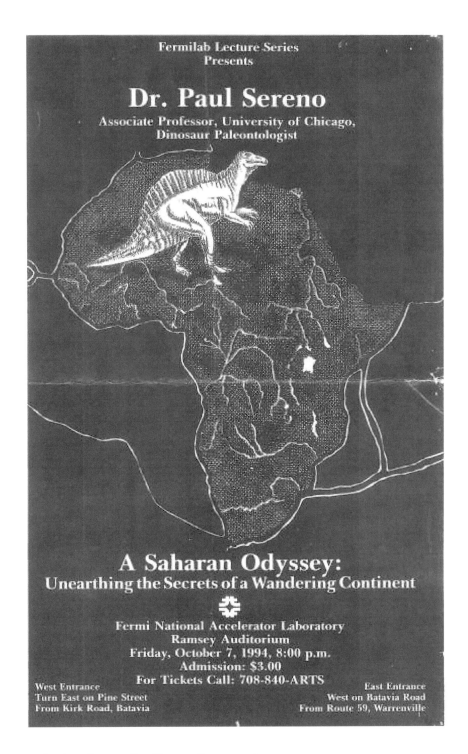

Author's collection See Chapter Forty-two for more.

Chicagoland's Paleo-World

Historically, going at least as far back as the 1880s, our Chicagoland area has been a trove of paleontological oddities, including museum displays, traveling exhibitions, paleo-conventions, and public lectures. Some displays such as those at the Field Museum became so *pivotal* to my younger, then maturing psyche! Decades later my wife and I reveled in such events during the mid-1980s through the early 2000s, venturing eventually outward from the Chicago metropolitan region to witness (and write about) these seemingly omni-present paleo-phillic lures ranging throughout the Midwest & beyond. No—I've never received formal journalistic training (as you may have noticed), but for a number of welcoming fanzines of the time, I accepted responsibility (sometimes with wife Diane's assistance) for documenting several of the more significant paleo-events transpiring in and around the Chicagoland area. Chapters in this section may someday be of historical interest, preserving a sense of Chicagoland's dinosaur fervor during this period, for posterity.

Incidentally, although not discussed here at further length, one of the most fantastic dinosaur-related shows appearing in Chicago during the past quarter century was 2000's amazing *DinoFest*, which was displayed expansively at Navy Pier. It was a jaw-dropping program that Diane and I (with Shannon Yeager) visited with glee!

A rare treat for Chicago north-siders is Dave's Down to Earth Rock Shop in Evanston, Illinois, a place I first visited as a college freshman during new student week in 1972. Over a quarter century later, Diane and I interviewed the owner, Dave Douglass, for the Fall/Winter 2000/2001 of *Dinosaur World* (issue no.9) (although not included in this selection). I also made a prehistoric animal sculpture that was exhibited in his downstairs museum fossil collection.

Museum exhibits come and go and then become generally forgotten. Yet here are just a few more examples that might retain historical interest.

These chapters originally appeared in the following publications:
Chapter Forty-two—Composited from articles appearing in *Fossil News: Journal of Avocational Paleontology*, vol.6, no.6, June 2000, pp.5-8; *Dinonews* (Museum of Western Australia, Perth),

no.9, March 1996, pp.17-18; *Prehistoric Times*, no.19, July-Aug. 1996, p.32.

Chapter Forty-three—Composited from articles appearing in *Fossil News: Journal of Avocational Paleontology*, vol.5, no.3, March 1999, pp.16-19; *Fossil News: Journal of Avocational Paleontology*, vol. 5, no.2, Feb. 1999, pp.7-8; *Prehistoric Times* no.79, Fall 2006, p.52.

Chapter Forty-four—*Dinosaur World* no.6, Spring/Summer 1999, pp.24-28 ... reprinted in *Fossil News: Journal of Avocational Paleontology*, vol. 7, no.12, Dec. 2001, pp.15-19.

Chapter Forty-two — *Paul Sereno Strikes ... again & again*

Fossils From the Dark Continent Shine Again!

Fossil discoveries from the African continent, found in places like Olduvai Gorge, the Karroo Basin, Tendaguru, and the Lumonya Hills, have often stirred human imaginations. Some of our most scientifically intriguing, if not downright thrilling episodes in the annals of vertebrate paleontology have sprung from sediments of Africa. Ranging through three eras of time, and spanning the history of our science, an array of important discoveries have come to light—including record of the mammal-like reptiles, tremendous dinosaurs and the earliest traces of human ancestors. Researchers who have conjured these creatures and the stories of evolutionary descent which he remains testify to, include an illustrious assembly of names—Bain, Broom, Jannensch, Fraas, Dart, the Leakeys, Johanson, Jacobs ... Now University of Chicago vertebrate paleontologist, Paul Sereno, has emerged from Africa's sands of time with a new trophy, further revealing the nature of Africa's Mesozoic prehistory.

Literally, on a whim, my wife and I attended a lecture presented by Dr. Paul Sereno at the Fermi National Accelerator Laboratory on October 7, 1994, concerning "A Saharan Odyssey: Unearthing the Secrets of a Wandering Continent." It was there that Sereno let the 'cat out of the bag' by announcing his description of ("relatively complete") *Afrovenator abakensis*. Sereno's address made national headlines and, in quick succession, followed televised news broadcasts, a television documentary—"Skeletons in the Sand"—and publication of his paper, "Early Cretaceous Dinosaurs from the Sahara," (*Science*, vol. 266, 10/14/94, pp.267-271).

Fortuitously it might seem, a skeletal cast of the *Afrovenator* skeleton (plus a thigh bone belonging to an as yet unnamed sauropod dinosaur) were exhibited in the Chicago Public Library, Harold Washington Library Center, where I often venture(d) during lunch breaks! This, as shall be discussed later, afforded exceptional opportunity to restore the *Afrovenator* in sculpture.

Dr. Sereno is no stranger to sensationalism. Still basking in the glow of recent successes, culminating in his description of the South American dinosaurs *Eoraptor* and new specimens of *Herrarasaurus*, Sereno literally willed yet another magnificent discovery to happen! This time, however, the fossils beckoned in another land south of the equator, but thousands of miles across the Atlantic, the country of Niger. Through shrewd foresight on the part of his field crew, perhaps reflecting the anticipation of no less than a lunar landing type event, and pitting dogged persistence against assorted adversities, Sereno and team prevailed to delight of many paleontology buffs scattered around the world.

Afrovenator joins a contingent of spectacular genera known from the dark continent. Some of the earliest known, Upper Triassic dinosaurs have been discovered in the country of Lesotho in southern Africa, including little *Lesothosaurus* and *Heterodontosaurus*. Such dinosaurs may represent the base of the ornithischian family tree. In the early years of this century, the Upper Jurassic dinosaurs of Tendaguru in Tanzania were excavated by German paleontologists, resulting in nearly complete skeletons of genera, such as the plated and spiny *Kentrosaurus*, and gigantic sauropods—*Dicraeosaurus* and *Brachiosaurus*.

The Tendaguru dinosaur fauna highly resembles that commonly encountered in North America's Morrison formation of similar age. A possible reason for this similarity is that by the middle Jurassic Period, juxtaposition of continental masses still allowed migratory routes of dispersal. Possibly by the Early Cretaceous Period, however, as the supercontinent Pangaea separated along rift zones, intercontinental access declined as oceanic barriers divided continental margins.

In the Early Cretaceous Period, approximately 100- to 135-million years ago, new fauna emerged and diversified, including peculiar fin-backed dinosaurs of northern Africa, the 40-foot long theropod, *Spinosaurus*, and the iguanodont *Ouranosaurus*. Although, until recently, restored with a tyrannosaur-ish sort of head, *Spinosaurus'* skull may have more closely resembled that of the British *Baryonyx*, which had a head described (superficially in appearance) as "crocodilian." Unfortunately, remains of *Spinosaurus* were destroyed during World War II, so *Afrovenator* is Africa's only 'surviving' theropod. Although it is not known exactly what the 'sails' of the two African genera were utilized for—what sort of adaptation—it is of interest that both happened to evolve in the same region and at about the same time. Do the sails represent a means of enhancing control of body temperatures in a hot dry climate, as has been suggested in the case of Permian 'sail-backed' reptiles *Dimetrodon* and *Edaphosaurus*?

A recently described fauna from the country of Malawi, near Tendaguru in southeastern Africa, has unveiled a new Cretaceous sauropod genus known as *Malawisaurus*, which grew to 30 feet in length. (A superbly written account of the associated fauna has been published in Dr. Louis Jacobs' *Quest For the African Dinosaurs*, 1993.) Jacobs' sauropod is an (armored) titanosaur, known on the basis of distinctive tail vertebrae. During the Cretaceous, titanosaurs flourished in the southern continent Gondwana. However, long after the camarasaurs, apatosaurs, and diplodocids suffered extinction on the north continent, titanosaurs such as the *Alamosaurus* and the *Hypselosaurus* migrated back into Laurasia. *Malawisaurus* enabled paleontologists to trace the migration of titanosaurs through geological time from Tendaguru through ancient Africa, across South America and then over a land bridge linking North and South America.

Skeletal remains belonging to *Afrovenator* were cleaned and cast at the Royal Ontario Museum in Toronto. An entire skeletal reconstruction resulted as technicians and students speculated how missing pieces of bone may have appeared. Surprisingly, skeletal anatomy evident in the new African dinosaur fauna reflects characterization of northern Laurasian dinosaurs, rather than other dinosaurs thought to be distinctive of Gondwana. It is possible that during the Late Jurassic, as Pangaea fragmented and separated, land connections extended between Gondwana and Laurasia, allowing migrations of dinosaurs through Europe. By the Early Cretaceous, with the land bridge severed, distinctive isolated populations emerged. Sereno concluded, "As a consequence, there is no recognizable … unity among dinosaur groups that persisted on Gondwanan continents during the Cretaceous. Rather these … dinosaur groups experienced different fates on each southern continent as isolation increased during the Cretaceous."

So on the evening of October 7th, I resolved to commemorate the successes of Sereno's expedition by making an *Afrovenator* sculpture. Relying on Sereno's published skeletal reconstruction and photographs of the skeleton itself, the work began. Although the published reconstruction expertly represented the actively posed predator two-dimensionally, it was necessary to observe (and photograph) the mounted skeletal cast to gain a proper three-dimensional perspective. A sense of how this powerfully built and dynamically balanced creature appeared in life steadily formed in the weeks ahead.

The remarkably short (skeletal) torso and long curved tail were aesthetically mounted. Most impressive, however, were the hips which were quite narrow when compared to the girth of the ribcage. *Afrovenator* was quite barrel-chested, perhaps especially after engorging itself with a good meal. The slender nature of the skeleton, the massive feet, and sharp talons connected to forearms no longer than a human's came alive in my imagination.

The (hopefully!) striking result is evident in the accompanying photograph (not here but printed in *Prehistoric Times* no.11). The 12-inch long sculpture (1/30 actual size) is poised in the position suggested by the mounted skeleton and Sereno's reconstruction. A pebbly skin texture was added in finishing touches, as has been found in association with a South American theropod dinosaur—*Carnotaurus*. The skin color is entirely conjectural. An urge to show *Afrovenator* stalking after a wounded (African) camarasaur, or even more stirring, in mortal combat with a spinosaur, was resisted because of added uncertainties associated with these lesser known dinosaurs. Also, although Sereno's new sauropod may have been contemporaneous with *Afrovenator, Spinosaurus* may have lived several million years later.

After its three month temporary display in Chicago, the skeleton will be shipped to Niger. Sereno has stated that he would like to design a more startling exhibit—mounted skeletons representing a herd of roving African dinosaurs! Funding for such a venture may be difficult to procure, particularly after furor over the discovery wanes. And yet, in coming months, even after its skeleton is delivered to officials in Niger, the excitement and inspiration *Afrovenator* instilled during its Chicago visit in the fall of 1994 won't be easily forgotten. For a relatively harmless, one-foot long restoration of the "African hunter from In Abaka" greets visitors daily in our living room. (12/26/94)

Sereno Strikes Dino-Treasure—Again!

With his usual dramatic flair, University of Chicago vertebrate paleontologist Paul Sereno concluded a lively presentation, "Tracking Dinosaurs on Wandering Continents," on the evening of April 30, 1996, at the University of Chicago campus, with the unveiling of a new theropod dinosaur skull, quite possibly the largest ever discovered.

Known as the "sharp-toothed reptile," the genus has a brain case smaller than its distantly related evolutionary cousin, *Tyrannosaurus*, and enormous size approaching 50-feet in length (up to ten feet longer than an adult tyrannosaur). The new skull's profile appears more highly arched toward the middle than in the case of a *T. rex* skull. An expansive antorbital fenestra, the opening in the skull in front of the eye socket, comprises nearly one third of the skull's overall head length.

Sereno captivated his audience with tales of adventure and hardship endured in the passionate quest for dinosaur fossils in forbidding and exotic places on the Asian, African and South American continents. He also divulged interesting details of his youth, including several unruly episodes prior to his fascination with a book (stolen from a library), entitled *The Fossil Book*, and his introduction to the world of art, instances which settled him down. Following an amusing moment while brandishing the pilfered volume, Sereno then tackled the mysteries of dinosaur evolution through a slide presentation. Biogeographical as well as evolutionary relationships between theropods and the splitting of continental masses during the Mesozoic Era were highlighted. Sereno hinted at the possibility that (undefined) catastrophes may have played prominent roles in the evolution and extinction of dinosaurs and mammals.

In addition to the gigantic skull, Sereno also spoke of another startling discovery, referred to as the "delta runner." This new creature known from a partial skeleton including limb, hip and shoulder bones had unusually long forelimbs and a tibia/femur ration suggestive of fast-paced movement. Sereno suspects both African dinosaurs were active predators. In life, the slender and cursorial dinosaur grew to about eight meters in length. Overall, the skeleton closely resembles that of a possible evolutionary coelurosaurian ancestor, the diminutive Upper Jurassic *Ornitholestes*. But hold on—Sereno also presented another predatory dinosaur too!

Their remains were found last year in Late Cretaceous deposits formed about 97-million years ago in a river delta (now situated in southeastern Morocco). Although described from bone fragments collected from an agglomeration of skeletons representing thousands of additional species, including a well preserved right maxilla belonging to, yes, another dinosaur, the "sharp toothed reptile." A team of artists and scientists already have fully reconstructed the skull. A resin cast of their impressive 'trophy' was shown to our astonishment at the lecture's conclusion. Sereno's graduate students assisted in the unveiling ceremony, displaying the skull cast, and some of the new fossils, a truly breathtaking culmination to an evening filled with lively anticipation. The specimen is thought to have weighed about 10 tons in life, perhaps outstripping South America's recently described and closely related South American *Giganotosaurus*. During the lecture, Sereno mentioned that bones belonging to the "sharp-toothed reptile" may have been described previously.

Artists are busily restoring the monster, framing flesh around cast skeletal pieces, and a full-sized skeletal reconstruction may eventually be prepared. Sereno also announced that his findings would be published in a forthcoming *Science* article, and that more details concerning description and phylogeny of these dinosaurs will be available shortly.

Two weeks later, we learned more of these discoveries and how the fossils fit into the distinctive African Cretaceous pantheon of super-predators. At his lecture, Sereno outmaneuvered a young male attendee who had asked what the scientific names of the new dinosaurs were. Politely refusing to yield on the matter, Sereno dodged in referring only to their monikers, "delta runner" and "sharp-toothed reptile." Mysteriously, Sereno mentioned that remains of the larger beast may have been discovered during the first half of the 20th century, although that collection had been destroyed in Germany during World War II.

Then, in his *Science* article (vol. 272, May 17, 1996, pp.986-991)—with a striking cover image comparing human and ddinosaur skulls—Sereno relieved the name mystery, assigning the "delta runner" fossils to a new genus and species, *Deltadromeus agilis,* and identifying the allosaurid "sharp-toothed reptile" as *Carcharodontosaurus saharicus*. It was the latter which had been described in the early 1930s. Perhaps the first flesh restorations of this dinosaur appeared in Don Glut's *The New Dinosaur Dictionary* (1982), in which the estimated length of the animal was prophetically reported as 50 feet!

If you shudder at the thought of lions, cheetahs, and hyaena prowling about the African plains, competing desperately for the flesh of hooved mammals, imagine the dread of a Cretaceous African herbivore with the likes of *Spinosaurus, Deltadromeus,* and *Carcharodontosaurus* roving about. No raptors apparently contributed to the crazed carnovorial cruelty, however. It is rather amazing to note that no fossils of armored or horned dinosaurs have been identified as yet on African soil.

What does it all mean? Quite simply, or so it would seem, in the Upper Cretaceous of Africa, faunal exchange between northern and southern continents was possible. Descendants of theropod fauna thought to be indigenous to North America from distantly related lineages, respectively, such as the closely related *Ornitholestes* and *Allosaurus* evolved into even more remarkable predators on the African continent, such as *Deltadromeus* and *Carcharodontosaurus*. Prior to geographic isolation and evolutionary radiation, ancestors of the African *Carcharodontosaurus*, the South American *Giganotosaurus*, and the North American *Acrocanthosaurus* may have ranged broadly across the globe during the Late Jurassic or Early Cretaceous through as yet unidentified land corridors. In the case of *Deltadromeus*, there is indication that Upper Jurassic coelurosaurs must have existed on southern continents.

Has Sereno found the new 'stars' of Michael Crichton's next novel, or Spielberg's next movie blockbuster? Or, possibly, have Sereno's raving successes on the paleontological frontier created the basis for a new sort of 'Indiana Jones' fictional movie character? (May 1, 1996 and June 5, 1996)

Joe Who? Jobaria!

You won't find your regular kind of "Joe" here, or even a cup of it. Where we talkin' about? The Crystal Palace, oops, we mean the Crystal Gardens solarium exhibit center at Chicago's Navy Pier. On the snowiest day of the new millennium so far experienced by Chicagoland, we tromped our way through endless snow drifts each countless feet high in our relentless pursuit to see (not only the newly opened Titanic exhibit at the Museum of Science and Industry), but also another remarkable sight, titanic in its own right, "Dinosaur Giants", featuring 135-million-year-old, *Jobaria tiguidensis*. What a pleasant experience this turned out to be, in spite of the foul weather!

Many curious associations can be made between the Titanic and *Jobaria*. They are both "extinct", each in their own way. Both were huge, by any standard, respectively, for 'manmade' and 'living' categories. Items in exhibited displays first both had to be found by resolute explorers with more than just a bit of serendipity involved, and then painstakingly pried from Nature's grasp. In both cases the splendid, and quite astonishing results were hard won. Of course, icebergs (even of the cometary kind) had nothing to do with *Jobaria's* demise, although both Titanic and "Joe" came to their "final" resting places in watery graves. Okay, so who is this "Joe" character?

Welcome to the land of technical journals. Paul Sereno and a band of stalwart co-discoverers/authors, admitted *Jobaria* into the annals of science by publishing the news of said dinosaur in a journal named, (what else?), *Science*.

(Our apologies for not mentioning all the numerous coauthors here by name.) The exact reference is *Science*, November 12, 1999, vol. 286, pp. 1342-47. It's all there really, although there's a lot to decipher in this paper.

Essentially, *Jobaria* was an Early Cretaceous sauropod, "long neck" type of dinosaur which lived in what is now Niger 135 million years ago. It is known from the Necomanian age/epoch, the Tiouraren' Formation of rocks to be more exact. Although extinct, *Jobaria* represents sort of a "living fossil." This is because it had a number of distinguishing ancestral ("primitive" is such a politically incorrect term these days) characteristics, more recognizeable in sauropods which lived some 30-million years earlier during the Jurassic period than in other more contemporaneous Cretaceous sauropods from other continents, seemingly more "derived" in their skeletal structures.

Sereno and company believe *Jobaria* simply evolved more slowly than other dinosaur genera. In other words, this genus' evolutionary design must have been extremely well engineered in order for "Jobarian" traits to have persisted so long, geologically. So much for the notion of ponderous and evolutionarily backward sauropods, ultimately destined for extinction because of poor adaptational design. Hey, in fact, these guys were among the best adapted terrestrial animals of all-time. In evolution, long-term "stasis" is generally 'good,' until you suddenly need to 'adapt.'

The Crystal Gardens exhibit really brings these dinosaurs back to life too, in all their glory. In fact, one can almost imagine them *living* here!

Sereno published a flesh-outlined skeletal reconstruction of *Jobaria* in the November *Science* paper. However, you may recall seeing restorations of *Jobaria* years before its formal 1999 description appeared. For instance, in 1994, Sereno and colleagues alluded to previously undescribed sauropod remains discovered in beds where the carnivorous, *Afrovenator*, had been found. (See *Science*, vol. 266, 10/14/94, p. 268.) Then, for Sereno's January 1995, *Natural History* article, artist, Douglas Henderson, completed a painting showing the *Afrovenator* stalking a herd of sauropods. These long-necked individuals must have been the as yet undescribed, (yep, you guessed it!) *Jobaria*. Just shows you how long it can take to present the facts accurately when it comes to paleontological specimens.

Here's what we see in the luxuriant Crystal Gardens. A tableau from deep time, a single moment in prehistory has been reconstructed and restored in grand style. One doesn't even notice the dinosaurs at first. The spacious, glass lined Greenhouse-like structure is a tiny haven in a city teeming with human turmoil. We experience the sound of calling birds, and a sight for sore eyes, ivy-covered palms and ferns. Natural light filters playfully from above. Strangely programmed fountains playfully fire darts of water across our path. These pass harmlessly overhead without spraying or splashing visitors. There is recorded music, of sufficient taste to calm even the savage beast. And then emerges the Beast himself, in the form of the "African Hunter" *Afrovenator abakensis*!

Afrovenator's 27-foot long cast skeleton is a bit more familiar, having been exhibited at the Chicago Public Library in the Fall of 1994. But this mount is intended to menace the newcomers, two cast *Jobaria* skeletons, displayed a few feet over, across the tiled walkway. One of these is 60-feet long and the other, a juvenile, is a 'mere' 40-feet long. The high-vaulted ceiling is with good measure, for the 60-foot long skeleton is rearing to dizzying height, thwarting *Afrovenator's* bloody "attack." The *Jobaria* skeletons are artistically positioned in life-like stances. There is implied analogy to a herd of elephants.

Can you feel the tremor of pounding, tree-trunk thick forelimbs, hear the African Hunter's strident war cry, or the sauropod parent's defiant bellow? Can you see these massive forms struggling against death's inevitability? Well, in case you can't, Mark Hallett, has painted the dramatic scene. His restoration hangs on a nearby wall. The caption reads, "Driven by hunger, 135-million years ago, a pair of *Afrovenator* emerge from the underbrush to attack a young *Jobaria*. The scene is based in part on a graveyard site that preserved skeletons of juvenile and adult *Jobaria* as well as bones and teeth of *Afrovenator*. Jagged grooves on the bones of the ribs of the juveniles offer telling clues of a lethal encounter." The fossils reveal the gory outcome played out on the African continent long before the days of roving lions and wildebeest.

Another painting, by Michael Skrepnick, shows two *Jobaria* standing erect, dueling with the large thumb claws on their forelimbs. One has slashed its rival. Skrepnick's painting vividly shows the conifer forests which formerly were situated along the bank of the broad river where the giants lived and died, (i.e. in an ancient flash flood). The fact that juvenile and adult bones were found together suggests *Jobaria* was a social creature, existing in herds, or as one caption reads, "...like the large plant-eating mammals of today".

Then there is a large bronze 1/15 scale statue of the rearing *Jobaria*, sculpted by Michael Trcic. This 5 foot long bronze is posed bipedally erect like the 60-foot long skeleton. A caption states the, ".... rearing posture is controversial. We think *Jobaria* could rear because its heavy muscular tail shifts center of gravity (mass) closer to its hind legs. By bending slightly at the knees, it could balance its mass on its hind legs. By rearing, *Jobaria* could feed high in trees or challenge rivals with its thumb claws." Paul Sereno has illustrated this argument on signage adjacent to Trcic's bronze statue. In deciding the pose for the 60-foot long *Jobaria* skeleton, Sereno, in fact, made observations of a trained elephant rearing on its haunches.

Another artist, Gary Staab, has produced a life-size restoration of a portion of *Jobaria's* neck and head. Whereas most sauropods had more neck vertebrae, *Jobaria* only had 12 (a "primitive" or ancestral condition). Staab has sculpted the bones, air sacs and muscles on the left side, and on the right, we see the living, 'breathing' animal. Did we say "air sacs"? That's right. A nearby caption clarifies that, "Air sacs are very unusual features in a neck. Found in dinosaurs and birds, these balloon-like sacs extend from the lungs along each side of the neck all the way to the skull to reduce the weight of the neck." This revealing exhibit was suggested by John Lanzendorf.

There is a wealth of additional information about the African dinosaurs artfully blended within foliage of the Crystal Gardens. Just roam about and you'll find it. For instance, we read that *Jobaria* adults grew up to 70-feet in length and weighed up to 40,000 pounds. They were 15-feet high at the hips. *Jobaria* hatchlings weighed less than 5 pounds and grew up to 2 to 3 feet long. The exhibited 40-foot long specimen may have only been 10 years old at time of death. Yet, it weighed 12 tons (24,000 pounds). And it just seems plain common sense that *Jobaria* probably had good olfactory and hearing senses.

The *Jobaria* skeleton is 95% complete, while *Afrovenator* is 70% complete. An adult *Afrovenator* weighed 6,000 pounds. The real bones of the *Jobaria* will soon be returned to the National Museum in Niamey, capitol of Niger.

There are so many exhibits here, so skillfully tucked away, nestled amidst the greenery. There is a *Jobaria* trackway, a "Femur Photo Stand", isolated casts of the skulls of the featured dinosaurs, more expertly rendered reconstructions by Sereno and Skrepnick. There are also films showing how the bones were expertly sleuthed, collected and assembled.

A credits sign mentions that the skeletal designs are Paul Sereno's creation and that the skeletal mountings were prepared by P.A.S.T. Educational materials and outreach are a product of Gabrielle Lyon's and Dr. Sereno's efforts. The exhibit was created by Project Exploration, of which more will be stated below.

Jobaria and the Crystal Gardens are immediately adjacent to the Children's Museum where now resides a cast skeleton of another African dinosaur, the geologically younger, 120-million-year-old *Suchomimus tenerensis*.

While many smokers would still walk a mile for a Camel, Paul Sereno will run 26 miles for a dinosaur! He participated in the 1999 Chicago Marathon to help raise funding for this (free) exhibit. Running ain't easy, (as Allen, who before his knees gave out used to run "mini" 10 K marathons, can attest). So, if you would like to help out and ease the strain in Dr. Sereno's weary legs before inauguration of his next fossil exhibit, you can help the cause by contributing to Project Exploration.

One of Project Exploration's principal missions is to make new dinosaur discoveries, such as this free "Dinosaur Giants" display, accessible to the public, especially to children. For information, or to read about the mission of this dedicated organization, contact, "Info@projectexploration.org", or phone (773)-643-3014. Next year, through Project Exploration's efforts, we'll be treated to the *Nigersaurus*, a geologically younger, 110-million-year-old, African sauropod described in the aforementioned November 12, 1999 *Science* paper.

Unlike the Titanic exhibit which cost us an arm and a (sea)leg to get into, the *Jobaria* exhibition is free to visitors. "Dinosaur Giants" was on display at Navy Pier from January 14 through March 19, 2000. Afterward the African "longnecks" and company will tour the country. Try to catch it when it passes through your 'neck' of the woods.

Chapter Forty-three — *Dinosaurs Come to Life*

By today's electrified standards, 1960s dinosaur paleontology was comatose. Ask teachers and professors about the possibility of growing up to study dinosaurs as a career back then, and all you'd ever hear was whining about how one could never earn a living out of dinosaurs. If only I hadn't listened to the naysayers in those dismal times. For dinosaurs have become worthy of scientific attention. Furthermore, they now have stellar ambassadors, such as Paul Sereno, who direct interests in paleontology toward our most essential audiences, tomorrow's scientists, medical practitioners, environmentalists and educators—today's grade school children.

Talents of premier paleoartists and visionary scientists, notably Dr. Paul Sereno of the University of Chicago, have brought the vital cutting edge of dinosaur popularization and science to Chicago's Navy Pier. For youngsters can be entertained, instructed and fully absorbed by two marvelous exhibits, the "Colossal Claw" Dinosaur Discovery exhibit displayed in the Children's Museum and just across the hall, the IMAX 3-D movie feature—*T. rex: Back to the Cretaceous*.

Sereno's Class Act:

Sereno is a clever, high-spirited phenom, commanding the artistic sense of Robert Bakker, the adventurous nature of Roy Chapman Andrews, and the uncanny luck of Barnum Brown. With a flair for the dramatic, children flock to him during guided tours at the Chicago Children's Museum. His spectacular field successes have paraded through Chicago's universities, libraries and museums in recent years ... *Sinornis*, *Eoraptor*, *Afrovenator*, *Carcharadontosaurus*, *Deltadromeus*, and now the simply awesome *Suchomimus*, found in late 1997.

The Dinosaur Discovery exhibit has been the Chicago Children's Museum's most striking success to date. Situated near the entrance to Navy Pier, the Museum's mission is to stimulate children's intellectual and creative capacities through constructively guided exposure to the world around them.

What it's like to go fossil digging in the Sahara Desert is an emphasis of the new exhibit, relying on *Suchomimus* as a central attraction. The displays resulted from a partnership between the Chicago Children's Museum, Dr. Sereno and his wife and colleague, educator Gabrielle Lyon of the University of Illinois at Chicago. The exhibits are supported by Amoco.

The exhibit consists principally of a 36-foot long fiberglass skeletal cast of the sensational new dinosaur and a cast of its large thumb claw, which visitors can handle. A videotape of the National Geographic Society documentary film, "Colossal Claw" (originally broadcast on November 15, 1998, three days after the discovery was formally announced with great fanfare), a mural-sized photographic enlargement of a 3' by 2' painting by Michael Skrepnick, now in the John Lanzendorf collection, vividly restoring the *Suchomimus* in its natural setting, and a 1/7 scale sculptural restoration by Stephen Czerkas can all be seen here.

There are also instructional diagrams and figures showing the position of continental land masses during the Mesozoic Era and how fossils are collected. By next summer, another exhibit, "Dinosaur Expedition," will be added, recreating the Sereno field camp setting and dinosaur excavation site for children to romp through. As Dr. Sereno stated, "Paleontologists live for the moments of discovery like we had on this expedition, and we want to share these moments with visitors of Chicago's Children's Museum."

When the exhibit opened on December 10, sponsors didn't just drop off a pile of bones at their doorstep. Museum official Elizabeth Lach explains that Sereno's graduate students have helped not only in erecting the exhibits but in helping children to understand what they're about. Elizabeth believes the exhibit merges the interests of children and their parents. For children, "… just knowing the name of the big dinosaur gives them a handle on things." Studying dinosaurs gives them an activity for parents to work closely with their kids. Elizabeth cited some of the more charming questions asked by awe-struck grade schoolers. "Is it a boy or a girl?", "Are those the real bones?", "What did it eat?", "Is it a 'brontosaurus-raptor'?" And of course, "Where is the bathroom?"

The real bones still reside at the University of Chicago, along with those of several other as yet undescribed species, still wrapped in burlap and plaster jackets. Sereno and students sculpted missing portions of bones to complete the likeness of *Suchomimus'* skeletal structure. They even molded and cast each bone into fiberglass themselves. After making a model of how the skeleton should be assembled, they sent the miniature skeleton and fiberglass bones to a company named "PAST" in Drumheller for mounting. Eventually the original fossils will be returned to Niger soon after a natural history museum sufficient to house the great collection has been built.

Suchomimus tenerensis, described in *Science* (vol. 282, 11/13/98) is a 120-million-year-old theropod found in the *Gadoufaoua* desert region of Niger. As Jay Leno quipped during his November 16th 1998 monologue, it is little wonder that a great big dinosaur like this might go extinct searching for fish in the middle of the Sahara Desert! Hah!

Ah, but during the mid-Cretaceous, this wasn't a desert region, as indicated in the Skrepnick mural. It was a lush, tropical environment. Bipedal *Suchomimus* is thought to have dined on fish and may have even tangled with the local varieties of crocodile. At least seven other kinds of dinosaurs also inhabited the region, but authors of the *Science* paper believe *Suchomimus'* snout and jaws were especially adapted for a piscivorous diet. Once you see the enormous skeleton, though, you may find this hard to believe. Imagine the titanic struggles for prey in this incredible paleo-world!

Features of the dinosaur are impressive. The 4-foot long head is extremely elongate and narrow. *Suchomimus* carried an arsenal of about 120 teeth in its mouth, including 7 premaxillary teeth. Teeth in the anterior of the jaws are sharply recurved and have a 'hook-like' shape. On its thumbs were 12-inch long claws, sheathed sharply in life. The latter two characteristics are shared with ancestral torvosaurids and are postulated to have evolved in the middle Jurassic.

Although nearly as long and massive as the Late Cretaceous African genus, *Spinosaurus*, the authors of the paper believe *Suchomimus'* enlarged dorsal and sacral neural spines supported a sail-like crest. One difference here though is that in *Suchomimus* the largest neural spines are in the sacral region, whereas in *Spinosaurus* the largest spines are situated over the mid-dorsal vertebrae.

Sereno believes "crocodile mimic" *Suchomimus* is more closely related to the Wealden age (Early Cretaceous) *Baryonyx*—the original "big claw"—than to the Late Cretaceous genus, *Spinosaurus*. Connections across the former Tethyan Seaway promoted migration of diverging clades of spinosaurs during the Early Cretaceous on the former continents, Laurasia (northern) and Gondwanaland (southern).

IMAX Dinosaurs:

T. rex: Back to the Cretaceous showcases several spectacularly recreated dinosaurs, the handiwork of Michael Trcic, around a soap opera-ish story concerning young "Allie" who is being neglected by her sometimes stern father (fictional paleontologist "Dr. Haden"). Does she imagine her mystical trip back to the time of the dinosaurs? Or is it 'real'?

Allie's school paper hypothesis that *Tyrannosaurus* laid clutches of eggs is doubted by her father, who ironically has recently discovered a strangely shaped egg in Canada. Then Allie is transported to the past where she has conversations with Charles R. Knight, who is seen painting dinosaurs from life *in* the Mesozoic, and Barnum Brown who is dynamiting fossils out of the Red Deer formation. Both encourage Allie to not give up too easily on her hypothesis until she has given it a fair shake.

Ultimately, Allie 'proves' her hypothesis when she confronts a *Troodon* stealing eggs from what turns out to be a *T. rex* nest. Then the 6-mile diameter K-Pg asteroid collides, *T. rex* dies, and the North American continent is covered in impact ash. When she returns to the 'present,' her father mysteriously has decided that her idea isn't so outlandish after all—even for a school report. The 'live' egg Allie rescues turns out to be the unhatched egg that her father later finds as a fossil.

Overall, I give this show 3.5 stars out of 4. As far as the absolutely incredible 3-D holo-effects, it's got *Jurassic Park* beat. The virtual reality gizmo involves donning a visor which transmits the stereo-optical image to your eyes.

Have loads of primeval fun at Chicago's Navy Pier.

(Coauthored with Diane E. Debus)

A "Gilded Cage" of Significant Slabs:

Near the hollow gaze of a brachiosaur "peering" over the second floor railing, like the animated 'brachs' in *Jurassic Park*, I glimpse several of the 20th Century's most famous fossil-bearing slabs. "Hey those are dinosaurs!" a boy yells gleefully to his parents as he sprints down the museum corridor. "Lucky guess, Kid", I muse. How could anyone, especially a child, identify these specimens as dinosaurs?

On October 1st, a small but highly significant exhibition of fossils from the lower Cretaceous, Yixian formation from Sihetun in China's Liaoning Province, opened at Chicago's Field Museum of Natural History. The National Geographic display features six stone slabs bearing exquisitely preserved specimens of extinct birds, bird-like dinosaurs, and two life sized sculptural interpretations. They will be exhibited until January 3, 1999, near the entrance to "Life Through Time" and adjacent to where the bones of specimen "PR2081", (otherwise known as "Sue"—the world's most famous *T. rex*) are undergoing preparation.

There, visitors will spy two fossils of *Sinosauropteryx prima*, two *Caudipteryx zoui* specimens, a fossil of *Protarchaeopteryx robusta*, and a slab bearing two specimens of *Confuciusornis*. Two remarkable sculptures of feathered dinosaurs, *Sinosauropteryx* and *Caudipteryx*, made by Canadian sculptor, Brian Cooley, greet approaching visitors along the aisle, where that precocious boy made his startling proclamation. (Yes—at first, Brian's sculptures looked for all the world like birds to me, but they are indeed dinosaurs!)

Discoveries such as these are threatening to render books on dinosaur and avian evolution, 'dead as a dodo' because the book publishers can't keep pace with the new discoveries. Formal descriptions of *Protarchaeopteryx*, *Caudipteryx* and (separately), *Sinosauropteryx*, all were published in English just this year. Not since the announcement of *Archaeopteryx*, first extracted from Bavaria's Lithographic limestone deposits in the previous century, have there been fossil discoveries bearing such profound implications for our understanding of the evolution of dinosaurs and birds. Thomas Henry Huxley, Charles Darwin and Edward Cope must all be enjoying victory cigars up there in paleo-heaven!

Perhaps *the* specimen on hand is that of the first specimen announced in late 1996 by paleontologist, Phil Currie, as the first "feathered dinosaur" specimen. It scored sensationally at the 1996 Society of Vertebrate Paleontology October meeting. A quickly prepared restoration of the dinosaur by artist, Michael Skrepnick, was rushed to print in the New York Times, and things rocketed from there! This bird, excuse me, DINOSAUR, is closely related to a more familiar dinosaur, *Compsognathus*. Now here before my eyes, was that very *Sinosauropteryx* specimen (or, really, half of the specimen, because two slabs resulted from the cleaved rock), responsible for enlivening debate over the evolution of feathers and flight!

Another exhibited *Sinosauropteryx* specimen is identified as the only dinosaur ever discovered along with the remains of a small mammal "....inside - apparently one of its last meals." Here we have Gouldian contingency at its finest! For in an alternate universe that mammal *could* have been MY ancestor! Thankfully these crow-sized, hungry dinosaurs, living in the real live "Jurassic Park" age of "nature red in tooth & claw," left some mammals alive to survive the K-Pg event 66-million years ago.

Caudipteryx steals the show, however, with Brian Cooley's masterful restoration continually attracting children and weary parents. But the *Caudipteryx* slabs are equally enticing. What would Sir Richard Owen, who "invented" the dinosaurs in 1842, and who throughout his brilliant career opposed evolutionary principles, murmured to himself if he were standing beside me gazing at the *Caudipteryx* specimens?

Well, Owen is dead and couldn't attend the show, so instead someone else murmured to me. "How on earth can they tell these are dinosaurs?" a Chinese gentleman asked. "They all look like big turkeys!" They did, and I had to reflect on how much study went into those scientific papers. Amazing that so much can be revealed from a jumble of bones. The feathers are there to behold of course, but it would be difficult for the ordinary "Joe" to regard them as feathers, unless, perhaps, you spied them from the corner of your eye! Thanksgiving is just around the corner, and the *birds*, Damn!—confusion reigns. I mean those DINOSAURS, were making me hungry.

Actually there *are* some birds too. There was a slab containing two *Confuciusornis*. What a beautiful fossil! A caption informs viewers that *Confuciusornis* had no teeth in its beak and that it apparently lived in colonies. Would it have alighted on my hand if I offered it some stale bread? At least it didn't have teeth. I wouldn't have the courage to feed birds with teeth! Birds armed with teeth wouldn't be satisfied with bread.

All these specimens are thought to have perished during eruptive volcanic episodes. The animals died and were quickly covered in tissue-preserving ash in a lake bed. This is why so many body details, such as ingested mammal bones, developing eggs and feathers can now be discerned by paleontologists.

With exception of *Confuciusornis*, none of the feathered creatures could fly. The reasons are elaborated in the scientific papers listed below and I won't digress. Authors of the scientific paper on *Caudipteryx* and *Protarchaeopteryx* believe these dinosaurs represent a lineage ancestral to *Archaeopteryx*, even though they lived at a later time.

The specimens are generously on loan from the National Geological Museum of China. Many other Liaoning fossils have "flown the coop" however, illegally falling into private hands. Unauthorized sales to museums have also been made. Authorities are cracking down on such exchanges.

Unless you are naturally generous with your hard earned dollars, plan your visit on the Field Museum's free admission day, which is Wednesday, because the $7.00 normal price of admission for adults isn't exactly "chicken feed." Catch the exhibition before this flock of rare birds, or, excuse me, DINOSAURS and birds, migrate on to their next destination!

Patagonian Dinosaurs Invade Chicagoland:

Well not quite to Chicago but … they made it to Aurora, IL anyway, which is about 25 miles as the proverbial bird flies west from Chicago's Field Museum's revamped dinosaur halls. This display of Patagonian fossils has proven so successful since its late March 2006 debut that it's even been extended through January 2007! What's all the fuss about? A bevy of dinosaur skeletal casts with a few restorations detailing former life of Mesozoic South America, that's what! This exhibit is downright sensational, yet compact as it all somehow fits inside the confines of a former U.S. Post Office building converted into a Children's Museum known as the Sci-Tech Hands On Museum located at 18 West Benton Street. They have a website too which you can check for the necessary logistical details in case you're planning to visit.

Upon entry, after you've paid your $12 fee, most imposing are the mounted 45-foot long skeletons of a pair of *Giganotosaurus* stalking a 56-foot long *Rebbachisaurus* sauropod. Those Giganoto-skulls are 6.5 feet long; the carnivorous genus was gigantic indeed. These animals lived between 100 to 108-million years ago, becoming extinct some 30-million years before *Tyrannosaurus rex* evolved in North America. From signage we learn that "Rebbachi-" happens to be the only sauropod to have been found in association with gastroliths.

You'll also soon be amazed by another dramatic 'bare bones' scene featuring 25-foot long, puny-armed *Carnotaurus* pursuing 30-foot long spiny-necked *Amargasaurus*, whose remains date from 116 to 121-million years ago. Off center stage you'll spy skeletons of the 225-million-year-old Triassic dinosaur, *Herrerasaurus* (15-feet long), tiny *Gasparinisaura* (3-feet long) and recreated dinosaur hatchling scenes (*Mussaurus* and a bowling ball-sized sauropod variety known from Nequen). There are also two marine reptiles represented, a 10-foot long elasmosaur, *Tuarangisaurus* and a 17-foot long tylosaur, *Lakusaurus*, both from the Upper Cretaceous seas.

There's so much more to see inside the museum like the 90-million-year-old *Megaraptor* sickle-shaped claw—now known to be a hand claw rather than a foot claw, as well as outside the museum premises. And for the collectors, there's a well written $15 souvenir booklet which you simply must have to preserve those important memories. Well worth the time.

References for Chapter Forty-three: (1)"An exceptionally well-preserved theropod dinosaur from the Yixian Formation of China", Pei-je Chen, Zhi-ming Dong, & Shuo-nan Zhen, in *Nature*, vol. 391, Jan. 8, 1998, pp. 147-152; (2) "Dinosaurs of a Feather", Michael Lemonick, in *Time*, July 6, 1998, pp. 82-3; (3) "Two feathered dinosaurs from northeastern China", Ji Quiang, Phillip Currie, Mark Norell, & Ji Shu-An, in *Nature*, vol. 393, June 25, 1998, pp. 753-761; (4)"Dinosaurs Taking Wing", Jennifer Ackerman, in *National Geographic*, July 1998, pp. 74-99.

Chapter Forty-four — *Say "YES" to Michigan's Paleo-Treasures!*

Would you be surprised to learn that not one but several of the country's finest exhibited paleontology collections reside in the Midwest? Among others, we midwesterners claim the Field Museum of Natural History, the Burpee Museum in Rockford, the Milwaukee County Museum, the University of Wisconsin's Geology Museum, and another venerable institution often overshadowed in standard listings of dinosaur places, the Exhibit Museum of Natural History of the University of Michigan in Ann Arbor.

The Exhibits Museum in Ann Arbor offers a stately collection of fossils, prehistoric dioramas, sculptural restorations, and reconstructed dinosaur skeletons sufficient to satisfy all enthusiasts of "prehistoria."

(Allen) made a 'quickie' visit to Ann Arbor's Exhibits Museum during the fall of 1984, at the time realizing this first-rate collection was well worth revisiting. Procrastination is a terrible thing, for 14 years sailed by before we could return (i.e. in August 1998). As have the continents of our globe, exhibit cases had shifted around in the intervening period, and there were some spectacular new acquisitions. This is well within the spirit of the place however, for the words a former Curator, Charles C. Adams, "A finished Museum is a dead museum," appear in an exhibit found in the lobby.

This is an institution with near ancient history, considering its natural history collections date to 1837. Many excellent specimens were procured from the Michigan Geological Survey and a generous donation of over 60,000 items made by a former museum director and naturalist, Joseph Beal Steere in the 1860s. There have been public admissions since 1882, although the present Exhibits Museum wasn't created as a distinct component of the University until July 1, 1956. Directors of the Exhibits Museum have included Irving G. Reimann, Dr. Robert S. Butsch (a former Exhibits Curator), Dr. William A. Lunk and Dr. Thomas E. Moore.

Representing merely two of the Museum's 80,000 annual visitors, we met the present Exhibits Museum Director, Bill Farrand, in mid-August for our tour of Michigan's greatest trove of fossil treasures.

Bill Farrand did graduate work at the University in the mid-1950s. He specializes in sedimentology of archaeological sites of the Paleolithic, Quaternary and classical Old World civilizations. He has done excavations on Crete Bronze Age (Minoan) sites, as well as sites in ancient Egypt, Jordan and Greece. Farrand has yet to find a Neanderthal fossil, but he has had many other successes. His enthusiasm for the institution and its continually expanding exhibit array is contagious.

Farrand explains that the Hall of Evolution we are about to enter is organized on the basis of evolutionary groups, and is not strictly a "life through time" representation of Life's history. But we were first drawn to two of the more recently created displays.

One would never dream of hunting Moby Dick in present day Michigan. But, avast Mateys! Prehistoric whales be here (har, har).... sperm whales, bowhead whales, and fin-backed whales, in fact! Paleontologists do not believe that bones of these creatures were carried as decorative objects by Native Americans from coastal regions, only to be re-deposited in Michigan's postglacial sediments. Instead, it is suspected that the seafaring mammals swam upriver from either the Atlantic or Gulf Coast regions. Fluvial waterways, (such as the Illinois River) swollen with glacial meltwater after the last glacial period ended, would have allowed whales to migrate into the prehistoric midwest, on toward the Great Lakes.

So it is fitting that perhaps the finest exhibition of fossil whales anywhere in the world now resides at the Exhibit Museum in Ann Arbor. This exhibition is of a cosmopolitan nature (not being restricted to fossil whales of Michigan) and probes early mysteries of whale evolution.

Events leading to the mounting of a 19 foot long skeletal cast of the prehistoric whale, *Dorudon atrox* went "swimmingly", one might say. *Dorudon* is suspended from the ceiling in the Hall of Evolution over an array of remains representing 5 other genera in the whale lineage. Only one of these had previously been placed on exhibit elsewhere. By the spring of 1996, artist Darryl Leja had painted a restoration of the animal swimming after fishy prey.

Dorudon lived in the middle to late Eocene epoch, 40 to 38-million years ago. Dozens of specimens have been found during expeditions to western Egypt, led by Philip D. Gingerich, director of the University of Michigan's Museum of Paleontology.

In order to financially support part of the effort, a museum-spirited "Buy A Bone" campaign successfully raised $40,000. This allowed Jennifer Moerman to have a 'whale' of a time preparing and casting a total of over 200 fossilized *Dorudon* bones. The most difficult task involved reconstruction of *Dorudon's* enormous, toothy skull because the original fossil had been compressed by overlying sediment. So Jennifer had to cast the original squashed specimen, accurately reconstruct the compressed resin "bones" into a skull shape, then remold and recast the lightweight resin reconstructed skull. Moerman's labors allowed exhibition without damage to the original fossils. It took great skill to build a great skull!

Completed *Dorudon* and fellow cetaceans were unveiled the weekend of October 18, 1997. A dinner was given to celebrate the opening of "Back to the Sea." On the menu were items such as "Primordial Soup," "Silurian Stuffing" and "Devonian Dessert." Lectures and other activities throughout the weekend made this a gala affair.

Dorudon's contemporary, 60-foot long *Basilosaurus*, also greets visitors here. *Basilosaurus* is known from many regions of the globe, including Egypt and the southeastern United States. The magnificent skull and the first fossils of its hind feet ever found (in Egypt) are there to behold, but the rest of the skeleton is still in the Egyptian desert. These vestigial hind limbs and feet were barely noticeable on the animal's exterior. A few years ago Gingerich thought the tiny limbs might have aided reproduction, allowing the males to clasp onto females more effectively during copulation. But he has abandoned that theory. "Just So" stories are human concoctions offering little indication of Evolution's hidden mysteries.

Did *Basilosaurus* prey on the smaller *Dorudon*, like killer whales of today attack other whales and porpoises? Possibly. Although from different ancestry, these early whales occupied the niche of the long extinct mosasaurs and ichthyosaurs.

Another fossil animal which is thought to be in the lineage of modern whales is, the dog-sized mesonychid, *Sinonyx*, (a 56-million-year-old, 4 legged cetacean ancestor, discovered in the Paleocene of China and described in 1995). "What!" you say? Whales evolved from dogs? Well, not quite, but wouldn't Melville have been amazed! As astonished as we would have been to contemplate Moby Dick cruising up the Illinois River! You can spy land-lubbing *Sinonyx* in Ann Arbor today. (Not all paleontologists are happy with the postulated mesonychid link. Some believe whales are more closely related to even-toed ungulates – (e.g. hippos, cows, and pigs, known as artiodactyls, instead.)

You will also see an assortment of Eocene genera at the Exhibit Museum, including *Pakicetus*, *Ambulocetus* and *Rodhocetus*, each far more recognizable as possible whale ancestors (or, in the latter case, perhaps full-fledged whales). Even with these additions to our evolutionary arsenal, we're still a long ways from "Flipper"! Readers are encouraged to seek out a wonderful book by Carl Zimmer, *At the Water's Edge: Macroevolution and the Transformation of Life* (1998) to learn more about the story of whale evolution.

Glancing up and down the Hall of Evolution, one can see it's a veritable Prehistoric Zoo in here. Best to climb the stairs to a higher vantage point. But when we do, we readily see something else apparently trod here first! Elephants!

Tracks of prehistoric mammals are not very common, and rarely publicized. Perhaps the most famous Cenozoic trackway is the 3.6-million-year-old Pliocene trackway at Laetoli in East Africa in which tracks of hominids were preserved in volcanic ash. Besides those of human ancestors, those of other species tracks were preserved, including those of prehistoric elephants.

In 1992, word spread that a Mastodon trackway bearing 20 footprints had been discovered by Harry Brennan on his farm in the town of Saline, Michigan, only 10 miles from Ann Arbor. The 10,000 year old trackway, now 2 to 3 feet below the present surface, had originally formed in the bottom of a pond. A sandy horizon was covered by a thin layer of muck known as "marl." A Mastodon (or two) walking in the muck, passed its feet into the soft sediment, leaving behind the longest and best preserved Mastodon trackway known to science!

Brennan had contacted museum scientists, and University of Michigan paleontologist, Dr. Daniel Fisher, took charge of the investigation. Yes, Mastodon tracks, not mammoth tracks. How can they possibly know this?

Prior to discovery of the footprints, in the course of excavating a lake on his farm, Brennan had discovered one clue, several Mastodon bones. It is unlikely that the bones Brennan discovered are those of the animal which made the trackway. However, according to the exhibit text, "based on the spacing between consecutive footprints relative to their size, in mammoth trackways, the prints are closer together." Therefore, odds are this is a Mastodon trackway.

The original trackway was too fragile to excavate. So instead the prints were stabilized in the field and marked. A mold of the section was made, and from this a fiberglass cast of the trackway was made. This section, bearing 10 of the 20 footprints, was the centerpiece of a gala opening on November 22, 1996.

Based on the Mastodon's stride length, the walker stood about 10 feet at the shoulder, weighed up to 6 tons, and was 12 to 15 feet long. We spy a mounted skeleton of another Mastodon below us in the exhibit hallway, belonging to a 10,400-year old female known as the "Owosso Mastodon," found in Owosso, Michigan, 21 years ago. A painting by Robert Thom (modified by John Klausmeyer) provides a dramatic glimpse into the not so ancient history of Michigan, when these great animals actually lived their furry lives.

What a strange place Michigan was not so long ago. The Mastodon roamed with Musk oxen, Giant Beaver, Caribou, Bison, Wolves, and Moose. (Don't forget those whales swimming upriver!) Did prehistoric humans contemplate their disappearance? Can we anticipate extinction of so many members of our favorite zoo animals in the natural wild?

Seeking a tonic for these sobering thoughts, we turn to other exhibit cases. This museum seems to have one of the highest proportions of displayed miniature sculptural restorations to fossil material we have ever seen. Numerous exhibit cases have sculptures situated alongside the fossil bones. As just one example an (uncredited) artist sculpted the Miocene pig, *Archaeotherium*, which stands by its fossilized skull. This two foot long (unpainted) model is a splendid rendition. Whoever thought a pig could look beautiful in sculpture? Just visit the Exhibit Museum and you'll see why.

An alcove lures us in to see the work of a master artist and sculptor, George Marchand. There are seven dioramas in this alcove, representing geologic time from the Cambrian through the Cretaceous periods, each one offering a highly detailed panorama of an ancient landscape. Marchand, who died in October, 1996, made these between 1961 and 1969. (Allen whimsically asked, "Which diorama is 'My Favorite Marchand'?" (i.e. martian-ed, get it?) Alas, as we shall explain, we haven't seen enough of them yet to make a decision.) It was left to Robert Butsch to subsequently complete the span of more recent geologic history during the 1970s.

Butsch completed five marvelous dioramas of North American Cenozoic scenes inside glass partitions featuring the prehistoric life of North America. He has restored the paleoecology of Southern Michigan, basically where we are standing, after retreat of the Wisconsinan ice sheet where American Mastodons capture the eye. He has also done dioramas of Wyoming 45-million years ago, South Dakota 30-million years ago, Nebraska 10-million years ago and a La Brea Tar Pits scene, 15,000-years ago. All are in living color and a wonder to behold! The recreated scenery in each diorama is exceptional, and the sculpted animals, few of which you will find in other museum exhibits, are expertly done.

As we headed home, I realized we had seen Marchand work elsewhere, namely the Field Museum in Chicago. Checking further with Museum representatives, (Nina Cummings of the Field Museum and Linda VanAller Hernick of the New York State Museum), we came to realize there were 3 Marchands at large.

Focusing solely on their paleo-dioramas, (i.e. because they made many other kinds of dioramas too) there are (or have been) Marchand paleoscenes also exhibited at the New York State Museum, the Buffalo Museum of Science, the Smithsonian, the Cincinnati Museum of Natural History, the American Museum of Natural History, and quite likely the Royal Ontario Museum. We suspect there were also Marchand dioramas formerly exhibited at the Science Museum of the St. Paul Institute in the 1930s. There are possibly many other places we are presently unaware of where Marchand dioramas are (or were) exhibited!

The Field Museum dioramas made during the early 1950s by George Marchand were removed from service when the paleontology halls were renovated. These are now in storage at the Chase Studios in Cedar Creek, Missouri. One wonders why they were banished from the 'kingdom'? Sadly, our knowledge of how the prehistoric realm has been perceived historically is fading as extraordinary exhibits such as these are taken from service.

If you don't think you're familiar with these dioramas, just find almost any circa 1960s- 1970s book on paleontology (e.g. *Modern Earth Science* (1965); *Evolution of the Earth* 2nd ed. (1976); *Handbook of Paleontology for Beginners and Amateurs* (1965)). Figures of these scenes were a welcome sight in such books, perhaps too often taken for granted, but you'll recognize them instantly. Most were underwater, aquarium view scenes, in which the former life of the Paleozoic periods was accurately and vividly recreated in sculpture. Let us briefly consider the Marchand family history, although only a carefully researched book could do this in sufficient detail.

First of all, there were three individuals at large creating these wondrous scenes, a father and two sons. The father, Jules Henri Marchand (1877-1951), was a pioneer of diorama building. Henri mastered the art of sculpting the living form of Paleozoic invertebrate animals and fishes. His work, often performed in collaboration with his two gifted sons, Paul and George (1903-1996), was principally done for the New York State Museum, the Buffalo Museum of Science, and the American Museum.

Eventually, they opened a studio in East Schodack, New York, offering their talents to many natural history museums through the decades. As noted by Linda VanAller Hernick, a New York State Museum paleontologist, the "crowning achievement" of their work at the New York State Museum was a reconstruction of the Gilboa Forest, a Devonian age scene completed in 1924. Paul Marchand served as a coordinator of Exhibits at the Buffalo Museum of Natural Sciences between 1925 and 1943.

George was responsible for the Ann Arbor dioramas and the eleven formerly exhibited Field Museum dioramas. George loved fishing and perhaps his underwater prehistoric scenes hint at what he contemplated during these relaxing moments. Would a muskie or a Eurypterid be reeled in next?

George also painted an impressive 40' long by 10' tall mural situated in the Exhibit Museum at the far end of the Hall of Evolution, behind the half-mount skeleton of an *Edmontosaurus*, showing a 1950s style dinosaur scene. The posture and habits of the painted dinosaurs are decidedly 'traditional.' Bill Farrand suggested the mural may have to be removed eventually. "Hey, we'll take it," I blurted instinctively. But as Bill noted, unless severed, or (if possible) rolled up, due to its sizeable dimensions, the mural wouldn't even fit through the Hall of Evolution's entrance.

Another remarkable wall sized mural, "Life Through Geologic Time," is fading into obscurity on the fourth floor above the Hall of Evolution. This was completed in 1950 by artists, Robert M. and Jane Mengel, under direction of Irving Reimann. The evolution of life is portrayed elegantly in rows ascending from the floor. Colorful "scenes from deep time" representing each of the major geological periods and epochs are shown as a stratified layering. This is a beautiful mural, distinctively dated in style and detail of certain restorations. (We realize it's old, but please don't paint over it! It's another one of Michigan's "paleo-treasures.")

And there are many other wonderful exhibits as well to captivate your imagination. A 200-million-year-old skeleton of the herbivorous "spiny, armored crocodile," or thecodont, *Desmatosuchus*, is on exhibit. This happens to be the most complete specimen ever found. It was found in 1917 in Crosby County, Texas. There are many fossils of other Permian and Triassic reptiles exhibited here too, including *Dimetrodon* and *Edaphosaurus*.

Visitors will notice an awesome head of the Devonian fish, *Dunkleosteus*. Museum artist, Dan Erickson, relied on this specimen in restoring a life sized model of this genus, which looks scarier than "Jaws." Replicas were shipped to the American Museum and the Cleveland Museum of Natural History. Too bad they couldn't cast another one of those 16 foot long monsters for the Exhibit Museum too.

Oh yes, they also have a fantastic dinosaur collection. The chief exhibit was mounted over 30 years ago. From the famous Cleveland Lloyd Quarry, are two of North America's most famous Jurassic dinosaurs—*Allosaurus* and *Stegosaurus*! *Allosaurus* is fully mounted over the reclining fairly complete *Stegosaurus* skeleton. This is a really 'cool' exhibit, even on a hot day! More recent acquisitions include the cast of an *Anchiceratops ornatus* skull, and a *Deinonychus* skeleton acquired through another "Buy-A-Bone" campaign first displayed in 1995. An "irregular pit" situated behind the right eye of the former may be a puncture wound.

Amy Harris is the coordinator of the Museum's "Dino Club," which is some 400 strong, primarily comprising grade school kids. Amy has also been highly involved in the successful "Buy-A-Bone" fund-raising events.

So come on down! This is a wonderful place to learn about the history of life. Our visit was short, but if you really want to absorb the exhibits, plan for a half day visit. Take notes, photographs and, most importantly, take time to become excited about the world we live in.[1]

(Coauthored with Diane E. Debus, 1998.)

Note to Chapter Forty-four: The University of Michigan Exhibit Museum of Natural History is located at 1109 Geddes Avenue, Ann Arbor, MI and is open seven days a week. Donations are requested for admission. Call (734) 764-0478 for further information. With special thanks to Bill Farrand who supplied us with several past issues of the museum's newsletter, "The Display Case," and other materials facilitating preparation of this chapter, and also Linda Van Aller Hernick who offered her knowledge of the Marchand family.

TEN

(Top) Lynne Clos, 2007 (used with permission) (Bottom) Punch, July 27, 1861. "The Antediluvians recognize an old acquaintance." Note the fiery comet streaking Earthward, viewed through the telescope!

Exo-paleontology

There is a profound fascination, at large, with paleontological mass extinctions theories and scenarios—*especially* those invoking cosmic forces that link astronomy with the fossil record. Such theories may 'read' or present like (paleontological) science fiction, yet they're delightfully founded in fact!

My predilection for both the science fictional and factual aspects is deeply ingrained, going back to my early college years while taking geology courses at Northwestern University. For example, to augment our understanding of the geological field, one professor at the time encouraged students to subscribe to the journal *Geotimes*. I read many articles therein over the two years of issues arriving at my doorstep, but now—as a senior citizen—the only contributions that left lasting impressions were those by an intriguing geologist named Steiner (who you will read considerably more about in Chapter Forty-seven).

A theory that gamma-ray bursts may have decimated marine life in the Ordovician seized my attention a few years ago, resulting in my *Fossil News* review, presented here as Chapter Forty-six. Interestingly, I recently learned it's now thought that 'only' 2.5-million years ago a 'nearby' supernova may have irradiated Earth, causing widespread floral and faunal changes—mutations. The resulting gamma-ray burst may have damaged the ozone layer, leading to these Pliocene catastrophes. The theory was televised on *How the Universe Works* (SCI Channel, first aired on Jan. 15, 2019—season 7, episode 3). Of course, we must take such matters with a grain of salt!

Meanwhile, with Chapter Forty-five in mind, fictionalized accounts of impacts from asteroids and comets still abound in the literature. Perhaps the very best science 'pen-picture' composed thus far outlining how and what happened sequentially during the Cretaceous-Paleogene (K-Pg) mass extinction of 66-million years ago may be found within pages 309-315 of Steve Brusatte's acclaimed 2018 popular book, *Rise and Fall of the Dinosaurs: A New History of a Lost World*. Any science geek or sci-fi nerd will relish his 'story,' founded in fact. And Steve's ecological subtext on extinctions is clear, a 'message' so well-stated beforehand by paleontologist Warren Allmon in *Rock of Ages Sands of Time* (p.303), who stated: "Humans may not be the measure of all things, but we may turn out to have an immeasurable effect on all things."

The final two chapters in this book allow intrepid readers to ponder a little geology, along the way.

Lastly, I should note key passages within Max Hawthorne's *Kronos Rising* 2013 sci-fi novel—hyped by *Prehistoric Times* as the 'Book of the Year.' Hawthorne offers an interesting scenario for why his titular and powerful, demonic *Kronosaurus* survived the K-Pg extinction, explaining how a few of its kind survived the half-mile high tsunami churned-up in aftermath of the bolide impact, details which aren't ordinarily apparent in such crypto-paleontological writing.

These chapters originally appeared in the following publications:
Chapter Forty-five—*Prehistoric Times* no.125, Spring 2018, pp.46-49.
Chapter Forty-six—*Fossil News: Journal of Avocational Paleontology*, vol.14, nos. 9-10, Sept. 2008, pp.4-8; Oct. 2008, pp.6-9.
Chapter Forty-seven—*Earth Science News*, (Earth Science Club of Northern Illinois) Vol. 39 (Parts 1 – 3), January 1988, pp.19-22; Feb. 1988, pp.19-23; March 1988, pp.19-22; Reprinted with additions and new information in *Fossil News: Journal of Avocational Paleontology,* vol. 13, Jan. 2007, pp.6-9; Feb. 2007, pp.6-9; March 2007, pp.4-7.

Chapter Forty-five — *Against the Grain: Off a Comet Into a Valley of Exo-dinosaurs*

There have perhaps been three historical periods in science and popular culture characterized by heightened intrigue concerning comets and satellite or planetary collisions with Earth. Such times have invoked dinosaurs as well. The first seems to have been during the 1870s, when Jules Verne's careful readings led to his 1877 novel, *Hector Servadac* (aka English title, *Off on a Comet).* A second period precipitated during the 1930s, when authors Philip Wylie and Edwin Balmer published their fantastic novel, *When Worlds Collide* (1932)— and a sequel thereafter, *After Worlds Collide.* Of course, the most recent period came during our lifetimes (i.e. as those of you who are reading this current *PT* issue must recall), during the 1980s when a comet (or an asteroid) collision was causatively linked to Late Cretaceous dinosaur extinctions (for which little more will be stated here). Dinosaurs didn't factor in quite as much—only tangentially during the second phase, yet oddly they *did* in relation to that *first* phase, albeit only derivatively—directly based on Verne's novel.

So how did dinosaurs hop onto a comet and then subsequently invade an outer space "valley"? And how did Verne get us all, well, off on comets (later to be 'inhabited' by dinosaurs)? The short answer is simply—*movies* based on Verne's novel! And what's an "*exo*-dinosaur" (a term coined by paleontologist Jose Luis Sanz), as in my title? Simply, any hypothetical non-terrestrial 'dinosaur,' possibly even from planets outside our Solar System. (After all, we're talkin' sci-fi here—remember.) Verne's 1877 novel is partly intended as social satire and contemporary political commentary, however those tendencies (its weakest points) need not concern us here.

Despite analogies that may be made to his *Journey to the Center of the Earth* (1864), Verne's 1877 *Comet* novel didn't incorporate prehistoric life. Nevertheless a short summary is needed concerning this imaginary 'trip around the solar system.' To prepare himself for this new writing venture, Verne read popular astronomy books by his friend (astronomer) Camille Flammarion, such as *Mondes imaginaires et mondes reels, Recits de L'Infiniti* and *Astronomie populaire.* Interestingly, a decade later, Flammarion published a massive illustrated book concerning evolutionary development of life through geological time, titled *Le Monde Avant La Creation De L'Homme: Origines De La Terre, Origines De La Vie, Origines De L'Humanite'* (1886), conceptually an update to 1860s French editions of Louis Figuier's *The World Before the Deluge*—a text Verne relied upon for writing chapters of his fictional *Journey*. (Was Flammarion's 1886 book in turn partly inspired by Verne's geological novel?)

So, briefly, in Verne's *Hector Servadac* a large comet "grazes" Earth, thus turbulently whisking away portions of the Mediterranean region onto its surface. However, this near 'collision' is sufficiently graceful such that protagonist Hector Servadac (note that this last name spelled backwards in French reads 'cadavres,' or corpses) doesn't even perceive what has happened, at first. Eventually other people are found to have also made the interplanetary transit. Together—with help of an 'on board' astronomer—they comprehend their plight … but they also consider how it might be possible to safely return to Earth many months later, since (providentially) the comet's orbit is elliptical.

Servadac and the astronomer have deduced that they and fellow passengers are aboard a 450-mile diameter comet (named "Gallia"), compositionally enriched in gold, that during a surprisingly close encounter with Earth, gravitationally attracted and attached sea and land, including the Rock of Gibraltar. *Providentially*—they encounter an active volcano, still lit by Gallia's "internal fires." First sailing onward past Venus, then they're outbound, fearfully, toward immense Jupiter … and then back toward Earth over the course of two Earth-years. During the portion of their journey outward beyond Earth's orbit, as the relatively thin atmospheric temperature declines and their ocean freezes, they seek refuge deep within bowels of said volcano—shades of Verne's *Journey to the Center of the Earth*. Again—providentially the volcano's geochemical fire (ignited by reaction of oxygen, supplied from Earth's atmosphere with minerals within the volcano) ceases just as the comet plunges to its farthest distance from the Sun, yet still retaining sufficient heat—thus allowing them to safely inhabit a warm interior chamber. Meanwhile, readers discern Verne's central (theologically themed) message, "… it is not for man to disturb the order of the universe …"

Even at a time when understanding of the physical mechanics of asteroid impact remained in its infancy, Verne's passengers on Gallia still worry whether it will be possible to return to Earth when the comet returns to its closest juncture with Earth. Will the comet directly collide at a relative speed of 21,000 mph? (Remember that the K-Pg Late Cretaceous asteroid was 'only' six miles in diameter—not a whopping 450 miles.) So they construct a hot air balloon out of sea vessel sails sufficiently voluminous to drift back onto the home planet through near space, presuming that the comet will only 'graze' our atmosphere like before, thus permitting safe passage. "None but a Divine Pilot could steer them now." And yet against all odds, they survive!

Although in his *Jules Verne On Film* (McFarland Publishers, 1998) Thomas C. Renzi claims that while 13 films have been based on Verne's *Comet* novel, only two of these incorporated dinosaurs into the plot. One of these, Karel Zeman's *On the Comet* (1970), is considered sublime for its cherished use of stop-motion animation, while the other is often regarded as rather abysmal, due to its reliance upon *One Million B.C.'s* (1940) stock footage of live lizards posing as 'dinosaurs.' But as we'll shortly see, I'll offer reasons as to why I'm 'against the grain' here, because I rather enjoyed 1961's *The Valley of Dragons*, despite its extensive use of stock footage. (And before continuing to discuss these films, a *Classics Illustrated* comic book was published in March 1959 featuring Verne's comet story, that remained true to the original tale, that is, without mentioning dinosaurs.)

Cometary collision theory began to take hold in public consciousness, with Ignatius Donnelly's *Ragnarak: The Age of Fire and Gravel* appearing shortly after in 1883. Chapter Three in this oddity was titled, for instance, "Could a Comet Strike the Earth?" Donnelly, not a scientist but a lawyer and politician writing in popular vein—opposed to the theory of continental glaciation—explained glacial "drift" deposits in terms of fallout from a comet impact occurring in an early historical period, as recorded by ancient civilizations in legends and lore. (He also published a book the previous year concerning the 'real' *Atlantis: The Antediluvian World*.) Today, *Ragnarok* remains interesting, but scientifically misguided. In his approaches, Donnelly perhaps anticipated mid-20th century writings of Immanuel Velikovsky, who also wrote of interplanetary catastrophes, while invoking paleontological evidence in books such as *Earth In Upheaval*—soundly denounced by scientists as 'crackpot.' However, the extent to which Verne's *Hector Servadac* novel influenced Donnelly's contemporary writings remains unknown.

As mentioned above, the second period of popular intrigue concerning planetary and cometary bodies on collision course with Earth dawns in the early 20th century, capped with publication of Philip Wylie's and Edwin Balmer's astounding novel *When Worlds Collide* (1932). (Yes—there had been angst too over the pending approach of Halley's Comet in 1910! But see Carl Sagan's and Ann Druyan's *Comet* book published in 1985 for more.) Interestingly, Paramount movie producers considered projecting *When Worlds Collide* and its sequel *After Worlds Collide* (1933) onto film under an ominous working title "The End of the World," during the 1930s. Notes found in a Hollywood file indicate an early tie to an early filmic foray into exo-dinosaurs that would have otherwise been lost to the ages. Fortunately, author Bill Warren recently discovered that there was then "… popular interest in planetary collision." Furthermore, the two novels were to be 'merged' more or less together in the movie, with some key modifications.

Warren noted, "Another suggestion was that the destruction of the world occupy only about the first third of the film, the remainder dealing with the survivors' adventures on the new planet. A further idea proposed that instead of two wandering worlds, one of which destroys the Earth and the other of which replaces it, only one passes through the solar system, taking out the

Earth, while the survivors head for Venus, then presumed 'to be in a geological state not unlike that of *The Lost World*—a place of steaming jungles, huge dinosaurs and primitive Tarzan-like men.'" (Warren, *Keep Watching the Skies,* Vol. 2 (2010), pp.895-896) Interestingly, today, concepts like rogue and wandering exo-planets are in vogue today, based on valid astronomical observations, as witnessed popularly via televised programs like the Science Channel's *The Planets* "Riddles of the Hell Planets" (2017). What an extravaganza that 1930s version of "The End of the World" might have been with its exo-dinosaurian fauna!

Which leads us first to Czech, Karel Zeman's impressive 1970 film, *Na komete'* (English title- *On the Comet*), which employed stop-motion 'exo-dinosaurs.' Zeman of course was no stranger to animating prehistoric animals, the appearances of which clearly were often inspired by Zdenek Burian's artistic restorations. Zeman had already animated dinosaurs and a host of other fossil creatures for his amazing *Journey to the Beginning of Time*. (See Chapter Five "Filmic Illustrations of Life Through Geological Time," in my *Dinosaurs Ever Evolving* (McFarland 2016) for more on that film.) According to Mark Berry's *The Dinosaur Filmography*, Zeman also produced two shorter films (unseen by this author) *The Black Diamond* and *Excursion to the Universe,* the latter employing a cometary gimmick, both of which also incorporated restorations of prehistoric animals. (Yes—Zeman excelled in projecting Verne's novels onto film, such as 1958's *Vynalez Zkazy* (*The Fabulous World of Jules Verne*).

Viewers were treated to a fantastic Vernian adventure in Zeman's 1970 entry! The English (dubbed) translation of *On the Comet* (film) is only loosely based on the novel, and due to a delicious "palette of cinematic trickery" (per Berry), at times looks, well ... kinda 'psychedelic.' (I mean you gotta see those *colors*, Man!) Clearly though, movie plot differs considerably from novel. For instance, following the "space catastrophe," Lieutenant Servadac immediately realizes that they're on a comet, headed beyond Earth's orbit for Mars. Unlike in the novel though, the comet's atmospheric temperature never declines, so its ocean never freezes, and hence there is no need of a warming volcano. And then of course there's that herd of rampaging exo-dinosaurs, not appearing in the novel, staging an attack on a village bazaar and fortress (transferred from Earth) unfolding over a generous ~5 minutes beginning at the ~ 41-minute mark! And there's even another evolutionary/primeval scene later in the film. So let's delve into these prehistorical scenes.

That surrealistic herd of evidently indigenous dinosaurs ("prehistorical cattle") comes inexplicably out of nowhere, implying that the comet has a primeval, or "antediluvian" bearing. We see theropods, sauropods and some camptosaur-like creatures. These exo-dinosaurs (modeled after real dinosaurs) rove around, snarling at the French cavalry. After cannon fire is unsuccessful in driving them off, ludicrously, they're repelled by the sound of clanging pots and pans! Most of the dinosaur scenes are stop-motion animated, but other means of animation were employed too.

Later, on a 'prehistoric island,' a short (tongue-in-cheek) evolutionary sequence is witnessed from aboard ship at sea, (The sailors must ward off a 'Nessie'-like sea dragon with those pots and pans.) Then after pterodactyls are spotted flitting about the ship, an *Edaphosaurus* is noted emerging from its "primeval" Coal Age, swampy "two million-year-old" forest, toward the shore. Servadac narrates the "Paleozoic" nature of their circumstances, but then suddenly there's a (Late Cretaceous) Burian-esque styracosaur on screen. And then we watch millions of years of evolution speeded up such that a walking lungfish type creature transforms in lateral view before our eyes, on land, into a boar sprinting on its hind legs! Zeman must have had fun with this phantasmagorical 'insular' feast. By film's end, however, as Servadac awakens on Earth, viewers are left wondering ... was it all just another Vernian dream? For me, kind of a letdown!

As Berry stated, "Zeman did not use stop-motion in the same manner as the great American animators. He employed it merely as one of many techniques on his varied palette of cinematic trickery, without necessarily worrying about duplicating nature.... Zeman might legitimately be compared to ... pioneer Georges Melies ... freely (melding) all manner of elements—stop-motion, cel animation, puppets, models and miniatures, matte shots, woodcuts, paintings—to create imagery in a style that is his alone." (p.292) Aligning Zeman's movie, *On the Comet,* with Verne's *Journey to the Center of the Earth* (novel and movie) may we detect an implicit message that when one probes down deep within Earth's strata, or (akin to what the astronomers say) further outward into the remoteness of outer space, mysteries of the deep Past will be revealed?

Although our second movie entry, *Valley of the Dragons,* was also founded upon Verne's 1877 novel, beyond opening credits and the first five minutes of *Dragons*, key novel elements are barely recognizable. Of course Verne's core idea is intact: a large comet grazes Earth, thus projecting humans onto said comet, to a place where there be 'dragons'—pseudo-dinosaurs and enormous paleo-mammals. Unimpressed author, Bill Warren, remarked "The storyline of *Valley of the Dragons* is that of *One Million B.C.* adding dialogue, two duelists from Earth and fake Morlocks, and taking place on Verne's comet." So much stock footage used too, hardly fooling anyone as to sources! And despite Warren's objections ... well ... at least there's that erupting volcano this time. (The comet also has been pre-populated by tribes of Stone-Age people.)

Right from the get-go, we spot an error, a model Earth is rotating in the opposite direction. The characters never discern how large this comet is, but neither film can 'show' viewers the diminished surface gravity while humans stride about—in both cases implying a far more massive comet than that envisioned by Verne. The characters don't immediately comprehend that they're on a comet, instead determining that they're on a "world of the past—100,000 years past." Following insertion of that famous gigantic-gator versus huge lizard battle, swiped from 1940's *One Million B.C.,* Servadac proclaims that the reptile vanquished in this scene is a (doubtful) "plateosaur." Later on, the two sole (modern) humans carried by the comet (i.e. on *their* recent interplanetary event) from Earth facilitate dispatch of another great reptile by formulating an explosive, in much the same way that Captain Kirk figured things out in a 1967 *Star Trek* episode where he fought the dinosauroid "Gorn." Three short scenes of Toho's "Rodan" (the flying monster) were also spliced into the 'Valley's action. And I could've sworn I also spied a gigantic version of that Geico insurance gecko peeking at me in one of the swiped scenes—lol.

Columbia Picture's *Dragons* was intended as another valid 'Vernian' experience, although with the addition of dinosaurs, perhaps its main focus of interest today! However, this 1961 entry wasn't the first on *film* to introduce outer space "exo-dinosaurs." (By the way, annals of paleo-fiction—published stories and novels, are rife with alien dinosaurs and space-faring dinosauroids.) Don Glut mentions several filmic examples in chapters of his popular *The Dinosaur Scrapbook*. For example, other exo-dinosaurs captured on camera include those seen in 1951 and 1952 televised episodes of *Tom Corbett, Space Cadet,* and also in segments of both *Flash Gordon* (1936) and *Flash Gordon Conquers the Universe* (1940)—serials incorporating live lizards enlarged via trick photography as well as an alien theropod-like "Gocko."[1]

Mark Berry flags another such example—1955's *King Dinosaur*—as "…a strong contender for the title of Worst Dinosaur Movie of All Time," so 'nuff said' here on that one. And still later examples include *Planeta Burg* (*Planet of Storms*), a 1962 Russian entry in which cosmonauts discover prehistoric life on Venus, and then in 1964/65 Japan's Toho Studios brought us "Ghidrah," a Venusian/Martian winged, three-headed drago-dinosauroid. Then there was a 1970s television show—*Land of the Lost* incorporating 'stop-mo' dinosaurs stemming from another space-time dimension. And kudos to Stephen Czerkas for his work on 1978's stop-motion animated *Planet of Dinosaurs*. And by no means was that aforementioned 1930s saga, "End of the World" *When Worlds Collide/After Worlds Collide* aborted movie adaptation, the first to *propose* filmic exo-dinosaurs. No, that milestone was instead achieved by Herbert M. Dawley's 1921 silent film *Along the Moonbeam Trail*, with its stop-motion animated "Lunarian" dinosaurs. (See Chapter Twenty-six.) This once 'lost' short film, resurrected partly through the devotion of Stephen Czerkas (with wife Sylvia), is now available on DVD.

It is curious that 1961's *Valley of the Dragons* was made *prior* to Zeman's *Comet*. Was the latter then a 'remake'? Hardly! In consideration of his other earlier, 'opus' film, *Journey to the Beginning of Time,* Zeman may have been just another 'dino-buff,' using his talents in revivifying prehistoric life—favoring the classical edifying element of evolution through time. Conversely, *Valley of the Dragons* would simply seem an opportunity to contort a movie around existing stock footage, while relying on Verne's then highly popular style of story-telling.

Dragons received a 2 out of 4 stars rating from Mark Berry in his *Dinosaur Filmography* and has a 5.0 out of 10 rating at IMDB (as of 12-31-17). *On the Comet* fared better, however, earning 3 out of 4 stars from Berry and 6.9 out of 10 at IMDB. Why—probably because of the way in which prehistoric monsters are projected. Zeman's approach was clearly superior. Okay, contrary to these ratings, I haven't shown my cards yet as to why I liked *Valley of the Dragons* nearly as much as I did *On the Comet*.

What works for me is the wacky, yet original pseudo-science woven into *Dragons'* plot, missing from both *On the Comet* and Verne's 1877 novel. It's not too elaborate, but at least represents an honest effort to suspend disbelief. I liked how Servadac deduces that this world of 100,000 years old they're exiled upon, suggests that their comet's highly elliptical orbit must have made a close approach with Earth *exactly that long ago* during our prehistory (oooh, eerie, chills!!), thus allowing the comet to carry off "an envelope of Earth's atmosphere" and a "fragment of the distant past" that remained unaltered by evolution whilst aboard the comet ('explaining' how cave peoples and prehistoric monsters got there). And as to why the comet collision wasn't more violent, it's explained in *Dragons* that upon the comet's close approach, Earth "rolled with the blow like a boxer taking a hard punch" ... or some such mumbo jumbo. (Hmmm—maybe that's why Earth is shown rotating the wrong way during the opening credits?) But in the end, Servadac and his 'modern' dueling antagonist from Earth contradict themselves by proclaiming that the comet should return to Earth's vicinity next—not in another 100,000 years, but only in 7 years.

Such a silly tale is Verne's novel *Hector Servadac*. But neither Verne, nor Zeman a century later, and certainly not Edward Bernds (Director of Columbia Picture's *Valley of the Dragons*) could've ever anticipated how intertwined dinosaurs and comets would later become—*scientifically*—during that spectacular 'decade of the dinosaur'—the 1980s! And finally, thanks to research conducted during the 1980s, we now realize that a 450-mile comet "grazing" Earth's atmosphere wouldn't be a farcical, fanciful or providential affair, but instead would prove catastrophic beyond all human experience and comprehension!

Note to Chapter Forty-five: (1) After this chapter ran as an article in *Prehistoric Times* no.125 (Spring 2018), on May 5, 2018 Don Glut e-mailed an interesting correction notice regarding the name Gocko. "FYI – the critter pictured ... identified as the fire-breathing 'Gocko' is misnamed. Although there were dinosaur-like creatures called Gockos in the early Alex Raymond *Flash Gordon* newspaper strip, their more or less counterpart in the 1936 movie serial is referred to only as a great beast (played by Glenn Strange). And that beast doesn't breathe fire (although the suit was revamped for another monster, one that *did* breathe flames, in a much later episode. Two monsters, one suit, neither of them called a 'Gocko.' (Supposedly the producers didn't think, after weeks had passed, that anyone in the audience would recognize that the outfits were really the same. As I recall, the 'Gocko' mistake was started by Forrest J Ackerman in one of the early issues of *Famous Monsters of Filmland* or *Spaceman* magazines." Don's note—information out of the deep fanzine vault—was also printed as a Letter in issue no. 126 of *Prehistoric Times*.

Chapter Forty-six — *A Cosmic Ice Age Extinction: (Learning by 'O-S'-mosis)*

Sheesh! Nearly the worst Ice Age of all-time, or at least the Phanerozoic, and nearly as lousy as a Chicago winter! Where was Al Gore when he was *really* needed to thwart climate change? What's all this mad sputtering about? Let's learn through what I'll refer to as "O-S-mosis" (i.e. where the 'O' stands for Ordovician, and the 'S' stands for Silurian).

Some of you may have seen the *History Channel's* "Mega-Disasters" 2007 broadcast on large Gamma-Ray Bursts ('GRBs'), that episode showcasing the latest causal theory for Late Ordovician mass extinctions, occurring some 444 million years ago. Given that, as a teenager, I collected numerous Ordovician fossils along the mid-west fossil corridor, my interest in this televised program was certainly piqued. And so I wondered whether those brachiopod and cephalopod fossils, so enthusiastically recovered decades ago from quarries and road cuts, represent organisms extinguished in a celestial catastrophe, as indicated in the Mega-Disasters program series. To place this intriguing program in perspective, let's probe far beyond what was presented in between commercials.

The earlier and middle Ordovician is generally known as the period of 'great biodiversification.' While during the Great Ordovician Biodiversity Event of approximately 465-million years ago warmer greenhouse conditions prevailed, within the ensuing 25-million years Mother Nature ('Gaia') was poised to take back many of her 'experimentations.' Traditionally, Late Ordovician extinctions, in sum—the second most severe as recorded in the fossil record (e.g. 60% loss in genera overall; 40% of trilobite genera; and nearly all graptolite genera), are considered the result of widespread glaciation.

Strangely, this extinction event wasn't even recognized until after Norman D. Newell identified a pattern of decreased biodiversity corresponding to the Late Ordovician, as portrayed graphically in his 1967 paper titled "Revolutions in the History of Life." (*Geological Soc. America Spec. Paper 89*, pp. 63-91) Several years later, geological evidence of former glaciation dating from this time was detected in North African deposits, indicating a probable cause for declining biodiversity. Meanwhile, researchers determined there were actually *two* phases of extinctions corresponding both to the beginning and end of the Late Ordovician ice age (the second more recent phase ('Hirnantian') was estimated to have endured from 20,000 up to 1-million years). By the 1980s, J. John Sepkoski quantified the severity of the Late Ordovician event using 'clade diversity' diagrams.

Circumstances contributing to the Late Ordovician extinctions are complex in nature. However, by 2003 the Ordovician-Silurian (or the "O-S") extinction boundary was thought so well documented that Peter M. Sheehan of the Milwaukee Public Museum felt sufficiently assured to proclaim of the Late Ordovician extinction, "...How it Became the Best Understood of the Major Extinction Events." (*Geophysical Research Abstracts*, vol. 5, 07188) And so paleontologists recognized a plausible paleo-geographical 'trigger' (with contributing geological factors), as outlined in the next paragraph, now considered conventional and founded on terrestrial factors.

When the former continent Gondwanaland occupied a South Polar position, oceanic circulation could no longer effectively convey heat parcels from the equator, globally. As ice sheets formed on Gondwanaland, increasingly more sunlight was reflected back into space. Also, throughout warmer Ordovician intervals, considerable carbonate had been sequestered in reef deposits, thereby depleting atmospheric carbon dioxide levels. Accordingly, warmer global 'greenhouse' conditions prevailing throughout earlier, warmer Ordovician stages could no longer be sustained. Sea floor spreading rates and plate tectonism are thought to have been more quiescent during this interval as well, contributing (along with the glacial eustasy ice buildup) to sea level decline (~ 50 to 100 meters). Plunging temperatures and adverse climatic effects wrought havoc on scores of marine fauna. All in all, marine organisms adapted to life in shallower continental shelf waters were subjected to a cascade of adversity; loss of shelf habitat even doomed *tropical* organisms.

Well, at least that's what happened during the Late Ordovician, more or less, according to convention. Puzzlingly though, during the ensuing extinction, open ocean (pelagic) marine fauna were more adversely effected than near-shore fauna which lost critical habitat. And in refining a computer model explaining long term fluctuations in atmospheric carbon dioxide levels through Phanerozoic time with effects on global climate, geochemist Robert A. Berner claimed in 1990 that the Late Ordovician ice age just didn't fit in so well with the pattern and particulars. (*Science*, vol. 249, pp.1832-36) Four years later, after enhancing the GEOCARB model, Berner decided that a "...short period of glaciation at the end of the Ordovician represents an *exception* ..." (*Science*, vol. 276, 4/25/97, p. 545) So were there other factors at large, driving the Late Ordovician ice age? Was this an unconventional extinction affair after all?

Further geochemical (stable isotopic) data sharpened details of this two-phase extinction. In 1995, P. J. Brenchley (et. al.), for example, interpreted his data as representing a "short-lived end Ordovician glaciation in a 'greenhouse' world." (*Modern Geology,* vol. 20, pp. 69-82) To them, it was the 'suddenness' of the second wave of environmental changes which contributed most significantly to Late Ordovician extinctions. While ice age conditions descended gradually, most extinctions occurred following that ice age. But icy conditions, lowered ocean levels and global cooling weren't in themselves first-phase killing 'agents.' Instead, resulting sea level decline may have disrupted marine currents, causing (eutrophic) over-productivity leading to the first extinction phase.

Subsequent to the ice age phase, a second 'rapid' pulse of global greenhouse warming restored oceanic levels, but with a deleterious twist. During deglaciation, as waters rose, toxic oxygen-poor waters were 'suddenly' re-circulated upward onto marine shelf platforms at a rate organisms could not adapt to, causing oceanic anoxia along continental shelf zones, and hence the second major extinction wave. This second main event left its mark in the form of widely distributed black shale deposits dating from this transitional time. Can mass extinctions and geological changes recorded be explained by an asteroid impact? After all, as Brenchley noted (doubtfully), "The end-Cretaceous and Ordovician extinctions share in common a phase of sudden extinction."

During those crazy yet intellectually inspiring 1980s, when extraterrestrial scenarios for causes of mass extinctions were so hotly debated, some paleontologists searched for evidence of Late Ordovician, extinction-triggering bolide impacts. At the time, three highly weathered impact craters were recognized, approximating from upper Ordovician age (bearing in mind there's usually a *wide* error in the accuracy of impact crater age-dating especially with older craters). However, they each had relatively small diameters, none larger than 8 miles, implying insufficient energies for causing widespread extinctions of global magnitude.

Yet indeed, for the 'comet camp,' some initially promising iridium geochemical data (yes—*iridium*, that incriminating metal irrevocably and insidiously associated with Late Cretaceous mass extinctions 65 mya) turned up at the Late Ordovician 'disconformity.' By 1989, for example, after examining critical paleontological and geochemical markers, Chinese researchers concluded that "... a high iridium anomaly has been discovered at the ...O-S boundary layer... If this result is confirmed ... it will provide more evidence that all of the abrupt changes mentioned... are probably related to the astrogeological factor rather than to the Ice Age at the end of the Ordovician or other terrestrial agents." (*Astrogeological Events in China*, p. 140) They were referring to a detected iridium anomaly of 0.64 parts per billion (ppb) found at the O-S boundary, markedly above background levels in sediments representing earlier and later stages. Results published independently that same year attributed this high detection (albeit speculatively) to an impact.

But rather than supporting a sensational Paleozoic asteroid impact scenario, that aforementioned 'promising' evidence found in Chinese sections where the Ordovician-Silurian ('O-S') boundary is well developed was instead eventually related to mundane sedimentation anomalies. For by 1992, the O-S impact 'bubble' burst following interpretation of a 0.23 ppb iridium anomaly detected in a Chinese clay layer representing the O-S (e.g. Hirnantian brachiopod fauna) transition (corresponding to the base of the graptolite *Glyptograptus persculptus* biozone), or the second wave of the two-step O-S extinction. In trimming enthusiasm for impacts on a grand scale, authors Wang, Chatterton, Attrep and Orth (*Geology*, Jan. 1992, vol. 20, pp.39-42) noted that the O-S boundary represented an age of extensive Gondwanaland glaciation as well as global glacio-eustatic regression.

Such circumstances were unlikely to have preserved a fuller, continuous sediment record. So because material in the studied sections presumably accumulated during a time of reduced sedimentation rate, this meant that any iridium (representing continuous background fallout from meteoritic dust entering the atmosphere) would have been naturally concentrated, (e.g. elevated), in the O-S boundary layer. Or as they stated, "... it is our conclusion that the latest Ordovician extinction in the Yangtze Basin does *not* appear to be linked with an extraterrestrial impact..." But so influential was the paradigm of comet impacts then, that the authors retained a bit of wishy-washiness by cautiously adding, "...although merely on the basis of iridium *we cannot preclude* (my italics) impact by a comet with very low abundances of platinum group elements."

Curiously and rather paradoxically however, researchers now suggest that an increased (100-fold) 'shower' of meteorite debris actually *contributed* to the aforementioned Great Ordovician Biodiversity Event of 25-million years *earlier*. Although a cause-and-effect explanation is lacking, at least there's 'temporal coincidence' for impact-driven *evolution* corresponding to the diversification event. (*Science*, vol. 318, 12/21/07, p.1854)

But besides the possibility of bolide impacts aren't there alternative factors driving the extreme fluctuations witnessed in Ordovician biodiversity (e.g. vibrant evolution suddenly followed by mass extinction)? Yes—and this takes us back to the theme of the aforementioned *History Channel's* Mega-Disasters program. For a theory of interfering energetic 'rays' from outer space was featured in 'Mega-Disasters,' which shall therefore be presently summarized.

Given that the past may sometimes be viewed as a reliable 'key' to the present, scientists focused on Late Ordovician marine fauna corresponding to the prolonged ice age major extinction interval of 444-million years ago. Were those spectacular trilobites, tentacled cephalopods, graptolites and early jawless fishes of the time zapped by gamma rays! 'Yes,' according to paleontologist Bruce S. Lieberman and astrophysicist Adrian L. Melott, both with the University of Kansas. For it is the nature and pattern of the Late Ordovician extinctions and, importantly, statistical probability of occurrence within the past billion years which led them to consider a gamma-ray burst (GRB) possibility for the ultimate cause of Late Ordvocian extinctions.

Melott and Lieberman (et. al.) elaborated on the pattern of Late Ordovician extinctions in their 2004 paper, (*International Journal of Astrobiology*, 3(1), pp. 55-61). The Ordovician was predominantly a greenhouse warmed condition, and so *sudden* onset of an ice age and sea level decline suggested that such rapid global cooling was ultimately powered by an "external forcing mechanism." Without disputing evidence for a textbook O-S ice age event, Lieberman and Melott suggested that the Late Ordovician ice age was triggered *extraterrestrially*, when Earth bathed in a mere *ten-second* gamma-ray burst.

Gamma-rays? Yes.

Gamma-rays are the highest energy wave form of *radiation* known, far more energetic even than x-rays. Discovered in 1900, gamma rays are a form of 'radioactivity' and are emanated during atomic bomb explosions. But remarkably, in 1967, understandably during the height of the Cold War, it was discovered that gamma-rays are intermittently beamed right to Earth from *outer space*. Were aliens waging atomic war somewhere within our Milky Way galaxy as a scientist speculated ... or were strange happenings, possibly even beyond our galaxy and imagination, signaling unknown cosmic properties and universal processes? As outlined in the *History Channel's* fascinating 'Mega-Disasters' 2007 program, it took four decades of rewarding astronomical research to find sensible answers to the mystery of large interstellar Gamma Ray Bursts, (or 'GRBs'), which have been described as the "most violent phenomena in the universe... shining with energies equal to all the light of stars in the observable universe."?

Gamma-rays are directed from all sky coordinates—all regions of the heavens, from each and all of the universe's 'billions & billions' of galaxies—not just from within our Milky Way galaxy. The GRBs appear 'beamed' linearly because they're ejected as narrow 'jets' from the poles of rapidly spinning stars. GRBs are thought either to be the 'birth cries' of black holes known as 'collapsars'—immense stars undergoing rapid gravitational collapse, or to signal collision of an orbiting pair of (much older and dense) bodies known as neutron stars, forming a black hole.

Calculations show that a GRB occurring within 6,000 light years of our Sun, beamed toward Earth, would be equivalent to 3,000 megatons of atomic bombs exploded simultaneously, thus packing sufficient punch to cause wide scale extinctions. Yes, just like the Late Ordovician event. As S. E. Thorsett profoundly stated in 1995, "It is tempting to speculate that some of these (gamma-ray) bursts may have led to mass extinctions which have occurred at similar intervals ... the furthest reaches of the universe may have close ties to life here on Earth." (*The Astrophysical Journal* vol. 444, pp. L53-L55)

How did consideration of 'rays' suggest an alternative, possible cause for the Late Ordovician ice age and extinctions?

The mechanism is conceptually simple. After the GRB irradiated Earth's upper atmosphere 444-million-years-ago, much of the ozone layer (which had presumably already formed by the Late Ordovician), protecting Earth's surface by absorbing the Sun's ultraviolet (uv) radiation, was chemically degraded, forming nitrogen-oxide compounds. Increased amounts of harmful uv light would kill vast amounts of exposed organisms such as planktonic forms floating within a few feet of the thin sunlit ocean surface. Furthermore, increased production of nitrogen oxides in the atmosphere (eventually precipitating as 'acid rain') would blanket the atmosphere in an "ominous brown haze" blocking visible sunlight. Prolonged global cooling would lead to the Late Ordovician ice age. Liebermann stated in *History Channel's* Mega-Disasters program focusing on large gamma-ray bursts, that only such a scenario provided the one-two, 'double-whammy' punch needed to trigger an ice age and exterminate so many species at once.

Too much, all at once—there were simply more intense stressors acting simultaneously than the biosphere could altogether sustain. Furthermore, Melott claimed such a scenario could have likely devastated Earth not just once, but perhaps several times in its geological history, on the order of once every few hundred million years. Authors of the 2004 *Astrobiology* paper concluded, "...even the most conservative rate estimates indicate that a (GRB) catastrophic event within the geological record is likely... every 500-million years." From a paleontological perspective, such a theory would seem difficult to falsify.

The pattern of extinctions is at least consistent with their theory. Planktonic forms floating near the surface of the water column would be more exposed to deadly uv light relative to organisms adapted to burrowing below in squooshy sediment. And, generally, marine organisms that underwent planktonic larval stages of development were cut down most swiftly in the Late Ordovician. These included graptolites, cephalopods and tiny creatures known as acritarchs. Certain trilobites from whose ontogeny (i.e. individual developmental life histories) it may be inferred went through planktonic larval stages, were more likely to be exterminated than other trilobite species which instead passed through *benthic* (living in ocean sediment) larval stages. Organisms whose larval stages persisted longer in planktonic phases of life were more vulnerable than species which transformed out of the planktonic phase earlier in life. Among bivalves, those which burrowed into sediment were conceivably more protected from uv light irradiation and acid rain than those which formerly prospered nearer sunlit surface waters. Also, the *Astrobiology* authors noted that most ensuing Silurian marine forms seem to be mostly descended from late Ordovician species that survived in more protected habitat—deeper water and higher latitudes—furthest from solar (uv) radiance.

Lieberman and Melott are by no means the first to suggest otherwise undetected instances of extraterrestrial irradiation as the cause of mass extinctions in the fossil record. Perhaps the first to do so was esteemed German paleontologist Otto Schindewolf, who in several papers published between 1950 and 1962 considered whether, for instance, Late Permian extinctions (aka the so-called Great Dying) resulted from widespread biological mutations caused by irradiation from a relatively nearby supernova. And, although claiming there's "...no compelling fossil evidence for past 'biological cataclysms'..." - a statement decidedly since overruled—physicist Malvin A. Ruderman tackled the supernova-promoting/ozone layer depletion angle, via uv-light and organismal scorching, in a 1974 paper (*Science*, vol. 184, pp.1079-81). Ruderman suggested that "the effects of ultraviolet deluges on past and future evolution may be more subtle."

Not surprisingly, throughout the Cold War late 1950s thru 1970s period, elegant cosmic radiation theories of extinction founded on ingenious celestial mechanisms, such as for explaining in *ad hoc* fashion how the last dinosaurs expired, continued to spring up. However as Steven D'Hondt commented in 1998, "Cosmic-radiation theories of mass extinction have clearly engaged the interest of both paleontologists and astronomers in the late twentieth century. Nonetheless, the intermittency of their appearance in the professional literature suggests that cosmic-radiation hypotheses of mass extinction have generally been of only marginal interest to both scientific communities." (*Earth Sciences History*, vol. 17, no.2, p. 167)

So should the History Channel be accused of breeding sensationalism, pandering for ratings? Is the Lieberman/Melott gamma-ray extinction scenario merely a throwback to earlier (pre-1980s) times, before people generally accepted reality of asteroid and comet impact events? Is the idea sound: can we believe it? Well, superficially, their new gamma-ray explanation for the Late Ordovician ice age and mass extinction pattern does seem rather *ad hoc*. Plausible alternatives remain; their gamma-ray model is merely another *dramatic* idea, or as they phrase it, a "credible working hypothesis." For example, in his considerations of biodiversity cycles as statistically documented in the fossil record, Melott has also entertained a *cosmic*-ray possibility. (Unlike gamma-rays, cosmic-rays are high energy charged *particles* formed in shockwaves of exploding stars known as supernovas—often precursors of black holes). It is generally believed that interactions of cosmic-rays with Earth's atmosphere can cause global cooling.

Is Melott, in effect, apparently hedging his bets in case gamma-rays don't explain key fossil record features to his satisfaction? Well, not quite. So then, how does this alternative cosmic-ray stuff fit into the Late Ordovician picture? The answer to this riddle may be resolvable through comprehending Paleontology's 'holy grail,' those mysterious 'mega-cycles' of the fossil record.

Ready for the really BIG picture?

Indeed, researchers do have a tendency to explain mysterious cycles of the geological record through the starry heavens. And astrophysicist Adrian Melott, lead author of a 2004 theory for extraterrestrial cause of the Late Ordovician ice age with ensuing mass extinctions, certainly isn't the first to be so tempted. On one occasion championed in the popular press a quarter century ago, then University of Chicago paleontologists David Raup and J. John Sepkoski documented a 26-million year recurrence pattern in mass extinctions since the Late Permian. They concluded their 1984 paper, "Periodicity of extinctions in the geologic past," provocatively. "If the forcing agent is in the physical environment ... are the extraterrestrial influences solar, solar system, or galactic? Although none of these alternatives can be ruled out now, we favor extraterrestrial causes for the reason that purely biological or earthbound physical cycles seem incredible ...." (*Proc. Natl. Acad. Sci.*, vol. 81, p.805) This influential paper launched the as yet unfulfilled search for the postulated Solar companion 'Nemesis' star.

And so, a decade later, relying on J. John Sepkoski's massive data compilation (first and last stratigraphic appearances of 36,380 marine fossil genera ranging across 67 stratigraphic intervals), in their 2005 paper (*Nature*, vol. 434, March 10, 2005, pp.208-210), Robert Rhodes and Richard A. Muller noted a strong 62-million year cycle in Phanerozoic (i.e. since the beginning of the Cambrian Period) biodiversity. At a lesser degree of statistically interwoven confidence they re-detected a ~ 140-million year cycle signal in biological diversity which researcher N. Shaviv had previously (yet controversially) attributed in 2002 to variations in cosmic-ray intensity.

Presuming the veracity of such statistical signals, what could have caused these ultra long-term cyclical changes?

Accordingly, it turns out there may be another set of cosmic 'culprits' out there capable of causing severe as well as cyclical changes in Earth's biodiversity, linked to the Solar System's oscillatory dance through spiral arms of our Milky Way galaxy. First, with Mikhail V. Medvedev, in 2007, Melott devised (*Astrophys. J.*, 664, pp.879-889) a celestial model accounting for the 62-million (plus or minus 3-million) year biodiversity cycle which correlates amazingly well with the 64-million year cyclical motion of the Sun (and planets) bobbing up, down and up again like a cork cast adrift on its way through our cosmic 'whirlpool'—the galactic plane—as our Solar System revolves around the Milky Way core. The statistically valid *exactness* of the two periods (62- vs. 64-million years) prompts further consideration.

While on its upward ('northward') motion *above* the galactic spiral arm plane, it seems the Solar System receives much more cosmic ray flux from the Local ('Virgo') Supercluster of galaxies toward which we are moving at an incredible rate. Given that cosmic-rays wield an energy punch 100-million times greater than that of our particle accelerators, a cosmic-ray stream would certainly be a most imposing force. When cosmic ray intensities increase on furthest displacement from the galactic plane, then Earth's atmosphere is zapped by intergalactic cosmic radiation emanating from the approaching Virgo Cluster, with resultant adverse consequences very similar in nature to what would be anticipated from a gamma-ray burst (GRB) incident on Earth. An important difference between gamma and cosmic ray scenarios though, is that the more intense cosmic ray *particle* stream would affect *all* organisms, even those ensconced in deepest ocean waters that would have been protected from the uv radiation caused by an ozone layer-degrading GRB. Awful genetic, chromosomal mutations would be expected to viciously cut a swath across all paleo-ecological levels, presumably driving mass extinctions.

While Melott and Medvedev didn't go so far as to proclaim that ice age epochs would result from exposure to cosmic rays every 62- to 64-million years, instead, researcher Shaviv boldly correlated measured ages of ice age epochs with recurrences of heightened cosmic-ray intensity. In a very simplified sense (and in contrast to Melott's and Medvedev's model), Shaviv's mathematically complex model projects greatest cosmic-ray exposure when the Sun (and Earth) crosses *through* the Milky Way's spiral arms where recurrences of supernovas (and hence cosmic rays) are more likely. Shaviv didn't associate times of terrestrial exposure to heightened cosmic ray flux with *mass extinctions*. But he suspected that cosmic-ray bombardment increases prolonged phases of atmospheric cloud cover and, therefore, global cooling on the order of 10 degrees—sufficient to trigger ice ages. Ultimately, relying on his detailed cosmic ray model, Shaviv crudely correlated a body of data and evidence to demonstrate a 140-million year periodicity in ice age epochs! (Compare Shaviv's theory to Johann Steiner's as described here in Chapter Forty-seven.)

And so, to summarize. Philosophically, a premise of paleontologist Bruce Lieberman's and Adrian Melott's 2004 gamma-ray theory for a Late Ordovician ice age plus mass extinction is basically, 'Given sufficient time, really BAD stuff is certain to happen,' while the two cosmic-ray models outlined above, together, purportedly explain recurrent (some might even claim 'periodic') episodes in ice age epochs through geological time, and long term cycles in biodiversity (including major Ordovician fluctuations). I've just delved into all this at a much deeper level than did *History Channel* through their entertaining 2007 'Mega-Disasters' broadcast on GRBs.

So far (i.e. as of 2007), neither Lieberman and Melott nor Shaviv have attempted combining both the cosmic- and gamma-ray celestial scenarios into an all-encompassing periodic irradiation theory of mass extinction. Although, geologists Craig Hatfield and Mark Camp conceptually already did the honors way back in March 1970. ("Mass Extinctions Correlated with Periodic Galactic Events," *Geol. Soc. Amer. Bull.*, v. 81, pp. 911-914)

But these are down-to-Earth intricacies of nature we're dealing with, (as well as, maybe, the cosmos). Admittedly, it's easy to become swept away by all the elegant mathematics and complex celestial mechanics. Yet it's unlikely that causes of real world mass extinctions will be solved by astrophysicists armed with calculators. So, instead, how about geochemists armed with, er, uh ... calculators? Can we trust geochemists more than astronomers to press the right buttons and to make more accurate modeling assumptions? Well, geochemist Robert Berner (mentioned in Part 1) has since decided that his refined GEOCARB III model (*Am. Journal of Sci.*, vol. 301, Feb. 2001, p.201) won't reflect, or explain, relatively short-lived ice age epochs such as the Late Ordovician occurrence, because the model averages (e.g. smears out) data over ten million year chunks of time. So at least the intricate geo-carbon dioxide model isn't inconsistent with a Late Ordovician ice age event after all, for which there is hard supporting geological evidence.

Furthermore, geochemists have also been modeling evolution and changes in atmospheric oxygen levels throughout Phanerozoic time, noting that intervals of declining oxygenation correspond to times of ecological adversity. During the Late Ordovician, respiratory systems of many species were ill-adapted to the worsening, increasingly anoxic, marine conditions and so they struggled: biodiversity decreased. Conversely, throughout the Phanerozoic, during times of increased atmospheric oxygen levels, biodiversity increased. Therefore, can broad scale shifts in biodiversity be explained in terms of changing atmospheric oxygen levels? Paleobiologist Peter Ward, author of two fascinating accounts, *Out of Thin Air*, (2006) and *Under A Green Sky* (2007), believes so. But even if terrestrial, *geochemical* explanations are proposed to explain 'sudden' changes in biodiversity, ultimately astronomers would still question whether celestial events somehow triggered those longer-term atmospheric fluctuations.

But—hey, what about those impacts? There's an extensive body of literature available on rays effecting biodiversity, little of which, other than through statistical correlations and astrophysical modeling, decisively *proves* a mechanism for extinction. But besides the reality of GRBs, we do know that asteroids and comets interact with planets. And there are now nearly a dozen proven impact craters scattered across North America (and the mid-west especially) whose ages are at least not inconsistent with the O-S (Hirnantian) boundary. Couldn't some of those have 'teamed up' and transformed Late Ordovician paleo-ecology in a dramatic and sudden 'one-two' (or three) punch fashion?

While realizing that motions of our solar system through the Milky Way and cosmos could pose spectacular yet awful consequences for life on Earth, (several of which may have been recorded geologically), for now, I just count myself as more of an 'impact' person. I'm just not 'rad' enough to be swayed by the Late Ordovician gamma-ray (GRB) death proposal, no matter how elaborate it may seem. *History Channel's* presentation was, as *Star Trek's* Mr. Spock might say, 'fascinating.' But is it logical?

Chapter Forty-seven — *Steiner's (and Grillmair's) Unorthodox Ice Ages Theory*

"Whether the solar system is a closed system or there are some galactic influences remains uncertain. Geologists may well keep in mind that future work may yield valuable and perhaps more valid galacto-geologic correlations." (Johann Steiner, "Possible Galactic Causes for Geologic Events: Reply," *Geology*, June 1974, p.280)

I was a fossil-loving college undergraduate back in the early 1970s when I became enchanted (yes, 'enchanted' *would* be the word) by the rather unusual writings of an Austrian geologist who claimed to have solved the Ice Age riddle, as well the recurrence of a host of geological/paleontological episodes and phenomena throughout Earth history. You see—I started as a geology major, then subscribing to *Geology* and *Geotimes* magazines, where I read articles by Johann Steiner, now a largely forgotten figure in the annals of the geological sciences. Steiner's theories haunted me for years—for some reason they were rarely referred to. Then over a decade later, after reading Steiner's obituary notice, I resolved to learn more. Who was this individual and where and how did he concoct his peculiar ideas? Could a scientist's published theories, which carried such weight for understanding Life's history, be so utterly wrong? Intriguingly, Steiner's motive seems allied to that of earlier scientists who sought to place paleontology on an exact mathematically elegant 'Newtonian' plane (or rather, in this case—a galactic one).

Since the 19th century, numerous theories have been proposed for recurrent Ice Ages, invoking a wide variety of causative factors. Some of these involve terrestrial factors. Others are founded in astronomical occurrences. The most successful theory, conceived by German meteorologist Milutin Milankovitch in the early 20th century, is based on secular changes in Earth's orbital geometry. Despite its highly acclaimed successes, a (more recently modified) Milankovitch theory doesn't predict, or pinpoint all ice age episodes throughout geological history. For instance, looking back over a considerably longer scale of time—way, way before the more familiar recent Ice Age—geologists have discovered evidence for other periods of continental glaciation. One such early phase of widespread glaciation was recently measured to have endured *continuously* (yet nearly unimaginably!) for 12-*million* years, beginning with onset of the Ediacaran age 633-million years ago. (*Science* 4/8/05, pp.181-182) Needless to say, there were no Charles R. Knight-stylized woolly mammoths alive then to frame the wan, preternatural dawn.

And a Late Ordovician mass extinction may be correlated with a prolonged ice age and sea level decline, events set in motion when the supercontinent Gondwanaland became 'land-locked' over the South Pole. Interestingly, some scientists now believe the Late Ordovician mass extinction was (instead) triggered by an intense cosmic gamma ray burst. (*Astronomy*, Nov. 2005, pp.34-41) A better-known, enduring ice age interval, known as the Permo-Carboniferous Glacial Age, occurred between 300 and 240-million years ago. However, the most recent age of widespread continental glaciation began about 10-million years ago. In spite of how much we complain about costs of running our air conditioners in summer, (and neglecting the looming factor of human-caused climate change) we currently live in a warmer interglacial stage expected to subside in a few thousand years before the next age of glacial advance, when perhaps woolly mammoths cloned from fossil DNA may once more roam across North America. (That was the long-term theoretical 'plan,' anyway—until tech-savvy humans interfered with the atmosphere, exacerbating global warming while inducing climate change!)

But Johann Steiner, with coauthor geologist Elfriede Grillmair, cited a host of ancient glaciations, for which geological evidence extending back to 2.2-billion years is perhaps less demonstrative.

Our current understanding of the cause of recurrent ice ages during the last 2 to 3-million years represents a great achievement in Earth science investigation. Widespread glaciations can be largely influenced by Earth's orbital geometry, as explained through Milankovitch's elegant and intriguing astronomical model. Here, temperature variations believed to have occurred throughout the past 3-million years (or since the Late Pliocene) seem to be rhythmically attuned with three independent secular factors influencing Earth's orbit with respect to the Sun's position. These factors include: 1) a 100,000-year rhythm corresponding to the period of time in which regular changes recur in Earth's orbital eccentricity (i.e. deviation from perfect circular orbit); 2) a 22,000-year period corresponding to a precession of the equinoxes along Earth's orbit; and 3) cycling of the obliquity of Earth's axis with a period of about 41,000 years. Theoretically, past (as well as future ice ages) should be 'predictable' too. (I know this paragraph introduces complex ideas, but for much more on this topic, see James D. Hays. John Imbrie and N. J. Shackleton, "Variations in the Earth's Orbit: Pacemaker of the Ice Ages," *Science*, 12/10/76, vol. 194, pp.1121-1132, and *Science,* 2/27/87, p.974 – in which 200 to 240-million-year-old Newark Basin strata doesn't fully vindicate Milankovitch's cycles, and John and Katherine Imbrie's *Ice Ages—Solving the Mystery*, Harvard University Press ed., 1986.)

Development of the modern ice age theory has been hailed as a triumph of modern investigative science. However, there are dissenters who think additional variables and factors—some terrestrial in nature and others astrophysical—have been overlooked, especially when it comes to occurrences of ice ages before the Pleistocene. Thus, we hear of other factors thought to be significant in this regard, which may have facilitated onset of ice ages, such as volcanic eruptions, asteroid impacts, drastic changes in the Sun's radiance, interactions between the solar system and interstellar dust cloud matter dispersed in the Milky Way Galaxy's spiral arms, cosmic-ray storms (which may peak every 140-million years on basis of meteorite evidence), and even a solar dust veil created by mass-extinction-inducing comet swarms.

Some geologists advocate gradually occurring, uniform, terrestrially-founded explanations for cause of the longer range pattern—like secular rifting of continents. Although exact details remain unclear, "... a reasonable case can be made that these changes in climate were caused by continental drift—the process that results in slow but continuous change in the geographic position of the continents." (Imbrie and Imbrie, *op. cit.*, p. 189) Continental ice sheets accumulated during these times when land masses were located in polar regions. As may have been the circumstance during the long Permo-Carboniferous event, "... surface reflectivity increased and climate cooled. (Imbrie and Imbrie, p. 191)

The notorious Russian psychologist Immanuel Velikovsky incorporated a recipe for explaining the cause of the recent Ice Age pattern in his outlandish theory of global catastrophe. According to Velikovsky, in his interesting though scientifically appalling book, *Earth in Upheaval* (1955), the recent Ice Age was a manifestation of cosmic disturbances creating geological upheaval and revolution. Despite Velikovsky's enthusiastic preoccupation with cataclysmic themes, his "working hypothesis" for cause of catastrophically-derived ice ages (pp.124-139) is outrageous (and shall be further neglected here). Velikovsky's ramblings couldn't be quashed quickly enough.

Velikovsky's ideas on origin of ice ages remain unsubstantiated today ... but do Steiner's?

Another fascinating hypothesis represented a wholesale attempt not only to explain the long term recurrence of ice ages since the early Precambrian, but also many of the major events known to have recurred throughout Earth's geological history. A fundamental tenet of this theory, perhaps expressed in its most elaborate format by geologist Johann Steiner between 1966 and 1973, was that (harkening back to high school physics class), Isaac Newton's 'universal gravitational constant,' expressed in the notation of physics as "G," really isn't so 'constant' after all, as it was purported to change as a function of both space and time.

(A quick note insertion—According to a theoretical prediction, "G" should be steadily decreasing at an infinitesimally small rate as a consequence of the expansion of the universe—which we'd never even notice during the course of our lifetimes. This is known as the "Dirac-Jordan effect" (see *American Scientist*, vol. 47, no.1, 1959, pp.25-40, "Gravitation – An Enigma," by R. J. Dicke).

Hereafter, we'll refer to Steiner's use of "G" as the near-Earth galactic force, which is dependent on the localized distribution of matter in the Milky Way galaxy. To Steiner, the combined gravitational effect of universal cosmic expansion with Milky Way matter interactions on "G" is additive.

Steiner (and others) stressed the importance of cosmic factors capable of causing localized disturbances in gravitational intensity that could be experienced recurrently by Earth and Solar System, besides cosmological expansion of the universe,, such as the Solar System's spatial relation to other matter concentrated within portions of the Milky Way galaxy itself, as within its spiral arms. Steiner's theory was rather complex. However, his 'bottom line' was that when localized galactic force has a greater absolute value than in the present, the effect of gravity is higher than today: stars 'burn' hydrogen fuel more intensely and planetary orbits and radii shrink. And when the galactic force has a smaller localized value, opposite effects should be noticeable. For instance, by 1977, Steiner claimed—on basis of sea-floor spreading rate measurements—that during the Jurassic Period, Earth's radius was 90% of its present-day size. (Steiner, "An Expanding Earth based on the basis of sea-floor spreading and subduction rates," *Geology*, vol. 5, pp.313-318, May 1977)

Steiner's selection of cosmic, fluctuating gravitational forces correlated with geological events echoed idealized time-honored philosophy. "In dealing with changes of the order of hundreds of millions of years, the long-term trends are commonly not known, for geology and related sciences are still largely in a descriptive phase ... For over half a century many earth scientists have cherished the hope that the cause of most major geological phenomena might be found in astronomy. The mechanics of the Solar System do not furnish periodicities of the order of hundreds of millions of years, but the dynamics of the Milky Way galaxy do." (Steiner, "The Sequence of Geological Events and the Dynamics of the Milky Way Galaxy," *Journal of the Geological Society of Australia*, vol. 14, no.1, 1967, p.100)

According to Steiner and others who considered the possibility, the value of the galactic force could depend upon localized distributions of visually observable matter (e.g. stars and nebulae) in our galaxy. Since the stellar density of our galaxy from its black hole core region to its wispy spiral arms has been mapped somewhat reliably, Steiner was able to develop a relation between the value of a galactic force as encountered by our solar system during different phases of its long-term orbit about the Milky Way core. This expression of how the galactic force varied over the past 400-million years as a function of distance from the galactic core was referred to as the "Galactic Model."

In his peculiar—if not downright 'cosmic—1967 paper, Steiner listed "Apparent Discontinuities of Biological Evolution (pp.119-121) among the host of geologically-recorded phenomena which could be freshly understood in light of a secularly-fluctuating galactic force intensity. According to Steiner's sister, Elfriede Grillmair, this was a topic he'd been professionally interested in since 1964. Steiner prefaced his argument in terms which no modern paleontologist would dare. "If a changing 'G' and resulting expansion and contraction causes the main marine transgressions and regressions … it should result in vast changes in the relative proportions of terrestrial and marine ecological niches available at a given time. These changes could conceivably be reflected in major apparent bursts of biological evolution, as indicated by the admittedly incomplete fossil record …" (p.119)

In his 1967 paper, "The Sequence of Geological Events and the Dynamics of the Milky Way Galaxy," Steiner ambitiously correlated appearances of major land floras with peaks of presumed changes in climate (e.g. long-term cooling phases when the alleged galactic force is decreasing (i.e. the Devonian-Carboniferous and Cretaceous-Tertiary boundaries), amphibians, mammals and birds—each in their respective 'ages'—enter the geological record. Due to planetary radial 'expansion' (under a prevailing "G" force fluctuation, these are marked as periods of marine regression. However, at the Permian-Triassic boundary, when the Galactic Model "… predicts a changeover from … terrestrial to a predominantly marine environment …" bipedal reptiles appear and certain reptiles (forms such as ichthyosaurs and plesiosaurs) 'return' to a marine environment. This—to Steiner—was an indication that the terrestrial surface area of Earth had shrunk to a considerable degree (as had the planetary radius).

Another curiosity surrounded Steiner's discussion of flying Mesozoic reptilians—linked to prolonged episodes of galactic force as experienced on Earth most distantly from the galactic core (at "apogalacticum"), or during the solar system's closest approach to the Milky Way core ("perigalacticum"). "It may be noted that the first flying reptiles of the Order Pterosauria appeared in the Jurassic and become extinct at the end of the Cretaceous. This period of time straddles the 'apogalactic' position. According to the galactic time-scale, the Mesozoic, which straddles the apogalactical position, predicts maximal values of "G" and presumably a denser atmosphere" (p.120). Such atmospheric conditions would 'buoy' up flyers such as pterosaurs, although Steiner didn't discuss why the evolutionary history of birds wouldn't follow a similar pattern. And finally, after attributing the Devonian-Carboniferous, Permian-Triassic and Cretaceous-Tertiary extinctions to "critical changes of the acceleration of expansion or contraction" based on the Galactic Model, Steiner correlated Man's "explosive" appearance "…with the geologically … impending perigalactical position which, according to this model, represents maximum availability of terrestrial ecological niches" (p.121). Well, maybe, until the Antarctic ice sheet melts due to human-induced global warming!

Considerable geological evidence has been amassed to support a theory that geotectonic events have been caused by gravitational changes, alleging an Earth which expands in a steadily decreasing cosmic gravitational field, for instance. For instance, observation of the matching—yet widely separated—coastlines of South America and Africa has been cited as a fracture zone resulting from planetary expansion—although the plate tectonics model provides a much stronger explanatory framework. Australian scientist Sam Warren Carey, who until 1956 supported Alfred Wegener's theory of continental drift, has staunchly advocated an expanding Earth model. (*Theories of the Earth and Cosmos*, Carey, 1988, and *Revolution in Science*, I Bernard Cohen, 1985, pp.563-64)

Steiner also made temporal correlations between recurrent geological events and factors influencing the galactic force effect throughout geological history. Later, he determined that intervals of major oceanic transgression and regression and 'craton-wide' (extended across continental regions) unconformities corresponded to Earth's most rapid episodes of radial contraction or expansion caused by the effect of a rapidly changing galactic force field. ("Possible Galactic Causes for Synchronous Sedimentation Sequences of the North American and Eastern European Cratons," *Geology*, Oct. 1973, pp.89-92) Consequently, as explained previously, he noted times of marine transgression and emergent continental areas had resulting impacts on biological evolution.

By 1972, however, one of my former Northwestern University professors—L.L. Sloss—had studied episodes of marine transgression/regression for six craton-wide sedimentary sequences, that were found to alternate with unconformities. Sloss related these in terms of episodic variation in global plate tectonics and sea-floor spreading instead of how Steiner did—relying on a hypothetical pulsating gravitational effect of our Solar System's motion through space. ("Sequences in the Cratonic Interior of North America," by L. L. Sloss, *Geological Society of America Bulletin*, vol. 74, Feb. 1973, pp.93-114)

In another paper, Steiner and colleague/sister Elfriede Grillmair correlated evidence for ice ages extending as far back as 2.3-billion years ago! In August 2006, I had the good fortune to correspond with Grillmair about her recollections on the galactic force problem as tied to the pattern of ice ages on Earth. Geologist Grillmair, who has participated in digs for dinosaur bones in the Alberta Badlands, stated that since her brother's death in 1982, she has labored alone to prove that their ideas are right.

Grillmair recounted how their collaboration began:

"In 1972, when I combed the University library for books to find the latest astrophysical data, I came upon a new set of galactic rotation rates. I combined these during my calculations with the given ellipticity of the Sun's galactic orbit and, just for the fun of it, let this ellipticity slightly increase going back in time, under the assumption that just like with cometary orbits, any elongated orbit will tend to become more circular the longer the object is exposed to the main central force. Well, the resulting curve was just stunning—it correlated almost exactly with the 'discontinuities of biologic evolution' (author's note—which her brother had written about in his 1967 paper), that is, with the geological subdivisions. Not only between Eras, for example Mesozoic and Cenozoic, or Periods, such as Jurassic and Cretaceous, but also between Epochs—Paleocene and Eocene, etc. This discovery really reinforced my conviction that my brother's theory was right, namely that the dynamics of the Milky Way galaxy had an undeniable influence on the geological events here on Earth. It clearly showed the oscillations of the function "G" (the gravitational constant, and the supposed alternating expansion and contraction of the Earth) perfectly correlating with the geological record. I showed this curve to my brother (i.e. Steiner) and he was equally stunned and his excited comment was something like 'Terrific! But first we have to get people used to a variable "G"—then we can publish this curve and go into more details!'"

The mathematically-gifted Grillmair also described how their working relationship led to the crafting of their 1975 paper:

"My brother was teaching at the University of Edmonton and I was raising a family in Calgary, as well as working occasionally for the oil industry to pay the bills. When I finally got around to studying it (i.e. Steiner's 1967 paper) I was instantly sold on his ideas. So he asked me to collaborate with him on (a) paper. I enthusiastically agreed as the subject was for me much more challenging than routine oil exploration work and I could easily contribute my own ideas. We communicated mainly by mail and he occasionally visited me in Calgary ... he gave me all the relevant material he had collected and left it up to me to do ... most of the footwork and all the calculations..."

Nonetheless, today some of their evidence now seems highly tentative. Steiner and Grillmair also correlated many other geological phenomena with fluctuating galactic conditions. They didn't completely reject the Milankovitch theory, but relegated its role in effecting ice ages to shorter, 1,000 to 100,000 year intervals. They cited Milankovitch cycles as representing "... a reinforcing mechanism which operates all the time; but which by itself does not cause glaciations." By contrast, their unobserved (as yet undocumented) galactic force effect was viewed as the "basic trigger mechanism." According to their model, ice ages are most probable when the Solar System is closest to the galactic core on its cosmic journey about the Milky Way. These periods correspond to a period of low gravitational effect on the solar system, corresponding to low galactic force values and hence minimal solar radiation.

Johann Steiner and Elfriede Grillmair claimed occurrences of ice ages during the past 600-million years have taken place when the solar system occupied a position closest to the galactic core (which they named the "perigalacticum"), and for Precambrian glacial events, when the Solar System occupied a position farthest from the galactic core (named the "apogalacticum"). (The Milky Way galaxy is believed to have a diameter of 80,000 to 100,000 light-years. A 'light-year' is the distance light travels in one earth-year or, roughly, six trillion miles.)

Furthermore, they claimed, "… the orbital position of the Sun is fastest at the perigalactica and slowest at the aopgalactica. In this model, the Phanerozoic (past 600-million years) is thus characterized by long warm periods representing the near apogalactic position (galactic summers) and comparatively shorter, cool periods representing the perigalactical positions (galactic winters) … The reverse is true for the Precambrian when shorter, warm periods alternate with longer, cool periods …" (Steiner and Grillmair, "Possible Galactic Cause for Periodic and Episodic Glaciations," *Geological Society of America Bulletin*, vol. 84, March 1973, p.1012)

But doesn't this all sound a bit, well, 'crackpot'?

On basis of their model, Steiner and Grillmair predicted that forthcoming evidence would be discovered substantiating occurrences of additional Precambrian ice ages (*Ibid.*, p.1010) Unfortunately, their model suffered from poor or inaccurately-known geological ages for successively older periods. "…Two very fuzzy curves are being compared and the degrees of agreement between them is surprising …" Yes—'fuzzy' indeed! However, despite obvious shortcomings, they felt that, "… the above facts and uncertainties should encourage geologists to pursue possible galactic-geological correlations further, until a clearer picture of their validity emerges." (*Ibid.*, p.1012)

Steiner and Grillmair eventually extended the central galactic force curves back to 800-million years in another paper ("Possible Galactic Causes for Synchronous Sedimentation Sequences of the North American and Eastern European Cratons," *Geology*, Oct. 1973, p.90), which I would have read during my college sophomore year. Even though they'd already criticized aspects of their own model, "This model is not based on quantitative … data prior to 350-million years ago … Thus, this galactic model must be considered as unsubstantiated prior to 350-million years." (*op. cit.*, March 1973, p.1009)

Like the continental ice sheets of recent geological history, Steiner's and Grillmair's elaboration has dissipated from mainstream geology. However, it must be admitted that the Galactic Model is, to say the least, intellectually intriguing for its overall cosmic scope and universal intent. (Bad puns, I know.) Unfortunately, perhaps because of its all-encompassing nature, it would be extremely hard to validate, especially when (per Occam's Razor) alternate theories suffice in explaining these geological phenomena—ice age occurrences included—particularly plate tectonics.

During the 1980s, some 'fringe' geologists claimed that geological evidence exists in support of the 'expanding Earth' model (which is rather consistent with Steiner's concept of a fluctuating galactic force variable "G"). Geologist S. Warren Carey, for example, rejected validity of the global plate tectonics model (which assumes Earth's radius has remained constant for billions of years). To him, the absolute cause of Earthly expansion, whether cosmological or terrestrial (e.g. radioactively-fueled internal heating resulting in increasing terrestrial diameter, with resulting geochemical phase changes in less dense rock throughout geological time as Earth's gravitational field decreases), remains unsettled and unproven. Whereas Carey advocated a continuously-expanding Earth model, Steiner instead had proposed an Earth radius which could *either* expand or shrink, depending on the galactic force field prevailing in any geologic era. There are, however, few remaining advocates of Steiner's 'pulsating Earth' model. The opposite of Carey's concept, that is, of a steadily, 'directionally' shrinking (e.g. 'cooling') Earth—as sea-floor spreading subsides in conveyor belt-like fashion into Earth's mantle—has been all but abandoned for decades.

Astronomer Howard E. Bond commented on Steiner's Galactic Model in the June 1974 issue of *Geology* (pp.279-280). Bond noted several problems with the approach used, as well as a calculation error, difficulties noted with orbital parameters, inconsistencies between predictions of the model and Solar System history, and lack of correlation between geological events and the gravitational force function derived by Steiner and Grillmair. Furthermore, Bond claimed, when alternative (perhaps) more accurate, values for the 'cosmic year' (i.e. the time for the Solar System to revolve about the Milky Way core once—or 220-million years) were substituted for those used by Steiner (i.e. 280-million years per galactic revolution) the alleged 'synchronicity' disappeared. Bond concluded, "In summary, it has been shown that Steiner's postulation of galactic causes for geological events was based on both physically implausible and erroneous assumptions. One may add further that there are no known astronomic reasons to have expected such a correlation in view of the extreme degree of isolation of the Solar System from other galactic objects. Geologists appear to be justified in continuing to regard the Solar System as an essentially closed system."

Note—the Sun has only traveled about the Milky Way's black hole core about 20 times, with an average period of about 230-million years per more recent estimates. The Milky Way galaxy with its ~200 billion stars—give or take a few billion—is one of many billions of galaxies in the universe, Andromeda being our closest galactic neighbor. Surely, more than plenty of opportunities for evolutionary forces to produce organic life forms … somewhere else!)

But Steiner did also have supporters. For in that same 1974 *Geology* issue, geologist J.N.J. Visser contributed information concerning two additional Precambrian glacial episodes (2.2-and 2.3-billion years old), coinciding with predictions of the Galactic Model, offering "… fairly good proof that a global event such as glaciation did not have to be of local origin and that one has to look for extraterrestrial forces to explain it."

According to Steiner's Galactic Model, the Solar System has occasionally encountered phases of space-time characterized by increased gravitational force effect. During these episodes, Earth's orbital radius and the Sun's radius should shrink, causing the Sun to 'burn' its nuclear fuel at a much faster rate, while Earth revolves in closer proximity to the Sun's more brilliant glare. (Astronomers have suggested that merely a 1% increase in the Sun's radiance could change Earth's climatic conditions significantly.) Because, at such times, Earth would also be closer to the Sun, one would anticipate evidence of increased heating of our planet during these (prolonged) phases of prehistory. Opposite effects and phenomena would be observed during periods corresponding to decreased galactic gravitational force.

At one extreme, geologists know that the ocean-atmosphere system hasn't, for example, on occasion boiled away into space, thereby precluding a possibility of life during such 'hothouse' episodes. Perhaps then, epochs characterized by warm paleoclimatic conditions should be accounted for in more 'conventional' terms such as meteorological variations, greenhouse effects due to volcanic emission of carbon dioxide or release of methane-enriched clathrate oceanic deposits, or even as the result of an intensified period of solar radiance (unrelated to possible changes in the value of a galactic force). Only several phases of galactic variable 'maxima and minima' have occurred since the origin of the Solar System 4.6-billion years ago, so it is truly difficult to correlate episodic geological phenomena with changes in the galactic variable, that is, with any degree of statistical significance. For this reason, Steiner himself admitted that correlation with certain geological events, especially those furthest back in time, becomes imperfect. (Note—An added difficulty is that extrapolation to a timespan beyond 1.25-cosmic years, roughly 335- to 400-million years ago, would require further consideration of the controversial 'Dirac-Jordan effect,' or a theoretical decline in the universal gravitational constant due to cosmological expansion of the universe.)

Michael Rampino and Richard Stothers, who in 1984—*Science*, 12/21/84, vol. 226—developed their own original hypothesis correlating a series of geological and tectonic phenomena including mass extinctions of life (exclusive of the recurrence of ice ages, though), and with both a 33-million year periodicity and a 260-million year periodicity, whose origins and cause lie in the complex motion of the Earth and solar system through the galaxy—also denounced the Steiner/Grillmair theory of 1973. This is because Steiner and Grillmair offered, "… no real geologic or astronomical evidence to support their hypothesis… Even less credible … is the speculation that the spatially varying gravitational potential of the Galaxy somehow induces a significant change of the strength of gravity in the Solar System…" ("Geologic Periodicities and the Galaxy," by Rampino and Stothers, in *The Galaxy and the Solar System*, 1986, eds. R. Smoluchowski, et. al., pp.254, 258.)

In personal correspondence dated March 17, 1988, Professor H. A. K. Charlesworth of the University of Alberta stated that Steiner's controversial ideas came to him during his graduate days at the Australian National University in Canberra. In an obituary notice that he co-authored regarding Steiner's life and career, Charlesworth also claimed that, "...In the repetition through time of apparently related geological events, he (Steiner) saw the possibility of extraterrestrial control and of discerning the record of the Earth's voyages around our galaxy. He threw all his resources into advancing this concept and charting its consequences. Had his papers on the Galactic Cycle been published slightly earlier, Dr. Steiner's insights and perception would have had more influence. As it was, Earth scientists of the 1960s and 1970s found his concepts difficult to blend into prevailing models."

Yet, as Elfriede Grillmair distinctly recalled in August 2006:

"As far as I can remember, we were in the beginning stages of several other papers, but my brother (Steiner) hinted that the professional journals refused any further publications because the mere suggestion that the cherished Newtonian gravitational 'constant' was varying in both time and space (and therefore just a 'function') had caused such an uproar in the scientific community. My brother's papers were denounced. However, there was some vindication when our 1973 paper on ice ages was reprinted in 1981, in a volume called *Benchmark Papers in Geology 57—Megacycles Long-term Episodicity in Earth and Planetary History*, edited by G. E. Williams (Hutchinson Ross Publishing Co.) I have no doubt that the rejection of his revolutionary new idea by his peers contributed greatly to my brother's eventual fatal illness and his untimely death at 56."

It surely didn't help Steiner's and Grillmair's cause either, that their Galactic Model conflicted with Geology's newfound paradigm—global plate tectonics, which in turn had soundly defeated lingering proponents of the expanding Earth model.

The peculiar, intriguing papers by Steiner (sometimes collaborated with Grillmair) weren't the last word in scientific attempts to explain and correlate long-term climate changes with other secular factors. In 1977, paleontologist Alfred Fischer and then-graduate student Michael A. Arthur published a paper titled "Secular Variations in the Pelagic Realm." They proposed that episodes of marine mass extinctions of the past 250 million years have been regularly spaced at a 32-million year periodicity. A large agglomeration of data, including temporal fluctuations of stable carbon isotope rations and fossil fauna was statistically correlated in an effort to demonstrate periodicity of mass extinction cycles. The data and conclusions were poorly received because the only indication of cycles evident in their work "... could be explained by the subjective way in which the data were chosen." (*The Nemesis Affair*, David

Raup, 1986, p.109.) Fischer and Arthur considered the possibility that fluctuations in the Sun's brightness provided an ultimate driving force. Even more likely was a possibility that variations in Earth's interior, related to convective heat patterns in the mantle, could have been responsible for the 32-million year cycle they observed in the pelagic realm.

Fischer later advocated a longer term, two-phased, 300-million year "super-cycle" initiated by increased rates of thermal convective cycling in the mantle, leading to continental breakup, marine transgression, and intensified volcanism, with resulting emission of carbon dioxide and an ensuing greenhouse effect. One hundred fifty million years later, reduced sea-floor convection rates promoted continental accretion, marine regression and atmospheric cooling due to reduced carbon dioxide emissions. Then cold, polar ice age conditions prevailed during climatic 'phase changes,' when meteorological conditions shifted from greenhouse-heated to colder, more temperate global conditions (e.g. late Cambrian, late Devonian, late Triassic, and late Eocene stages).

> "In summary, I take the crisis in late Cambrian time (at about 500 mya), the one in the late Devonian time (at about 355 mya) and that in the late Eocene (40 mya) to mark the crossover points between the major climatic episodes in earth history ... the regularity of these intervals ... suggest(s) the possibility that the inciting cause—plate activity, and ultimately, mantle convection—is a cyclical process with a period of about 300-million years: *roughly the periodicity that is commonly attributed to the period of galactic rotation* (italics mine)." (Fischer quoted in *Catastrophes and Earth History*, W. A. Berggren and J. A. Van Couvering (eds.), 1984, p.146.)

Fischer also offered evidence correlating epochs of maximum plate tectonic activity, continental dispersion, volcanism and eustatic sea level.

Paleontologist David Raup later stated:

> "... that Fischer and Arthur were a little vague on mechanism is understandable. There is no theorem in science that says that the description of a phenomenon must be accompanied by a mechanism for how the phenomenon works, although a viable mechanism is always preferred. The absence of a mechanism is often used as a weapon against research conclusions that we don't like. For example, the idea of continental drift was put down for many years partly on the grounds that nobody had come up with a mechanism." (*The Nemesis Affair*, 1986, pp.109-110.)

And yet, today, while most paleontologists have at least heard of Alfred Wegener—lauded as 'inventor of the Continental Drift theory—probably few to none of you have prior heard the names Johann Steiner or Elfriede Grillmair!

Until now.

In proposing his Galactic Model, it is suggested here that Steiner was rather seduced into a traditional pattern set decades before by earth scientists who ambitiously strove to place a unified paleontology-geology on equal footing with mathematically exact Newtonian sciences. It's more or less analogous to the physicist's quest to solve the 'theory of everything' (although in this—'everything,' geologically speaking) with mathematical elegance.

Certain scientists of the 18th and 19th centuries held this lofty ideal, like Baron Georges Cuvier with his "law" of the correlation of anatomical parts, as well as through proposed theories interrelating geological revolutions with extinction boundaries. We also find an allusion to celestial Newtonian mechanics in the sentence concluding Charles Darwin's *The Origin of Species* (1859, 1st ed.), where he stated, "… whilst this planet has gone cycling on according to the fixed law of gravity, from so simple a beginning endless forms most beautiful and most wonderful have been, and are being evolved …" We even find traces of cyclical astronomy embedded into geology's Huttonian foundation. For it was Scottish (uniformitarian) geologist James Hutton who in his *Theory of the Earth* (1788), stated, "For having, in the natural history of this earth, seen a succession of worlds, we may … conclude …there is a system in nature … in like manner as from seeing revolutions of the planets …" Therefore, Hutton concluded, "… we find no vestige of a beginning, no prospect of an end."

Although for the most part, 20th-century scientists disregarded the idea that the cosmos could have had much influence on repeated events in geological history, fully over a decade before the birth of the "Nemesis" theory (fostering a celestially-triggered 26-million-year mass extinction cycle), Craig Hatfield and Mark Camp published their article, "Mass Extinctions Correlated with Periodic Galactic Events," *Geological Society of America Bulletin,* vol. 81, pp.911-914, March 1970). Here they suggested, "If mass extinctions do occur with a degree of temporal regularity, their periodicity should be considered in any postulated cause for the extinctions. Known regular events with such long periodicity are not numerous and, so far as the authors are aware, may be expected only in the realm of astronomy."

In retrospect:

> "Johann Steiner was a distinguished scientist who accomplished much despite turning to science rather late during his lifetime. During World War II he served in the German air force, and was interned as a prisoner of war in the United States. He earned a first-class honors degree in geology from the University of Alberta in 1961, followed by a M.Sc. the

following year. His Ph.D. was obtained from Australian National University in 1967. Dr. Steiner completed thorough and 'calm and reasoned' work on Devonian depositional environments, meteorites and Cretaceous coal seams, although his passion clearly was in charting ... large-scale patterns in the evolution of the Earth. In the repetition through time of apparently-related geologic events, he saw the possibility of extraterrestrial control and of discerning the record of the Earth's voyages around our galaxy. Steiner's Galactic Model received publicity in *Nature* and in (the aforementioned) *Benchmark Papers in Geology* volume ... earning him a permanent place in history on the subject." (Excerpted from "Memorial to Johann Steiner," by H.A.K. Charlesworth and J. Sebastian Bell.)

Placing matters into modern context, in July 2006, Grillmair's son. Caltech astrophysicist Carl J. Grillmair, informed me that his uncle, Johann Steiner:

"...took the rotation velocities of stars around the Galaxy and, rather than inferring dark matter (as most astronomers now do) to provide the force necessary to contain them, he postulated a gravitational constant potential that varied with Galactocentric radius. I know from conversations with my mother (E. Grillmair) that they subsequently expanded this variable gravitational potential (variable, that is, compared to the inverse square law $1/r^2$ Newtonian form) to depend not only on radius but on absolute position and perhaps (if I remember correctly) also on angular velocity. I believe that other researchers have correlated a variety of geologic and climatic indicators with the Sun's movement through the spiral arms (for example), which doesn't sound implausible to me. It would certainly be interesting to be able to make such connections ..." (But see brief note at end.)

For the most part, Steiner's treatment of the geological occurrences of ice ages received a cold reception from the scientific community. At a time when the Earth sciences seemed in upheaval, Steiner and Grillmair added further fuel to the fire—or was it food for thought? By the time of Steiner's death (1982), it was becoming clear that while Newtonian science represented mathematical 'ideality' for a host of scientific and academic professions, the Earth sciences— paleontology included—must sometimes proceed on basis of *historical* lines of evidence (which could still potentially lead to startling new paradigms).

Note—In her August 18, 2006 correspondence to me, E. Grillmair dispensed with the "utterly ridiculous Big Bang theory of cosmological expansion and other concepts such as 'dark matter' and 'dark energy'." Today, the latter two concepts are very much in vogue, even in popular science.

ABOUT THE AUTHOR

Allen A. Debus has written numerous articles on paleo-topics and science fiction—usually intertwining subjects of science fiction and paleontology, several for refereed journals ... and many more for a variety of remarkable fanzines—the sort of writing he prefers. The latter variety includes *Prehistoric Times, Modeler's Resource, Dinosaurus Magazine, G-Fan, The Baum Bugle, The Dinosaur Report* (U.S. Dinosaur Society), *Filmfax Plus, Mad Scientist, DinoPress, Castle of Frankenstein, Dinonews*—(Dinosaur Club of the Museum of Western Australia), and *Scary Monsters*. He's a former co-editor of *Dinosaur World* (1996-1999), and a former contributing editor of *Fossil News: Journal of Avocational Paleontology*. Allen has also written eight other books each emphasizing prehistoric animal popular culture and imagery, including *Dinosaur Memories: Dino-Trekking for Beasts of Thunder, Fantastic Saurians, 'Paleo-people,' 'Dinosaurabilia,' and other 'Prehistoria'* (with Diane Debus, 2002); *Paleoimagery: The Evolution of Dinosaurs in Art,*(with Diane Debus, 2002); *Dinosaurs in Fantastic Fiction: A Thematic Survey* (2006); *Prehistoric Monsters: The Real and Imagined Creatures of the Past That We Love to Fear* (2010); *Dinosaur Memories II: Pop-Cultural Reflections on Dino-Daikaiju & Paleoimagery* (2017); *Dinosaur Sculpting: A Complete Guide, 2nd ed.* (with Bob Morales and Diane Debus, 2013); and *Dinosaurs Ever Evolving: The Changing Face of Prehistoric Animals in Popular Culture* (2016). Allen is a self-taught dinosaur sculptor and a retired environmental chemist, having earned a bachelor's degree from Northwestern University in 1976, (with graduate and post-graduate work completed afterward, respectively, at the University of Illinois-Champaign, and Chicago).

ABOUT THE COVER

The **Front** cover was designed by artist Todd Tennant, under direction of the author. Lettering for the main title was borrowed from a 2012 advertisement for Monster Bash's "Prehistoric Monster Bash"; permission to incorporate the lettering style—as modified was generously granted by Ron Adams of Monster Bash and Creepy Classics Magazine. Readers are encouraged to check out the annual Monster Bash Conference event at www.monsterbash.us. At **top left** is a *Pteranodon* sketched by Jack Arata for *Dinosaur World* and at **bottom** is Othenio Abel's c. 1920 *Stegosaurus* restoration. Visible on the **Back** cover, also designed by Todd Tennant, (**clockwise from top-left**) are seen (1) detail from 1960 *Dinosaurus!* movie poster—copyright Universal Studios (2)) detail from 1960 *Dinosaurus!* movie poster—copyright Universal Studios (3) *Brontosaurus* detail from Palmer Plastics 52-piece model kit skeleton box, circa 1960 (4) Gfantis sculpture by Allen A. Debus, 2012; see Chapter Eleven (5) Movie still from *Reptilicus* (1961), as published on back cover of Dean Owen's 1961 Monarch Books novelization, copyright American International Pictures (6) detail from *The Land Unknown* movie poster, copyright Universal-International Pictures, 1957 (7) Image from *Boy's Magazine,* a British publication no longer printed or circulated, 1933, as reprinted in James Van Hise's *Hot-Blooded Dinosaur Movies*, 1993, p.65 (8) **Middle column – top,** Tyrannosaur and *Triceratops* images from Martha McBride Morrel's *Cuando El Mundo Era Joven*, Argentina 1944, p.189 (9) **Middle column – middle**, Jack Arata's *Torosaurus* restoration author's collection, and as published in *Dinosaur World*, no.3, Oct. 1997, p.60.